A History of Military Morals

History of Warfare

VOLUME 138

The titles published in this series are listed at *brill.com/hw*

A History of Military Morals

Killing the Innocent

By

Brian Smith

BRILL

LEIDEN | BOSTON

Cover illustration: Lot-6112-5: "German Schutte-Lanz Airship SL2 bombing Warsaw, Poland, in 1914" Translated from German publication, "Our Air Fleet in the World War." Artwork by Hans Rudolf Schulze. Courtesy of the Library of Congress. (2016/10/28).
https://flickr.com/photos/127906254@N06/29992553564 Last accessed 29 December 2021.

The Library of Congress Cataloging-in-Publication Data is available online at https://catalog.loc.gov
LC record available at https://lccn.loc.gov/2022007715

Typeface for the Latin, Greek, and Cyrillic scripts: "Brill". See and download: brill.com/brill-typeface.

ISSN 1385-7827
ISBN 978-90-04-51343-3 (hardback)
ISBN 978-90-04-51548-2 (e-book)

Contents

PART 3
The Twentieth Century: Aerial Bombing and a Shift in Norms

Figures

Introduction

1 Introduction

In his landmark book *Just and Unjust Wars*, Michael Walzer sets out the central tenets of the post-World War II war convention, the "set of articulated norms, customs, professional codes, legal precepts, religious and philosophical principles, and reciprocal arrangements that shape our judgments of military conduct."[1] One of the foundational principles of this implicit/explicit code of conduct is that "noncombatants cannot be attacked at any time. They can never be the objects or the targets of military activity."[2] Unfortunately, however, battlefields are never so tidy, and noncombatants are frequently killed. For this reason, militaries must decide what their relationship to nonfighting populations will be. Walzer goes on to clarify that the international community largely agrees that while civilians cannot be deliberately killed, they might be killed unintentionally or accidentally as a byproduct to a legitimate strike. This moral calculus is commonly referred to as the doctrine of double effect.

First articulated by the Scholastic tradition, double effect permits the unintentional killing of noncombatants under certain conditions. Summarizing this view, Walzer explains that one can pursue a military strike that will lead to civilian death if:

> 1) The act is good in itself or at least indifferent, which means, for our purposes, that it is a legitimate act of war; 2) The direct effect is morally acceptable – the destruction of military supplies, for example, or the killing of enemy soldiers; 3) The intention of the actor is good, that is, he aims only at the acceptable effect; the evil effect is not one of his ends, nor is it a means to his ends; 4) The good effect is sufficiently good to compensate for allowing the evil effect.[3]

Well-directed strikes that are directed only at legitimate targets can tolerate some civilian casualties, that is, so long as those deaths are proportional to the value of the thing being targeted. A technically legitimate military target that would lead high civilian casualties and produce little strategic benefit

---

1 Michael Walzer, *Just and Unjust Wars: A Moral Argument with Historical Illustrations* (New York: Basic Books, 2006), 44.

2 Ibid., 180.

3 Ibid., 153.

 | DOI:10.1163/9789004515482_002

is illegitimate. According to Walzer's framework, acting on legitimacy alone may evince bad intentions, a cavalier indifference to indirect harm. Not every legitimate target is a permissible one. For this reason, Walzer argues that it is essential for militaries to convey their good intentions by taking steps to show that they are actively seeking to minimize unintended harm to noncombatant populations and that in doing so they willingly take additional risks onto themselves. Taking on additional risk while seeking to mitigate civilian casualties demonstrates, at least in principle, a commitment to proportionality. Since it is often quite difficult to quantify the value of civil death in relation to a military outcome, it is now common for militaries to deploy casualty mitigation tools to show that they have made a reasonable effort to minimize harm to non-fighting populations. Within the framework that Walzer establishes, one can infer good intentions when the strike is directed at a proper object, a legitimate military target, and when the attacker makes a reasonable effort to minimize harm.

Walzer's formulation of the war convention summarizes with admirable precision the ascendant beliefs about killing in war since World War II. As one can see, *intentions* carry a great deal of the moral load of this ethical discourse. Militaries must not only have good intentions but demonstrate them through in their targeting protocols. Indeed, many scholars have noted this peculiar feature of twentieth-century just war theorizing. Maja Zehfuss argues, "It is difficult to overstate the centrality of the prohibition against the intentional killing of civilians in scholarship that positions itself as on the ethics of war."[4] In a similar vein, Neta Crawford writes,

> The dominant paradigm of legal and moral responsibility in war stresses both intention and individual accountability. Deliberate killing of civilians is outlawed, and international law blames individual soldiers and commanders for such killing, because … the dominant view of agency is a theory of individual autonomous moral agency. Killing civilians is legal and may be forgiven if it was unintended, incidental, and considered proportional to a militarily necessary operation.[5]

4 Maja Zehfuss, "Killing Civilians: Thinking the Practice of War," *The British Journal of Politics and International Relations* 14 (2012); 423–440. 423.

5 Neta Crawford, *Accountability for Killing: Moral Responsibility for Collateral Damage in America's Post-9/11 Wars* (New York: Oxford University Press, 2013). 6–7.

Along these same lines, Major General Charles Dunlap explains,

> It is not surprising ... that there is no legal or, indeed, moral imperative to spare the sentient adult population of a belligerent from the vicissitudes of war, short of refraining from direct targeting. To be clear, while it is plainly wrong to target civilians or to conduct military operations – bombing or otherwise – for the "sole or primary purpose of spreading terror among the civilian population," it is nevertheless also fully expected – and tolerated – that military operations could have psychological and other unpleasant consequences for civilians.[6]

While it is wrong to deliberately target civilians, it is permissible for the noncombatant population to be forced to bear up under the hardships of war, which typically includes the possibility of indirect death.

Some rightly point out that if there were no mechanism to permit the killing of the innocent, war would be impossible. Jeff McMahan expresses these stakes quite poignantly:

> Given the conditions of modern warfare, it is virtually impossible to fight a war without harming innocent civilians. Thus, to avoid pacifism, one must concede the permissibility of harming or killing innocent civilians in war. But terrorism, which virtually everyone condemns, is distinguished by the fact that it harms or kills innocent civilians. Must we choose between embracing pacifism and condoning terrorism?[7]

He continues, "Common sense morality avoids this dilemma by insisting that the harming or killing of innocents counts as terrorism only if it is intended. Acts of war that have the harming of the innocent as a foreseen but unintended effect are not terrorism."[8] Without some moral standard, like the intention to commit harm, confusion would arise, and one might be forced to construe acts of war which lead to civilian death as morally indistinguishable from terrorism. Identifying an explanatory or exculpatory bright line helps to distinguish between terrorism and an excused but tragic military action. To be sure, such

---

6 Charles Dunlap, Jr., "Clever or Clueless?: Observations About Bombing Norm Debates," in *The American Way of Bombing: Changing Ethical and Legal Norms, From Flying Fortresses to Drones* eds. M. Evangelista and H. Shue (New York, 2014). 117.

7 Jeff McMahan, "Revising the Doctrine of Double Effect," *Journal of Applied Philosophy* 11:2 (1994). 201–212. 201.

8 Ibid.

a line is essential to the moral permissibility of war. Without such a framework one would be forced to either morally permit or condemn both.

In the broadest terms, this book seeks to explain how we have arrived at the ascendant normative framework that prohibits the intentional harm of noncombatants described by Walzer and the many subsequent commentators above. As such, this project is mainly about how militaries rationalize or explain the killing of noncombatants with weapons as part of the direct fighting logic of war, but as will be seen, the subject matter will sometimes digress into other kinds of violence that have a negative impact on civilians: genocide as an object of war, rape, pillage and plunder, and other unauthorized uses of violence. At many points in the history of warfare, a tactic or act of violence might fall well outside of the ascendant norms of the period. Sometimes those outliers became trends and help to shape the conduct of future wars. This project seeks to chart the most broadly accepted norms in the conduct of war, particularly in the Euro-American tradition.

As noted by Walzer, the twentieth-century double effect framework finds its origins in Scholastic discourse. While this is certainly true, this book seeks to show that one must be cautious when tracing historical lineages. Linking twentieth-century discourse to that of the thirteenth century gives the misleading impression that if one looks back over history, one will find an unbroken line tracing the gradual development and broad acceptance of something like double effect arguments for killing the innocent in war. This is not the case. In many respects, the role double effect plays in permitting the killing of noncombatants is quite new. Its resurgence in military ethical discourse can be traced back to just around World War I, but as a more formal principle in international law it arose after World War II. The following chapters seek to paint a nuanced picture of how we came to speak in the language of double effect, intentional harm, and noncombatant immunity in the twentieth century.

As Geoffrey Best notes, “Humanitarian commentators in our own time too easily fall into the good-natured error of transferring their universalist principles on to ages and people whose views of their own species were strictly discriminatory.”[9] The conflict Best is referring to can be described as the tension between antiquarianism and anachronism. It is of course quite useful to transpose concepts and terms back onto history. These are, at times, quite necessary tools of historical demarcation, shorthand for the historian to cluster and categorize in ways that make the past intelligible to the present. However, these descriptions at times conflict with the language and self-understanding the people of that period used. In this sense, the transposition of concepts

9 Geoffrey Best, *War and Law Since 1945* (Oxford: Clarendon Press, 1994), 24–25.

onto history can be anachronistic. Making the past intelligible runs the risk of overlooking the language and self-understanding of those theorizing in that historical period. Of course, trying to reconstruct the intellectual architecture of the past runs the risk of antiquarianism. Ideas can be situated in such a foreign ecosystem that they appear too alien to be meaningful or, indeed, intelligible. This book, admittedly, has certain antiquarian sympathies. It seeks to show that the relevance of older ways of talking not in terms of their immediate applicability to contemporary discourse but in terms of how certain ideas gave way to others. In other words, rather than seeing military ethical discourse as a continuous, progressive development, this project seeks to show the disjunctions and breaks, how one moral vocabulary was replaced by others. By retrojecting universalist principles onto the past, we obscure these meaningful differences and discontinuities.

Indeed, many commentators writing on the development of noncombatant protections paint compelling pictures of broad historical continuity of military ethical discourse. On this view, over many centuries, there has been a slow accrual of protections extended to discrete classes of people. In this development, however, it is implied or assumed that these antecedent elements were interminably building up to a more robust notion of noncombatant immunity. As Best rightly cautions, however, one should be wary of retrojecting twentieth-century humanitarian terminology and categories onto the past. In doing so there is some risk to constructing oversimplified narratives of ethical development which see unbroken lineages from the thirteenth to twentieth century. What these well-meaning attempts to reconstruct a progressive view of military history obscure is the fact that as soon as tacticians and philosophers started arguing that certain people should be exempted from direct harm in war, there simultaneously emerged arguments showing how the innocent could still be permissibly killed. This book seeks to show alternative stories about the permissibility of killing. It is quite true that in different periods there were discursive traditions which claimed certain people were exempt from harm. What unfolds in the following chapters, however, is a neglected story, the underbelly of those arguments – *how the rules protecting the innocent could be violated.* In different periods fighting communities established discrete rules prohibiting certain harm. Simultaneous to the establishment of these rules there emerged excuses and justifications for violating them. This is a book about the different ways militaries rationalized to themselves the killing of those people they believed should be exempt from killing.

From the sixteenth to eighteenth centuries there were three discursive traditions, all grounded in some conception of the natural law, that rationalized killing the innocent in different ways. The late-Scholastics gave early expression to the tradition of double effect, largely in response to the emerging realities of siege warfare. Specifically, double effect arguments were deployed to excuse the bombardment of cities in which some unspecified number of innocent people lived. Secondly, the martialist tradition, a tradition largely borne out of the realities of the Eighty Years' War, focused on military discipline and the bridling of soldiers' passions, transforming them into a perfect instrument of violence. This literature emphasized martial virtue and the wisdom and judgment of a judicious prince. Villages and towns could be destroyed only when command to do so by a legitimate authority. Finally, the humanist or republican tradition focused largely on the duty and right to punish certain wrongs. Sovereigns had a right to defend themselves and to discipline aggressors. This discourse typically included various accounts of what might be necessary to attain those ends, including the killing of the innocent. Certain negative side-effects may be necessary when pursuing one's right of punishment.

By the nineteenth century, this complex and polyvocal discourse abruptly became monotonal. After the Napoleonic wars, and the Congress of Vienna, virtually every commentator on war abandoned the natural law theorizing of the prior 300 years and instead focused exclusively on the principle of necessity interpreted through a utility framework. The emerging discourses of national sovereignty and utilitarianism helped to establish the concept of "military necessity." Since the central duty of states came to be seen as self-preservation, they were given a great deal of latitude in determining what was necessary for their survival, which meant that noncombatants could be intentionally killed under certain conditions. Throughout the nineteenth century, the battlefield was seen as spatially delimited. The closer to a zone of hostility one was, the more likely one was to be characterized as a combatant. Under these conditions, it was the responsibility of non-fighting populations to avoid zones of conflict. Finally, as the twentieth century saw the rise of the devastating reality of aerial bombing, the discrete spatial relations that had governed the utility calculations of the nineteenth-century battlefield were obliterated. It is during this period that one begins to see modifications to the ascendant ethical discourse of military necessity. In this period many theorists started to reassert intention as the moral bright line delimiting acceptable and unacceptable civilian death.

In many respects, this book seeks to understand how ethical arguments transform and gain credibility. In the rough sketch indicated above, it is striking that moral vocabulary tends to adapt to changes in institutional and material conditions. The Scholastics deployed the principle of double effect to address the tactic of bombing cities during sieges. In the nineteenth century, the logic of necessity arose in conjunction with the idealized attempts to professionalize militaries as the tool to defend national sovereignty. Twentieth-century theorists rediscovered double effect when aerial bombing rendered obsolete, or at least deeply morally questionable, the strict appeal to necessity. The dissonance between interwar bombing realists, who attempted to cast air warfare in terms of necessity, and what the international community was willing to accept could not have been greater. Pinpointing these shifts provides a plausible framework to understand the development of ascendant military ethical vocabulary. Indeed, understanding ethical discourse as a response to shifts in institutional and material conditions has striking implications for our own twenty-first-century context. As will be seen, contemporary revolutions in technology are placing a great deal of strain on the ascendant vocabulary of intention described by Walzer above.

To be sure, understanding the ways in which military technologies have historically influenced arguments permitting the killing of the innocent is significant, especially since the rise of twenty-first-century weapons, e.g. sophisticated casualty-mitigation software, precision-guided munitions, autonomous weapons, and AI, have placed new strain on military ethical vocabulary. It is becoming increasingly clear that the ascendent moral framework that privileges intentions is poorly suited to our technological landscape. As more and more decision-making is outsourced to complex algorithms and autonomous systems, as militaries come to rely on sophisticated modeling software and other casualty mitigation protocols, it is less clear how individual moral responsibility and "bad intentions" will apply to killing in war. If the history of the development of moral arguments is to be a guide, one should be attuned to these tensions. History suggests that shifts in ethical language arise under just such conditions.

2 Theory of Moral Development

In a late eighteenth-century treatise on international law, the Scottish jurist Sir James Mackintosh (1765–1832) observed that the history of the laws of nations is a history of successive attempts at system building. What makes these successive attempts so difficult is that in one age "the language of science is so completely changed" that "his moral reasoning would be almost unintelligible to some of his hearers or readers."[10] Mackintosh continues: "But these reflections, while they moderate the pride of invention, and dispel the extravagant

---

10 Sir James Mackintosh, *A Discourse on the Study of Law of Nature and Nations* (London: Printed for T. Cadell ..., 1799), 22.

conceit of superior illumination, yet serve to prove the use, and indeed, the necessity, of composing, from time to time, new systems of science adapted to the opinions and language of each succeeding period. *Every age must be taught in its own language*."[11] As Mackintosh suggests, the self-flattery inherent to the idea of moral progress obscures the fact that discourse seeks to adapt itself to different environments. In a similar vein, the great nineteenth-century historian, Henry Thomas Buckle (1821–1862) argues that "it is evident, that if we look at mankind in the aggregate, their moral and intellectual conduct is regulated by the moral and intellectual notions prevalent in their own times."[12] Like Mackintosh, Buckle also seeks to show the variability and change from one generation to the next, and that history is the project of tracing language and the prevalent intellectual and moral notions of each period.

One of the key assumptions of this book is that each generation develops its own moral vocabulary, and the ethical arguments and assumptions that underwrite international law and military ethical reasoning in one age will likely be transformed and repurposed in another. In this sense, moral arguments function like adaptive technologies.[13] Quite obviously, ethical arguments do not occur in a vacuum. The history of military legal and ethical reasoning suggests that moral arguments are calibrated to address conditional needs and emergent crises, where the moral vocabulary of one age may be ill-suited to the next. In short, moral arguments are constructed to meet historically proximate demands. While they are in some sense contingent, this is, of course, not to say that moral arguments are groundless or that they can be changed haphazardly, at will. What appears to be the case is that emergent moral judgments from one age produce a kind of scaffolding for subsequent judgments.[14] The moral arguments and intuitions of one era disclose possibilities (and foreclose on others) for the arguments and intuitions that come after. While in a certain sense relative, moral judgments are not produced at random. They certainly cannot be whatever we want them to be. In a real sense, moral judgments are

11 Ibid., 23. Emphasis mine.

12 Henry Thomas Buckle, *History of Civilization in England, Volume 1* (London: Parker, Son, and Bourn, 1861), 162.

13 For more on language as cognitive technology, see Michael Frank, Daniel Everett, Evelina Fedorenko, and Edward Gibson, "Number as a Cognitive Technology: Evidence from Pirahã Language and Cognition," *Cognition* 108:3 (2008); 819–824; and Michael Frank, Evelina Fedorenko, and Edward Gibson, "Language as a cognitive technology: English-speakers match like Pirahã when you don't let them count," *Proceedings of the Annual Meeting of the Cognitive Science Society* 30 (2008); 439–444.

14 See Val Turner & Marvin Berkowitz, "Scaffolding Morality: Positioning a Socio-cultural Construct," *New Ideas in Psychology* 23 (2005); 174–184.

contingent, that is, responses to historical, cultural, institutional, and technological changes.[15]

This book seeks to describe how we have received the moral vocabulary that dominates the Western ethical discourse. In short, this is a story about how norms emerge and what constraints they impose upon us. As Quincy Wright explains, "War has objective and subjective, universal and particular, aspects which may be differently stressed. One aspect emphasizes its violence, another the conflict out of which it arises, another the attitudes it engenders, and another the conditions which define and regulate it."[16] What Wright is pointing to is the fact that, as many theorists on the evolution of war often forget, the character of war is not determined merely by the implements of violence and revolutions in military technology, conflict and the rise of threat environments, bureaucratization and the capacity of political organizations, and so on.[17] Attitudes, beliefs, norms, and dispositions can also influence how war is practiced. Neta Crawford, for instance, shows how shifts in emphasis between norms and practices that privilege noncombatant immunity versus those that give greater weight to force protection can change the character of war, specifically by determining the amount of tolerable risk to which civilians are exposed.[18] Simply put, beliefs about what counts as acceptable practices in war have nontrivial effects. The ethical language in which military behavior is described functions as a soft limitation.

The point here is not to naively assume that moral language is a perfect restraint; nor is this to suggest that our ethical arguments are progressing in some interminably linear and objectively better sense. Quite the contrary, as the dynamics around changes to the language of intention will show, certain moral technologies that are adapted to one moral environment can have pernicious side effects in others. For instance, at their outset, moral arguments grounded in the Scholastic and utilitarian traditions sought to limit the

---

15 Richard Rorty, "Human Rights, Rationality, and Sentimentality," in *Truth and Progress: Philosophical Papers* (New York: Cambridge University Press, 1998); 167–185; and Bruce Friesen, *Moral Systems and the Evolution of Human Rights* (New York: Springer, 2015).

16 Quincy Wright, *A Study of War* (Chicago: University of Chicago, 1942), 5–6; John Ulric Nef, *War and Human Progress: An Essay on the Rise of Industrial Civilization* (Cambridge, MA: Harvard University Press, 1950); and Geoffrey Best, *Humanity in Warfare: The Modern History of the International Law of Armed Conflict* (London: Methuen, 1983).

17 For instance, see Jack Levy and William Thompson, *The Arc of War: Origins, Escalation, and Transformation* (Chicago: University of Chicago Press, 2011).

18 Neta Crawford, "Targeting Civilians and U.S. Strategic Bombing Norms: *Plus ça change, plus c'est la même chose?*" in *The American Way of Bombing: Changing Ethical and Legal Norms, from Flying Fortresses to Drones* eds. Matthew Evangelista and Henry Shue. (Ithaca: Cornell University Press, 2014).

amount of harm inflicted on the innocent. Over time, these traditions evolved in such a way that permission for excessive cruelty derived from their premises became formulaic.

Thinking about military ethical discourse as adaptive and evolutionary also assumes that the words and terms that underwrite our moral landscape will also to some limited extent determine how that landscape develops and how we ultimately interface with the phenomenal world. The words we use to describe military phenomena matter. At the very least, new words not only signal changes to the ethical landscape but they also allow us to delimit and frame sites of moral concern. Hannah Arendt writes, "One way to date the actual birth of such general historical phenomena as revolution … is, of course, to find out when the word which from then on remains attached to the phenomenon appears for the first time. Obviously, each new appearance among men stands in need of a new word, whether a new word is coined to cover the new experience or an old word is used and given new meaning."[19] As Arendt suggests, by looking at word origins, one gets a sense of both the dynamic antecedent moral landscape out of which subsequent principles and norms were built and the expanded moral framework the new word opens. Sometimes old words are repackaged and given new meaning. This process is important, but often overlooked in military ethical discourse, which is striking since there is growing evidence to suggest that language frames our moral experiences in crucial, even if imperceptible, ways.[20]

It is essential to recognize the degree to which vocabulary helps to define our sense of the world and how we navigate it. Again, referring to Best's admonition of well-meaning historiographers who retroject twentieth century notions onto the past, one should be much more attuned to changes in terminology, as this can often signal the emergence of a new site of moral conflict. As noted above with the antiquarianism/anachronism debate, there is something imprecise to conflating "civilian casualties," a term used to denote inappropriate killing in war in the twentieth and twenty-first centuries, with the violence inflicted on the "innocent," "inhabitants," or "women and children," all terms used to describe protected groups in earlier periods. To be sure, it is difficult *not* to refer to those innocent townspeople who were killed in the sixteenth and seventeenth centuries as "noncombatants" or "civilians." After

---

19 Hannah Arendt, *On Revolution* (New York: Penguin, 1990), 35.

20 There is some evidence, for instance, that moral judgments are influenced by the language they are framed in. See Janet Geipel, Constantinos Hadjichristidis, and Luca Surian, "The Foreign Language Effect on Moral Judgment: The Role of Emotions and Norms," *PloS One* 10:7 (2015); doi: 10.1371/journal.pone.0131529.

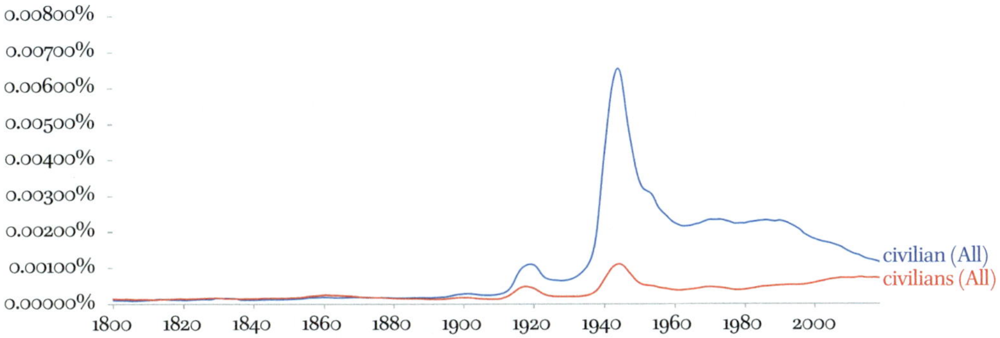

FIGURE 0.1 Ngram search of 'civilian' and 'civilians'

all, as conceptual tools, our words are expedient bundles of information; they work well enough. The challenge is that even though our words work and are perfectly intelligible, these subsequent classification schemes were informed by context-specific conditions that cannot be so easily conflated: the "non-combatant" emerged as part of the military professionalization discourse of the nineteenth century, and the "civilian" arose as part of twentieth century discourses which sought to delimit one's status in the industrial nation state.

To put this problem a bit more directly, there cannot be "civilian casualties" before there are "civilians" (see Figure 0.1). While it makes intuitive sense to think of the peasants and townsfolk who were killed in sixteenth- and seventeenth-century wars as "civilian casualties," there is reason to be cautious here. Presuming that continuity exists, that the words we have invented to describe twentieth century war can be unproblematically cast back onto history obscures the important differences that have existed in the many terminological contexts throughout time. One must consider that *how* a phenomenon is described changes our experience of it.[21] Rather than thinking of civilian casualties as a singular or unified phenomenon that can be traced back through history, it is better to think of this term as seeking to grapple with a phenomenon that has been described and redescribed in different contexts and milieus. These variations in description are important because they help to illuminate disjuncture between moral vocabularies.

---

21 In other words, it is not a question of whether we should transpose "anachronistic" terms and concepts onto the past – there are good reasons for doing so at different times – but one should also be cognizant of what is obscured in the process. Branko Mitrović, "Attribution of Concepts and Problems with Anachronism," *History and Theory* 50:3 (2011); 303–327.

To illustrate the power of vocabulary, take Guy Deutscher's analysis of William Gladstone's commentary on the use of color in Homer's work. Specifically, he noticed that "there was something seriously amiss with Homer's description of color."[22] After an exhaustive study, Gladstone shows that Homer used only a handful of words to describe a variety of – what we would today think of as – colorfully distinct things; indeed, Gladstone demonstrates "how remarkably colorless Homer's vibrant verse is" – he uses black 170 times, white 100 times, red thirteen times, yellow ten times, and violet six times.[23] As Deutscher concludes, "In those instances when colors are mentioned, they are often vague and highly inconsistent: his sea is wine-colored, and when not wine-colored, it is violet, just like his sheep. His honey is green, and his southern sky is anything but blue."[24] What is even stranger is that these same peculiarities extend to other Greek writers of this period.

The conclusion Deutscher derives from this example is not that the ancient Greeks were unable to differentiate between colors – many early commentators, including Gladstone, surmised that the Greeks were color-blind or had not yet developed the ability to differentiate color – but rather their available color vocabulary points to a particular cultural expression which curated, framed, limited the ways in which the ancient Greeks experienced and talked about the world. While there may be no phenomenal difference between a world without the color-concept "blue" and a world with it, in a very real sense it is only within a world with the word-concept "blue" that certain meanings and values can accrue. And these meanings and values open new analogies, metaphors, representations, and ways of being.

It is significant that no language attempts to delimit or account for every discrete phenomenon. Many things go unnamed, and some might even be unnamable inasmuch as they are supersensible.[25] This raises important questions as to why and for what reason language-groups delimit some phenomena and not others. The choices and accidents that lead to their being some

---

22 Guy Deutshcer, *Through the Language Glass: Why the World Looks Different in Other Languages* (New York: Metropolitan Books, 2010), 31.

23 Ibid., 34, 35.

24 Ibid., 36.

25 Kant, for instance, argues that abstract concepts – like freedom, justice, and the state – need effective analogies and symbols in order to be rendered sensible, i.e. to generate an intuition about the concept. In this sense it is significant that languages employ a "scaffolding" of symbolic and analogical language for the purposes of rendering otherwise inscrutable concepts intuitive, what he calls "hypotyposis." The implication is that without effective analogies, tokens, and symbols supersensible phenomena would remain unintelligible. See Immanuel Kant, *Critique of the Power of Judgment* trans. Paul Guyer and Eric Matthews (New York: Cambridge University Press, 2009), §59, 225.

word-representations and not others are consequential, and they point to the profound ways that words can come to influence our experience of the world. Some of the significance of this position can be summarized in the following thought experiment that Deutscher asks us to imagine. He invites English speakers to try to view the experience of color through the Russian language, and to "look at *siniy* [dark blue] and *goluboy* [light blue] as two separate colors."[26] In doing so, one can begin to understand what is at stake in the ancient Greeks language's lack of differentiation between blue and green:

> Just as English lumps *goluboy* and *siniy* under one "blue" concept, other languages extend this lumping principle to the whole green-blue range. And if you happened to grow up in a culture where this chunk of the spectrum has just one label, let's say "grue," wouldn't it seem silly that some languages treat leaf grue and sea grue as two separate colors rather than as two shades of the same color?[27]

This thought experiment is designed to show that color vocabulary does not automatically reduce to a universalist, biological, or phenomenal necessity. After all, there is no relevant biological difference between the British, Russians, and the ancient Greeks; they all could make, if asked, the same discriminations between colors (unless they were color blind). However, it is significant that these linguistic-cultural groups make the color distinctions that they do, and it is also significant how these differences allow the speakers of those languages to interface with the world in specific ways.

Recent studies show that Russian speakers, when presented with timed, fast-paced color discrimination tasks were more effective in distinguishing between dark and light blue colors than English speakers completing the same task.[28] Not only is there evidence that the "salience of color categories changes across languages," but that cultural variables influence the formation of these categories.[29] There is strong experimental evidence to support, at the very least, a "weak relativity hypothesis," which suggests that even though there are

---

26 Deutshcer, *Through the Language Glass*, 69.

27 Ibid.

28 Jonathan Winawer, Nathan Witthoft, Michael Frank, Lisa Wu, Alex Wade, and Lera Boroditsky, "Russian Blues Reveal Effects of Language on Color Discrimination," *Proceedings of the National Academy of Sciences* 104:19 (2007); 7780–7785.

29 Ian Davies and Greville Corbett, "A cross-cultural study of colour grouping: Evidence for weak linguistic relativity," *British Journal of Psychology* 88:3 (1997); 493–517; and M. Pilling and I.R. Davies, "Linguistic relativism and colour cognition," *British Journal of Psychology* 95:4 (2004); 429–455.

remarkable similarities in color concepts between language groups, there are also non-trivial differences.[30] Experimental evidence demonstrates that "new color terms are added to convey hard-to-name colors that have become salient in a language/culture."[31]

Of course, linguistic evolution is a very complex and obscure process, likely linked to biology, accidents of etymology, cultural contingencies, and other aesthetic and utility-based considerations that gave rise to some terms and not others. The point is that this implicit/explicit selection process – i.e. the process of repurposing old words and the creation of new words to demarcate phenomena – is suggestive of a complex interplay between bio-linguistic evolutionary constraints on the one hand, and cultural valuation and selection on the other. And while it may be difficult to quantify the exact cultural significance or value that accrues to discrete terms across languages, recent experimental evidence shows that the language we use impacts our ability to interface with the world in crucial ways that vary across language groups.[32]

In short, the language we use matters; it influences the way we interface with the world, and in this sense, there is an analogy to be made between color-vocabulary and military ethical terms and concepts. The ways words demarcate, delimit, and frame phenomenal experiences allow us to interface with the world in new and unprecedented ways. Again, the point is not that words can mean whatever we want them to mean or that all of reality is merely a cultural-linguistic construct but rather that words help to create new pathways of thought and action. They help us to frame the possible and permissible. A world with "civilian casualties" is different than a word which has a passing interest in the death of "townspeople." These terms help mediate and curate our experience with the world. In short, the claim here is that the historical contingencies which give rise to certain terms are important, and the development of moral terms not only delimit a phenomenon, but they allow us to interface with the world in new ways.

---

30 In other words, one does not need to commit a strong version of the Sapir-Whorf hypothesis or linguistic relativity to accept this argument.

31 Galina Paramei, Yulia Griber, Dimitris Mylonas, "An online color naming experiment in Russian using Munsell color samples," *Color: Research and Application* 43:3 (2018); 358–374. 359.

32 For instance, recent studies show that language can influence how one experiences the visual world and time. See S. Chabal and V. Marian, "Speakers of different languages process the visual world differently," *Journal of Experimental Psychology* 144:3 (2015); 539–550; and Bylund and Athanasopoulos, "The Whorfian time warp: Representing duration through the language hourglass," *Journal of Experimental Psychology* 146:7 (2017); 911–916.

3 Methodology

The arguments in this book are derived from the textual analysis of an array of contemporaneous source material spanning from the sixteenth to twentieth century: military manuals, histories, treatises on international law, treaties and conventions, conference proceedings, memoirs, journal and newspaper reportage, philosophical texts, war correspondent literature, private records, peace activist reports, academic articles, and works of fiction.[33] Instead of focusing specifically on the nuances of singular arguments, the survey of this literature looks at trends, broad continuities, and clusters in thought that inform judgments about the killing of the innocent in war. In short, a reader does not need to be specifically familiar with each point of reference. Since this book seeks to link commentators on this subject across time and genre of writing, the assumption is that most readers will not be familiar with many of the authors addressed in this book. While some space is dedicated to situating and contextualizing individual arguments, this book seeks to identify shared assumptions and broadly ascendant norms across time.

To aid in this effort, interspersed throughout this book one will find Google Ngram Viewer search results.[34] Drawing on Google's herculean effort to create a digital library of all the books in print – as of 2017 they had digitized somewhere on the order of 30 million of the roughly 130 million books in existence, scanning billions upon billions of pages of printed text[35] – the Ngram Viewer allows researchers to make comma-delimited word searches in English, French, German, and several other languages.[36] The results of these searches

33 For all primary source material see the Bibliography.

34 *Gram* comes from the Greek *gramma*, signifying something has been scratched, drawn, or written, such as alpha-numeric characters, signs, or symbols of any kind. *N* stands for the number of such characters, signs, or symbols linked together in sequence. An *Ngram*, therefore, is a search for any sequence of *grams* strung together, be they words, numbers, symbols, phrases, etc. http://books.google.com/ngrams.

35 On 14 July 2020, the catalogue was updated to include material through 2019. Google SearchLiaison, 2020, 14 July. https://twitter.com/searchliaison/status/1282797986863386624.

36 For background on the methodology and use of Google Ngram Viewer for historical analysis, see Jean-Baptise Michel, Yuan K. Shen, Aviva Aiden, Adrian Veres, and Matthew Gray, "Quantitative Analysis of Culture Using Millions of Digitized Books," *Science* 331:6014 (2011); 176–182; A.C. Sparavigna, and R. Marazzato, "Using Google Ngram Viewer for Scientific Referencing and History of Science," *airXiv* Preprint, airXiv:1512.01364 [cs.DL] (2015); and Thomas Lansdall-Welfare, Nello Cristianini, "History Playground: A Tool for Discovering Temporal Trends in Massive Textual Corpora," *airXiv* Preprint, airXiv:1806.1806.01185[cs.CL] (2018). See also, Erez Aiden and Jean-Baptiste Michel, *Uncharted: Big Data as a Lens on Human Culture* (New York: Penguin, 2013).

chart the frequency of a *gram* used (in print, obviously) for each year out of a desired range (but becomes noticeably spotty the further beyond the nineteenth century one goes). These searches provide a striking macroscopic view of cultural-linguistic trends.[37] Some have even suggested that embedded within the frequency of word-usage may be profound clues about cultural or normative trends, touching on an array of social or political issues.[38]

Early advocates of this methodology were quite optimistic that Ngrams would not only give a much more robust insight into cultural-linguistic change, but that it might be used to generate predictions about where culture was moving.[39] It turns out, however, that it is quite difficult to draw robust (and sometimes even reasonable) inferences from Ngram searches. They are remarkable macroscopic snapshots of word usage, but it is not always clear how to interpret word-use variation over time. After all, these searches do not capture variations in meaning or sense, and in this regard the graphic representations of word frequency may give the illusion of (dis)continuity that does not always exist. This is in part because the more frequently a word is used, the more likely it is to have different meanings and usages. Word counts bundle all word-usages together, so the most frequently used words will capture far too many meanings to make reasonable differentiations.[40]

There are many other pitfalls that one must consider, as well. Search results do not give us an indication as to whether a word or phrase is positively or negatively used. Certain texts were reprinted in multiple sources, which may lead to redundancy and overrepresentation in the digital archives. Some have also noted the selection bias in Google's strategy of scanning university libraries. Ngram searches will reflect the purchasing and collection patterns of academic book repositories. Academic literature may be overrepresented, and

---

37 See Shawn Graham, Ian Milligan, Scott Weingart, *Exploring Big Historical Data: The Historian's Macroscope* (London: Imperial College Press, 2016).

38 See Seth Roth, "Fashionable Functions: A Google Ngram View of Trends in Functional Differentiation (1800–2000)," *International Journal of Technology and Human Interaction* 10:2 (2014); 34–58; Jamie Trinidad, "'Culturnomics' and International Law Research," in *International Law and Select Proceedings of the European Society of International Law* eds. August Reinisch, Mary Footer, and Christiana Binder (Oxford: Bloomsbury, 2016), 261–276; and Ted Underwood, *Distant Horizons: Digital Evidence and Literary Change* (Chicago: University of Chicago Press, 2019).

39 See Kalev Leetaru, "Culturomics 2.0: Forecasting large-scale human behavior using global news media tone in time and space," *First Monday* 16:9 (2011); doi:10.1093/llc/fqv037.

40 For some recent methodological guidance on using Ngrams, see Nadja Younes and Ulf -Dietrich Reips, "Guideline for Improving the Reliability of Google Ngram Studies: Evidence from Religious Terms," *Plos One* 14:3 (2019); 1–17. https://doi.org/10.1371/journal.pone.0213554.

popular media and literature not to mention other non-academic sources are less likely to be accounted for.[41] Finally, and perhaps most problematically, this approach assumes that *frequency* in print is a signal of cultural value.[42] There are many reasons to be cautious and indeed skeptical about privileging frequency in print when assessing a word's cultural value.

The outspoken polemicist, Christopher Hitchens, provides a humorous anecdote that hints at what the omission of certain words in print might signify. After completing his famous dictionary, Samuel Johnson, the eighteenth-century moralist and lexicographer was, as Hitchens explains,

> waited upon by various delegations of people to congratulate him, of the nobility, of the quality, of the Commons, of the Lords – and also by a delegation of respectable ladies of London, who tended on him at his Fleet Street lodgings, and congratulated him.
>
> "Dr. Johnson," they said, "we are delighted to find that you have not included any indecent or obscene words in your dictionary."
>
> "Ladies," said Dr. Johnson, "I congratulate you on being able to look them up."[43]

This is a very clear example of an instance in which the cultural value of – and one might even say psychosexual obsession with – certain words militate against their registry in Ngram word counts. Words can have immense cultural value in their *absence*. That-which-shall-not-be-named can be a source of deep devotion, fear, anxiety, repugnance, and so on. Sacred and profane words and phrases have typically been governed by different norms in writing. Furthermore, things get much more complicated when one considers the role of (self)censorship in writing, i.e. how word usage might be legislated, not

---

41 Eitan Adam Pechenick, Christopher Danforth, and Peter Sheridan Dodds, "Characterizing the Google Books Corpus: Strong Limits to Inferences of Socio-Cultural and Linguistic Evolution," *PLoS ONE* 10:10 (2015).

42 For more on the problems and promises of "word-frequency," see Ioan-Iovitz Popescu, *Word Frequency Studies* (Berlin: Mounton de Gruyter, 2009). For some of the many challenges of making valid inferences on Ngram view data, see Alexander Koplenig, "The impact of lacking metadata for the measurement of cultural and linguistic change using the Google Ngram data sets – Reconstructing the composition of the German corpus in times of WWII," *Digital Scholarship in the Humanities* 32:1 (2015); 169–188.

43 "Christopher Hitchens Defending Free Speech Debate," Hart House Debating Club, University of Toronto, 2006-11-15. https://www.youtube.com/watch?v=H_ohvc_ZqzA (Accessed 20 May 2021).

to mention the multiple strategies oppressed communities employ to bypass censorship laws through creative or coded phrasing.[44]

Finally, there is an immense literature dedicated to Martin Heidegger's strategic devise of *sous rature* – writing "under erasure" – where an "erased" term remains visible though crossed through. Being struck-through signified both the term's inadequacy but functional necessity.[45] A frustratingly deficient word may be used frequently. The point is that certain words are, in themselves, suggestive of a lacuna. Even though language falls short, there is ultimately no way to escape the limitations in word use. As such, words may be deployed as a placeholder or mere filler. This analysis subsequently gave rise to deconstructionism. Following Heidegger, Jacques Derrida emphasized the ultimate indecipherability of words, where embedded in each gram is a kind of linguistic play.[46] The idea is that words entangle with their negations and binary contradictions to produce a kind of indeterminacy of meaning. Under these conditions, it would be impossible to extract individual words from the web or matrix of references in which they are embedded. On this view, it would be difficult to know what the frequency of discrete grams would signify at all, that is, extracted and denuded from their ecology of reference.

To further illustrate the challenges of strong inferences in Ngram searches that are more relevant to this present analysis, refer to Figure 0.2 below. There one can see the historical-linguistic trajectories of the word-phrases "international law" and "law of nations." In the early 1870s, international law overtook and started to rapidly outpace the much older expression law of nations. At first glance, one could take these trends to confirm Mark Janis's thesis that Jeremy Bentham's utilitarian-linked conception of international law over time slowly started to overtake the common and natural law theories which typically referred to the law of nations.[47] After all, for reasons that will be seen in subsequent chapters of this book, there is good reason to think that a

44 See Keith Allan & Kate Burridge, *Forbidden Words: Taboo and the Censoring of Language* (New York: Cambridge University Press, 2007); and Bryan Trabold, *Rhetorics of Resistance: Opposition Journalism in Apartheid South Africa* (Pittsburg: University of Pittsburg Press, 2018).

45 Martin Heidegger, *The Fundamental Concepts of Metaphysics: Word, Finitude, Solitude*. Trans. William McNeill and Nicholas Walker (Bloomington, IN: Indiana University Press, 1995).

46 See Jacques Derrida, *Of Grammatology*. Trans. Gayatri Spivak (Baltimore, MD: Johns Hopkins University Press, 2016).

47 See Mark Weston Janis, "Blackstone and Bentham: The Law of Nations and International Law," in *America and the Law of Nations 1776–1939* (New York: Oxford University Press, 2010).

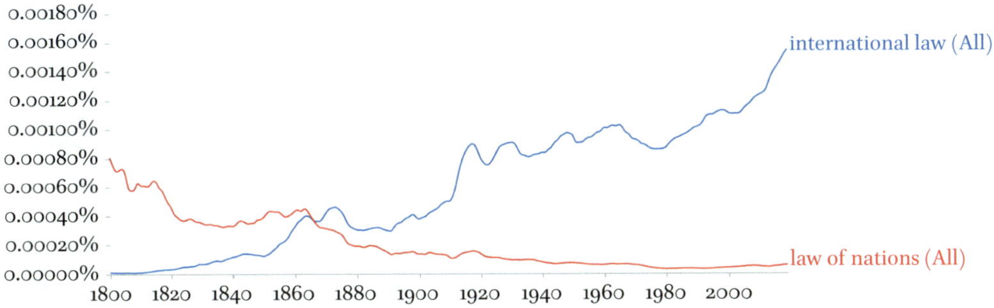

FIGURE 0.2 Ngram search of 'international law' and 'law of nations'

Benthamite influence was responsible for a significant shift in ethical reasoning. Superficially at least, one could conclude that this graph captures the ideological clash that took place between these conflicting traditions.

The difficulty with this analysis, however, is that in common usage, as Janis admits, "Bentham's term 'international law' soon became synonymous with the classical term 'the law of nations,'" that is, even though "Bentham's restriction of the subject to inter-state relations was neither faithful to the more generous scope of the classical law of nations nor a fair approximation of state practice."[48] In short, commentators with different ideological persuasions often used these terms interchangeably, without much thought as to the conceptual origins or traditions they ostensibly refer to. In fact, many people who employ the term "international law" may be entirely unaware of or indeed indifferent to its utilitarian origins.[49]

The preceding paragraphs have enumerated the challenges and problems with this methodology in order to stress the high degree of caution needed when making inferences from linguistic metadata. Many of the purported claims that Ngrams elucidate "objective" cultural-linguistic trends are deeply problematic, to say the least. There are, however, three uses of word frequency that are employed in this book: 1) identifying terminological origins, 2) spotting linguistic or norm cascades, and 3) showing word or term obsolescence.

48 See M.W. Janis, "Jeremy Bentham and the Fashioning of 'International Law,'" *American Journal of International Law* 78:2 (1984); 405–418.

49 Edward Gallaudet argues that the term can be traced to Richard Zouch's 1650 treatise *Juris et judicii decialis sue juris inter gentes*: "The jurist was the first who designated our science by words *jus inter gentes*, and it is from this expression that the term international law has been derived." Edward Gallaudet, *A Manual of International Law* (New York: A.S. Barnes & Co, 1879), 17.

3.1 *Terminological Origins*

Ngrams can capture terminological origins and what might be called linguistic coalescence, where a term is coined, or an existing word is repackaged. While a concept may have existed in some antecedent sense prior to the term or phrase that eventually arises, the emergence of new terms or phrases can represent a normative focus, or an emerging cultural significance that did not exist prior to its formation. Take the example of "genocide" (see Figure 0.3). The word genocide was a neologism coined by Raphael Lemkin in 1944.[50] He explains that "New conceptions require new terms" and that "This new word, coined by the author to denote an old practice in its modern development, is made from the ancient Greek word *genos* (race, tribe) and the Latin *cide* (killing, thus corresponding in its formation to such words at tyrannicide, homicide, infanticide, etc.)."[51] As Lemkin acknowledges, the creation of the word genocide was not meant to imply that violence directed against certain groups *qua* groups was new. Rather it was the unprecedented qualitative and quantitative novelty of the multiple crises of the early twentieth century that required terminological and indeed moral precision in delimiting the phenomenon.

Lemkin argues, "The above-described techniques of genocide represent an elaborate, almost scientific system developed to an extent never before achieved by any nation. Hence the significance of genocide and the need to review international law in the light of the German practices of the present war."[52] While in some sense related to "denationalization" and "mass murder" – the terms historically used to speak of related, antecedent practices[53] – the

50 As is the case with many Ngram searches, there is some misdating of texts. According to the Ngram search, it seems as though "genocide" was coined around 1940, but upon closer examination of the texts the search brings up, they were actually printed at later dates. Some of the earlier hits are for different words that have similar spelling. For more on the problem of misdating in Ngram, see Laura Miller, "The Trouble with Google Books: How Rampant Errors Threaten the Scholarly Mission of the Vast Digital Library," *Salon* 9 September 2010. https://www.salon.com/2010/09/09/google_books/ (Accessed 20 May 2021).

51 Raphael Lemkin, *Axis Rule in Occupied Europe: Laws of Occupation, Analysis of Government, Proposals for Redress* (Washington D.C.: Carnegie Endowment for International Peace, 1944), 79. Lemkin had also toyed with the word *ethnocide*.

52 Ibid., 90.

53 Lemkin argues that "Denationalization was the word used in the past to describe the destruction of a national pattern" (Ibid. 79). Incidentally, in a preliminary session of the Treaty of Versailles, the Commission on the Responsibility of the Authors of the War and on Enforcement of Penalties ruled that in WWI Germany and Turkey had violated the 'laws of humanity' regarding Turkey's Armenian and Greek speaking population. They had been guilty of "murders and massacres; systematic terrorism," "deportation of civilians," "internment of civilians under inhumane conditions," and "attempts to denationalize the

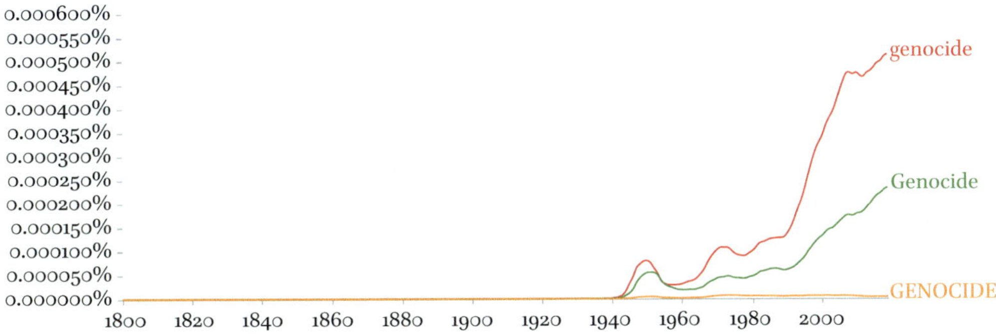

FIGURE 0.3 Ngram search of 'genocide'

Holocaust exploded these historical categories. While one could quite reasonably point to precursors that in some sense approximated the phenomenon, it was only with the formulation of the terminological category *genocide* that those acts acquired the moral dimension they now do. Armed with this neologism, we have been able to cast our eyes back onto history to morally engage with antecedent cases, though, as noted above, it is not without problems that we do so.

3.2 *Linguistic or Norm Cascades*

Ngram searches can capture linguistic or norm cascades – where an existing word or phrase goes from relative obscurity to prominence in a short period of time.[54] While genocide works for this category, take another example: "human rights" (Figure 0.4). Around the middle of the nineteenth century, one finds only a smattering of references to human rights, which was sometimes expressed as "natural human rights." In 1845, Elisha P. Hurlbut, a nineteenth-century jurist who served on the New York Supreme Court, published *Essays on Human Rights, and Their Political Guaranties*, in which he traces the origins of human rights to the revolutions of the late eighteenth century, which were

inhabitants of occupied territory." "Commission on the Responsibility of the Authors of the War and on Enforcement of Penalties," *The American Journal of International Law* 14:1/2 (1920); 95–154. 114.

54 For the development of norm cascades, see Cass Sunstein, "Social Norms and Social Roles," *Columbia Law Review* 96:4 (1996); 903–968; and Cristina Bicchieri, *Norms in the Wild: How to Diagnose, Measure, and Change Social Norms* (New York: Oxford University Press, 2017). See also, Kathryn Sikkink, *The Justice Cascade: How Human Rights Prosecutions are Changing World Politics* (New York: Norton, 2011).

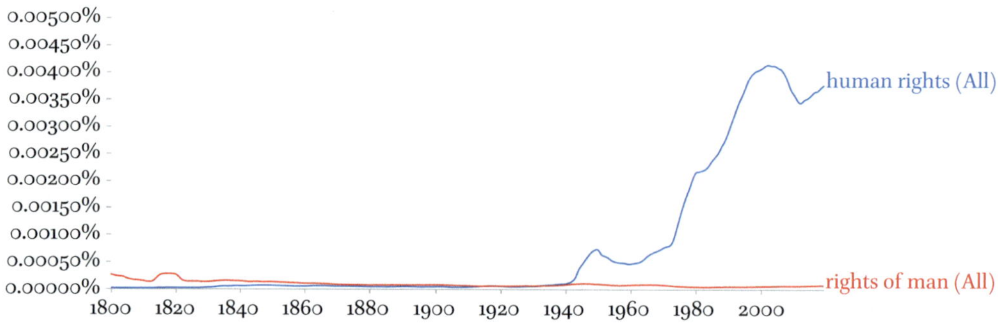

FIGURE 0.4 Ngram search of 'human rights' and 'rights of man'

articulated in the language of natural rights, specifically the "Rights of Man."[55] Despite this relatively early expression of the concept, Figure 0.4 shows that for about 100 years, from 1840 to 1940, references to human rights were almost non-existent – they were, in fact, slightly less frequent than "rights of man" for that same period. In the years following World War II, however, there is a noticeable uptick. Much like the quick coalescence around the word "genocide," the rapid surge in the usage of "human rights" signifies the emergence of a moral and conceptual focus that lacked a sustained formal or theoretical dimension prior to the 1940s. In a very real sense, the crisis of the Holocaust and World War II generated a historical moment where the analytic technology "human rights" found greater conceptual and moral purchase.

Of course, the cascade in word use, as the examples of "genocide" and "human rights" illustrate, is not meant to imply consensus or unanimity, that there is something like a uniform and agreed upon usage of these words that can be traced through time. Undoubtedly, the vociferous disagreements that continue to inform the debates around these words attest to the difficulty in precisely pinning down meanings. But this indeterminacy is part of what makes the cascade a point of methodological interest. These cascades ultimately demonstrate that these terms themselves have quickly become sites of debate and moral concern, in all their polyvocality and contention. And as a site of new or newly inflected concern, novel and unanticipated judgments begin to emerge.

55 Elisha Hurlbut, *Essays on Human Rights, and Their Political Guaranties* (New York: Fowlers and Wells, 1848). See also Lynn Hunt, *Inventing Human Rights: A History* (New York: Norton, 2008).

3.3 *Word Obsolescence*

Finally, Ngram searches can illustrate the obsolescence of words or terms that used to be quite common, either in the form of a slow decline or a reverse cascade that is also suggestive of changes in the ethical landscape.[56] Take, for instance, "letters of reprisal" and "retorsion" (Figure 0.5). Both concepts are now quite obscure. Historically, letters of reprisals could be obtained when an individual was aggrieved by a foreign nation, but he could not obtain justice through the foreign nation's legal system. The individual would petition his own sovereign for letters of reprisal in order to take matters into his own hands.[57] Going back to the Middle Ages, this was seen as a legitimate form of private war. It is perhaps significant that this mechanism of justice started to go out of fashion around the end of the Napoleonic Wars and the subsequent Congress of Vienna (1815) which gave rise to the Concert of Europe. While it was increasingly seen as a problematic institution – in the late eighteenth century, for instance, Benjamin Franklin had advocated for its abolition[58] – it was not until it threatened the stability of the balance of European powers that they were formally abolished by international agreement in the Paris Declaration (1856), the agreement which ended the Crimean War (notice the final significant uptick in usage of "letters of reprisal" occurs around this time).

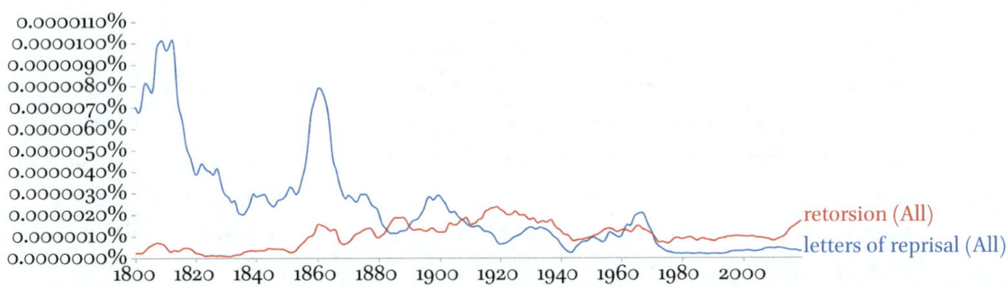

FIGURE 0.5 Ngram search of 'letters of reprisal' and 'retorsion'

56 Despite some research that attempt to identify the mechanisms of word death, it is not clear that this is a settled issue. See Tim W. Machan, *What is English?: And Why Should We Care?* (New York: Oxford University Press, 2013); pp. 57–84. See also, Alexander Petersen, Joel Tenebaum, Shlomo Havlin, and H. Eugene Stanley, "Statistical Laws Governing Fluctuations in Word Use from Word Birth to Word Death," *Scientific Reports* 2:313 (2012); doi:10.1038/srep00313.

57 Article 1, Section 8, Clause 11 of the US Constitution, for instance, gives congress the power to "declare war, grant letters of Marque and Reprisal, and make rules concerning captures on land and water."

58 Baron Russell, *The French Corsairs* (London: Hale, 1970), 34.

The decline and eventual abolition of "letters of reprisal" charts an important shift in the formalization and, indeed, professionalization of international politics. No longer were states believed to be permitted to outsource their foreign policy to private individuals, especially as it pertained to issues of crime and punishment. Grievances between the private citizens of different states needed to be addressed through the formal channels of their governments. It is significant that in this period of decline for letters of reprisal we witness a simultaneous rise of retorsion (i.e. in-kind retaliation). Even though it was a morally challenging tool of foreign policy, nevertheless letters of reprisal represented, at least in theory, a kind of pressure-release valve. It was a mechanism of perceived justice that was just short of war. In the moral and legal vacuum that ensued from this ethical and institutional shift, retorsion started to emerge as a conceptual alternative to what had formally been the domain of private war. States could justifiably engage in limited retaliations to avenge wrongs that the offending state was unable or unwilling to prosecute. And while it never gained the prestige that letters of reprisal did (this is perhaps because, as will be seen, retorsion was routinely conflated with terms "reprisal" and "retaliation") its rise was suggestive of important tectonic shifts that were taking place in the international arena at that time. Of course, both words have largely disappeared from public discourse – a kind of word death.

3.4 *How Ngrams Supplement the Central Argument*

The central argument of this book is not inferred or derived from Ngram key word searches. Rather, Ngram searches are provided to quantitatively dramatize the trends unearthed through careful analysis of a trove of primary source materials from the thirteenth to twentieth centuries. These graphs illustrate points of interest in the rise of new ethical concepts that began to calcify and were operationalized in altogether new ways.

For instance, take the principle of noncombatant immunity mentioned above (Figure 0.6). Though military ethicists and legal theorists have argued

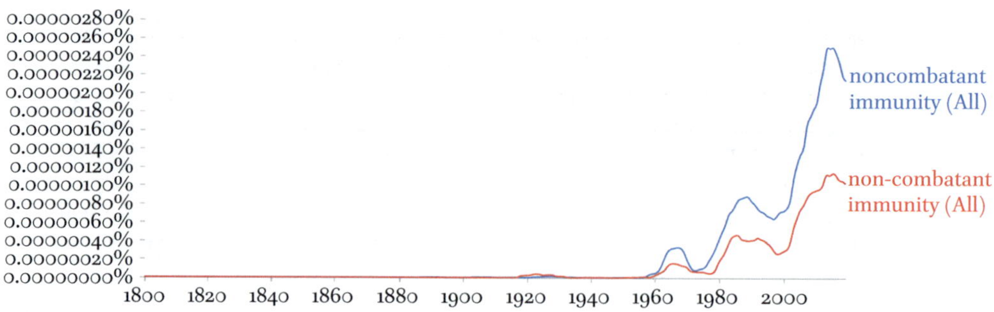

FIGURE 0.6 Ngram search of 'noncombatant immunity' and 'non-combatant immunity'

for certain protections to be extended to different classes of people, the formal notion of "noncombatant immunity" did not emerge until the twentieth century. At least those words do not appear together until then. It is perhaps not surprising that with the growing focus on securing and protecting human rights after World War II that noncombatant immunity would partake in this upsurge of interest. Indeed, it appears that most of the writing on noncombatant immunity follows the burst of writing on human rights, around the same time as the human rights "breakthrough" took place in the 1970s and with it the rise of the human rights culture.[59] Furthermore, as the formal category of "human" – the individual holder of rights – expanded (through abolitionist, civil rights, and postcolonial movements) we begin to see an increase in theorizing about the kinds of protections that should be extended to those not engaged in war-making.[60] The right to life quite reasonably translated into immunity from the arbitrary and accidental harms stemming from war-making.

Once the concept of noncombatant immunity achieved more formal conceptual status, even if actual protections lagged, one can see the concomitant focus placed on the intentionality of targeting civilians. It is only after Protocol I of the Geneva Convention in the late 1970s that the bulk of the writing on intentionally or deliberately targeting civilians occurs. Again, as will be seen in the subsequent chapters, throughout the eighteenth and nineteenth centuries militaries routinely believed they were justified in perpetrating what we today would call "intentionally targeting civilians," that is, under certain conditions. Quite simply, intention was not the morally significant variable in determining whether civilian death was legitimate or not. It is important to recognize that it would be anachronistic to locate the concepts of "civilian casualties" and "noncombatant immunity" in military ethical discourse prior to the twentieth century. As noted above, the idea of the civilian was only slowly acquiring a formal character throughout the eighteenth and nineteenth centuries. Terms like "inhabitants," "residents," "subjects" or "noncombatants" were much more commonly used to describe innocent victims of war. But what is more, it is problematic to retroject the static conception of noncombatant immunity back onto the nineteenth century. As will be seen, while noncombatants were gaining more and more protections in the military ethical literature of this period, 'noncombatant immunity' is more precisely a product

59 Jan Eckel and Samuel Moyn (eds.), *The Breakthrough: Human Rights in the 1970s* (Philadelphia: University of Pennsylvania Press, 2013).

60 For more on the evolution of humanitarian norms, see Neta Crawford, *Argument and Change in World Politics: Ethics, Decolonization, and Humanitarian Intervention* (New York: Cambridge University Press, 2002).

of the human rights culture of the twentieth century. Indeed, throughout the eighteenth and nineteenth centuries there were many ways for non-soldiers to become legitimately entangled in conflict. The moral frameworks governing combat in those times produced their own incentives and prohibitions.

4 Overview of Chapters

This book is divided into three sections. The first section, "Killing the Innocent: Three Discursive Traditions from the Sixteenth to Eighteenth Centuries" examines the rationalizations for killing the innocent that emerged in late Scholastic, martialist, and republican/humanist discourses. While there have been many attempts to classify the various strands of just war thinking on the Continent, this approach looks exclusively at justifications for killing the innocent.[61] Chapter 1, "The Origins of Double Effect: The Scholastic Tradition," shows that Aquinas's original formulation of double effect was ambiguous, and that it did not apply to killing the innocent in war. By the late sixteenth century, as innovations in siege technology progressed, putting women and children in fortified cities in much greater danger, Scholastic commentators adapted and repurposed Aquinas's original formulation of double effect to excuse the killing of the innocent in war. While there were permutations and variations within this tradition, the consensus was that militaries could bomb fortified cities as long as they did so for legitimate strategic reasons and had no intention of killing the noncombatants who resided inside.

Chapter 2, "Martialists: Soldiers and Tacticians," sets out the professional discourse of soldiers and tacticians who had returned to England from the Eighty- and Thirty-Years' Wars. In contrast to the Scholastic tradition, these texts do not explicitly thematize civilian casualties. The emphasis is almost exclusively on the necessity for princes to be steeped in the Art Militarie and the subsequent order and discipline of his military. This conceptual focus had

61 Richard Tuck argues that the Grotian tradition synthesized Scholastic elements, but the principal distinction was between Scholastic and humanist thought. See Richard Tuck, *The Rights of War and Peace: Political Thought and the International Order from Grotius to Kant* (New York: Oxford University Press, 1999). Karma Nabulsi argues that there were three principal traditions of just war from the early modern era to the present: Martialist, Grotian, and Republican. See Karma Nabulsi, "Conceptions of Justice in War: From Grotius to Modern Times," in *The Ethics of War: Shared Problems in Different Traditions* eds. Richard Sorabji and David Rodin (Burlington, VT: Ashgate, 2006). See also, Erica Charters, Eve Rosenhaft, and Hannah Smith (eds.), *Civilians and War in Europe, 1618–1815* (Liverpool: Liverpool University Press, 2012).

an indirect impact on the question of killing the innocent. Soldiers should not be governed by inordinate passions, for instance, indiscriminately killing women and children when storming a fortress. Instead, they should strictly follow lawful orders, the authoritative judgments of their princes and commanders. Within this tradition, killing the innocent was not a priori prohibited. What mattered was whether the killings had been authorized by a lawful command.

The final chapter of the first section, Chapter 3, "Humanists and Republicans," presents the arguments from an array of commentators, from Machiavelli to Vattel. While this chapter clusters together a rather divergent group of thinkers, what unites these theorists is the appeal to the principles of necessity and right in their justifications for killing the innocent. There are, of course, many nuances and particularities within this tradition, but in contrast to both the Scholastics and martialists, humanists and republicans routinely elided the moral exculpatory force of good intentions or the simplistic appeal to right authority, instead favoring prudential wisdom and the calibrated use of necessary violence to pursue just ends. In other words, in certain difficult cases one could deliberately kill the innocent on the basis of right and necessity.

The second section, "The Nineteenth Century," seeks to explain how the robust, polyvocal discourses around killing the innocent of the prior three centuries consolidated into the virtuously unanimous and homogenous framework that dominated the nineteenth century. By the early nineteenth century, virtually every prominent commentator on the laws of war and international law more broadly justified the death of innocents by way of necessity or some rudimentary utility calculus. Chapter 4, "Killing Noncombatants in the Nineteenth Century," explores several significant variables that explain this shift, for instance, the rise of concepts like the "noncombatant" and "military necessity." Both ideas gain conceptual ground during this time. What is more, this chapter also looks at the legacy of utilitarianism and the post-Westphalian discourse on self-preservation that dominated the recourse to violence in this period.

Contrary to the popular claim that noncombatant immunity was a central tenant of international law of the nineteenth century, Chapter 5, "Deliberately Killing Noncombatants," shows that from the Napoleonic Wars to World War I there were four ways that militaries could deliberately target and kill the innocent. First, it was widely accepted that states had a right to lay waste to border lands in order to eliminate threats or to punish enemies. Ravaging, looting, and burning were all permissible. Secondly, the enemy could be starved into submission through blockades. These policies were pursued with the express knowledge that civilians would bear the brunt of these hardships. Third, civilians living in or around fortified places could be explicitly targeted in order to expedite surrender. While it was believed to be honorable for besieging armies to permit women and children to leave the surrounded city, this was not required. And finally, if an enemy nation was believed to have deliberately killed the innocent, the wronged country had the right to repay the perceived violation through reprisals, retaliation, or retorsion. These were terms that meant different things at different times throughout the history of military ethics, but, in essence, the norms that emerged around these practices permitted states to justify the killing of the innocent for other people's crimes.

The third section, "The Twentieth Century: Aerial Bombing and a Shift in Norms," provides a detailed account of how the ascendant norms of military necessity that were operative in the nineteenth century gave way to those of the twentieth century. Chapter 6, "A New Problem: Inventors and Future-War Fiction on Aerial Bombing," is a forecast of the transformation that was slowly taking place. Inventors, aeronaut adventurers, and military tacticians all weighed in on how weaponized air ships would change the way war was waged in the twentieth century. Many believed that the terror it would inflict on the noncombatant residents of enemy nations would make the consequences of war so unpalatable that nations would be forced to find alternative ways to solve international disputes. Bombers would be a source of peace. While the technology was still in its infancy, future-war fiction writers also started to imagine the impact of aerial bombing. While some suspected that these terrible new weapons would be a source of peace, others envisioned apocalyptic futures.

Chapter 7, "Interwar Approaches to Bombing: Two Discursive Traditions," follows the interwar debates on aerial bombing. The all-consuming destructive power of conventional land forces and the negligible impact of aerial bombing during World War I failed to resolve the central debates that had persisted in the early twentieth-century, pre-war period. Many continued to argue that air warfare would be even more devastating in the next war, while bombing optimists believed that it would make war *less* devastating or of a much shorter duration. This chapter follows these two discursive traditions. The first, international liberals, argued that the laws of war should be modified to specifically address the impact bombing would have on civilian populations. In the many interwar conferences on revising the laws of war and disarmament, one begins to see intention-based arguments for killing the innocent reemerge, i.e. bombing could be permitted as long as pilots do not intentionally target civilians. Quite in contrast to this tradition, bombing realists, still very much under the influence of nineteenth-century military ethical norms, argued that limited numbers of civilians should be swiftly and deliberately targeted to expedite victory.

In the interwar period, it became common to characterize the moral stakes of the aerial bombing of civilians in terms of intentionality. It was becoming a moral liability to intentionally bomb civilians. What is so unusual about this moral formulation is that at no point throughout the nineteenth century had accusations like this been made. The laws of war, in fact, had permitted the deliberate bombing of residential districts in siege warfare. Even detached suburbs could be bombarded. Chapter 8, "The Return to Intention: Post World War I" seeks to explain this shift to intention-based language. There were three principal factors that contributed to this transformation. First, it is significant that German war crimes were characterized as intentional. Despite its incoherence – technological limitations largely vitiated any realistic appeal to intention – the documents from the Paris Peace Conference repeatedly characterized German guilt in these terms. In the debates at this conference, one can see in vivid relief the tension between the nineteenth-century logic of necessity and the emerging norm of intention, which was largely grounded in ethical formulations based on the "law of humanity." Furthermore, as international jurists were looking to (re)build an international order in this period, it is also noteworthy that a renewed interest in the late Scholastic work of Vitoria and Suarez emerged. Moreover, given the rise of intention-based language used to justify bombing in war, a number of prominent Catholic writers vociferously pushed back against what they saw as a misapplication of the principle of double effect. All these trends further popularized and reinforced the turn to intentions in governing the morality of killing in war.

Chapter 9, "Postscript," follows in a very general way the legacy of intention-based arguments post-World War II, especially as it pertains to the rise of nuclear weapons and terrorism. Over time, this ethical language became more enshrined in post-World War II international agreements. What is more, intention evolved into the central moral criterion differentiating tragic but necessary civilian casualties from terrorism – militaries kill civilians incidentally or accidentally whereas terrorists kill civilians intentionally. However, recent trends in moral psychology concomitant with emergent military technologies – precision-guided munitions, automated systems, and artificial intelligence – all place significant strain on the language of intention. It is unclear whether this ethical language is well suited for the twenty-first century. Understanding how ethical justifications for killing the innocent have been adapted and reformulated in response to new technological environments will provide a useful basis for understanding the impending shifts that are unfolding in the twenty-first century.

PART 1

Killing the Innocent: Three Discursive Traditions from the Sixteenth to Eighteenth Centuries

CHAPTER 1

The Origins of Double Effect: The Scholastic Tradition

1 Recovering the Primitive Christian Tradition

In the early nineteenth century, various peace societies started springing up throughout the West.[1] This timing was not coincidental; it corresponded to a moment in history when international law was in its infancy and incipient humanitarian movements (abolitionist, male suffrage, labor rights) were making tenuous progress. The optimism that war and other violent cruelties might be eradicated was a product of a general confidence in a humanitarian trajectory in global politics.[2] The ascendant just war tradition of that time condemned wars of aggression, and the emerging concept of state sovereignty had started to make provisions against wars of retribution, e.g. the idea that states could be punished for violating the natural law. Justified war was increasingly being pressed into the much more limited corner of self-defense, which often included the defense of weaker states or one's own ethnic minority who lived within neighboring states.[3] Broadly speaking, states were justified in taking only those actions believed necessary to deter or repel an invader in order to preserve the state.

---

1 The Massachusetts and New York Peace Societies were founded in 1815; the London Peace Society was founded in 1816; the International Peace Congress was formed in London in 1843; and the Ligue Internationale de la Paix et de la Liberté was founded in 1861. See Peter Brock, *Radical Pacifists in Antebellum America* (Princeton: Princeton University Press, 1968); Meredith Weddle, *Walking in the Ways of Peace: Quaker Pacifism in the Seventeenth Century* (New York: Oxford University Press, 2001); and David Cotright, *Peace: A History of Movements and Ideas* (New York: Cambridge University Press, 2008).

2 For a clear expression of the lost opportunity following the Treaty of Paris (1815), see John Spencer Bassett, *The Lost Fruits of Waterloo: Views on a League of Nations* (New York: Macmillan, 1919). For contemporary analysis of post-war peace regimes, see John Ikenberry, *After Victory: Institutions, Strategic Restraint, and the Rebuilding of Order After Major Wars* (Princeton: Princeton University Press, 2001).

3 For a more detailed look at the changing norms on military intervention and particularly humanitarian intervention in this period, see Martha Finnemore, *The Purpose of Intervention: Changing Beliefs about the Use of Force* (Ithaca: Cornell University Press, 2013); and Alexis Heraclides, "Humanitarian Intervention Yesterday and Today: A History," *European Review of International Studies* 2:1 (2015); 15–37.

 | DOI:10.1163/9789004515482_003

What is perhaps most striking about these nineteenth-century peace movements is that they challenged war-making's most fundamental supposition. Pacifists rejected the premise that states possess an inherent right to self-defense. In his 1815 treatise on Christianity and war, David Low Dodge explains that war is a process of "returning evil for evil" and this is "a fundamental principle in the system of self-defence."[4] Pacifists like Dodge had concluded that the principle of self-defense was far too elastic and permissive. If international peace were truly to be attained, a much more critical eye would need to be directed at the allowances self-defense afford states. Despite its apparent restrictiveness, almost all manner of violence could still be justified through the appeal to self-preservation. Pacifists argued that it merely served as a pretense for injustice; it exacerbated conflict, precipitated reprisal, and led to cruelty and excess. A contemporary of Dodge, Samuel Whelpley, likewise concurs that "it is … this vain and haughty principle of self-defence, which has filled the world with blood and slaughter. It induces preparation, carries hostility on its front, provokes aggression, and after all, generally fails to accomplish its object; for it affords no safety."[5]

Beyond the negative practical effects that stem from the concept, these pacifists also challenged the right of self-defense on theoretical grounds. Jonathan Dymond, for instance, argues that,

> The lawfulness of defensive war, is commonly simplified to the right of self-defence. This is one of the strong holds of the defender of war, the almost final fastness to which he retires. *The instinct of self-preservation, it is said, is an instinct of nature; and since this instinct is implanted by God, whatever is necessary to self-preservation, is accordant with his will.* This is specious, but like many other specious arguments, it is sound in its premises, but, as I think, fallacious in its conclusions. That the instinct of self-preservation is an instinct of nature, is clear – that, because it is an instinct of nature, we have a right to kill other men, is not clear.[6]

As Dymond notes, just because an instinct is natural does not imply that it is authoritative, and it certainly does not follow that one can do anything

---

4 David Low Dodge, *War Inconsistent with the Religion of Jesus Christ, as it is Inhuman, Unwise, and Criminal* (New York: Dodge and Sayer, 1815), 77–78. See also Adin Ballou, *Christian Non-resistance* (London: Charles Gilpin …, 1848).

5 Samuel Whelpley *Letters Addressed to Caleb Strong, Esq. Late Governor of Massachusetts* (Providence, RI: Printed by Miller and Hutchens, 1818), 28.

6 Jonathan Dymond, *An Inquiry into the Accordancy of War with the Principles of Christianity* (Philadelphia: Printed by I. Ashmead & Co., 1834), 86. Emphasis in original.

perceived necessary to survive. After all, even natural instincts can be deceiving. Humans also have the instinct to eat and reproduce, but these natural impulses are also heavily circumscribed in many ascetic, religious, and moral traditions. As Dymond explains, Christianity explicitly requires its followers to "love your enemies better than yourself."[7] In this respect, satisfaction of primal instincts is pitted against obedience to God's law. Dymond, in fact, continues, it "is not whether we should love our enemy better than ourselves, but whether we should sacrifice the laws of Christianity in order to preserve our lives [...], whether we should be willing to 'lose our life, for Christ's sake and the gospel's.'"[8] Christians should not only strive to make love the primary object of their behavior, but they are also commanded to submit to and obey God's law. Fundamentally, these pacifists understood that no matter how justified self-defense might seem, it still demands an evil to be committed, and, as such, much as Socrates had pointed out several centuries prior, it is morally safer to suffer rather than to actively commit an injustice.[9]

These nineteenth-century arguments against self-defense find their origins in primitive Christian thought. These pacifists were not iconoclastically overturning and unsettling rigid dogma; they were harkening back to and seeking to rehabilitate the pre-Thomist Christian tradition of Justin Martyr, Athenagoras, Irenaeus, Tertullian, and Ignatius which had expressly forbidden individual self-defense.[10] In the early days of Christianity, embattled Christian communities faced a great deal of persecution and oppression. In this context, the question of self-defense was highly relevant and the focus of much theoretical discussion.[11] What one finds in these early Christian commentaries is the emphasis on God's role as a defender, the primacy of the eternal over the temporal, and the corrupting effects of sin. Quite literally, assessments about repelling external threats deploy the trope of "lambs to the slaughter" – that is, Christians have a duty to follow the Good Shepherd even if it means being

7 Ibid., 89.

8 Ibid.

9 Plato, *Gorgias* trans. Donald Zeyl (Indianapolis: Hackett, 1987), 469a–c, 31–32.

10 See Josiah Leeds, *The Primitive Christians' Estimate of War and Self-Defense* (New Vienna, OH: Peace Association of Friends in America, 1876).

11 For more on the early Christian attitudes and traditions, see C. John Cadoux, *The Early Christian Attitude to War: A Contribution to the History of Christian Ethics* (London: Headly Bros. Publishers, 1919); Roland H. Bainton, *Christian Attitudes Toward War and Peace* (New York: Abingdon Press, 1960); David Kopel, "From the Early Apostles to the Fall of Rome," in *The Morality of Self-Defense and Military Action: The Judeo-Christian Tradition* (Santa Barbara, CA: Praeger, 2017); Jan Arno Hessbruegge, "Judeo-Christian Tradition," in *Human Rights and Personal Self-defense in International Law* (New York: Oxford University Press, 2017), 31.

led to their deaths. Since God was charged with the care and protection of His flock, any harm inflicted on His sheep was ultimately His concern. The sheep should remain content to rest in His wisdom and leadership, even if this meant that they would be sacrificed in the process.

For instance, in explaining Zechariah 9:15–16, which speaks of God's role as a defender of His flock, Tertullian writes, "No one gives the name of sheep to those who fall in battle with arms in hand, and while repelling force with force; but only to those who are slain, yielding themselves in their own place of duty with patience, rather than fighting in self-defense."[12] The very posture of God's flock should be one of patience and acquiescence, even in the face of violent death; His sheep must wait upon the wise and benevolent Good Shepherd. In His own time and in His own way, God will defend His people and, ultimately, avenge all wrongs. In the meantime, however, one should accept and endure any temporal hardships that come, including physical death. Accordingly, the orientation of the supplicant should not be towards the external world of injustice but towards one's inner depredations and need for grace. Hence, in the work of the early Christian Fathers, devotees were "enjoined to the defense of their soul from the wiles of the devil," which meant that "the preservation of the body is a matter of secondary importance."[13]

There is a noteworthy disjunction between this early Christian tradition, which made it exceedingly difficult to justify personal self-defense, much less killing in war, and the much later Christian just war tradition(s) that arose in continental Europe, indeed, as Christianity was becoming more entangled with the affairs of state. One of the purposes of this chapter is to show how the Christian tradition evolved to accommodate not only killing in war but more specifically the killing of the innocent.[14] At some point, especially when confronted with the realities of siege warfare, European Christians developed ways to rationalize killing in self-defense. Under certain conditions, killing of the innocent became not only tolerable but permissible. The story of this transformation in thinking is suggestive of a theory of the subtle changes in ethical reasoning about killing in war. Moral arguments tend to be the product of attempts to cope with the changes in the material conditions of war. Perhaps the most consequent of these ethical adaptations is Thomas Aquinas's notion of double effect – the belief that unintended harms can be morally acceptable

12 Quoted Leeds, *The Primitive Christians' Estimate of War and Self-Defense*, 29.

13 Ibid.

14 Of course, Christian Just War theory has non-Christian antecedents which took hold and evolved differently outside of Europe. See Rory Cox, "Expanding the History of the Just War: The Ethics of War in Ancient Egypt," *International Studies Quarterly* 61 (2017); 371–384.

if the original action is aimed at a morally good end. The development of this formulation has had a profound effect on the changes in permissions of the innocent in subsequent centuries.

One of the primary claims of this chapter is that Aquinas's ethical language of double effect was not initially developed as a conceptual tool to permit the killing of the innocent in war. Rather, this later formulation emerged as a sixteenth-century attempt to confront the shifting realities of siege warfare and, to a lesser extent, hostage taking. Late Scholastic writers repackaged and redeployed double effect arguments to help explain and justify the violence meted out to the innocent who lived in fortified cities. Just as more formal protections and immunities were being extended to the innocent, simultaneously there emerged arguments that allowed for the violation of those protections. What these developments show is that ethical formulations, like that of double effect, function as a kind of adaptive technology. Militaries needed some way to reconcile excess killing in war with the perceived justness of their cause. While moral arguments are never perfect restraints in war, these ethical formulations deployed to rationalize killing the innocent, to a limited extent, helped to determine the character of war and the extremes of violence to which combatants were willing to resort.

2 Before Double Effect

As indicated by the above nineteenth-century pacifists, up until the thirteenth century it was by no means self-evident that Christians had the right to defend themselves against injustice, much less outright violence.[15] For instance, in a fourth-century Christian discourse, the Church Fathers Augustine and Evodius debated "whether an attacking enemy or an ambushing murderer can be killed without any inordinate desire, for the sake of preserving one's life, liberty, or chastity."[16] Augustine begins by noting that the civil law "permits a traveler to kill a highway robber in order to keep from being killed himself, or that permits anyone who can, man or woman, to kill a sexual assailant, before he or she is harmed. The law also commands a soldier to kill the enemy; and if he

15 See Suzanne Uniacke, *Permissible Killing: The Self-Defence Justification of Homicide* (New York: Cambridge University Press, 1996); Fiona Leverick, *Killing in Self-Defence* (New York: Oxford University Press, 2006); and Whitley Kaufman, *Justified Killing: The Paradox of Self-Defense* (New York: Lexington Books, 2009).

16 Augustine, *On Free Choice of the Will* trans. Thomas Williams (Indianapolis: Hackett, 1993), 8.

refuses, he is subject to penalties from his commander."[17] Evodius responds to Augustine with the standard early Christian view: "I don't blame the law that allows such people to be killed; but I can't think of any way to defend those who do the killing."[18] Those who kill in self-defense, Evodius concludes, are "condemned by a more powerful, hidden law."[19]

Augustine, however, seeks to persuade Evodius that while private individuals cannot kill out of self-defense, soldiers are permitted to do so.[20] Specifically, they can kill out of self-defense if they maintain the proper mental state, i.e. they are not attacking out of base desires, like revenge or rage.[21] They must defend themselves with the correct motive, i.e. "not for the love of transitory goods (which can be taken against our will)" but "the proper love of transcendent goods, which are invulnerable to attack," and when they do so in order to prevent a greater evil.[22] Furthermore, princes are permitted to punish certain wrongs, a right which extends to soldiers, as the hands and feet of the prince's justice.

These subtle shifts in moral reasoning were significant. They anticipate the centuries of interior movement toward the internal motives of individuals when locating the factors relevant for evaluating the morality of killing in war, a movement which was just beginning to unfold. As a result of this shift, unfortunate outcomes could be justified by an appeal to a soldier's or prince's proper mental state or right spirit. Unlike Tertullian, who had argued that baptized Christians could not become soldiers and remain consistent in their faith, Augustine provided a defense of Christian violence, that is, for soldiers and princes.[23] At that time, however, private individuals still lacked the right

17 Ibid.

18 Ibid. 9.

19 Ibid.

20 For more on Augustine's approach to this problem: see Richard Shelly Hartigan, "Saint Augustine on War and Killing: The Problem of the Innocent," *Journal of the History of Ideas* 27:2 (1966); 195–204; and Louis Swift, "Augustine on War and Killing: Another View," *The Harvard Theological Review* 66:3 (1973); 369–383; and Robert Holmes, "St. Augustine on the Justification of War," in *On War and Morality* (Princeton: Princeton University Press, 1989).

21 For more on Augustine's role in the development of this interior world, see Phillip Cary, *Augustine's Invention of the Inner Self: The Legacy of a Christian Platonist* (New York: Oxford University Press, 2000).

22 J. Warren Smith, "Augustine and the Limits of Preemptive and Preventive War," *Journal of Religious Ethics* 35:1 (2007); 141–162. 144–145. See also, N. Vorster, "Just War and Virtue: Revisiting Augustine and Thomas Aquinas," *South African Journal of Philosophy* 34:1 (2015); 55–68.

23 Incidentally, as John Mattox argues, "Augustine's *jus in bello* doctrine does not provide anything approaching a list of rules either for identifying or, once identified, for

of self-defense in the Christian tradition. This is because Augustine held the view that:

> private self-defense can only proceed from some degree of inordinate self-love or 'wrong-headed desire.' In defending himself a man's egoism either manifests or gains control over his action, and the passion of selfishness, concupiscence or libido warps his moral judgment so far as to render him totally incapable of deciding rightly between himself and his neighbor.[24]

For Augustine, the instinct of self-defense was problematic because even those who are guilty naturally seek to defend themselves. Self-love will override one's ability to accept the consequences of guilt.[25] Rather than giving in to an impulse, it is better to yield to an authoritative judgment that had been issued by a proper authority about where guilt actually lies. Even if the law permitted individuals to defend themselves, self-defense was, at best a temptation to sin. Individuals, in the heat of the moment, were simply ill-equipped to make a proper assessment. Even if one's instinct for self-defense just so happened to align with justice, and the truly guilty party were killed, such an outcome would merely be *accidentally* good and not, properly speaking, just. Augustine concludes that all except princes and those explicitly commanded to kill by God are implicated in the crime of murder when they kill.[26]

3 The Rise of Double Effect

We do not find a robust justification for individual self-defense in the Christian tradition until the thirteenth century, when Aquinas gives expression to what is now commonly referred to as the doctrine of double effect.[27] In the *Summa*

---

safeguarding non-combatants." John M. Mattox, *St. Augustine and the Theory of Just War* (London: Continuum, 2006), 63.

24 Paul Ramsey, *Basic Christian Ethics* (Louisville: Westminster John Knox Press, 1950), 175.

25 As Michael White explains, "The assumption seems to be that it is extremely difficult, and perhaps impossible to kill in self-defense without acquiescing in the affect of anger or some other passion implying the preference of one's own life to the life of someone else." Michael White, *Political Philosophy: An Historical Introduction* (New York: Oxford University Press, 2012), 166.

26 Augustine, *The City of God Against the Pagans* ed. R.W. Dyson (New York: Cambridge University Press, 2002), Book 1, Chap. 21, 34.

27 As will be seen, it is not no means clear that in its early formulation it was a 'doctrine.' Indeed, there is a great deal of scholarly debate as to when it became doctrine, if it ever did.

Theologica Aquinas specifically addresses the question as to whether or not it is lawful to kill a man in self-defense (Q64, Art.7).[28] To this he explains: "nothing hinders one act from having two effects, only one of which is intended, while the other is beside the intention. Now moral acts take their species according to what is intended, and not according to what is beside the intention, since this is accidental as explained above."[29] Furthermore, "though proceeding from a good intention, an act may be rendered unlawful, if it be out of proportion to the end"; he continues, "it is lawful to repel force by force, provided one does not exceed the limits of a blameless defense."[30] A secondary harm is morally neutral, i.e. permissible, as long as it is not expressly intended, the unintended harm being "accidental" or "incidental," and as long as it is proportional to a just end. Aquinas stipulates that self-defense must not intend the death of the perpetrator and it must be "moderate."[31]

While originally employed in the service of private individuals engaging in one-off acts of private self-defense, in twentieth- and twenty-first-century military ethical discourse, this novel formulation has taken on a life of its own.[32]

---

28 This issue was less controversial among Roman-inspired medieval jurists who typically believed individual self-defense was permitted in the canon law tradition. For an extensive discussion of self-defense among medieval Romanists, see Frederick Russell, *The Just War in the Middle Ages* (New York: Cambridge University Press, 1975).

29 Thomas Aquinas, *Summa Theologica* Volume 3, trans. Fathers of the English Dominican Province (New York, 2007). Pt. II-II, Q. 64, Art.7.

30 Ibid.

31 T.A. Cavanaugh argues that Aquinas's double effect is the synthesis of two earlier arguments; "he follows Alan [of Lille] in prohibiting an intent to kill and Alexander [of Hales] in requiring an intent to preserve one's own life. In his predecessors' works, one sees the resources upon which Aquinas relies in discussing two effects, one of which the agent intends, the other of which is not intended or is, in Aquinas's terms, *praeter intentionem*." T.A. Cavanaugh, *Double Effect Reasoning: Doing Good and Avoiding Evil* (New York: Oxford University Press, 2006), 5. The rudiments of Aquinas's formulation certainly can be traced back to Aristotle's *Rhetoric*, which states, "Equity will not appreciate alike, but make all due distinction between errors and crimes on the one hand, and between errors and accidents on the other. Accidents are things that happen unforeseen and unexpected, and that imply no malice of intention in him who occasions them. Errors take place, not indeed without thought and expectation, but without any deliberate design of pravity of purpose. Crimes are committed not only with consideration and knowledge, but with a premeditated design of mischief: and to this class is to be referred the gratification of all inordinate passions." Aristotle, *A New Translation of Aristotle's Rhetoric*, trans. John Gillies (London: Printed for T. Cadell ..., 1823), Book 1, Chap 8, 236.

32 Subsequent chapters of this book will address this evolution in greater detail. As Peter Knauer has pointed out, originally the principle of double effect was a marginal feature of Scholastic moral theory. Over time, it came to be seen as much more fundamental to understanding all of morality. See Peter Knauer, "The Hermeneutic Function of the Principle of Double Effect," *Natural Law Forum*. Paper 127 (1967), http://scholarshilaw

Some variation of this argument is routinely featured in discussions about the unintended consequences of justified military operations in war. For instance, John Finnis writes, "If, as Aquinas seems to assert and never denies, one can spear an assailant's heart in self-defense without intending to kill, it is possible to wage war too – lethally and often successfully – without that intent."[33] Robert Holmes also claims, "If, therefore, the prohibition against killing innocents applies only to intentional killing, and one can knowingly kill innocents without intending to do so, the way is open to saying that it is absolutely wrong under all circumstances to kill innocent persons and at the same time to supporting wars that kill them without limit."[34] Aquinas's double effect appears to provide a convenient way out of a paradoxical dilemma, one where war, or any punitive violence for that matter, is permitted but the death of innocents is strictly prohibited.

However, it is by no means clear that Aquinas's double effect, as originally formulated, had anything to do with the unintended killing of civilians in war.[35] Many commentators have noted how underdeveloped and inchoate the concept is in his work.[36] Some interpreters advise against the temptation to

.nd.edu/nd_naturallaw_forum/127?utm_source=scholarshilaw.nd.edu%2Fnd_natural law_forum%2F127&utm_medium=PDF&utm_campaign=PDFCoverPages. Along these same lines, Daniel Schwartz argues that orthodox readers of Aquinas knew that what was at stake with private killing was not the issue of intentionality but "the moral limits that membership in a community sets on the pursuit of private ends, including the private end of staying alive." In other words, Aquinas himself was ambivalent about double effect. Daniel Schwartz, "Thomas Aquinas and Antonio de Córdoba on Self-Defence: Saving Yourself as A Private End," *British Journal for the History of Philosophy* 26:6; 1045–1063. 1047.

33 John Finnis, *Aquinas: Moral, Political, and Legal Theory* (New York: Oxford University Press, 1998), 287.

34 Robert Holmes, *On War and Morality* (Princeton: Princeton University Press, 1989), 196. More recently, see, Lawrence Masek, *Intention, Character, and Double Effect* (Notre Dame, IN: University of Notre Dame Press, 2018); and David Černý, *The Principle of Double Effect: A History and Philosophy Defense* (New York: Routledge, 2020).

35 This has been pointed out by Germain Grisez, "Toward a Consistent Natural-Law Ethics of Killing," *The American Journal of Jurisprudence* 15:1 (1970); 64–96; Daniel Montaldi, "A Defense of St. Thomas and the Principle of Double Effect," *The Journal of Religious Ethics* 14:2 (1986); 296–332; and Bradley Gershel, "Applying Double Effect in Armed Conflicts: A Crisis of Legitimacy," *Emory International Law Review* 27:2 (2013); 741–754.

36 John Kenneth Ryan argues that the entire section on war is "not complete, that it contains gaps that must be filled in by later scholastics, and that some of its solutions are too general and need further distinction." John Kenneth Ryan, *Modern War and Basic Ethics* (Milwaukee: The Bruce Publishing Company, 1933), 23. See also Joseph Mangan, "An Historical Analysis of the Principle of Double Effect," *Theological Studies* 10:1 (1949); 41–61. Ghoos argues that the principle was not formalized until the seventeenth century.

attribute a theory of cognition to Aquinas that he just did not possess.[37] Others have cautioned against the temptation to "reason backwards to the private individual from the other uses allowed only to the sovereign."[38] Conversely, it may also be problematic to read into Aquinas's justification of individual self-defense an argument equally applicable to sovereign authority.[39] While the logical formulation of double effect seems to work as it is framed in contemporary ethical discourse, Aquinas never makes the connection between private and public self-defense explicit. The issue of war is taken up in Question 40 of the *Summa*, whereas murder and self-defense are taken up in Question 64. Aquinas never expressly links the moral formulation of double effect to war, which he could have easily done had he wished to do so.[40] In fact, Aquinas conspicuously does not appeal to double effect in those instances in which it would have seemingly been most applicable, for example, when Samson commits suicide by pulling down the temple on top of himself and his Philistine enemies.[41] All of this seems to imply that Aquinas developed the double effect formulation as a conceptual tool to address a very specific problem, and that it was not originally meant to be so broadly applied.[42]

When Aquinas does take up the question of war, he makes no mention of secondary or unintended effects. Instead, he focuses entirely on the proper procedure leading up to the use of violence. Aquinas's just war theory builds on the proceduralism of the Roman fetial system, which "lay down three conditions for a just war: There must be a formal declaration by proper authorities;

Josef Ghoos, "L'Acte à double effet: Etude de théologie positive," *Ephemerides Theologicae Lovanienses* 27 (1951), 30–52; and Eric Rovie, "Reevaluating the Historical Evolution of Double Effect: Anscombe, Aquinas, and the Principle of Side-Effects," *Studies in the History of Ethics* 2 (2006).

37 See Robert Pasnau, "Intentionality made mysterious," in *Theories of Cognition in the Later Middle Ages* (New York: Cambridge University Press, 1997).

38 James Turner Johnson, "Aquinas and Luther on War and Peace: Sovereign Authority and the Use of Armed Forces," *The Journal of Religious Ethics* 31:1 (2003); 3–20. 10.

39 At the very least, as Leget argues, "When the self-defense of the community is considered, the intentionality becomes more complicated." Carlo Leget, *Living with God: Thomas Aquinas on the Relation between Life on Earth and 'Live' After Death* (Leuven: Stichting Thomasfounds: 1997), 185.

40 The point here is not that it is unreasonable to use the one section to interpret and explain the other, but that it is curious that these justifications are not explicitly linked given the nature of Aquinas's discussion on war.

41 Gregory Reichberg, *Thomas Aquinas on War and Peace* (New York: Cambridge University Press, 2017), 175. For the passage on Samson and suicide, see Aquinas, *Summa Theologica*, Pt II-II, Q. 64, Art.5.

42 See also Thomas Cavanaugh, "Aquinas's Account of Double Effect," *The Thomist: A Speculative Quarterly Review* 61:1 (1997); 107–121.

this must include a charge, which must be one of two things, either an attempt to resist injuries or an attempt to avenge them; there must first be a demand for reparations, and the guilty party must be given a chance to satisfy this."[43] Likewise, Aquinas sets out the three criteria for a just war:

> In order for a war to be just, three things are necessary. First the authority of the sovereign by those who command the war is to be waged. For it is not the business of the private individual to declare war ... Secondly, a just cause is required, namely that those who are attacked, should be attacked because they deserve it on account of some fault ... Thirdly, it is necessary that the belligerents should have a rightful intention, so that they intend the advancement of the good, or the avoidance of evil.[44]

An authorized ruler cannot attack out of greed or a sense of ambition; he can only do so retributively, to punish clear violations of the natural law. As Doyne Dawson explains, "A just war must repel or avenge injuries or recover goods."[45] In this sense, a just war was always in some sense punitive; a justified attack was directed at those who had violated the natural law. As such, the guilty party deserved to be disciplined. Even in repelling an attacker, one was not merely engaged in self-defense but also the punishment of an aggressor. Indeed, even in ostensibly defensive wars, the defending army was routinely believed to be justified in exacting a punishment on the aggressor, invading their lands in search of compensation for what had been lost or destroyed. In this respect, a just war was virtually always characterized as a punitive response to a violation of the laws of nature, even if this offensive action was first aroused by enemy aggression and initially took the form of repelling an assault. While political communities certainly have the right to self-defense, this right could be lost. Much like criminals breaking into a home, belligerent nations that have violated the natural law lose the right to self-defense.

The procedural focus of Question 40 in the *Summa*, namely, that a just war must follow a process beginning with an authoritative judgment from an authorized sovereign, raises additional questions as to whether double effect applies to war-making. To begin with, when an individual kills accidentally or incidentally in self-defense, this act appears to be governed more immediately

43 Doyne Dawson, *The Origins of Western Warfare* (Boulder, CO: Westview Press, 1996), 125; James Turner Johnson, *Ethics and the Use of Force: Just War in Historical Perspective* (Burlington, VT: Ashgate, 2011), 8.

44 Aquinas, *Summa Theologica*, Pt II-II, Q. 40. Art. 2.

45 Dawson, *The Origins of Western Warfare*, 174.

by situational variables and contingencies that are simply not the same for military conflicts. The procedural emphasis of Aquinas's formulation suggests that military violence ought to be the product of careful deliberation and planning, a clear assessment of guilt followed by a measured use of punitive violence. Aquinas explains that retribution must be moderated by a sense of proportionality (the punishment must be suited to the crime), and it must ultimately aim at peace. In this respect, a ruler's intention to wage war is grounded in clear thinking, foresight regarding the potential harms, and prudential wisdom. The amount of force to be used must be reasonably well calibrated, i.e. appropriate to punish the violation and to bring the offender into a posture of submission.

In stark contrast, the private citizen engaging in self-defense is both desperate and, presumably, weaponless. He is an individual engaging in the ad hoc weaponizing of things at hand to defend himself, not someone prepared and trained to kill.[46] It is the ad hoc-ness of the weapons and the uncertainty around the use of force that makes death truly accidental and uncertain.[47] In other words, while some amount of risk may be knowingly extended to the assailant, there remains no reasonable foresight that he will be killed, much less intentionally killed.[48] To be sure, the strictures governing individual self-defense bear little resemblance to sovereign rulers who could explicitly intend to kill those whom they judged guilty, a right that Aquinas believed,

46 *Pace* McMahan, self-defense will likely not involve a flame thrower! See Jeff McMahan, "Revising the Doctrine of Double Effect," *Journal of Applied Philosophy* 11:2 (1994); 201–212. 202.

47 As Schwartz explains, among some Scholastics, like Antonio de Córdoba, "there might be something morally problematic about this specific sort of private defensive killing, a killing with involved cool, calculated lethal choices." Schwartz, "Thomas Aquinas and Antonio de Córdoba on Self-Defence: Saving Yourself as A Private End," 1050.

48 There is some ambiguity here since, as Joseph Boyle argues, "The use of *'quandoque'* to describe the frequency of the deadly consequences following from an act of self-defense suggests that the assailant's death is not a natural and totally predictable consequence of the defensive act as such. But there appear to be types of self-defense in which the use of the minimum force needed to preserve one's life does have the assailant's death as a natural and certainly foreseeable consequence." To Boyle, there is a difference between rare events and recurring events. In the former, one might not be able to properly calibrate the requisite amount of force. In the later instance, when one had a better sense of the possible outcome, one would be required to employ the minimum amount of force to preserve one's life. Joseph Boyle, "*Praeter Intentionem* in Aquinas," *The Thomist: A Speculative Quarterly Review* 42:4 (1978); 649–665, 658. John Makdisi argues that "foreseen death" is included in Aquinas's definition of intention. In this sense, Aquinas would reject conventional readings of the Doctrine of Double Effect that permit unintended but foreseen death. John Makdisi, "Aquinas's Prohibition of Killing Reconsidered," *Journal of Catholic Legal Studies* 57:1 (2018); 67–128. See also Reichberg, *Thomas Aquinas on War*, 178.

as Augustine before him, extended to soldiers as well.[49] Again, for killing to be truly accidental or incidental, one may be limited to simply resisting an attacker in such a way that might only indirectly lead to death. Since war is predicated on the implements of violence, this suggests that it is already governed by a different moral rubric.

4 Killing the Innocent

There remains a great deal of difficulty in discerning exactly what Aquinas believed about the killing of the innocent in war.[50] As Frederick Russell contends, the decretalists – a thirteenth-century canon law tradition that favored papal interpretations issued by way of letters – paid little attention to antecedent medieval expressions of noncombatant immunity, but nevertheless took it for granted that "the indiscriminate violence of the *routiers* and marauding knights" was illegitimate.[51] In Gratian's twelfth-century *Decretum*, he had declared that pilgrims, clergy, monks, women, and the defenseless poor should be immune from harm, under pain of anathema.[52] All of this is to say that it is reasonable to assume that Aquinas held some view of immunity rooted in the canon law, or that he too simply assumed that some classes of individuals

49 Aquinas, *Summa Theologica*, Pt II-II, Q. 64. Art. 3.

50 For a spirited defense of noncombatant immunity in Aquinas's thought, see D.H. Weiss, "Aquinas's Opposition to Killing the Innocent and its Distinctiveness within the Christian Just War Tradition," *Journal of Religious Ethics* 45:3 (2017); 481–509. 482. Likewise, Richard Miller argues that the framework of Aquinas's question "Is it always a sin to go to war?" is meant to emphasize the centrality of nonviolence. Richard Miller, "Aquinas and the Presumption against Killing and War," *Journal of Religion* 82:2 (2002); 173–204. 183. However, as Frederick Russell points out, "Aquinas's own views straddle the issue, for while he clearly did not approve of the killing of innocent men, neither did he unqualifiedly prohibit it." And later, "It is difficult, if not impossible to attribute to Aquinas a clear doctrine of non-combatant immunity; he seemed to remain safely within the confines of Augustine's theory of war guilt whereby innocent men might legitimately be punished even by an unjust war for sins they had committed in other connections." Frederick Russell, *Just War in the Middle Ages*, 275. In a similar manner, McKeogh writes, "There are hints in Aquinas's writings of a wish to distinguish between the innocent and the guilty. There are statements that it is never lawful to kill the innocent but they do not lead to any idea of non-combatant immunity nor a moral basis for one. Aquinas does not address the issue of the killing of innocents in the context of war but rather in relation to the judicial process." McKeogh, *Innocent Civilians*, 63.

51 Russell, *The Just War in the Middle Ages*, 186.

52 Gratian, *Concordia discordantium canonum* [c. 1150] (Lipsiae: Ex Officina Bernhardi Tauchnitz, 1879), Causa 24, Q. 3.25, 997.

should not be killed in war.[53] However, as Lucy Lynch points out, even if Aquinas was appealing to medieval canonists, they did not always provide "sufficient regulation for non-combatant immunity, nor was it specific enough in determining exactly who was classed as the non-combatant."[54] In other words, just because canonists argued that certain classes of people should be protected, this did not mean that there was universal agreement about who met these criteria and how fixed that innocence was.[55]

While Aquinas claims that a just war will be fought proportionally, the specific requirements of proportionality were not well developed. This issue does not appear to be central to his theory. However, even if this idea is underdeveloped, he clearly assumes that war should be directed at only those who are guilty. As noted above, he believes that by its very nature a just war – a properly calibrated application of punitive violence – would not be "inordinate and perilous," and it would not lead to the slaughter of innocents or a breakdown of military discipline.[56] Just wars require not only strategic preparation but also a culture of strict martial discipline. A wise prince will be competent not only in his ability to judge when the natural law has been violated but also in the execution of his judgment in the form of a military campaign. On this account, one wonders if Aquinas envisioned the death of innocents to be possible in

53 Though, in contrast, Hartigan argues that Aquinas "seemed strangely oblivious to contemporary custom and thought, which was concerned with proper conduct in war. It is obvious that neither the customary limitations on war, such as the Peace of God, with had developed since Augustine's time, nor the work of the canon and civil lawyers had any influence on him." Richard Shelly Hartigan, *The Forgotten Victim: A History of the Civilian* (Chicago: Precedent Publishing, Inc.,1982), 44. Similarly, Lawrence Duggan points out that Aquinas ignored antecedent council rulings that prohibited the founding of religious institutes for military service. Lawrence Duggan, "The Evolution of Latin Canon Law on Clerical Armsbearing," in *Between Sword and Prayer: Warfare and Medieval Clergy in Cultural Perspective*. Eds. Radosław, Jacek Maciejewski, and John Ott (Leiden: Brill, 2018), 511.

54 Lucy Lynch, "'Protecting the Non-Combatant': Chivalry, Codes and the Just War Theory," *Ex Historia* 6 (2014), 59–80. 67.

55 Trevor Russell Smith shows how these protected classes routinely were forced to take up arms and how combatants frequently attributed combatant status to these groups of people for various reasons. Trevor Russell Smith, "Willing Body, Willing Mind: Non-Combatant Culpability According to English Combatant Writers, 1327–77," in *Killing and Being Killed: Bodies in Battle: Perspectives on Fighters in the Middle Ages*. ed. Jörg Rogge (Bielefeld: Verlag, 2017).

56 In this respect, it bears noticing that on this view war is any punitive violent military action. What today we see as armed conflict just short of war would be included in the Thomist conception of war. See Johnson, "Aquinas and Luther on War and Peace," 3–4.

just war at all.[57] In fact, one gets the impression that innocent people likely would be nowhere near the battlefield where military engagements were taking place.[58] These implicit assumptions about the nature of war and how it would be executed also help to obviate the need for a conceptual mechanism justifying the incidental or accidental death of innocents.

The spatial realities of thirteenth-century battlefields may have limited the urgency of questions about the appropriate object of punitive violence.[59] Nevertheless, even if one presumes some rudimentary notion of blanket noncombatant immunity in Aquinas's thought, much as many have attributed to the medieval canonists, this does not entirely clarify the question of *who* possesses innocence. Rather than an appeal to double effect, there are at least two other ways of justifying killing the innocent in Aquinas's thought. First, since the sovereign is authorized to punish those who have offended the natural law, some of the radiating effects of war might legitimately extend to the innocent, which could be both foreseen and permitted. For example, if a single father is justly put to death for murder, his children may suffer severe repercussions because of their father's crime and subsequent punishment. They may find themselves ostracized and in extreme penury, which may lead to illness, starvation, and death. Since the guilty party ostensibly deserved the punishment, the suffering of his family does not count as an accidental effect, but rather a feature of the punishment. The consequences of the father's sin that are felt by the children are not justified by way of double effect; they are the extension of a just punishment. By forsaking the law, the guilty man also forsakes

57 This is especially true if combat resembles the military tournaments and contests that he mentions in Aquinas, *Summa Theologica*, Q. 40. Art. 2. Pt II-II, 1354.

58 There may be some truth to this characterization. The Italian historian, Francesco Guicciardini, notes that "Before the year 1494, wars were protracted battles bloodless, the methods followed in besieging towns slow and uncertain; and although artillery was already in use, it was managed with such lack of skill that it caused little hurt. Hence it came about that the ruler of a state could hardly be disposed." Quoted in Geoffrey Parker, *The Military Revolution*, 10. Colm McKeogh also argues that in the thirteenth century, casualty rates for war were low, and that "noncombatants as long as they were not inhabitants of a town under siege, could remain apart from hostilities." Colm McKeogh, *Innocent Civilians: The Morality of Killing in War* (New York: Palgrave, 2002), 48. See also John France, *Western Warfare in the Age of the Crusades, 1000–1300* (London: Taylor & Francis, 1999). For an account of casualty counting in this period, see John Gagné, "Counting the Dead: Traditions of Enumeration and the Italian Wars," *Renaissance Quarterly* 67:3 (2014); 791–840.

59 Of course, this had a lot to do with where you were located. For the devastation of the French Peasantry during the Hundred Years War (1337–1453), see Nicholas Wright, *Knights and Peasants: The Hundred Years War in the French Countryside* (Suffolk, UK: The Boydell Press, 1998), Chapter 3.

his family and his community, and as a result they suffer too.[60] Since human communities are a fabric of social ties, it is difficult to imagine that *any* punishment could be isolated and imposed only on a singular individual without some radiating effects. The same might also be true of indirect harm in times of war. If a sovereign authority believes that an enemy nation has violated the natural law or poses such a threat to its neighbors that it must be confronted, the consequences of this punishment might also radiate to the innocent population living within that guilty nation.

If the first alternative to double effect looks at the natural radiating effects of legitimate punishments, the second looks at certain ambiguities in Aquinas's moral categories of innocence and guilt. The way he conceptualizes guilt may be expansive enough to include a range of people who today would be considered innocent (that is to say, women, children, and those not directly participating in combat). To understand how expansive this category of guilt might be, it is worth analyzing the section in the *Summa* where Aquinas explicitly addresses the problem of killing the innocent (Q. 64. Art. 6). There he makes a striking caveat: "an individual man may be considered two ways: first, in himself; secondly, in relation to something else. If we consider a man in himself, it is unlawful to kill any man, since in every man though he be sinful, we ought to love the nature which God has made, and which is destroyed by slaying him."[61] Aquinas continues, however, by saying that "the slaying of a sinner becomes lawful in relation to the public good."[62] The relevant prohibition with regard to the second way a man might be considered innocent stems from whether or not he is sufficiently "righteous" or "just," since "the life of righteous men preserves and forwards the common good."[63] The phrasing here is a little oblique, but what Aquinas is suggesting is that all men are sinners stemming from the effects of the fall and original sin, but some of these sinners may attain higher states of grace, and with that grace they exert a positive influence on

---

60 This is a point reiterated later by Grotius, who writes, "the offence of one man may sometimes occasion inconvenience or loss to another." Hugo Grotius *The Laws of War and Peace* trans. A.C. Campbell (New York: M. Walter Dunne, Publisher, 1901), Chap. 21, Sect. 11, 265. The great nineteenth-century military commentator, Francis Lieber, also justifies harm to the innocent during war in this same way: "for there is hardly a single sentence inflicted upon society, upon a criminal, which does not affect in the infinite catenation of human connexions, morally or physically, some innocent person – children who depend upon the father, a mother who mourns the disgrace of her son, or an innocent wife or friend." Francis Lieber, *Manual of Political Ethics, Part II* (Boston: Charles C. Little and James Brown, 1839), 644.

61 Aquinas, *Summa Theologica*, Pt II-II, Q. 64. Art. 6.

62 Ibid.

63 Ibid.

their peers.[64] Of course, others may be not only sinners in the general sense of original sin, but active criminal agents who undermine the public good, and are therefore explicitly guilty in light of the natural or positive law. A nation made up mostly of these actively guilty members could be liable to collective punishment, presuming their deaths would be necessary to punish that community's crimes. In contrast, he argues, cities that are inhabited with enough righteous men should be spared, for instance, a state with demonstrable moral opposition to the criminal agency of its leaders. A just prince should be wary of too much punitive violence against a city with sufficient righteous members. This is because killing a righteous man is not only wrong but also counterproductive. Aquinas is concerned that such killing will have profoundly negative practical consequences for the moral status of the enemy state. Without these moral exemplars, the state will fall further into degradation.

With this formulation in mind, it is worth noting that if a sovereign sees that an enemy nation repeatedly violates the natural law, say, by routinely engaging in wars of aggression, this may be due to pervasive moral corruption and the lack of righteous men in that community. Indeed, such a city would likely be populated not only with passive sinners who have been tainted by original sin, but with active sinners, criminals bent on violating the natural law. Residents of a corrupted city are likely entangled in the moral guilt of their leaders. To illustrate this point, recall Abraham's pleading with God prior to the destruction of Sodom and Gomorrah: "And Abraham drew near, and said, Wilt thou also destroy the righteous with the wicked? Peradventure there be fifty righteous within the city: wilt thou also destroy and not spare the place for the fifty righteous that are therein?"[65] In the end, God could not find fifty righteous people in either city. The inhabitants had become so corrupted by sin and idolatry that this opened them up to collective punishment. On this account, everyone – men, women, and children – were tainted and therefore could be killed. If the multitude generally persists in sin, guilt could be collectively distributed.[66]

64 Nigel Biggar argues this passage shows "Aquinas was not at all consistent in arguing … that one may intend to kill sinners on the grounds that they have forfeited human dignity, and there is therefore no evil in their deaths." Nigel Biggar, *In Defence of War* (New York: Oxford University Press, 2013), 108.

65 Genesis 18:23–24, KJV.

66 Jude Dougherty writes, "St. Thomas's position on collective guilt might startle the modern reader. He writes, 'When the whole multitude sins, vengeance must be taken on them.'" Jude Dougherty, "Collective Guilt," in Metaphysics *as Foundation: Essays in Honor of Ivor Leclerc* eds. Paul Bogaard, Gordon Treash, and Ivor Leclerc (New York: SUNY Press, 1993), 229. R. Charli-Carpenter argues that civilian immunity based on gender categories were only inchoate in Aquinas's thought. See R. Charli-Carpenter, *"Innocent Women*

As Abraham's haggling with God suggests, this formulation is ambiguous inasmuch as it is negotiable; it is not clear exactly how many righteous men are needed to make a political community sufficiently good, not to mention the fact that the criteria for who counts as a righteous person are not always obvious or formulaic. Discriminating between the just and unjust or the righteous and wicked was not self-evident, especially since these moral categories did not perfectly map onto one's role in governance or in warfare. To further illustrate this point, in an earlier question dealing with the permissibility of killing sinners, Aquinas notes that lurking in the decision to kill a sinner is a rather painstaking assessment.[67] Sinners are often embedded among the righteous.[68] And while sinners can be killed, the innocent should not be harmed.[69] Clearly, there is a duty to discriminate even if it is not always possible to do so. To complicate matters, however, he also notes that sinners eat away at the common good: "a little leaven corrupteth the whole lump," and elsewhere he explains that "the happiness of sinners, which feeds a punishable licentiousness and, like an internal enemy, strengthens an evil will," all of which is suggestive of the deeply corrosive moral threat the unrighteous pose to the righteous innocent.[70] So the exact proportion of sinners to righteous members, and the speed at which this corrosive potential plays out, all must be factored into a prince's decision about lawful punishment.

One can perhaps see how this formulation could both prohibit but also help make a case for forceful intervention: it would be unjust if a city populated with enough righteous people were attacked for the sins of their corrupted leaders. Nevertheless, one can also imagine Aquinas affirming the view that an enemy nation is sufficiently corrupted, full of so many unrighteous sinners

---

and Children": Gender, Norms and the Protection of Civilians (Burlington, VT: Ashgate, 2006), 41.

67 Aquinas, *Summa Theologica*, Pt II-II, Q. 64. Art. 2.

68 Here Aquinas refers to the parable of the tares and wheat (Matthew 8:29). Since weeds are interspersed with the harvest, it would be ill-advised to sacrifice the wheat just to purge the tares. This too is suggested of some unarticulated ratio of tares to wheat. If the field was mostly tares, then it would make sense to cut it all down rather than seek to salvage a harvest. Incidentally, Francisco Suarez argues that this parable does not apply to justifiable killing of innocents in war because in that case "the pulling up the tares was inexpedient to the end sought by the head of the household" which seems to imply that in the case of the tares it would be inexpedient to pull out tares in that particular case, but that it may not be inexpedient in every case. Francisco Suarez, *A Work on the Three Theological Virtues: Faith, Hope, and Charity* in *Selections from Three Works* trans. Gwladys Williams, Ammi Brown, and John Waldron, ed. Thomas Pink (Indianapolis: Liberty Fund, 2015), 970.

69 Aquinas, *Summa Theologica*, Pt II-II, Q. 64. Art. 2.

70 Ibid., Pt II-II, Q. 40.

(including women and children) that there is no hope for regeneration (like Sodom and Gomorrah); there are so many tares that the entire field should be plowed. A city ostensibly run by a sinful and unrepentant regime that is widely supported by the actively sinful, indeed, degenerate public, by this formulation, could be collectively punished. According to Aquinas's criteria this may open them up to the direct and indirect effects of punishment. While the sinful public may not be directly responsible for the specific crimes of their leaders, their deaths may be permissible in relation to the common good.

The point here is not to suggest that Aquinas was seeking to construct such a narrow criterion of innocence that practically anyone could be killed at will; neither is this to argue that lingering in his just war theory is a bloodthirsty militarism that could easily justify the murder of whole cities or even genocide. To the contrary, all of Aquinas's moral formulations are carefully designed as restraints against excess and cruelty, especially when wielded by a mindful and introspective Christian prince seeking to live a righteous and dutiful life. Instead, the purpose of this section is to show that Aquinas did not expressly employ double effect reasoning to excuse killing the innocent in war, and that within his formulations there are other logical pathways to permit the killing of those whom we would today consider to be noncombatants. That these elements are severely underdeveloped in his thought pays tribute to the wisdom and prudence he believed just authorities should wield.

5 Killing the Innocent after Aquinas

Aquinas's thought provides different ways of permitting the killing of the innocent in war *besides* the appeal to intentions. A just punishment may rightfully impact members of the community as a feature of the punishment. In this sense, indirect effects are themselves a feature of the punishment. Guilt may also spread collectively among an enemy nation, especially those where a thoroughly sinful public supports a criminal regime that engages in unlawful military incursions in neighboring countries. Collective degeneracy may open the entire nation up to punishment, including nonfighting populations. Both variants are consistent with the proceduralism Aquinas emphasizes in his theory of just war.

There is additional evidence that Aquinas did not mean to apply intention-based arguments to explain the killing of the innocent in war. It is striking that among those scholars who came after Aquinas, in many cases those directly influenced by him, *none* deploy double effect formulations to explain killing in war. This is further indication that double effect had not yet been adapted to explain killing in war. In fact, like Aquinas, most writers in this period focus on the just causes and proper procedures for waging war. Quite conspicuously none of these commentators make appeals to intentions or indirect effects to justify killing the innocent in war.

For instance, in *De Monarchia* (1312), Dante Alighieri (1256–1321) never speaks of the risk war poses to the innocent. Instead, he simply focuses on the proper procedure for going to war, albeit with a slightly different emphasis than Aquinas. He writes, "There are obviously two fixed rules of single combat … that not in hatred, nor in love, but in pure zeal for justice, the contestants or champions should enter the field by common consent."[71] Dante emphasizes the duty to fight with the right spirit and the justness of the cause. For William Ockham (1287–1347), the primary tension around killing the innocent in war was not whether such killing was accompanied with a good intention, but whether the prince was making demands in line with God's law. He believed soldiers should resist ungodly commands to kill the innocent:

> We offer our hands against any enemy, but we think it wicked to stain them with the blood of innocents. Our right hands know how to fight against the wicked and against enemies, but not to butcher good men and fellow citizens. We remember that we took up arms in defense of citizens, not against citizens. We have always fought for justice, goodness, and the safety of the innocent.[72]

The presumption here, as was seen in Aquinas's thought above, is that the prince's authorization for a just war would not lead to inordinate slaying. Of course, Ockham's formulation does not escape the same categorical difficulties of distinguishing between the good and the wicked, and neither is there any attempt to use intention-based arguments to justify killing the innocent.

In a similar manner, the Neapolitan jurist, Lucas de Penna (1325–1390), affirms the procedural approach to war, which includes an emphasis on just causes.[73] He does, however, argue that soldiers who commit murder can be put to death, but he does not specify whether the soldier's mental state factors into his guilt or if unintentional killing counts as murder.[74] In fact, there is no

71 Dante Alighieri, *De Monarchia*, trans. Aurelia Henry (New York: Houghton, Mifflin, and Company, 1904), 117.

72 William of Ockham, *A Short Discourse on the Tyrannical Government*, trans. John Kilcullen, ed. Aurthur Stephen McGrade (New York: Cambridge University Press, 1992), 84.

73 Walter Ullmann, *The Medieval Idea of Law as Represented by Luas de Penna: A Study in Fourteenth-Century Legal Scholarship* (New York: Routledge, 1946), 170–171.

74 Ibid., 170.

mention of intentions at all in his discussion of killing in war. When Lucas does speak about punishing soldiers for murder, rape, robbery, and other acts of violence against non-fighting residents, one gets the impression that it is not that the soldiers have violated the immunity of the innocent but that they had become undisciplined and carried away by inordinate passion.[75]

The Italian jurist Giovanni da Legnano (1320–1383) employs a variation of the Thomist justification of private self-defense. He explains that killing an assailant in self-defense or defense of one's property is justifiable since the man only "commits irregularity by killing or wounding, even without a guilty intention, as appears in the case of a judge."[76] This can be seen as a straightforward recapitulation of Aquinas's more limited original use of double effect for private self-defense. However, Legnano employs no such formulation to justify the killing of the innocent in war. In fact, using the same categories of collective guilt, Legnano argues that enemy nations can be punished *en masse* due to the "excesses of the inhabitants."[77] Like Legnano, the Scottish theologian, John Mair (1467–1550), also employs the language of intention in certain cases, for instance, in his assessment of Moses murdering the Egyptian.[78] He does not, however, deploy these arguments to justify killing the innocent in war. Rather, he affirms the standard procedure, that a just war must be waged by a legitimate princely authority, have a just cause; and be pursued with the right intentions.[79]

Finally, in *Utopia* (1516), Thomas More's (1478–1535) Utopians only go to war for good reasons, "to protect their own land, to drive invading armies from the territories of their friends, or to liberate an oppressed people, in the name of compassion and humanity, from tyranny and servitude."[80] Furthermore, they only go to war after the right procedures have been met. Rather than a rush to battle for minor offences, the Utopians "enter a conflict only if they themselves have been consulted in advance, have approved the cause, and have demanded

75 As will be seen in the following chapter, this is a major theme among martialist writers in later periods.

76 Giovanni da Legnano, *De Bello, De Represaliss et De Duello* [1360], trans. Thomas Holland (London: Oxford University Press, 1917), 300.

77 Ibid., 210. For fourteenth-century debates on self-defense between Giovanni da Legnano and John Wycliffe, see Rory Cox, "Natural Law and the Right of Self-Defence According to John of Legnano and John Wyclif," in *Fourteenth Century England, VI*, ed. Chris Given-Wilson (Rochester, NY: Boydell & Brewer, 2010).

78 John Mair, *In Quarum Sententiarum quaestiones utilissimae suprema ipsius lucubratione enucleatae* ([Paris?]: Venundantur a sui impressore Iodoco Badio, 1519), Dist. XV, Q.19.

79 Ibid., Dist. XV, Q.20.

80 Thomas More, *Utopia*, trans. G. Logan and R. Adams (New York: Cambridge University Press, 2003), Book II, 89–90.

restitution, but in vain, and only if they are the ones who begin the war."[81] Here, too, More privileges the proper motive in going to war and the right procedures in carrying out an attack. More also explains that the Utopians are quick to declare war against unjust nations. He writes, "those who devote themselves to doing injury are punished by death or slavery."[82] The presumption seems to be that guilt rests on the entire enemy nation that is devoted to doing injury, and that enemy nations might be enslaved; though, given the satirical nature of the book, it is unclear if More truly supported this approach. At the very least, this formulation partakes of the ambiguous categories of guilt and innocence seen in Aquinas's thought above.

It is quite noticeable that *none* of the commentators in the centuries immediately following Aquinas employ a version of double effect to justify the killing of the innocent in war. This connection is even missing where one would most expect to find it, in the commentaries on the *Summa* that emerged in the late fifteenth and early sixteenth centuries. Take for instance Thomas Cajetan (1469–1534), who was perhaps one of the most prominent theologians of that time.[83] In his authoritative commentary on Aquinas's work, he conspicuously differentiates the relevant domains concerning intentional and accidental death. He does this first by delineating between public and private killing. Summarizing Aquinas, he explains:

> When a public person orders a soldier to slaughter the enemy, he does so as a means to an end, which is subordinate to the common good. As it says in the *Summa*: A private person cannot intend to kill someone in order to save himself; he can merely intend to save himself by way of self-defense. Self-defense is permitted even if it leads to the death of another. And in this way a man is not killing; the death is an accident. In fact, the attacker has killed himself. Public authority is not needed to authorize these cases of self-defense.[84]

In this concise passage, Cajetan sums up the distinct moral considerations facing public and private individuals. Princes can intentionally order soldiers to

81 Ibid., 90.

82 Ibid., 91.

83 Cajetan led the defense of the Church against Martin Luther and the Reformation. See Thomas Cajetan, *Cajetan Responds: A Reader in Reformation Controversy* trans. Jared Wicks (Eugene, OR: Wipf & Stock, 1978).

84 Thomas Cajetan, *Commentarium super Thomae Aquinatis Summa Theologica, Volume 4* (Rome: Polyglot Press 1882), II-II, Q. 64, Art. 7.

slaughter the enemy, but the private person does not have the same right to deliberately kill.[85]

Notice that Cajetan's formulation is devoid of any indication that permissible harm applies to innocents' deaths if they are unintended.[86] The common good authorizes a public person to punish evildoers. In securing his own life through self-defense, however, the private individual can only accidentally kill his assailant. In fact, for Cajetan, it was the criminal who put his own life at risk in committing a wrong. It is striking that even at this late stage in the development of the discourse on the justification of killing, the function of double effect was still quite narrow having not yet evolved to address the killing of innocents in war.

6 Repurposing Double Effect in the Sixteenth Century

Immediately following Cajetan, in the mid- to late-sixteenth century, technological developments in war were beginning to transform the spatial relationship of the battlefield in profound ways. If Aquinas could build a theory of just war on the assumption that the innocent would be nowhere near the battlefield and that war mostly consisted of soldiers squaring off in pitched battle, the rise of siege warfare demanded new moral formulations, especially since urban warfare was beginning to threaten the wellbeing of all a city's inhabitants.[87] Aquinas's double effect formulation was a ready-made

85 Daniel Schwartz argues that one of the major moral dilemmas among Scholastic writers was the degree to which soldiers were culpable for killing the innocent in an unjust war. Killing the innocent, which included enemy soldiers, was only permissible when committed with "invincible ignorance." Soldiers could only be excused if they were convinced that they were fighting a just war. As seen in Ockham's argument above, some argued that soldiers had a duty to disobey unjust commands. See Daniel Schwartz, *The Political Morality of the Late Scholastics: Civil Life, War, and Conscience* (New York: Cambridge University Press, 2019).

86 Though, as Mangan points out, Cajetan does employ a version of double effect to explain the case where a judge knows a man is innocent by way of some private source and yet condemns him to death based on the evidence presented in court. But it is not clear that these cases were appealing to a formal principle of double effect. In fact, the conceptual weight seems to be the public good. See Mangan, "A Historical Analysis of the Principle of Double Effect," 54. Mangan also points out the degree to which double effect became more central to neo-Scholastic formulations of morality (56). See also, Christopher Kaczor, *Proportionalism and the Natural Law Tradition* (Washington D.C.: The Catholic University of America Press, 2002), 36–37.

87 Ryan also argues that the Scholastic just war theory was evolving in the sixteenth century due to the rise of new problems such as "the sack of cities, the capture of ships and the

conceptual tool that could be adapted to this purpose. Military action that led to indirect death could still be prudent and just if it were both necessary and unintended.

The threat context in which the late Scholastics were writing was quite different than Aquinas's. There was a major shift in the centrality and function of fortified cities. As the nineteenth-century historian, Charles Oman, explains, it was the sheer dominance and strategic effectiveness of fortifications that had determined the character of war in the medieval period. He writes:

> A land could not be conquered with rapidity when every district was guarded by three or four castles and walled towns, which would each need several months' siege before they could be reduced. Campaigns tended to become either plundering raids, which left the strongholds alone, or to be occupied in the prolonged blockade of a single fortified place.[88]

In the medieval period, the infrastructural demands of laying siege made it a costly and inefficient tactic. Indeed, in the early sixteenth century even Machiavelli speaks of the tactical limitations of siege warfare. He writes, "It is virtually impossible to keep an army encamped idly in a siege for a whole year," which is why a besieging army might more easily resort to "[burning] and [pillaging] the countryside when they arrive" in the enemy's lands.[89] During the Hundred Years' War,[90] Trevor Russell Smith explains that English military campaigns typically "moved through enemy territory on a broad front so that they could destroy as much as possible while avoiding enemy forces and lengthy sieges."[91] In the operational framework of the fourteenth and fifteenth

---

like, and the obligations of individual soldiers with regard to the justice of the war in which they engage." Ryan, *Modern War and Basic Ethics*, 25–26. Stanley Windass argues that "It was not St. Thomas … who worked out the complexities of the double effect theory, but the Spanish neo-scholastics of the sixteenth century; and they were dealing in the first place with cases of quite a different nature from that of killing in self-defense." Stanley Windass, "Double Think and Double Effect," *Blackfriars* 44:516 (1963); 257–266. 261. See also Adam Kosto who argues that in this period there was a shift from voluntary to involuntary hostages. See Adam Kosto, *Hostages in the Middle Ages* (New York: Oxford University Press, 2012).

88 Charles Orman, *The Art of War in the Middle Ages: AD. 378–1515* (Oxford: B.H. Blackwell, 1885), 59.

89 Niccolo Machiavelli, *The Prince* ed. Quentin Skinner (New York: Cambridge University Press, 1988), 39.

90 The Hundred Years' War was a territorial and dynastic conflict between England and France between 1337–1453.

91 Smith, "Willing Body, Willing Mind," 81.

centuries, sieges were logistically demanding enterprises, so they would be relatively uncommon in the broad scope of a state's military campaigning.[92]

By the sixteenth century, things had started to change. Christopher Allmond argues that at that time, "English kings abandoned raids in favour of a policy of direct conquest. No conquest could be affected unless all fortified towns and castles were brought under the control of the invader."[93] And as Geoffrey Parker notes, the military revolution around siege warfare led to important tactical changes in terms of how wars were waged: "A heavily defended fortress or town, sheltering perhaps 10,000 men and supported by lesser strongholds in the vicinity, was far too dangerous to be left in the wake of an advancing army: it had to be taken, whatever the cost."[94] Military fortresses, which were the economic and political nodes of power, needed to be destroyed or taken in order to guarantee dominance over the region.

The change in tactics was also facilitated by new technologies and feats of engineering which were revolutionizing how sieges were conducted.[95] In the early sixteenth century, bastion, or "star" forts, began spreading throughout the continent. Unlike medieval forts, which produced blind spots and dead zones, this new design proved less vulnerable to cannon fire.[96] Strategies and techniques to conquer fortified cities adapted and emerged alongside these novel architectural designs, including the use of incendiary bombs. What the focus of this escalation shows is that, as Martha Pollak argues, throughout the sixteenth and up to the seventeenth century the city was becoming the "privileged site of war."[97] Allmond concurs that "the civilian was no longer the accidental victim of war but was now becoming one of the chief targets of those who were waging a 'just' war with royal or princely authority."[98] If military discipline could reign in the excesses of cavalry raiding parties, it was much harder to see how one could avoid killing civilians who lived alongside of soldiers behind the walls of a fortified city. Quite simply, technological innovations were changing the kinds of encounters armies were having with those who might have otherwise been seen as innocent.

---

92 Richard Jones suggests that while sieges did occur, they were not the central focus of campaigns. Richard Jones, "Fortifications and Sieges in Europe," in *Medieval Warfare: A History* ed. Maurice Keen (New York: Oxford University Press, 1999), 164.

93 Allmond, "War and the Non-combatant," 262–63.

94 Parker, *The Military Revolution*, 16.

95 See Maurice Keen, "The Changing Scene: Guns, Gunpowder, and Permanent Armies," in *Medieval Warfare: A History* ed. Maurice Keen (New York: Oxford University Press, 1999).

96 See Sébastien le Prestre de Vauban, *The New Method of Fortification* (London: Printed for Abel Swall, 1691).

97 Pollak, *Cities at War in Early Modern Europe*, 109.

98 Allmond, "War and the Non-combatant," 263.

It is difficult to underestimate what a remarkable strategic shift this represented, especially in terms of its effect on the moral vocabulary being used to justify the death of innocents in this period. While women and children were certainly being killed in medieval wars, this shift in tactic required new ways of talking about the death of innocents. Given the emerging realities of siege war, it makes sense that sixteenth-century just war theorists needed to conceptualize civilian death in new ways. In this respect, emphasizing the intentions of belligerents became an effective moral adaptation that permitted, among other things, the bombing and burning of fortified cities, a tactic that would likely kill some indeterminate number of innocent inhabitants. While those deaths could be foreseen in some abstract or probabilistic sense, one could not always be certain exactly how many innocents resided in the fortress and how many would be killed in an artillery barrage. In fact, while the category of innocence was become more formalized in this period, the realities of siege warfare threatened to obviate these advances. As will be seen, some would argue that all the inhabitants of a fortified city had the potential to transform into combatants due to the necessities of self-defense. Assuming reasonable distinctions between the guilty and the innocent could be made in fortified cities, the fact that targeting was happening from greater distances also generated new uncertainties. At least early on, where those balls and bombs eventually fell much more closely resembled acts of chance. All these realities of siege warfare further insulated the good intentions of the besiegers from the negative consequences of just punitive action.

The rise of double effect arguments to rationalize the death of the innocent can also be explained by two other developments in the Scholastic just war tradition. First, several commentators simply no longer saw the need to defend the private right to self-preservation and, hence, the need to justify self-defense when facing an unjust aggressor. John Mangan argues that many late-Scholastics – Domingo de Soto (1494–1560), Luis de Molina (1535–1600), Gabriel Vasquez (1549–1604), and Leonardus Lessius (1554–1623) – simply assumed that Aquinas really thought one could kill an unjust aggressor in self-defense with foresight, a formulation that would weaken one's appeal to the unintentionality of the killing.[99] Francisco Suarez (1548–1617) expressly affirms the principle of "prudent self-defense."[100]

---

99 Mangan, "A Historical Analysis of the Principle of Double Effect," 45, 54. See Dominicus de Soto, *De justitia et jure libri X* (Venetiis: Apud Minimam Societatem, 1594); 414; Gabriel Vazquez, *Commentariorum, ac disputationum in tertiam partem Sancti Thomae, tomus tertius* [1598–1615] (Antverpiae, Apud Petrum & Ionnem bellerso, 1621), Q.80 Art. 4, 380.

100 Francisco Suarez, *A Treatise on the Laws of God the Lawgiver* in *Selections from Three Works* trans. Gwladys Williams, Ammi Brown, and John Waldron, ed. Thomas Pink (Indianapolis: Liberty Fund, 2015), Pt. I, Bk. II, Chap. XVI, 360.

In fact, in a polemical late-seventeenth-century publication assessing the morality of the Jesuit just war tradition, Nicholas Perrault[101] argues that Molina, Lessius, Tomas Sanchez (1550–1610) and Domingo Bañez (1528–1604) were arguing that "Nature gives a right of self-defence without need of demanding any toleration. And this natural right gives leave to send as well as receive a Challenge," and this includes "the right of killing an enemy in secret rather than by fighting with him."[102] Perrault argues that these late Scholastic authors were now arguing for such an expansive definition of self-defense that just about any type of cruelty and trickery was permitted. Of course, from the pen of such an antagonistic author, this is clearly an ungenerous and exaggerated reading, but the grain of truth that it intimates is that a noticeable distance had opened between the original use of double effect and how it came to be employed by late-Scholastic commentators. What some believed a cynical tool to justify any kind of cruelty, the Scholastics viewed as a reasonable adaptation of Aquinas's theory.

By the sixteenth century, private self-defense was no longer the moral conundrum it had been to thirteenth-century Christians. Two hundred years of moral evolution had changed the tenor of the ethical discourse. Aquinas's original formulation of double effect had become a vestige that traced back to a now defunct moral context borne out of the tensions between the early Christian tradition and Roman canon law. The incommensurability of these two traditions – between the quiet, devotional life of Christianity and the demands of empire – led Aquinas to a novel formulation that ultimately produced a moral allowance for what Roman law had long permitted. Aquinas was able to provide a sophisticated rationale for killing in self-defense in a Christian context that had originally forbade acting on the inordinate passion that comes when one kills out of self-preservation.

Secondly, in the late Scholastic period there was also a shift away from collective guilt. At that time, guilt came to more formally rest on an individual's capacity to contribute to the war effort.[103] As seen above, while Aquinas never explicitly formulates a theory of collective guilt (that is, outside of original sin), the collective guilt of a political community can be seen as the logical

101 This was the pen name of anti-Catholic agitator, Ezerel Tonge (1621–1680).

102 Nicholas Perrault, *The Jesuit Morals*, trans. Ezerel Tonge (London: Printed for John Starkey, at Miter in Fleet-street, 1670), 322.

103 Though the case may be slightly overstated, Marcus Schulzke argues that there was a shift from a guilt- to rights-based status of noncombatants in this period. See Marcus Schulzke, *Just War Theory and Civilian Casualties: Protecting the Victims of War* (New York: Cambridge University Press, 2017), 27–30. For similar claims, see also Colm McKeogh, "Civilian Immunity in War: From Augustine to Vattel," in *Civilian Immunity in War* ed. Igor Primoratz (New York: Oxford University Press, 2007).

inference of the degrading effects of sin. And in the commentators immediately following Aquinas, the blurred distinction between guilt and innocence gave the impression that enemy nations might be held collectively guilty. Sixteenth-century Scholastics were less inclined to use these kinds of formulations. Instead, they were increasingly focusing on the role one played in prosecuting the war.

These novel shifts in the understanding of innocence and guilt also likely stemmed from the historical realities of warfare, such as the Ottoman expansion into Central Europe which dramatically increased encounters between Christian and non-Christian communities. It is telling that the late Scholastic authors of this period specifically began to theorize the limits of Turkish innocence in response to these changes.[104] In these new guilt-innocence classification schemes, it was common for authors in this tradition to affirm the innocence of Turkish women and children, specifically since they were not involved in physical combat. It was no longer held that non-fighters could be justifiably killed due to their unbelief or some innate guilt or the corruption they accrued by virtue of their proximity to the guilty. This issue had very likely become more pressing since soldiers on the continent were increasingly finding themselves in situations where they had to decide whether to kill or spare Muslim women and children.

Perhaps the clearest example of this adaptation and repurposing of Aquinas comes from Francisco de Vitoria (1483–1546). He argues that "it is never lawful in itself intentionally to kill innocent persons."[105] And rather than appealing to an interior moral status to define innocence, he argues that protections should be extended to Turkish women and children, travelers and visitors, and clergy and monks categorically since they were not able to participate in the war-making effort.[106] Following Vitoria, Bañez also argues that Turkish women and children should be considered innocent, even though they lived among unbelievers.[107] Unbelief was no longer a sufficient criterion of guilt. Suarez also makes the case that not everyone living in and among the enemy deserves to

104 'Turkish' was a stand-in for Muslims in general. For an overview of these conditions, see L.S. Stavrianos, *The Balkans Since 1453* (London: Hurst, 2000); Daniel Goffman, *The Ottoman Empire and Early Modern Europe* (New York: Cambridge University Press, 2002); and Brian Davies (ed.) *Warfare in Eastern Europe, 1500–1800* (Boston: Brill, 2012).

105 Francisco De Vitoria, "On the Law of War," in *Political Writings*, eds. Anthony Pagden and Jeremy Lawrence (New York: Cambridge University Press, 1991), 314.

106 Ibid. 315. A later commentator extends this list to include "harmless farming folk, and other peaceful civilians (*gens togata*), who are all to be presumed innocent unless there is evidence to the contrary." Ibid., n.37.

107 Domingo Bañez, *Commentaria in secundam secundae D. Thomae* (Venetiis: Apud Bernardum Iuntam, 1587), 1377.

be punished. Like Vitoria, he argues that women, children, and all those unable to bear arms should not be intentionally killed; however, "all other persons are considered guilty; for human judgment looks upon those able to take up arms as having actually done so."[108] Those members of an enemy nation capable of fighting, however, would be treated as *prima facie* guilty.

Noting this general shift away from using broad conceptions of collective guilt, Colm McKeough argues that by the sixteenth century "War was now fought by and against people who bore no guilt for the unjust cause of their side."[109] The crimes of political leaders and the moral status of the public at large ceased to be the primary criterion of guilt.[110] Remember, for Aquinas, the category of innocence did not necessarily correspond to the individual's role in aggression or war-making.[111] While guilt or innocence could refer to one's legal standing in light of the natural or positive law, it also had a great deal to do with one's inner posture. Due to original sin, all were tainted with some degree of guilt, which not only predisposed them to degenerate behavior (like supporting criminal regimes) but also left them open to punishments that served the public good. Only the truly penitent and righteous had accrued the requisite degree of innocence, and even then, they were to be spared largely due to their ability to positively influence their communities. On this view, since everyone is tainted with the stain of original sin, no one, not even women and children, could be classified as *a priori* innocent. By the sixteenth century, however, many commentators were more formally relying on innocence as a categorical distinction.[112] People were starting to be seen as innocent by virtue of their social status and occupation. More specifically, they accrued guilt in relation to their ability to directly contribute to the war effort.[113]

---

108 Suarez, *A Work on the Three Theological Virtues*, 962.

109 McKeogh, *Innocent Civilians*, 77.

110 Though, some of these moral formulations can be seen in some commentators up to the eighteenth century.

111 Ideally, soldiers would fight soldiers, but even in the twelfth and thirteenth centuries, attacks on cities and towns put the public in harm's way.

112 In the tenth century there were nascent attempts to separate innocence from one's spiritual status. The Peace of God Movement, for instance, sought to demarcate classes of people as innocent in reference to their class, and this largely out of the practical necessities stemming from violence in France. See Thomas Head and Richard Landes (eds.), *The Peace of God: Social Violence and Religious Response in France around the Year 1000* (Ithaca: Cornell University Press, 1992). This is not meant to imply that the emerging norms seeking to prevent civilian casualties were borne solely out of theological traditions. For instance, James Turner Johnson points out the influence of the Chivalric tradition. See James Turner Johnson, *Just War Tradition and the Restrain of War: A Moral and Historical Inquiry* (Princeton: Princeton University Press, 1981), 131.

113 McKeogh links guilt with combatancy in this period. McKeogh, *Innocent Civilians*, 85.

7 Sixteenth- and Seventeenth-Century Scholastics on Siege Warfare

The late Scholastic writers were especially attuned to the shifts in siege warfare. In their writings one repeatedly finds what can best be described as a formulaic attentiveness to the innocent in the context of urban war, something that simply was not a feature of Aquinas's analysis. Virtually every Scholastic writer in this period, when contemplating the moral risk of killing the innocent in war, focuses on the storming or bombing of fortified cities. For instance, Vitoria explains that:

> it is occasionally lawful to kill the innocent not by mistake, but with full knowledge of what one is doing, if this is an accidental effect: for example, during the justified storming of a fortress or city, where one knows that there are many innocent people, but where it is impossible to fire artillery and other projectiles or set fire to building without crushing or burning the innocent along with the combatants. This is proven, since it would otherwise be impossible to wage a war against the guilty.[114]

Vitoria's contemporary, Bartolomé de las Casas (1484–1566), also argues that when attacking an enemy city in a just war "some innocents can be accidentally killed, without foreknowledge or intention."[115] Balthasar Ayala (1548–1584), a student at Vitoria's School of Salamanca, speaks of both the bombardment of cities and hostage-taking as features of modern war.[116] He writes:

> just as intentional killing of innocent persons, for example, women and children, is not allowable in war (if unintentional, as when a town is assaulted with catapults and other engines of war, the case is different, because such things are inevitable in war), so also in this hostage-taking, but only to confiscate their property, yet it may be that they are

114 Vitoria, "On the Law of War," 315. It appears that Vitoria derives this reading from Franciscus de Sylvestris Ferrariensis, *Commentaria in Summa contra Gentiles* (Romae: Sumptibus et typis Orphanotrophii a S. Hieronymo Aemiliani, ca. 1516).

115 Bartolomé de las Casas, *Brevissima relación de la destruycion de las Indias* ([s.n.] [n.], 1552), 75–76.

116 Vitoria did not thematically elaborate on hostage-taking. As Adam Kosto explains, Vitoria simply assumed "that if hostages are guilty, as if they had borne arms, they may be killed; if not, as in the case of children or women or other innocents, they may not be killed." Kosto, *Hostages in the Middle Ages*, 215.

> unintentionally killed without any liability arising, provided that the hostage-taking could not otherwise be carried out.[117]

Reiterating Vitoria's position, while some civilian death is to be expected, it is not, strictly speaking, intended. In a similar manner, Molina explains that "No matter how just the war is, killing an innocent person directly and intentionally is a crime."[118] He clarifies, however, that when a city is under siege in a just war, it is permissible to reduce it by cannon shot or fire. The besieging army would not be at fault if any of the innocent are killed because those deaths occurred accidentally and without intention.[119]

This point is also affirmed by Bañez, who agrees that "Having laid the foundation, let the first conclusion be that intentionally killing the innocent is intrinsically evil, that is, unless it is done by God's express order."[120] He continues, "Accidentally and unintentionally killing innocents is not a sin, for example, when a city is being conquered by one who is waging a just war and is not otherwise able to conquer (the city) unless by means of the torments of war. Although he knows that many innocent people must be killed, it is permissible for him to conquer the city."[121] And in a very similar manner, Suarez permits those tactics that:

> necessarily involve the death of innocent persons, as in the burning of cities and the destruction of fortresses. For, absolutely speaking, whoever has the right to attain the end sought by a war, has the right to use these means to that end. Moreover, in such a case, the death of the innocent is not sought for its own sake, but is an incidental consequence; hence, it is considered not as voluntarily inflicted but simply as allowed by one who is making use of his right in a time of necessity.[122]

117 Balthasar Ayala, *On the Law and the Duties Connected with War and on Military Discipline* [1582] trans. John Pawley Bate (Washington DC: Carnegie Institute of Washington, 1912), 33.

118 Luis de Molina, *De justitia et jure opera omnia* [1593–1609] (Coloniae: Sumptibus Marci Michaelis Bousquet, 1733), Tract II, Dis CXIX.2.

119 Ibid.

120 Bañez, *Commentaria in secundam secundae D. Thomae*, Q. XL, Art. 1, 1378.

121 Ibid., Q. XL, Art. 1, 1377.

122 Suarez, *A Work on the Three Theological Virtues*, 968.

Like Vitoria, he concludes that without such permissions "it would be impossible, through any other means, to end the war."[123]

The death of innocents in siege or urban warfare became a near universal concern among the Scholastic writers of this period. Arguments virtually identical to those above run throughout the late-sixteenth and early-seventeenth-century Scholastic commentaries on war.[124] Given the realities of siege war and the proliferation of fortified places, commentators in the Scholastic tradition faced a clear moral dilemma. If the guilty could safely retreat into secure places filled with innocent people, then the guilty could never be punished. Many commentators in this tradition emphasized the necessity of siege warfare and practical questions of liability. The Dominican friar, Joannes Baptista Corradus (1536–1606), concludes that when a republic has been invaded in a just war, the just invader is not liable for the accidental loss of life and property.[125] The Flemish Jesuit theologian, Leonardus Lessius (1554–1623), also argues that when "taking a defended tower the innocent population, which is crowded in the same place as the enemy, can be killed unintentionally, otherwise victory would be impossible."[126] Martin Becanus (1563–1624), a Dutch Jesuit priest, writes that "the man who attacks an enemy city is not guilty of waging war against the innocent (otherwise it would not be a just war), but only against those harming him, even if there are some innocent among them."[127] The Spanish Jesuit theologian, Gaspar Hurtado (1575–1647), argues that when conducting a siege one could not kill hostages to bring the siege to an end; however, when taking a castle the innocent can be killed incidentally and accidentally.[128]

It was quite common for commentators in this tradition to emphasize the tragic necessity associated with waging a just war. Juan Márquez (1565–1621), a Spanish professor of theology at the School of Salamanca, writes, "The reader is perhaps waiting for the case showing when it is lawful to kill the innocent. Killing the innocent is only permitted in a just war when victory cannot be

---

123 Ibid.

124 For more on the political morality of this period, see Daniel Schwartz, *The Political Morality of the Late Scholastics: Civic Life, War and Conscience* (New York: Cambridge University Press, 2019).

125 Joannes Baptista Corradus, *Responsa ad cujuscunque pene generis casuum Conscientiae, Volume 1* (Venetiis: Apud Societatem Minimam, 1596), Q. 108, 147.

126 Leonardus Lessius, *De iustitia et iure ceterisque virtutibus cardinalibus* (Lugduni: Sumphibus Ludovici Prost Haeredis Roville, 1622), Lib. 2, Ca 9, Dub. 7, §37. 82.

127 Martin Becanus, *Summa Theologiea Scholasticae* (Lugduni: Sumptibus Antonii Pillehotte …, 1621), Q. VII.5, 598.

128 Gaspar Hurtado, *Tractatus de iustitia et iure* (Matriti: Apud Ioannem Sanches, 1637), Dis XI., Difficultas VIII, 107.

obtained in any other way, for example, when a king has surrounded a rebellious city and there is no way to enter without fire and blood."[129] The Spanish Franciscan friar, Enrique de Villalobos (??–1637), states that "In a just war it is lawful to kill the innocent accidentally and without intention, for example, when one must set fire to a city. In such cases, the innocent find themselves at the mercy of the guilty, otherwise it would be impossible to wage a just war."[130]

Despite what may seem like an overly generous attitude to civilian death in pursuit of just ends presented in some of the passages cited above, remember that this ethical formulation was designed to be a check against the extreme permissiveness that would follow if military leaders could justify any cruelty by way of an appeal to good intentions and necessity. In reality, this framework was meant to represent a much more painstaking and, indeed, morally introspective process.[131] As already seen to a limited extent in the examples above, by the early seventeenth century, several Catholic theologians were providing formulations that were specifically designed to limit the ability of a just belligerent to take a city knowing that it would lead to the death of the innocent. Several theologians emphasized Vitoria's stipulation that if an attack on a city would lead to the death of many innocents, the attack had to be strategically significant. Such loss of life would not be appropriate for a minor victory. Take, for example, Don Pedro de Aragón (1538 ca.–1592), who argues that one must be cautious when storming a citadel that a great number of the innocent should not be killed for such a small victory; however, he continues, it is permissible to kill the innocent accidentally and without intention in a just war.[132]

129 Juan Márquez, *El governador christiano* (En Amberes: En casa de Jacobo Meursio, 1664), Ca XIV, 74.

130 Enrique de Villalobos, *Manual de Confessores* (Salamanca: En casa de Diego Cyssio, 1625), 241.

131 Nevertheless, these arguments were still controversial in this period. Blaise Pascal (1623–1662) for instance, rejects "directing the intention" type arguments, as they were derisively called. He complained that "according to this mode, everything will be permitted; it will stick at nothing" – by appealing to intentions, one could justify virtually any type of cruelty. To Pascal there was something disingenuous to the partitioning of outcomes and inner posture: "To men you grant the outward material effect of the action; and to God you give the inward and spiritual movement of the intention; and by this equitable partition you form an alliance between the laws of God and the laws of men." To this, he concludes, "I am by no means sure but a person would feel much less sorry to see himself brutally killed by an infuriated villain, than to find himself conscientiously stilettoed by a devotee." Blaise Pascal, "Letter VII" *Provincial Letters* [1656], trans. Rev. Thomas McCrie (New York: Derby & Jackson, 1859), 232, 233, 246.

132 Pedro de Aragón, *Ordinis eremitarums augustini, atrium, & Sacrae Theologiae Magistri, & in Clarissima Salmaticensi Academia publici professoris* (Venetiis: Apud Minimam Societatem, 1608), Q. LXIV, Art. VI, 245.

A similar argument is put forward by Gregory Sayers (1560–1602), an important, if now obscure, English moral theologian of this period.[133] In *Clavis Regia Sacerdotum*, an influential, posthumously published book that went through ten editions between 1605 and 1659, Sayers writes:

> We must consider, as Vitoria rightly advises, lest greater evils than the ones that are being avoided follow from such a war, it is only for the sake of the highest victory in war that one should attack a citadel or a town in which there are many innocent and few truly guilty people. Without such a purpose, it will not be permissible to overthrow the town. In fact, the evil which would follow from the death of so many innocents would not outweigh that small benefit resulting from the end of the war, which such a storming of the citadel would bring. Always beware that greater losses are not inflicted in war than the good that would follow.[134]

While Sayers believes that the innocent could be killed incidentally, their deaths must be factored into a wider prudential calculus that also takes into consideration how those deaths specifically contribute to victory, in effect, minimizing evil in the long run.

The Jesuit scholar, Luis de Torres (1562–1655), explains the moral caution behind the double effect formulation this way:

> The incidental killing of the innocent is lawful in a just war; however, one must be able to give an account when tested. Sometimes it is necessary to set fire to a citadel. This is permitted even if the attacking army knows that innocent children will be killed. The common good will not permit us to pass over the innocent, whose lives and property are linked to that of the guilty. They are liable to be attacked; otherwise, one would never be able to wage a just war against enemies.[135]

Torres suggests that an attacking army must be able to defend the decision to pursue a military action that will lead to the deaths of innocent children. The presumption appears to be that, like Sayers, killing the innocent for strategically insignificant victories would not be acceptable. Good intentions are

133 E.J. Mahoney, "Gregory Sayers O.S.B. (1560–1602): A Forgotten English Moral Theologian," *The Catholic Historical Review* 11:1 (1925); 29–37.

134 Gregory Sayers, *Clavis Regia Sacerdotum* (Monasteri Westfaliæ: Samptibus Michaelis Dalii, & Bernardi Raesfelt, 1628), Lib. VII, Ca XII, 459.

135 Luis de Torres, *Disputationum in secundam secundae D. Thomae* (Lugduni: Sumptibus Horathii Cardon, 1617), Disputation XC, Dub., III, 1151.

demonstrated through proportionate harm. And finally, Raimondo Capizucchi (1616–1691), a Dominican friar who would later become a cardinal and one of the leading theologians under Clement X, employs a near identical formulation to Sayers. Capizucchi appeals directly to Vitoria, and in the process presents a similarly restrictive version of double effect. He writes:

> As long as the killing of innocents must be observed, it ought to be avoided, according to Vitoria's *On the Law of War* number 31, lest greater evils follow from war than are being avoided by that war. For if at the end of the war either storming the enemy's citadel of the town where the enemy's garrison is located confers (only) a small victory and there are many innocent casualties, it does not seem that it would be permitted to kill many innocent men to fight against a few guilty ones by subjecting them to the fire of either siege weapons or by another method by which innocents would be indifferently oppressed along with the guilty. And finally, it never seems permitted to kill innocents even by accident and unintentionally unless when it is not possible to expedite a just war and carry on in any other way.[136]

Again, killing the innocent is permissible, but only when the war is just, the strategic demands are high, the ratio of guilty to innocent makes the attack plausible and proportional, and greater evil does not follow the decision to attack.

As can be seen in the above quotes, despite accusations to the contrary, the appeal to intentions was not designed to be a kind of rubber stamp that commanders could resort to cynically. Nor was it meant to permit an invading army to do whatever it believed necessary to win a just war. While the proper intention of a belligerent was a necessary component in the Scholastic moral formula, it was often ensconced in a much broader set of moral considerations that needed to be in place to assess the legitimacy of the action in question. For instance, the war must be just, and, depending on how risky the specific operation was in terms of the potential loss of innocent life, one must have a clear sense that its objective a) would not make matters worse, and b) that it was absolutely necessary to the effective prosecution of the war on the whole. In many instances, even good intentions would not be enough to cover disproportionate harm to the innocent. To be sure, slaughtering the innocent for

136 Raimondo Capizucchi, *Ordinis Praedicatorum Sacrae Theologiae Magistri. Controversiae Theologicae Selectae Scholasticæ Morales, Dogmaticæ Scripturales, ad Mentem D. Thomae Aquinatis* (Rome: Typis Angeli Bernabo, HH. Manelphi Manelphij, 1670), Para. 8, 713.

minor strategic victories is unacceptable, even when one is fighting a just war with proper intention.

By the late-seventeenth century, certain Scholastic commentators were posing more restrictive versions of the double effect formula, but this was not universal. At that time, the norms around bombing and setting fire to cities had been subject to their own evolutionary forces. Technological developments would eventually create new dangers for the non-fighting residents of fortified cities. For instance, up to that point in time, most of the literature on bombing fortified places focused on creating a breach large enough for a besieging army to storm. In fact, storming the breach represented perhaps the most desperate moment in all siege warfare, since, in the frenzy of a sack, it was especially difficult to differentiate the guilty from the innocent. And, of course, even if distinctions between the innocent and guilty could be made, it was not clear that soldiers could be sufficiently restrained in the heat of battle.[137]

By the mid to late-seventeenth century, innovations in artillery and the arts of siege warfare had started to make the breach and, with it, the frenzied sack much less necessary. Siege warfare had become an increasingly vast operation where a besieging army was confronted with sophisticated feats of engineering and complex fortification structures. With these infrastructural changes, sieges were not only transforming into massive duels between cities and their besiegers, but the numbers of soldiers and non-fighting inhabitants comingling and residing in these fortified places gave rise to new vulnerabilities. With the additional pressure points of a potentially weak residential population, an assaulting army could now feasibly direct enough force on a besieged place that surrender could be gained without a costly assault. These developments are largely due to the rise of more effective artillery and incendiary bombs, for example, the carcass and red-hot shot,[138] both of which allowed besieging armies to bombard and set fire to the interior, residential spaces of cities. Now that incendiary shells were being fired into city centers, one could expect that the innocent inhabitants would suffer a new degree of risk.

Of course, some earlier Scholastic commentators mention launching firing bombs into cities. The Dominican priest, Lodovico Carbon de Costacciaro (1532–1597), writes that "In a just war it is permissible to kill the innocent indirectly and by accident. For example, it is not unjust to kill the innocent when

137 This is a problem that will be more explicitly thematized by the martialist writers of the next chapter.

138 The carcass was a cast iron shell casing filled with a highly flammable liquid. It was specifically designed to set targets on fire. The red-hot shot was a solid ball that was superheated, also designed to set its target on fire.

firing bombs at an enemy who has been unjustly resisting."[139] Likewise, Rutilio Benzoni (1542–1613), who would serve as the Bishop of Loreto, argues that it is permitted to take a city with the use of machines that will knowingly kill the innocent, as long as those deaths are not strictly speaking intended.[140] The Jesuit scholar, Tomás Sánchez (1550–1610), also declares:

> One is not at fault for killing the innocent by accident and without intention. Just as when in a just war a city – in which everyone knows there are many innocents, even children – is either bombarded with missiles are reduced to ash with fire balls, and the innocent perish equally with the guilty.[141]

However, many of the earlier formulations appear to be focused primarily on using cannons as a kind of battery-engine for creating an opening for troops to storm in. The above sequence seems to imply that, after the walls had been breached, the town could be ransacked and set on fire by the onslaught of troops. This was permissible because the innocent, according to some, had acquired guilt due to their proximity to the guilty. And finally, the Dominican theologian, Pedro de Ledesma (1550–1616), writes:

> According to the Theologians [Vitoria, Soto, and Bañez], it is sometimes lawful to kill the innocent accidentally and without intention. For example, when a city cannot be taken with a battery of shots [against the walls] it is lawful to bombard a city in a just war even though it is understood that many innocents will die in the process. Since those deaths are not intended, it is permissible to take the city.[142]

Like, Sánchez, Ledesma's formulation also prioritizes the breach. One must first attempt to reduce the walls with cannon shot. However, if that proves unsuccessful, a besieging army may then proceed to bombard the city itself.

As implied by Ledesma's formulation, many commentators were beginning to grapple with the strategic effectiveness of urban bombing. In fact,

---

139 Lodovico Carbone de Costacciaro, *Tractatus de omnium rerum restitutione amplissimus* (Venetiis: Apud Haeredes Ioan Baptistae Somaschi, 1592), Q. LXII, 441.

140 Rutilio Benzoni, *Speculum episcoporum et curatorum* (Venetiis: Apud Minimam Societatem, 1595), Bk. 3, Q. 3, 60.

141 Tomás Sánchez, *Disputationum de Matrimonii Sacramento* (Antverpiae: Apud Ioannem Hasrey, 1614), Bk. IX, Dis. XX, 14, 231.

142 Pedro de Ledesma, *Segunda Parte de la Summa* (Lisboa: Impresso por Pedro Crasbeeck, 1617), Trat. 3, §35, 76.

throughout the seventeenth century one can find numerous examples of this kind of transitional language, where commentators included both the tactic of artillery as battery to be used against the city's walls versus the bombardment of the city's residential districts. For instance, the Spanish theologian, Benito Remigio Noydens (1630–1685), writes,

> one can infer that it is lawful to use powder mines in order to demolish city walls, to throw bombs into cities, and to set fire to ships so that they do not come into the enemy's power, even though the innocent will be killed by these methods. This is because the death of the innocent is not intended by the demolition of the towers and walls; it merely accidentally follows.[143]

While the conceptual emphasis is on the traditional objects of attack – the walls and towers of fortified places – Noydens also includes the bombing of cities. In a similar manner, Pedro de Tapia (1582–1657), the Archbishop of Seville, writes, "It is not wrong to kill the innocent accidentally and without intention in a just war, for example, when a city cannot be taken except by storm or by launching balls of fire and sulfur. This can be done even if it is known that many innocents will be put to death."[144] For Tapia, a city could be taken by storm or through incendiary bombardment. And finally, Antonius Schiara (16??–1718) addresses both the traditional tactic of bombing the citadel and the city itself. He writes, "Laying siege to a citadel or city is just even if it is known that more than a few innocent people live there. In those places, women and infants are fused together with the guilty. Cannon balls can be shot into them without scruple in order to put them under or to capture them."[145]

As the seventeenth gave way to the eighteenth century, however, many commentators simply dropped the hybrid formulations that included the demolition of walls, and the bombing of towers or citadels. It became much more common to explicitly justify the bombing of the city itself. For instance, Hermann Busenbaum (1600–1668), a Jesuit priest from Westphalia, explains that when "driven by necessity to attack a city, it is permissible to direct the artillery to the place where innocent people will be, as long as the death of

143 Benito Remigio Noydens, *Decisiones practicas y morales para curas, confesores y capellanes de los exercitos, y armadas* (Madrid: Por Andres Gareta de la Inglesia, 1665), Ch. 9, §1.3, 77.

144 Pedro de Tapia, *Catenae moralis doctrinae* (Hispali: in aedibus archiepiscopalibus excussit Ioannes Gomez a Blas, 1657), Bk. III, Q. XIII Art. VIII, N.2, 222.

145 Antonius Schiara, *Theologia bellica* (Romae: Typis, & Fusoria Cajetani Zenobiii Typographi …, 1703), Bk. VII, Diff. XIII, 250.

the innocent is unintentional."[146] In a similar manner, the Jesuit moral theologian, Claude Lacroix (1652–1714), argues that "If it is necessary, one may direct artillery to the place where many innocent people are located, as long as their deaths are unintentional."[147] In his commentary on Aquinas's *Summa* (1751), the Dominican theologian, Charles-René Billuart (1685–1757), employs a similar formulation. It is permissible to use explosive ordnance to take a city, even though the innocent may be killed, i.e. as long as the war was just and those deaths were accidental and unintentional.[148] Along these same lines, the prominent Catholic bishop, Alphonsus Liguori (1696–1787), explains that "In the same manner, if in the pursuit of victory it is necessary to attack a city, one may direct artillery fire to the place where many innocent people are located. The death that follows is permitted because it is unintended."[149]

Take one final example of this trend: Father Fulgentius Cuniliati (1715–1759), one of the most prolific Dominican writers of the eighteenth century, quite provocatively writes:

> I said that it is not wrong to kill the innocent accidentally and without intention. For example, there may be no other way of taking a city without launching incendiary projectiles into it, in which case the deaths of a great many innocents will result. Those deaths are for the common good; an opportunity to obtain victory supersedes the private interests of innocent persons.[150]

Though he employs a slightly antiquated term for the technology of his period (*ballista* is something closer to a catapult or large crossbow), Cuniliati clearly

---

146 Hermann Busenbaum, *Theologiae moralis* [1645] (Venetiis: Apud Hieronymus Albriccium ..., 1689), Bk. III, Tract IV, Ch. I.IV, 126.

147 Claude Lacroix, *Theologia moralis* [1704–1714] (Parisiis : Apud ludovicum vives ..., 1866), Lib. 3, Pars I, Dub. IV. IV., 59. Gabriel Groddeck (1672–1709) also rules on the permissibility of launching fire balls into cities *indirecte & praeter intentionem* during a siege. Gabriel Groddeck, *De eo quod justum est circa tormenta bellica* (Lispsiae: In Collegio Pavlino Prope Avditorium Theologicum, 1710), §XXVIII, 105–106.

148 Charles-René Billuart, *Summa S. Thomae hodiernis academiarum moribus accomodata* (Leodii: Apud Everardum Kints, 1751), Diss, VII, Art. III, §III, 287.

149 Alphonsus Liguori, *Theologia moralis* [1748] (Parisiis: Apud Adrianum le Clere et Soc, 1857), Lib. 4, Tr. 4, Dub. IV, §4, 208. While the German jurist Jakob Friedrich Ludovici (1671–1723) does not explicitly identify fire bombing, he does mention that the killing the innocent in a siege is permitted by custom. See, Jakob Friedrich Ludovici, *De limitibus defensionis in bello defensive* (Halae: Apud Christoph, Andr. Zeitlerum, 1706), §XXV, 19.

150 Fulgentius Cuniliati, *Universae theologiae moralis accurata complexio instituendis candidatis accommodata* (Venetiis: Bettinelli, 1772), Tr. IV. Ch. IX, XII, 162.

has in mind the firing of incendiary projectiles into city centers, which by the eighteenth century had become a pervasive and completely normalized tactic. He provides the conventional intention-based argument to support such attacks. Notice, however, his supporting arguments: without such attacks, punishing the guilty would be impossible, and that the common good supersedes private interests. This second supporting argument is a novel variation within this tradition. Prior commentators relied mostly on the fact that innocent death is accidental or unintentional, even if there was some disagreement about the degree of foresight necessary to make such death truly accidental. Others focused on the degree to which the innocent accrue guilt due to the company they keep. On this view, living among the guilty morally taints the innocent. Cuniliati's argument, however, rehabilitates a line of thinking that can be traced back to Aquinas. Since everyone is guilty through the effects of original sin, all non-righteous people can be killed for the common good.

8 English and Scottish Protestants

These same Scholastic formulations were also prevalent among certain English and Scottish Protestants.[151] In particular, these arguments acquired a new urgency in the period leading up to and through the 1640s, as the cruelties of the English Civil War (1641–1651) were beginning to unfold.[152] For instance, the English theologian and philosopher William Ames (1576–1633) employs the standard version of double effect in his *De Conscientia*, an influential book that went through about seven editions between 1631 and 1670. He writes, "The innocent can be injured unintentionally, by accident, for instance, when attacking a city or a ship which contains various innocent people. In war, when the innocents reside among the enemy, they become part of the guilty."[153] As seen in the Scholastic tradition, this belt-and-suspenders type reasoning was common. Not only did the attacking army have the right intention – they were not trying to kill the innocents – but the inhabitants of the city were, in the

151 Including bishops in the Church of England. See Lancelot Andrewes, *The Patterns of the Catechistical Doctrine at Large* (London: Imprinted by Roger Norton …, 1650), 412.

152 See John Kenneth Ryan, *The Reputation of St. Thomas Aquinas Among English Protestant Thinkers of the Seventeenth Century* (Washington D.C.: Catholic University of America Press, 1948); and Manfried Svensson and David Van Drunen (eds.), *Aquinas Among the Protestants* (Hoboken, NJ: Wiley, 2018).

153 William Ames, *De conscientia, et eius iure, vel casibus, libri quinque* (Amstelodami: Apus Ioan Ianßonium, 1643), Bk. 5, sec. 33, 355.

end, not entirely innocent. They accrued some guilt for living among the guilty defenders of the city.

Likewise, John Weemse (1579?–1636), a member of the Church of Scotland, writes:

> To kill an innocent man accidentally, and besides his intention, when he is exercised in his lawfull calling, this is not a sinne to him; but if he should of purpose kill an innocent man, that indeed were a sinner contrary to the law; and even as in just warre, when the victory cannot be had otherwise unlesse there be innocent men killed, as well as the guiltie, yet they must be safely killed, because the warre is just warre, and secondly because it is not their intention directly to kill the innocent, but because otherwise the victory could not be obtained. So a Iudge is bound to proceed according to that which is proved, and if he killed the innocent man, it is beside his intention; for his intention is here to doe justice, and not to kill the innocent, and he is bound to preferred the universall good, before the particular.[154]

Again, one sees in Weemse's formulation the tragic necessity of resorting to attacking cities. If the guilty hide among the innocent they would never be punished. When resorting to lethal force, however, one must make every effort to kill only the guilty. If the innocents are killed unintentionally, this is permitted in the same way judges must rule on the evidence at hand. But notice here too Weemse employs what could be interpreted as a "common good" argument. The universal good of punishing the guilty is to be preferred to the particular good of preserving the innocent.

Samuel Rutherford (1600–1661), another Scottish minister who was sympathetic to Thomist formulations,[155] argues that "There is no defensive warre different in spece and nature from an offensive warre, if we speake physically, they differ only in the event and intention of the heart, and it is most cleare that the affection and intention doth make one and the same action of taking away the life, either homicide, or no homicide."[156] While Rutherford doesn't explicitly argue that the innocent can be killed unintentionally in a just war,

154 John Weemse, *An Explanation of the Ceremonial Lawes of Moses* (London: Printed by T. Cotes for Iohn Bellamie ..., 1632), 67–68.

155 See Peter Judson Richards, "'The Law Written in Their Hearts'? Rutherford and Locke on Nature, Government and Resistance," *Journal of Law and Religion* 18:1 (2002); 151–189.

156 Samuel Rutherford, *Lex, Rex: The Law and the Prince. A Dispute for the Just Prerogative of King and People* (London: Printed for Iohn Field ..., 1644), 338.

his conceptual frame clearly pays homage to the Scholastic ethical formula. Finally, around this same time, the minister Francis Bland, writes,

> First, in killing, slaying and taking away the life of the enemy, is much regard to be had with what mind, intentions, and affections it is done. S. Bernard hath a distinction very profitable to this purpose. First, if thou prevailest, and perhaps killest a man out of desire of conquest and revenge, then thou livest a Murtherer; which name who is he that would not be loath to carry among men ... Secondly, there is nevertheless some which neither out of zeal, to revenge, nor desire of victory, kills a man, but only for remedies sake, of avoyding or eschewing some ill; as to suppresse Rebellion, to resist invasions, oppressions and the like; In which cases to kill an enemie is undoubtedly lawful.[157]

As Bland suggests, when killing in war is governed by base motives, like conquest or revenge, it is evil. Only when men are pursuing some proper end – righting a wrong or preventing some evil – can killing be justified. In those cases, the justness of the cause and the probity of the intentions permit the Christian soldier to kill.

As the urbanization of war continued, justifications for the death of innocents were routinely framed in the form of tragic but necessary trade-offs, rooted in the principles of justice and proper intentions. A wise prince would carefully assess the proportion of violence needed to obtain a just end and authorize his military to achieve those ends. For instance, in the late seventeenth century, Louis Du Pin's (1657–1719) *New Ecclesiastical History* (1699) which happened to be translated into English by "a learned Divine of the Church of England," reaffirms Vitoria's assessment of killing innocents. He writes,

> Is it lawful to kill the Innocent? Our author [Vitoria] answers that it can never be lawful to do it with an intention of doing it: So that it is unlawful to kill Children, Women, Husband-men, Clergy-men, or any of those that do not bear arms. But it may sometimes accidentally happen that it is not unlawful to kill the Innocent; as when a Fort or Town is besieged, in which there are many innocent Persons that Perish with the Guilty.[158]

157 Francis Bland, *The Souldiers March to Salvation* (York: [s.n.], 1647), 29.

158 Louis Ellies Du Pin, *A New Ecclesiastical History of the Sixteenth Century* (London: Printed by H. Clark, for Tim Child ..., 1699). 443.

The logic of double effect offered theorists of this period a conceptual tool to qualitatively differentiate offensive and defensive wars (even if they technically looked the same), and it also opened up tactical frontiers that would have been closed to the stricter adherents of early just war theory which adamantly protected innocents from the effects of war.

Writing in the mid-eighteenth century, Thomas Rutherford (1712–1771), the Regis Professor of Divinity at Cambridge, argues:

> But if an army, or any small parties of an army, come into the territories of the enemy, notwithstanding they are at liberty to treat the inhabitants, who are not in arms, as enemies, yet they have not a general liberty to kill them. The plain reason, why they have not this general liberty, is, because the death of the inhabitants, who are not in arms, is not necessary for obtaining the just ends of war: and all harm, which exceeds what is necessary, is causeless harm.[159]

He goes on to explain that "in the siege of a town, many of the inhabitants, who are not in arms, may happen to be killed. This is such a chance of war, as the law of nature will excuse. For if the act of besieging the town is lawful, *the accidental consequences*, which follow from taking the necessary steps to carry on this act effectually, are not chargeable upon the besiegers."[160] What is striking about this argument is that it more explicitly hybridizes the notions of necessity and intention. To begin with, necessity is limited by much more definite principles of justice. Typically, killing the innocent is not necessary to obtain just ends. In certain cases, like sieges, however, the innocent can be killed accidentally. Those deaths are permitted because they happened by chance. No innocent person is being targeted qua innocent. The nature of bombing as a matter of fact will lead to these ends. In this respect, the role of chance is a limiting factor in unintentional violence. For instance, Rutherford argues that while "putting a garrison to the sword for holding out long" during a siege might deter others from holding out so long, no "motive of mere utility, will render it just to take away the lives of those who are in our power, and have not deserved to lose them."[161] The same goes, he concludes, for killing prisoners of war in retaliation. Rutherford suggests that it is only when genuinely governed by the laws of chance that the killing can reasonably be seen as accidental

159 Thomas Rutherford, *Institutes of Natural Law* [1754–6] (Baltimore, MD: William and Joseph Neal, 1832), 522.

160 Ibid. (my emphasis).

161 Ibid.

and therefore permissible. While killing hostages and prisoners of war might be perceived as necessary, they are prohibited because they would necessarily be intentional.

Despite these limits, the norms of utility or necessity and proper intentions are clearly in tension. Rutherford goes on to explain that certain cruelties are permissible by "absolute necessity."[162] For instance, he explains that laying waste to a country may be permitted under certain conditions:

> if the progress of an enemy cannot otherwise be stopped, we may destroy the forage or other provisions that are in the way: or if towns or villages would afford them such posts as would give them a considerable advantage against us, and we cannot otherwise prevent them from gaining these posts, we may destroy such towns or villages. To waste a country, unless for some such necessary reasons as these, is causeless harm.[163]

Again, much like his formulation with siege warfare, the purpose of ravaging villages and towns is framed much closer to the language of double effect. Inhabitants are not despoiled by virtue of their 'enemy' status; they receive the accidental effect of an action directed at an advancing army, as plundering was not prohibited.

For Rutherford, the limits to cruelty in war are derived from the law of nature and they point to "what is intrinsically just."[164] And while many nations who engage in war will resort to the slaughter of prisoners and innocent inhabitants, including women and children, no matter how common or normalized such practices have become those states are guilty of violating the law of nature. The fact that they are not punished for transgressing the law has led to an atmosphere of "general license" and what he calls "impunity of fact," as opposed to "impunity of right."[165] In this respect, Rutherford's focus on intentions was designed to function as a moral limitation to the brutal excesses of necessity. In difficult situations, particularly those that will lead to the accidental death of innocents, it was especially important that one had an eye to intrinsic justice so as to not allow mere utility to dominate one's moral framework.

Arguments that explained or justified the death of innocents by way of proper intentions persisted up until the late eighteenth, early nineteenth

162 Ibid., 525.
163 Ibid.
164 Ibid., 526.
165 Ibid.

century.[166] For instance, John Daniel Gros (1739–1812), a German-born Protestant who had migrated to America and ended up teaching at Columbia University, makes a distinction between direct and indirect intentions. He explains that "actions done by direct intention are very considerably higher in morality or immorality than those which proceed from indirect intention."[167] To illustrate this point, he explains that "The destruction of a city or buildings consequent upon direct intention, is the height of cruelty and inhumanity, causing our detestation; if the strength of a place, the vigorous resistance of a garrison, renders it unavoidable, it causes pity and regret."[168] In his analysis, Gros is not suggesting that good intentions absolved one of any and all responsibility:

> No, if unintended consequences may as well flow from our actions as those which are intended, they certainly ought to be taken into consideration before we determine upon an end; they consequently, considered or not considered, come within the compass of at least indirect intention, and as intentional acts, actions of the most direful effects must have a high degree of immorality.[169]

Even with genuine accidents, one must still contend with the "direful effects" caused by the action. For instance, when assessing "the accidental blowing up of a powder-magazine, and the death of many persons occasioned thereby," one must weigh the complex mix of purpose, means, and the diligence to mitigate harm in order to formulate a complete moral diagnostic.[170] On this reading, sufficiently anticipated and weighted incidental harm could be permitted, as long as it was not directly intended, and the end was not full of "direful effects."

Like some of the late Scholastics had indicated, not taking sufficient care, or producing great harm for a minor strategic victory, would count as "indirect intention." Gros explains,

---

166 As will be seen in the following chapters, these arguments precipitously went out of fashion in the nineteenth century among non-Catholics and international law theorists.

167 John Daniel Gros, *Natural Principles of Rectitude: For the Conduct of Man in All States and Situations of Life* (New York: Printed by T. and J. Swords, 1795), 86.

168 Ibid., 87.

169 Ibid.

170 Ibid., 87–88.

> Should we suppose a nation so cruel as to make destruction, plunder and bloodshed their principal object? Should we derive all these calamities from direct intention? No, if there is yet a sense of honesty, justice or humanity, these things would not take place had they been avoidable: in all these proceedings the intention is indirect; it is to compel an enemy to terms of a lasting peace.[171]

Like many employing the double effect formula, the presumption in focusing on intentions is to instigate genuine moral deliberation. It is designed to avoid the temptation to resort to mere utility. This is because there was something morally dangerous about relying on utility to justify certain harms. As Rutherford had indicated, utility can precipitously lead to the ravaging of countries and the murdering of hostages. While one might, in the end, be forced to bomb towns or burn houses, one should resist these temptations for as long as possible. After all, there might be another way to attain just ends. Reluctantly, however, destructive methods are permitted if there is no other way to punish the guilty.[172]

9 Conclusion

Double effect justifications for killing the innocent were dominant among those commentators writing in the Scholastic or Scholastic-inspired intellectual traditions. From the late sixteenth to seventeenth centuries, Jesuit and Dominican scholars reflexively and ubiquitously utilized the language of intention to permit the killing of the innocent, specifically due to the emerging realities of siege warfare. In fact, the above references are hardly exhaustive.[173]

---

171 Ibid., 86.

172 Indeed, it was even common to argue the God was subject to double effect formulations when punishing the guilty. The Scottish minister, Neil Douglas (1750–1823), writes, "As the primary intention of divine judgments, or chastisements, is the reformation, not the destruction of sinners, this being only an incidental effect, in consequence of their willful blindness and obstinacy." Neil Douglas, *Britain's Guilt, Danger, and Duty* (Dundee: Printed in the year 1795), 25.

173 See also Grégoire de Valentia, *Commentariorum theologicorum tomus tertius*(Lugduni: Sumptibus Horatii Cardon, 1609), Dis III, Q. XVI, Punct, III, 271; Johannes Caspari, *Decas Qvaestionum Illustrium Juridico-Politicarum* (Jenae: Typis Tobiae Steinmanni, 1623), Q. 9; Martino Bonacina, *Sacrae theologiae* [1623] (Lugduni: Sumptibus Anisson & Joannis Posuel, 1700), Dis II, Sect. I, §11.17, 468; Joannis Wiggers, *Commentaria de iure et iustitia* (Lovanij: Sumptibus Georgii Lipsii, 1662); Johannes de la Cruz, *Directorium conscientiae in duas partes divisum* (Parisiis: Apud Edmundum Couteror, 1673), Art. 3, Dub. 3.2, 75;

These arguments were also pervasive among English and Scottish Protestants in the years leading up to and during the English Civil War.

While there is unanimous agreement that attacking cities can be permitted in a just war when the belligerent has the right intention, it is also worth noting that in the many examples presented above there are significant nuances and disagreements. For instance, even though it was much more common to link guilt to one's role in combat – Turkish women and children were no longer seen as inherently guilty – there were disagreements about how to characterize the status of the innocent residing in fortified places. Some appear to argue that the innocent lose their protected status due to their proximity to the guilty. By living in the same space as the enemy, they accrue some degree of culpability. Others leave their innocence intact and focus on other variables, like how accidental the killing really is or whether those deaths contribute to the common good. There are also disagreements about whether foresight of harm is compatible with good intentions, or if, for the harm to be permissible, it needs to be a genuine accident. Some argue that the foresight of harm is incompatible with good intentions; whereas others argue that one can both foresee harm and commit it unintentionally. Others place a stricter burden of foresight onto attacking armies; they must have some sense as to the overall military value of the strike. The innocent cannot be killed for minor strategic gains. And finally, some focus on the permissibility of punishing sinners, that is, those not proximately guilty for the crimes that led to war, if their deaths are for the common good.

It is also significant that these arguments represent a clear shift away from Aquinas's original use of double effect, which focused exclusively on the private right of self-defense. By the sixteenth century, the realities of siege warfare had placed new demands on just war theorists. What is more, the revolution in conceptions of innocence made it much more difficult to justify killing those who were not directly engaged in combat. Novel reformulations of double effect gave sixteenth-century moral theorists a critical tool to fight just wars even though they might bring about some unfortunate results.

The double effect formulation for killing the innocent in war (and all its subtle permutations) represents a powerful moral technology, derived from the demands of a particular threat environment. As has been seen, these

Anaklet Reiffenstuel, *Theologia moralis brevi* (Barcinone: Ex Officia, & Sumptibus Josephi Giralt Calcograph & Bibli, 1736); Konstanty Swięcicki, *De jure naturae et gentium in genere; et de jure belli et pacis in specie* (Venetiis: Apud Antonium Bortoli, 1763), Part II, Dis II. Ca I, 211; and Joanne Lercari, *Tentamina Theologico-Moralia* (Lucae: Ex Officina Jacobi Justi, 1778), Dis. XI, Ca III. Dub. V.5, 399.

arguments thrived up until the end of the eighteenth century – countless moralists and jurists formulaically appeal to intention-based arguments for killing the innocent. By the nineteenth century, however, these arguments almost entirely disappear from the wider discourse on killing in a just war. Among those writing in Catholic circles these arguments do persist, but they are no longer taken seriously in the wider international legal discourse. The rival intellectual traditions of the martialists and humanists, treated in the following chapters, eventually take on a much greater degree of influence. The strictures of martial discipline and the necessity-based arguments of the humanists begin to overtake those arguments that focus on good intentions. However, despite falling into disuse and obscurity for roughly a century, double effect arguments will surge back into prominence with the rise of aerial bombing in the early twentieth century where they would once again dominate ethical discourse.

CHAPTER 2

Martialists: Soldiers and Tacticians

1 Introduction

As the late Scholastics were adapting and repurposing Aquinas's novel formulation of double effect to deal with the transformations that were taking place in sixteenth-century warfare, there also was emerging an English tradition of military writing that partook of similar just war terminology but that did not employ intention-based justifications for killing the innocent. As David Lawrence explains, "The transformation of European warfare during the sixteenth century required that soldiers learn the new lexicon of warfare and those assigned the task of commanding armies were expected to adapt to the various changes to technology and tactics."[1] Professional soldiers and military tacticians developed their own literature that to some limited extent discussed the moral liability of killing in war.

In the wake of the Hundred Years' War (1337–1453) and during the Eighty Years' War (1568–1648),[2] soldiers and tacticians returning from battle wrote an impressive array of military manuals and guidebooks promising to instruct readers on various technical aspects of warfare (drills, battle maneuvers, how to defend and assault fortified places, etc.) with the ultimate aim of creating perfect soldiers.[3] Given the cruelties and excesses of war in this period, the

1 David Lawrence, *The Complete Soldier: Military Books and Military Culture in Early Stuart England, 1603–1645* (Leiden: Brill, 2009), 20.

2 The Hundred Years' War was a territorial and dynastic conflict between England and France. The Eighty Years' War, also known as the Dutch War of Independence, was a military conflict between the Seventeen Provinces and the armies of Philip II of Spain.

3 Jonathan Davies points out that prior to 1562 only six books of a military nature were published in England. After 1587, when thousands of English troops were returning from the continent after fighting in the war of Dutch Independence, there was a proliferation of military manuals. Jonathan Davies, "'A Combersome Tying Weapon in a Throng of Men': The Decline of the Longbow in Elizabethan England," *Journal of the Society for Army Historical Research* 80:321 (2002); 16–31. 17. For more on this technical military literature see, Geoffrey Langsam, *Martial Books and Tudor Verse* (New York: King's Crown Press, 1951); Henry Webb, *Elizabethan Military Science: The Books and the Practice* (Madison: University of Wisconsin Press, 1965); J.A. Holding, *Fit for Service: The Training of the British Army, 1715–1795* (Oxford: Clarendon, 1981); J.R. Hale, *War and Society in Renaissance Europe, 1450–1620* (Baltimore, Johns Hopkins University Press, 1986); Barbara Donagan, *War in England, 1642–1649* (New York: Oxford University Press, 2010); and Paul Scannell, *Conflict and Soldiers' Literature in Early Modern Europe: The Reality of War* (New York: Bloomsbury, 2015).

 | DOI:10.1163/9789004515482_004

commentators in this tradition were notably attempting to confront and challenge some of the practices that had become normalized.[4] In this reformative spirit, this literature rehabilitates elements of the early Scholastic just war tradition as well as that of chivalry in medieval England.[5] The result is not only a highly moralistic discourse, but one that treads very lightly in terms of theory, theology, or, for that matter, a broader sense of the strategy and politics of war.[6] There are sporadic references to classical republican literature, typically Machiavelli, and there was periodic interplay between English and continental martialist traditions. Most citations, however, are taken directly from scripture. When references are made to the just war tradition, they are simplistic and formulaic. For all its weaknesses, however, this set of books was attempting to cope with many of the same transformations in warfare that had led the late-Scholastics to employ the innovative justification of double effect for killing the innocent.[7] Surprisingly, while this literature is indebted to the just war

4 It was also not uncommon for some to argue that the art of war had been neglected since the fall of Rome. A fictionalized Thomas More makes this point: "The art of war, like every other art, ecclesiastical architecture along excepted, was greatly deteriorated during those years of general degradation, which preceded and produced the overthrow of the Roman empire. It would have revived about the time when gunpowder was introduced, without the introduction of any such new agent, as a necessary consequence of the systems of dominion which then began to be developed." Robert Southey, *Sir Thomas More; Or, Colloquies on Progress and Prospects of Society* (London: John Murray …, 1829), 208.

5 James Turner Johnson traces the fusion of the just war and chivalric tradition to the prior century: "In the period of the Hundred Years' War (midfourteenth through midfifteenth centuries), the chivalric code was absorbed into the developing tradition on just war, naming the same categories of people as noncombatants but adding provisions specifically concerning combatants." James Turner Johnson, "The Erosion of Noncombatant Immunity in Asymmetric War," *ASPJ Africa & Francophonie* 6:1 (2015); 53–64. 54.

6 Beatrice Heuser argues that very few manuscripts in the period "reflect on politics or make contributions to thinking about the political purposes of warfare, and thus of strategy." Beatrice Heuser, *Strategy Before Clausewitz: Linking Warfare and Statecraft, 1400–1830* (New York: Routledge, 2018), 91. Though, Ben Lowe points out that the literature on war in this period is part of a broader humanist tradition that sought to posit the commonwealth as an agent of peace. Ben Lowe, "War and the Commonwealth in Mid-Tudor England," *The Sixteenth Century Journal* 21:2 (1990); 170–192.

7 For the legacy of chivalry on this period, see D.J.B. Trim (ed.) *The Chivalric Ethos and the Development of Military Professionalism* (Boston: Brill, 2003). For more on the transformations in warfare taking place in this period, see Christopher Duffy, *Siege Warfare: The Fortress in the Early Modern World, 1494–1660* (London: Routledge and Kegan Paul, 1979); Geoffry Parker, *The Military Revolution: Military Innovation and the Rise of the West, 1500–1800* (New York: Cambridge University Press, 1988); Clifford Rogers (ed.), *The Military Revolution Debate: Readings on the Military Transformation of Early Modern Europe* (New York: Routledge, 1995); Mark Charles Fissel, *English Warfare, 1511–1642* (New York: Routledge, 2001); James Raymond, *Henry VIII's Military Revolution: The Armies of Sixteenth-Century Britain and Europe* (London:

tradition in many ways, it makes no reference to intention-based justifications for killing the innocent that had dominated Scholastic writing on this subject. These manuals, usually written from the perspective of the obedient soldier, almost exclusively emphasize the role lawful orders and authoritative judgements play in legitimizing killing in war.

One may recall that a crucial component of Aquinas's conceptualization of just war is the need for the public authority to be guided by a sense of prudence, right intention, and fidelity to the natural law. In doing so it was assumed that he would take all the necessary variables into consideration when planning and authorizing a punitive military strike. There are two significant implications of this line of thinking. The first is that all killing must stem from a proper authority. A soldier must be authorized to kill and to punish the enemy before he can do so. Unless the legitimate authority gives a specific order, killing was not justified.[8] The second implication is that effectively punishing the guilty requires that princes and their commanders have a good sense of their enemy's capabilities (in terms of men and materials), and, perhaps even more importantly, their own military's competency. Focusing on the technical proficiency of a prince's military led to an emphasis on drills, proper training, and strict military discipline.

The martial virtues of strict discipline and steadfast obedience to lawful commands are some of the most pronounced themes that run through these manuals. The killing of those not engaged in fighting is not thematically address. However, one does catch indirect glimpses of the threat war posed to non-fighting populations; plundering, raping, raiding, burning, and hostage-taking are all suggestive of the risks the innocent faced in times of war. When these issues are touched upon, they are almost exclusively addressed through this prism of restraint and dutiful obedience. An array of martialist writers sought to instruct princes and their commanders on how to maintain perfect control over their armies, which ranged from the much more mundane technical matters (the signals and signs for commanding troop maneuvers, where and how to set up camp, etc.) to much more idealized conceptualizations of martial virtue (preventing soldiers from drinking, swearing, gambling, or sleeping with prostitutes). Quite ambitiously these manuals sought to establish the criteria for the perfect soldier – a pious and obedient servant who is the

Tauris Academic Studies, 2007); Mark Greengrass, *Christendom Destroyed: Europe 1517–1648* (New York: Penguin, 2014).

8 For an elaboration of this point, see Daniel Montaldi, "A Defense of St. Thomas and the Principle of Double Effect," *Journal of Religious Ethics* 14:2 (1986); 296–332.

complete master of his passions, even in the heat of battle, and will unquestioningly follow orders.

Given the unprecedented degree to which killing the innocent was thematized in the late Scholastic literature, it is noteworthy that this subject receives very little direct attention by those writing in that tradition. When there is dedicated attention paid to the moral peril of killing the innocent, its analysis is theoretically underdeveloped. This is even more remarkable and worth discussion given the degree to which the city was becoming implicated in military conflict in this period.[9] As Olaf van Nimwegen explains, "Sieges dominated warfare in the early modern era, particularly in northern Italy and the Low Countries, two densely populated regions with many fortified towns. Administration, trade and industry were concentrated in towns, so their conquest yielded great political and economic advantages."[10] The professional soldiers writing on martial virtue would have been intimately aware of the new pressures nonfighting residents faced in war. When addressed at all, however, prohibitions against killing the innocent were largely subordinate to or entailed by the precise judgment of wise leadership in a just war, princes would direct soldiers to kill only the guilty, but, unlike the Scholastic tradition, the innocent-guilty distinction was neither thematized nor was it theoretically well developed.[11]

---

9 For an account of how the Hundred Years War impacted civilian populations, see Clifford Rogers, "By Fire and Sword: *Bellum Hostile* and 'Civilians' in the Hundred Years' War," in *Civilians in the Path of War*, ed. Mark Grimsley and Clifford Rogers (Lincoln: University of Nebraska, 2002); and James Gilbert, "A Medieval 'Rosie the Riveter'? Women in France and Southern England during the Hundred Years War," in *The Hundred Years War: A Wider Focus, Part 1* eds. L.J. Villalon, and Donald Kagay (Boston: Brill, 2005), 341. For more on how the Eighty Years' War impacted civilian populations, see Marco van der Hoeven (ed.), *Exercise of Arms: Warfare in the Netherlands, 1568–1648* (Boston: Brill, 1997); and Fernando González de León, *The Road to Rocroi: Class, Culture, and Command in the Spanish Army of Flanders, 1567–1659* (Boston: Brill, 2009).

10 Olaf van Nimwegen, *The Dutch Army and the Military Revolutions, 1588–1688* trans. Andrew May (Suffolk, UK: Boydell Press, 2010). 119.

11 Continental martialist traditions would routinely make nominal distinctions, appealing to the canon law. For instance, the Italian soldier, Pierino Belli (1502–1575), argues that protections should be extended to women and young boys. He also notes that "the canon law commands the sparing of priests, hermits, monks, pilgrims, traders, and farmers, with their stock." Pierino Belli, *A Treatise on Military Matters and Warfare* [1563], trans. Herbert Nutting (Oxford: Clarendon Press, 1936), 81. Belli recommends that Italian generals keep in mind the example of the Romans who, when a city was taken by surrender, did not plunder and slay its inhabitants. He writes: "I would that such examples were kept before the eyes of the generals, prefects, captains, and soldiers of our armies, who plunder the unhappy districts that without any fault on their part have fallen into the hands of the

As counterintuitive as it might seem, and despite nominal provisions in some manuals, the killing of women and children was not thought to be *prima facie* wrong; such killing was only immoral when driven by inordinate or private passion rather than following a lawful order. In many respects, the conferral of innocence represented a *judgment* rather than the recognition of an innate category. It is certainly true that military codes in this period, largely inspired by Maximillian I's military reforms of 1508 and Charles V's military penal codes of 1532, were increasingly relying on vaulted language that demanded soldiers live pure lives, not swearing or drinking, and that they be perfectly restrained in battle; they should not kill "priests, women and children, and the bringers of provisions; under pain of punishment for perjury or death."[12] It is quite difficult, however, to differentiate these strictly moral claims from the practicalities of war. These prohibitions could just as easily be seen as part of the necessity of making a fighting force so bridled that it could be commanded and directed even in the fury of battle. To be sure, these restrictions had a significant practical and one might say logistical component. Commanders needed to ensure that after victory plunder and the spoils of war were systematically collected and that local inhabitants were spared, if for no other reason than to produce regular contributions of food and money.

The limited scope of this so-called immunity can be seen repeatedly in these manuals when protected status of the innocent is posited conditionally. Stipulations that appear to ensure that certain classes of people should not be killed are immediately followed by some kind of caveat; one should not kill the innocent *unless so directed*. Very few commentators within this tradition focused on the innocent *qua* innocent. Most military manuals of this period make it clear that the moral bright line was not strict immunity or the innate innocence of certain populations, but the duty to obey the lawful command of an authorized authority.

enemy in fact sometimes even through the carelessness of our soldiers, and perhaps even by the treachery of some of them. But whether they retake these places by force of arms or even by surrender, they plunder as heartily, and abstain not a whit more from murder, debauchery, and pillage, than if they had taken a city of the Turks at the cost of much labour and bloodshed." (Ibid., 232). Of course, the implication of this last line seems to suggest that Belli believed that Turks were subject to a different criterion in war.

12 Colonel Eardley Wilmot, "Military Law in Germany," *Journal of the Royal United Service Institution, Volume 35* (London: Harrison and Sons, 1891), 1031, and William Winthrop, *Military Law, Volume 1* (Washington, DC: W.H. Morrison, 1886), 7.

2 Early Martialists

The first impression one gets when reading these early manuals and treatises on military discipline is the presumption that through the process of rendering warfare scientific, many of war's excesses and cruelties would simply be eradicated. When poorly trained and undisciplined, armies would commit all manner of atrocities. Take, for instance, Barnabe Rich's (1540?–1617) essay on martial discipline. He writes:

> as by the knowledge of warre and exercise of armes, Empires haue beene purchased, Kingdomes enlarged, Princes preserued, Iustice maintained, good Lawes protected, and the Common wealth defended, so in neglectinge martiall exercises and laying aside of their weapons, how many kingdomes hath beene brought to calamitie, howe many countries ruinated, and howe many florishinge Citties sacked, beaten flatte to the ground, couered ouer with moulde, and almost worne out of memorie.[13]

Notice that injustice – kingdoms being brought to ruin and calamity, cities being sacked – are the result of ignorance and poor leadership. Civilians are put at risk when leaders are unwise, and armies are undisciplined. In another essay on this topic, Rich writes "As the Warres cannot bee performed without Souldiers, so Souldiers cannot bee contained without pay: for where there is want of pay, there must needs be but a scambling warre, manie disorders committed, and as many opportunities permitted."[14] There was a growing sense that military order and discipline were to be found in the institutionalization of armies, and that atrocities and disorder would be eliminated only if soldiers were effectively trained and well compensated. Rich concludes: "Force that is not assisted by advice, is like a horse without a rider, of it owne selfe, it destroieth it selfe: and force and furie, without Discipline and Order, are easily vanquished and overthrown by Policie."[15]

On this same theme, Sir Roger Williams (1540?–1595) argues that war is maintained by three principles: "A good Chiefe; a good Purse and good Justice."[16] He explains that the "covetous and miserable," the "ambitious," and "proud and vaine glorious" leaders "weary others with wants and discourtesies, keep the Officers of his Armie in pickes and quarrels, cause them to disgrace one

13 Barnabe Rich, *A path-way to military practice.* (London: Printed by Iohn Charlewood, for Robert Walley, 1587), Preface.

14 Barnabe Rich, *A Looking-Glasse for Warre* (London: By Thomas Creede ..., 1604), 71.

15 Ibid. 34.

16 Sir Roger Williams, *A Briefe Discourse of Warre* (Londed: Printed by Thomas Orwin ..., 1590), 2.

another, lay al the fault on his felowes, spare neither friend nor foe to recover wealth; it is unpossible for such a Chiefe to end anie action honorablie."[17] The excesses of war can be traced directly to the character defects of leaders and their inability to purge the moral deficiencies of their men. Well-funded and disciplined armies with effective leadership produce fewer quarrels and conflict.

Sir John Smythe (1534–1607) explains that martial discipline not only prevents wars – i.e. long periods of peace leads to lax behaviors that can only be restrained through martial readiness – but it also leads to restraint in fighting. He writes, "Even so, an Armie in the field, being well payed, provided for and governed, doo live as orderly in towne, campe and field, without robbing, spoiling, or otherwise injuring any other, but the common enemie."[18] Poorly maintained and under-disciplined armies don't set up orderly encampments, "but dispersed and scattered by band in many Townes and Villages with great disorder," which ultimately leads to "the spoil, and misusing of the common people."[19] Smythe was critical of those "disorderlie proceedings, contrarie to all pietie, and discipline Militarie" that led to wanton destruction and cruelty.[20]

Bertrand de Loque (1550–1605) argues that it is crucial for a prince to learn to "overcome his owne passions and affections," and above all that he be reluctant to wage war.[21] This is because:

> Warre maketh all things deare, and causeth famine: for the soldiers treade downe the fruites of the earth under their feet, spill the vines, steale, burne whole townes and villages, spoile and rob mens houses, take away mens cattell, fourrage their barnes, lye in waite to set upon men in the high waies hinder the traffique and concourse of Marchants, and doe commit many such like mischiefes, which for the most part cause great famine and scarsity.[22]

17 Ibid. For similar arguments see Geffrey Gates, *The Defence of Militarie Profession* (London: By Henry Middleton, for Iohn Harison, 1579); Thomas Styward, *The Pathwaie to Martiall Discipline*. London: Printed by T.E. for Myles Jenyn …, 1581); and Justus Lipsius, *Politica: Six Books of Politics or Political Instruction* [1589] ed. Jan Waszink. (Assen, The Netherlands: Royal Van Gorcu, 2004).

18 Sir John Smythe, *Certain discourses, written by Sir John Smythe, Knight: concerning the formes and effects of divers sorts of weapons, and other verie important matters militarie* (London: Printed by Richard Iohnes …, 1590), Dedicatorie.

19 Ibid.

20 Ibid.

21 Bertrand de Loque, *Discourses of Warre and Single Combat*, trans. John Eliot (London: Printed by John Wolfe …, 1591), 19.

22 Ibid., 15.

It is crucial that armies are led by restrained Christian princes, who in turn instate disciplined captains. He rejects the idea that "the laws cease, & are of no force in warre"; such sentiment "is not of God but of the Devil."[23] Without dutiful obedience to the law of God and strict training in martial discipline, "There is no mischief, that soldiers do not put daily in practice against the poore people."[24]

Rejecting the "brute customes" of antiquity, Thomas Digges (1546–1595) argues that the monarchies of the world should embrace and train their subjects in the "Science Militare."[25] A wise ruler understands that the "corruption of Militare Discipline, and the licentious living of the Souldiour" opens them to all manner of errors of judgment, both in and out of combat.[26] Digges promises to guide the reader to an algebraically precise understanding not only of tactical artillery but the fundamental health and wellbeing of the military. And it is this mathematical precision that guarantees that the excesses of war will be reined in. Digges suggests that even the accidental death of the innocent can be avoided through proper discipline. For instance, he stipulates "That no souldier kill any woman or aged man, nor any impotent person, albeing in the Furie of Victorie, and this also on paine of life: nor shall lay hands on them, on paine to be punished according to the nature of the offence."[27] Proper training and strict martial law will check the soldier's excessive passions, even in the throes of victory.[28]

Digges's desire to restrain troops from the harm that might be inflicted on civilians in the "Furie of Victorie" is a subtle but important indication that the martialist writers, much like the Scholastics, were also beginning to cope with the same transformations of warfare, namely the shift away from raids and

23 Ibid., 22.

24 Ibid., 24.

25 Thomas Digges, *An arithmetical warlike treatise named Stratioticos.* (London: R. Field, 1590), Preface.

26 Ibid.

27 Ibid., 289.

28 In a similar manner, John Norden denounced ungodly warfare "wherein spoyles are helde lawfull prize, riots maintained as rightfull lawes, innocents murthered, maydens, wiues and most graue matrons violently abused, and forcibly deflowred, churches destroyed, religion confoũded, houses consumed with fire, all vertue and honest duety abandoned, vice exalted and imbraced, the seruice of God contemned, Gods children murdred, euery degree, sexe and qualitie grieued and oppressed, all trades and traffique hindred, the best men troden downe, and all this (with many other miseries) by a crewe of most mercillesse and wicked men." Like Digges, the implication is that men trained in discipline and martial virtue would not be prone to cruelty and excess. John Norden, *The Mirror of Honor* (London: By the Widowe Orwin for Thomas Man, 1597), 2.

dynamic troop engagements to sieges and the occupation of towns and villages. By the late-sixteenth century, the siege had arguably become one of the most pressing moral concerns for strategists and planners.[29] Not only did such assaults pose the greatest technical and logistical challenges, but they also represented the greatest risk to the restraint and order of the invading soldiers. Many manuals are quite clear that these sorts of engagements must be led by a properly instated captain, that is, not only someone formally designated to serve in this capacity, but someone who has the technical understanding of how to make a proper assault and who knows how to lead and restrain those under his command.

While sieges had been common in ages past, they were costly and difficult to maintain. Advances in engineering and artillery, however, made assaults on fortifications much more tactically plausible, hence much more frequent.[30] Some of these transformations can be seen in the martialist literature of this period. For instance, early in the sixteenth century, Machiavelli had noted that with the rise of artillery "no wall is so thick, but in a few days they will beat it down."[31] In a similar manner, Humfrey Barwick rhetorically muses, "was not our olde Castles and Towers, sufficient to defend against all Rams and engins in those daies used: whereas in these times there is hardly any devise of fortification that will serve against the force of the Cannons?"[32] It is not surprising that in this period many manuscripts were dedicated to the art of fortification, laying siege, and assaulting fortified places.[33] Mastering the art of the siege was now crucial to successful leadership. Edward Davies sums up this point of

29 Many commentators have pointed out the way changes in siege warfare from 1560–1660 affected the laws of war. See Randall Lesaffer, "Siege Warfare in the Early Modern Age: A Study on the Customary Laws of War," in *The Nature of Customary Law: Legal, Historical, and Philosophical Perspectives* ed. Amanda Perreau-Saussine (New York: Cambridge University Press, 2007), 179; and John Lynn II, "Honourable Surrender in Early Modern European History, 1500–1789," in *How Fighting Ends: A History of Surrender* eds. Holger Affelerback and Hew Strachan (New York: Oxford University Press, 2012), 102.

30 Some of these trends were anticipated in the fifteenth century. See Leon Battista Alberti, *On the Art of Building in Ten Books* [1452], trans. Joseph Rykwert, Neil Leach, Robert Tavernor (Cambridge, MA: MIT Press, 1988).

31 Niccolò Machiavelli, *The Art of War* [1521] in *The Works of the Famous Nicholas Machiavel* trans. Henry Neville (London: Printed for John Starkey ..., 1680), 353.

32 Humfrey Barwick, *A Breefe Discourse, Concerning the Force and Effect of all Manuall Weapons of Fire* (London: Printed for Richard Oliffe ..., 1592), "To all Skilful Captaines and Soldiers."

33 Besides the texts already cited, see Peter Whithorne, *Certain Wayes for the Ordering of Soldiours in Battleray* (London: by W. Williamson: for Ihon Wight, 1573); and Raimond Fourquevaux, *Instructions for the warres. Amply, learnedly, and politiquely, discoursing the method of militarie discipline*, trans. Paul Iue (London: Printed for Thomas Man ..., 1589);

view when he explains that effective captains must be proficient in "drawing platforms, in the Mathematikes, in the martiall Lawes, in besieging of townes, batteries, mines, and each thing else belonging to martiall discipline."[34]

To say that the storming of fortified cities was a central moral concern in the military manuals of this period is not to say that the risks this tactic posed to the noncombatants was explicitly theorized. Many completely elided the subject. William Fulbeck (1560–1603), for instance, simply reaffirms the longstanding view that "cities surprised may bee sacked" whereas "cities yielded upon condition may not."[35] There was a wide-spread belief in this period that a city obstinately defended opened itself up to certain reprisals by way of the invading troops. Even if manuals made nominal provisions about the duty to not harm women and children, the rules governing the sack of cities and town elided or overrode those protections. The permissibility of the sack was the product of a unique set of circumstances; indeed, these conditions were such that, practically speaking, one could not ensure that truly innocent people could be spared. Very few of these early manuscripts dealt directly with the plight of the civilian population in the process of a siege, and in no cases do they expressly argue that the status of the residents of besieged cities is unambiguously innocent.[36] In this respect, the nonfighting population was treated as an incidental component to the process of taking the city. Again, what prohibitions did exist were intimately linked to the issue of martial discipline and the persistent readiness to obey orders. Simply put, soldiers' passions should not get the better of them. It would be a mistake, however, to claim that there was an expansive and uncomplicated principle of noncombatant immunity at work in this discursive tradition. As seen in the previous chapter on the Scholastics, even though notions of collective guilt were waning by the

---

Thomas Smith, *The Art of Gunnery* (London: Printed for William Ponsonby, 1600); Robert Norton, *Of the Art of Great Artillery* (London: Printed by Edw. Allde …, 1624).

34 Edward Davies, *The Art of Ware, and Englands Traynings* (London: Printed by Edward Griffen …, 1619), 109.

35 William Fulbecke, *The Pandectes of the Law of Nations* (London: Imprinted by Adam Islip …, 1602), 51.

36 For instance, there is almost nothing about the proper treatment of the innocent in the statues governing the army of Henry VIII. One statute indicates that after victory, "no man presume to robbe nor pyll, or to take them prisoners," but the specific prohibition against sacking is not specified. *Statues and Ordynances for the Warre* (London: Fleetstreet by Thomas Barthelet …, 1544). And other texts provided only instrumental or pragmatic accounts of restraint. For instance, Jacopo de Porcia argues that soiling and pillaging should be prohibited "until thyne enemyes be conquered, and wholly brought under." Jacopo di Porcia, *The Preceptes of Warre* trans. Peter Betham (London: In the Olde Lewery by Edwarde Whytchurche …, 1544), Statue 146.

sixteenth century, it was not uncommon for enemy populations to still be seen as collectively guilty, especially those residing in a recalcitrant fort.

While many commentators looked down on the reckless violence that took place during a sack, the sacking of towns was still quite common in this period. Despite their resistance to crazed violence, martialists believed that certain pressures could justifiably be placed on civilian populations. Forcing residents back into a besieged place was a common strategy. Some even proposed slightly more aggressive variations of this tactic. The Italian captain, Onosandro Platonico, writes,

> But the women, the children, the olde folke, and those which by other occasion be weak, and not mere to faight, thou shalt inforce to enter into the Citye, as those whyche can not woorke, nor helpe in anye thing the same, which for the defence of a City is desired, so that they much soner consuming the vittailes, may be of so great harme to the besieged citizens.[37]

And in imitation of the "Italian method," Gyles Clayton explains:

> if you be minded to win any Cittie or strong Hold, then you shall first foresee which way is best to ouercome all the small Townes rounde about the same, and so ouercomming them with violence, shall driue them into the Cittie or strong Holde, for succour at their freendes and Countrimens hands, the which multitude of people beeing driuen into the Cittie, you may with more ease famish and ouercome them.[38]

Rather than merely forcing back in those seeking to leave, commentators like Platonico and Clayton argued that all noncombatants in the surrounding area could be corralled into the fortress just prior to the siege. But in general, these techniques, as proposed in the manuals, were designed to expedite surrender, not as a plot to slaughter the innocent. After all, as Clayton continues, "where discipline is used with lenitie and curtesie, more prevaileth then furie & rashness, without discretion or wisedom."[39] In fact, as the above passages from Digges's treatise indicates, much greater attention was beginning to be

37 Onosandro Platonico, *Of the Generall Captaine* trans. Fabio Cotta (London: By William Seres …, 1563), 134.

38 Gyles Clayton, *The Approoved Order of Martiall Discipline* (London: By I.C. …, 1591), 74.

39 Ibid., 75.

directed at the frenzy and confusion that was likely to come about when the fortress's walls had been breached and the city risked being sacked.

While many manuals addressed the three prominent siege tactics of surprise, deception, and blockade, much greater moral consideration was starting to be directed at the process of taking a fortress by force.[40] Artillery fire, as would soon notably change, was directed exclusively at the castle walls and battlements with the goal of creating a breach large enough for troops to storm. It was in this moment, when soldiers were pouring into the fortress, that nonfighting residents were most vulnerable. Many commentators started to emphasize the duty of restraint during an assault. For Sextus Frontinus, the presumption was that unless a specific proclamation had been made, "everye man must refrayne from sacking upon the same paine."[41] Furthermore, he generally stipulates that soldiers shall not "misuse the people of the country where the warre is made, either in body or deeds, except that they bee declared rebels to the kind."[42] William Garrard concludes that "if there be any citie, castle or other fort, yielded up by the enemie, withough expugnation: there shall no man be so bold to enter into the said place, to spoile or otherwise to kill or doe any outrage, without leave of the Generall, upon paine of the losse of his life."[43] As can be seen in the above formulations, the operative distinction was authorization. Soldiers were only permitted to sack the town if they had been given permission to do so, typically after the invading force had achieved its objective of taking the city.

40 Besides the texts already mentioned, see Flavius Vegetius Renatus, *The Foure Bookes of Flavius Vegetius Renatus* (London: Fletestreate …, 1572); Paul Ive, *The Practise of Fortification* (Imprinted at London by Thomas Orwin for Thomas Man and Toby Cooke, 1589); William Bourne, *Inventions or Devises Very Necessary for All Generalls and Captaines, or Leaders of Men* (London: Printed for Thomas Woodcock, 1590); Don Bernardiono de Mendoza, *Theorique and Practise of Warre* trans. Sir Edwarde Hobt (Middelburg: Printed by Richard Schilders, 1597).

41 Sextus Juliys Frontinus, *The Strategemes, Sleyghtes, and Policies of Warre*, trans. Rycharde Morysine (London: Thom. Bertheleti typis impress, 1539), 264. This same formulation is found in Robert Barret, *The Theorike and Practike of Moderne Warres Discoursed in Dialogue Wise* (London: for William Pasonby, 1598), Item 28, 147.

42 Frontinus, *The Strategemes, Sleyghtes, and Policies of Warre*, 265.

43 William Garrard, *The Arte of Warre* (London: Printed for Roger Warde …, 1591), 48. Edward Davies employs the same provision. Davies, Edward, *The art of vvare, and Englands trayn-ings* (London: Printed by Edward Griffen …, 1619), 67. Note the implication that a general might permit his soldiers to plunder and kill under certain situations. These conditions are not elaborated on in this literature. For an account of the formal procedure around plunder, see also, John Cruso, *Military instructions for the cavallrie* (Cambridge: Printed by the Printers to the University of Cambridge, 1632).

There were those in this literature who were gesturing towards something closer to noncombatant immunity, as the straightforward stipulations of the canon law suggest. Matthew Sutcliffe, perhaps the most expansive in the protections he extends, writes, "In executing of wars, this precept must diligently be had in remembrance, yet there be no crueltie used. There is moderation even in the execution of justice."[44] To this end, "Women, children, and old folks, by the orders of warre observed nowe [...] are exempted from the souldiers furie in the sacke of Townes."[45] Just wars, he continues, are "moderately prosecuted."[46] And though not explicitly articulated in reference to the siege, these same concerns were raised by Queen Elizabeth I (1533–1603) in her 1597 orders to the Earl of Essex. She commands him,

> give orders to all persons serving under your charge, that there be no slaughter of people, except such as shall with open force manifestly withstand you, but mercy given to all such as freely submit, detaining them prisoners, to be redeemed according to their estate. You shall also order all serving in our navy and army in no sort to spoil or destroy any church or place appointed for Divine service, nor any hospital for sick or poor people, upon pain of death; but if any of the enemy shall flee to any church or hospital to save himself, and shall be a person worth taking prisoner, it shall be lawful to recover him, without destroying the church or hospital. You shall also forbid all persons to kill or willfully hurt women, children, or aged or sick persons; all these directions you shall publish for the military discipline of the forces.[47]

One should notice that this formulation, as generous as it seems, essentially reproduces the norms of siege warfare. Those places that are obstinately defended – where a force manifestly withstands the besiegers – can be slaughtered. While these stipulations were certainly pushing against the established norms of this period, they could certainly be interpreted within the ascendant views of the period.[48]

---

44 Matthew Sutcliffe, *The practice, proceedings, and laws of armes* (London: by the deputies Christopher Barker ..., 1593), 11.

45 Ibid., 12.

46 Ibid.

47 Queen Elizabeth I, "Vol. CCLXIII," *Calendar of State Papers, Domestic Series, Elizabeth, 1695–1597*, ed. Mary Anne Everett Green (London: Longmans, Green, Reader and Dyer, 1869), 441.

48 There was a parallel discourse on military discipline among the Spanish in the sixteenth century that also prioritized the control of soldiers' passions in the heat of battle. For

While some manuals advocated categorical restrictions, the scope and intensity of protections varied in this discursive tradition. Overall, the primary focus of this moral concern was the duty of the soldiery to be restrained in the heat of battle. And as such, these prohibitions help to create a framework for, at the very least, a descriptive understanding of noncombatant death; the innocent die when leadership is unwise, and soldiers are undisciplined. Killing the innocent is a technical problem. The only killing that is permitted is that which is explicitly authorized by a legitimate commander.

3 Seventeenth-Century Martialists

By the seventeenth century, many of these formulations which were just starting to be employed in the sixteenth century continued, with some subtle though noteworthy shifts. Military manuals and treatises continued promoting restraint even as the passions and fury of war risked overwhelming soldiers, driving them to commit various atrocities. For instance, in John Bingham's *The Tactics of Ælian* (1616), he affirms the need for armies to restrain their passion: "At the first warre was made, and men fought (as wild beast grapple together) led with furie, and rage, and not with skill."[49] In order to avoid the desperation of the breach, he preferred to "provoke the enemie ... to towel him out of his advantage, and bring him into the field, where he may more easily be dealt withall."[50]

In *The Swedish Discipline* (1632), a summary of Gustavus Adolphus's *Articles of War* (1621), William Watts explains that if a soldier fires upon "any Churchman, or aged people, Men or Women, Maydes or Children" that is "unless they first take Armes against them," those soldiers would be subjected to the "pain

example, Sancho Londoño writes: "That no soldier kills a woman, a child, an old man, any innocent inhabitant, even if it is in the fury of victory, under pain of life, if I lay my hand on such people, I will be subjected to the same penalty, being punished according to the quality of the crime." Sancho Londoño, *El Discurso sobre la forma de reducir la disciplina militar* (Madrid: Por Luys Sanchez, 1593), 33. See also Diego de Alava y Viamont, *El Perfecto Capitan* (Madrid: Por Pedro Madrigal, 1590); Cristóbal Mosquera de Figueroa, *Comentario en breve compendio de disciplina military* (Madrid: Por Luis Sanchez, 1596); Francisco de Valdes, *Espeio, y disciplina military* (Brussellas: En casa de Roger Velpius ..., 1596). In the Italian tradition, Flamio della Corce makes similar arguments. See Flaminio della Croce, *Theatro Militare* (In Anversa: Appresso Henrico Aertssio, 1617).

49 John Bingham's *The Tactics of Ælian: Or, Art of Embattailing an Army After y Grecian Manner* (At London : for Laurence Lisle ..., 1616), 92.

50 Ibid., 96.

of punishment at the discretion of the Iudges."[51] John Cruso argues that a good soldier "depresseth those vehement passions," and that when taking a village it is a moral liability when soldiers proceed with "ignorance joined with a naturall furie."[52] In his manual on *Camp Discipline* (1642), Robert Devereux, the Earl of Essex, prohibits "ransoming of persons, spoiling, pillaging, parting of the prey, or wasting and burning by fire [...] upon the same paine of death."[53] Similar observations can be seen in King Charles I's *Lawes and Ordinances of Warre* (1640),[54] the New Model Army's *The Souldiers Catechisms* (1644) and many other manuals and treatises of this period.[55]

In fact, these idealized conceptions of the soldier were so common that they were routinely lampooned in mid seventeenth-century plays and literature. In Jasper Mayne's *The Amorous Warre* (1659), three humorous courtiers paint a very different picture of the soldier. One, named Callias, maintains that "Armies Plundering Townes and ravishing of Virgins, As naturally follow a good Face Stolne, as this was, as Aches doe you Wenchings, Or as your Taylor, Artops, Follows you With an old Bill unclear'd."[56] And later, Callias's friend Artops notes that "Some under Officers perhaps there may be, whose Trade & Occupation 'tis to Kill, And to grow rich by Slaughter; Vile *market spirits*, who doe not fight for Fame, or Cause, but thinke that is most ith' Right which gives

51 William Watts, *The Swedish Discipline, Religious, Civile, and Military* (London: Printed by Iohn Dawson ..., 1632), 57.

52 Cruso, *Military instructions for the cavallrie*, 2, 86.

53 Robert Devereux, *Camp Discipline, or The Soldiers Duty* (London: Printed for Jo. Rothwell ..., 1642), 5.

54 *Lawes and Ordinances of Warre* (London: Printed by Robert Barker ..., 1640). See also *Military Orders and Articles Established by His Majesty, For the Better Ordering and Government of His Majesties Army* (Oxford: By Leonard Lichfield ..., 1642); and *Lawes and Ordinances of Warre: Established for the Better Conduct of the Army* (London: Printed for Luke Fawne, 1643).

55 Robert Ram, *The Souldiers Catechisms: Composed for the Parliaments Army* (London: Printed for J. Wright ..., 1644); Robert Monro, *The Scotch Military Discipline Learned from the Valiant Swede* (London: Printed for *William Ley* ..., 1644). See also some earlier texts: James Achesone, *The Military Garden, Or Instructions for All Young Souldiers and Such who are Disposed to* Learne (Edinburgh: Printed by Iohn Wreittoun, 1629); Dieur Du Praissac, *The art of warre, or Militarie discourses of leavying, marching, encamping; and embattailing an armie* (Cambridge: Printed by Roger Daniel ..., 1639); Henri duc de Rohan, *The complete captain* (Cambridge: Printed by Roger Daniel ..., 1640); and Henry Hexham, *The first part of the principles of the art military practiced in the warres of the United Netherlands* (Delf, Holeand: [n.p.], 1642).

56 Jasper Mayne, *The Amorous Warre* (Oxford: Printed by Henry Hall ... for Ric. Davis, 1659), 2.

most pay."[57] Rather than the pious instrument of a godly prince, soldiers were seen as hired mercenaries who grew rich from raping and killing the innocent. Indeed, many of the manuals written in this period appear to have in mind exactly this kind of behavior where professional soldiers are driven by "market spirits" and who viewed the chaos of war as a money-making venture. These hired killers were all too willing to abandon ranks to slaughter innocent residents to collect the choicest pieces of plunder for themselves. To be sure, these lampooned descriptions are a useful corrective to the highly idealized literature of the military manuals of the period.

Much as it was in the sixteenth century, however, the moral focus in the military codes and articles of this period was not the inherent innocence of the nonfighting populations but the degree to which the prince was steeped in the "Arte Militarie" and his soldiers were restrained and disciplined.[58] A wise and just prince would, in theory, be able to produce men with "Strength, Courage, Judgement, Learning, and other qualities befitting a perfect Souldier."[59] And in this respect, the moral liability in war was not merely the murder of the innocent but the moral framework of restraint and obedience that circumscribed and controlled all violence in war. As one manual notes, it was paramount that soldiers learned to "governe their passions, and to bridle their insatiate appetite."[60] But as seen above, this did not always mean that women and children should not be killed. While some manuals entirely forbade the killing of certain classes of people, again we see that others prohibited only unauthorized killing. As Alexander Leighton (1570–1649) notes, "What the meanest soldier doth, hee must doe it by authority, which doth warrant him to doe many things, which otherwise were unlawful; as to kill, sack, and spoyl; which without authority, were murther, robbery, and cruell oppression."[61] Much as it was in the sixteenth century, pillaging, raiding, and killing were not *prima facie* prohibited; they were only forbidden when driven by private passions or when they occurred outside the lawful command structure. This can also be seen in Robert Ward's *Adimadversions of Warre* (1639) in

57 Ibid., 65–66. The characterization of soldiers as ravenous and greedy was common in the literature of this period. Diane Purkiss, *Literature, Gender, and Politics During the English Civil War* (New York: Cambridge University Press, 2005), 42.

58 Sir Thomas Kellie, *Pallas armata, or Militarie instructions for the learned* (Printed at Edinburgh: By the heires of Andro Hart, 1627).

59 Ibid., "To His Fellow Advocates".

60 Edward Cooke, *The Character or Warre, or The Image of Martiall Discipline* (London: Printed by Tho: Purfoot, 1626), Chap. XVI.

61 Alexander Leighton, *Speculum Belli Sacri: Or The Lookingglasse of the Holy War* ([Amsterdam]: Printed [by the successors of Giles Thorp], 1624), 20.

which he explains: "No Souldier shall set fire upon any Town or Village in the enemies Land, without he be commanded by his Captain: neither shall any Captain give any such command, unlesse hee hath first received it from us or our General" (ordinance 88); and, "No man shall set fire upon any Hospitall, Church, Schoole, or Mill, or spoyle them anyway, except he be commanded."[62]

Some authors pointed out the moral confusion that would arise if a soldier engaged in private or unauthorized acts of violence. For instance, Abraham Gibson explains that when soldiers take it upon themselves to engage in private violence they are "pressed with an unanswerable *Dilemma* on both sides. If he be *killed*, he hazardeth with his body his *soule*, because hee dieth in passion and distemper. If he *kill* hee hazardeth with his soule his *body*, because hee fighteth both against *God* and against *Gideon*. Against *God*, whose *image* he extinguisheth and against *Gideon*, whose *law* he contemneth."[63] Much as the Scholastic tradition had argued, these risks appear not to be entirely operative when one is under the command of a wise and just Christian prince. After all, when properly directed, soldiers could be the agents of divine justice. Richard Baxter explains:

> Q. When is killing Murder, or unlawful? A. When it is done without Authority from God, who is the Lord of Life. Q. To whom doth God give such authority to kill men? A. To the Supream Rulers of Common-wealths, and their Magistrates to whom they communicate it. Q. May they kill whom they will? A. No: None but those whose crimes are so great as to deserve death by the Law of God in Nature, and the just Laws of the Land: even such whose crimes make their death the due interest of the Republic, and needful to its good safety.[64]

While it was presumed that an authorized prince was limited by the "Law of God," in practice this gave leaders a great deal of latitude to identify and punish the guilty. As seen above, it was common within the Scholastic discourse to argue that the innocent, when in close proximity to the guilty, accrued a certain degree of culpability.

Of course, the Scholastics also presupposed that a wise prince embarking on a just war would definitionally be restrained. Just punishments are by nature proportional and limited. In his 1677 treatise on the *Art of War*, Roger

62 Robert Ward, *Adimadversions of Warre* (London: Printed by Iohn Dawson ..., 1639), 48.

63 Abraham Gibson, *Christiana-Polemica, or A Preparatiue to Warre Shewing the Lawfull Use Thereof* (London: Printed by Edw. Griffin ..., 1619), 23–24.

64 Richard Baxter, *The Catechizing of Families* (London: Printed for T. Parkburst ..., 1683), 316.

Orrery also emphasizes "circumspect command" and the duty of "right reason" in war-making:

> For though it is too usually said by some, and believed by others, That Success does cover all the Faults and Errors in War – Yet doubtless the wiser Portion of Mankind, had much rather owe their Victories to their good Conduct, than to (what is termed, by the World) their Good Fortune; Nay, had rather be less Successful, by stedily pursuing the Dictates of Right Reason, than be more Victorious, by declining them at any Time; Since Wise Councels are still within the Power of Wise Men; But Success is not.[65]

It can only be counted a victory if the war is prosecuted with good conduct. In the above passage, Orrery emphasizes the role of proper motive in deciding to go to war, that the prince have as his aim to be a force for good, which can be seen in the discipline of the soldiers and their conduct in fighting. However, as with the Scholastic just war tradition, the moral framework in which sacking a town might be permissible can be traced back to the justifiable reasons for war to begin with, i.e. to what extent the people themselves are complicit in the crimes of their rulers. None of this is explicitly theorized in these manuals, but the fact that leaders can authorize soldiers to sack towns under certain conditions accords with this tradition.

The level of restraint promoted in these late sixteenth- and early seventeenth-century treatises on war is sincere, but the cumulative effect, considering the historical realities that produced these texts, is one of hyperbole and exaggeration, much in the same way Mayne depicted in *The Amorous Warre*. As noted above, these manuals' demands for perfect, self-disciplined soldiers bear almost no resemblance to the realities of war as they were practiced in this period. The sacking of cities and the outright murder of the innocent was a common occurrence. The brutality of the Thirty Years' War (1618–1648) is legendary.[66] After the fall of Magdeburg (1631), for instance, "20,000 men, women, and children lay charred and blackened corpses amidst the ashes of the hapless city."[67] And this was hardly an isolated case. Villages

65 Roger Boyle Orrery, *A Treatise of the Art of War* (Savoy: Printed by T. N …, 1677). See introduction.

66 Geoffrey Parker estimates that the war killed between 3–4 million people, roughly 15 to 20 percent of the Holy Roman Empire's population. Geoffrey Parker, *The Thirty Years' War* (New York: Taylor & Francis, 1997), 178.

67 Thomas Walker, *The Science of International Law* (London: Cambridge University Press, 1893), 247.

that lay too close to the zone of conflict were repeatedly raided. Some were sacked dozens of times in the course of the war.[68] As hostilities were winding down, armies had become little more than "marauding bands of armed, hungry, and vicious men. Rape, murder, and pillage thus became the order of the day."[69] Similar atrocities followed the Irish Rebellion (1641) and the English Civil War (1642–1651), both of which took place in this period.[70] To be sure, the disjunction between the extensive commentary on proper military conduct and the practices of war could not be greater.

On a certain level, the dissimilarity between war manuals and military practice is easily explained. Commentators in this period set out to rebuke and restrain the clearly excessive practices and norms of the day. They were not necessarily designed as normative commentaries seeking to weigh in on the question of noncombatant immunity. Much like the manuals written at the end of the sixteenth century, early seventeenth-century manuals were reactionary and reformative in nature. Their aim was to restore the dignity and honor to a severely degraded profession, and, more importantly to gain greater degrees of control over their fighting forces.

Despite its oftentimes lofty rhetoric, one can catch inadvertent glimpses of the norms of war as they were practiced. For instance, Donald Lupton seeks to defend the soldierly profession despite the "many grosse and heinous Abuses dayly committed by Soldiers; as suppose, Sacriledge, Swearing, Murthers, Rapes, Stealing, Pillaging, Firing of Houses, Drunkennes, or the like."[71] Richard Ward also explains, "Warre spares none, neither man, women, nor child, neither young nor old: Virgins and Wives in warre are ravished and vitiated: Infants are trampled without pity or mercy under the horses feet, or tossed upon spears points: Women with child are often cut up and dissected."[72] And finally, Thomas Jordan promotes the virtues of a perfect soldier in contrast to a

68 Alexander Gillespie, *A History of the Laws of War, Volume 2: The Customs and Laws of War with Regard to Civilians in Times of Conflict* (Oxford: Hart Publishing, 2011), 140.

69 Cathal Nolan, *The Age of Wars of Religion, 1000–1650: An Encyclopedia of Global Warfare and Civilization, Vol. 2* (Westport, CT: Greenwood Publishing, 2006), 544.

70 For instance, see James Morgan Read, "Atrocity Propaganda and the Irish Rebellion," *The Public Opinion Quarterly* 2:2 (1938); 229–244; Barbara Donagan, "Atrocity, War Crime, and Treason in the English Civil War," *The American Historical Review* 99:4 (1994); 1137–1166; Pádraig Lenihan, *Confederate Catholics at War, 1641–49* (Cork: Cork University Press, 2001); Micheál Ó. Siochrú, "Atrocity, Codes of Conduct and the Irish in the British Civil Wars, 1641–1653," *Past & Present* 195 (2007); 55–86.

71 Donald Lupton, *A Warre-like Treatise of the Pike* (London: Printed by Richard Hodgkinsonne …, 1642), 3.

72 Richard Ward, *The Anatomy of Warre* (London: Printed for John Dalham and Rich Lownds, 1642), 8.

corrupted soldier who "glories in how many townes they burnt, how many women great with child were slaine, how many men blowne up at the springing of a mine."[73] A Christian soldier, on the other hand, must "not kill where he may save with his owne safety."[74] Such descriptions are poignant illustrations of the contemporary practice of war-making, which runs through virtually all of the manuals of this period.

By reasserting the martial discipline framework, commentators indirectly fed into certain descriptive justifications for the killing of the innocent. Women and children could be deliberately killed if, as an instrument of righteous judgment, the soldier had been expressly authorized to do so. One is left to infer that those killed are somehow implicated in their nation's crimes or that their deaths are simply a proportional necessity that tragically arise during a just war. The rationale is never fully articulated; though, as had been the case in Aquinas's framework, it seems likely that the presumption was that those women and children being killed were not, strictly speaking, innocent. In this respect, most martialist writers appear to follow this more exact Thomistic formulation that implicitly presumed some amount of collective guilt.

For instance, as Richard Ward explains, "*What if the cause of the Warriour be good, the end of the warre cannot be evill*; and contrarily the end of a fight is not judged to be good, except a good cause and a right intention did precede the fight."[75] This formulation leaves open the status of those killed in the process of waging a just war. Indeed, at its most extreme it is indicative of an inchoate ends-justifies-the-means or lesser evil approach to violence. If the means used to prosecute a just war cannot be evil if the cause underwriting military action is just, it becomes difficult to conceptualize what restraint would look like in that war. The laxity of this phrasing is suggestive of an extreme permissiveness. In all likelihood, however, this argument is not meant to be as expansive as it comes across at first blush. After all, the expression of justified violence will always be measured: "actions best shew the intentions: every tree is knowne by his fruite."[76] As Aquinas had argued, a truly just war will not give way to inordinate slaying and plundering. A just cause will produce a very particular fruit. Nevertheless, one can trace the ambiguity of authorized violence against the innocent throughout the martialist tradition.

---

73 Thomas Jordan, *The Christian Soldier, or Preparation for Battaile* (London: Printed for Edward Christopher, 1642), 4–5.

74 Ibid. 5.

75 Ward, *The Anatomy of Warre*, 21.

76 Gregory Thims, *The Protestant Informer, or Information to all Protestants: Shewing the Causes and End, of this Unjust Warre* (London: [s.n.], 1643), 9.

4 Shifts in the Martialist Tradition

Like the shifts that took place in the Scholastic tradition, by the late seventeenth century, one begins to see additional changes in the military manuals and technical literature, especially regarding the assault of fortified places. Unlike in the martialist literature of the late sixteenth century, where virtually every text lays out the proper techniques for creating a breach so that fortresses could be stormed, by the late seventeenth century one finds many more references to the bombing and burning of the residential districts of fortified cities.[77]

This shift is in part due to advances in artillery which had been steadily developing since the late sixteenth century. The English antiquarian, Francis Grose (1731–1791), traces the first use of incendiary bombs back to the siege of Vakterdonc, a town in Guelders (in present-day Netherlands), which took place in during the Eighty Years' War (1588). Grose explains that the besiegers used a novel invention, one designed for the "destruction of the human species"; the attackers launched "hollow balls filled with powder materials that could not be extinguished" onto the city.[78] This novel approach to siege warfare worked to terrible effect; the bombs, "falling on a house, beat in the roof and floors, and set fire, which communicating with the neighboring houses, burnt two thirds of the town, the fire being so violent, that it was impossible to extinguish."[79] The great Polish military engineer, Kazimierz Siemienowicz (1600–1651), in his highly influential *Artis Magnae Artilleriae* (1650), describes the effectiveness of the *grenado*, another novel explosive shell fired from a cannon developed in this period, and used to deadly effect at the Siege of Bois-le-Duc (1629).[80] He reports from eyewitness testimony that during this attack,

77 For instance, David Papillon's 1645 treatise on fortification exclusively focuses on creating a breach. He makes no mention of burning houses within besieged places. In fact, he recommends placing all essential provisions "in those houses seated round about the Market-place, as the most safest seat of the Towne." David Papillon, *A Practicall Abstract of Fortification and Assailing* (London: Printed by R. Austin ..., 1645), 51. See also Richard Elton, *The Compleat Body of the Art Military* (London: Printed for Robert Leybourn, 1650), and Alonso Zepeda y Adrada, *Epitome de la fortificacion moderna* (Brusselas: Por Francisco Poppens, Mercader de Libros, 1669).

78 Francis Grose, *Military Antiquities Respecting a History of the English Army, Volume 2* (London: Printed for S. Hooper ..., 1786), 327.

79 Ibid.

80 This was another of the many sieges during the Eighty Years' War.

> nothing was to be seen but Balls of Fire … which were sometimes thrown by Hand, and sometimes slung through the Air by Machines, and conveyed Fire and Dread wherever they fell, nothing more cruel or frightful was ever seen; and so Word can be more expressive of their sad Effects, than Death itself: For whenever they fell upon the Ramparts our Houses, the poor Inhabitants were put into the utmost Consternation; And almost as often as our Engineers throw any amongst them, we presently saw Household-goods, Furniture and Cloaths blown up into the air.[81]

And finally, by the 1670s, the French military engineer, Sébastien le Prestre de Vauban, had developed a new type of incendiary bomb, the *carcass*, a bomb which was soaked in oil and was designed to uncontrollably burn upon impact.[82] Though similar in design to the early incendiary bombs mentioned above, the *carcass* proved to be even more devastating. Indeed, one Spanish military engineer, Sebastián Fernández de Medrano (1646–1705), notes that the *carcass* was "the most infernal device that could be invented, since … villages can be reduced to ashes with it."[83] This, along with other types of fire-bombs, made the interior of cities much more vulnerable to the bombardments of besieging armies.[84]

These revolutions in military technologies, while not exclusively responsible for the transformation of military regimentation, certainly signaled the need to rethink the organization of military regiments and the strategic function of the artillery in war. It is therefore not surprising that in this period one begins to see the formation of the first dedicated artillery corps among the great European powers. The French military historian, Nicolas Edouard Duparcq, traces the origins of the French Royal Artillery to 1693.[85] And while the origins of the English Royal Artillery, in rudimentary form, can be found in certain reforms dating back to the 1680s, a dedicated artillery regiment was not

---

81 Kazimierz Siemienowicz, *The Great Art of Artillery*, trans. George Shelvocke (London: Printed for J. Tonson …, 1729), 243.

82 Vauban, *The New Method of Fortification*, 71.

83 Sebastián Fernández de Medrano, *El perfecto artificial, bombardero y artillero* (En Brusselas: En Casa de Lamberto Marchant, 1699), 77.

84 Jean-Denis Lepage claims that Vauban "recommended to limit their use [*carcass*] in order to spare innocent lives among the population of besieged cities." Jean-Denis Lepage, *Vauban and the French Military Under Louis XIV* (Jefferson, NC: McFarland & Company, 2010), 38. See also, Christopher Duffy, *The Fortress in the Age of Vauban and Frederick the Great, 1660–1789* (New York: Routledge, 1985).

85 Edouard le Barre Duparcq, *Elements of Military Art and History* trans. George Cullum (New York: D. Van Nostrand, 1863), 143.

officially established until 1716.[86] It is within this same timeframe, in 1710, that Spain founded its Royal Artillery, Real Cuerpo de Artillería.[87]

Along with these technological developments and the ensuing regimental restructuring that followed throughout continental Europe, one also finds novel justifications beginning to emerge in the martialist literature. As seen above, Richard Ward argued that the means to obtain just ends cannot be evil. While there is a great deal of imprecision in this formulation, it seems reasonable to assume that it was a somewhat clumsy attempt to give expression to the more or less conventional just war argument of authorized and proportional harm. If led by a wise prince who authorizes punitive violence, one could expect proportional and discrete authorizations to kill. By the late seventeenth century, however, some theorists were beginning to actively theorize the deaths of innocents by way of an embryonic principle of necessity – that the innocent can be killed instrumentally and proportionally for good ends.[88]

Take, for instance, Sir James Turner (1615–1686?), a soldier and scholar who participated in the Thirty Years' War. Ambitiously engaging with St. Paul, Machiavelli, and Hugo Grotius, he explicitly builds on the argument of necessity and the morality of committing the lesser evil in his 1683 tract *Pallas Armata*.[89] He explains that

> Nature teacheth all men, That it is better to kill than to be kill'd, because of two evils the least evil is to be chosen. It were to be wish'd, that War might be ended, and an Enemy overcome without killing; for a bloodless

86 Captain Francis Duncan, *History of Royal Regiment of Artillery* (London: John Murray, 1872), 81.

87 Don Antonio Vallecillo, *Ordenanzas de S.M. el Régimen, Disciplina, Subordinación y Servicio* (Madrid: Emprenta de los Señores Andrés y Deas, 1850), 10.

88 This will be developed more in the next chapter, but at this time the humanist tradition, which more actively theorized necessity as a justification for civilian death had gained ascendency through thinkers like Alberico Gentili, Hugo Grotius, and Samuel Pufendorf. It is not surprising that this much more theoretically robust literature was beginning to filter into the martialist writers' manuals and treatises. For a summary of the interplay between early martialist and humanists, like Gentili and Grotius, see K.R. Simmonds, "Some English Precursors of Hugo Grotius," *Transactions of the Grotius Society* 43 (1957); 143–157.

89 Daniel Defoe was scathingly critical of Sir James, especially his willingness to plunder and murder villagers in Western Scotland. He writes, "It is impossible to give the Detail of the Cruelties and inhuman Usage the poor People suffered from this Butcher, for such he was rather than a Soldier." Daniel Defoe, *An Ecclesiastical History of Scotland* (London: Printed for Charles Marsh, at Bible in Angel-Court …, 1734), 160.

> Victory is the most honourable: But if that cannot be done, it is folly, nay, madness to say that Souldiers should not kill their Enemies.[90]

He does admit, however, that "too many Souldiers are cruel in putting to death those whom they might spare, as such who yield themselves Prisoners ... or old people, Women, Children, in assaults and storms; this is the Souldiers excess, this is their fault, this is their crime."[91] When necessary, and when commanded to do so, soldiers should be prepared to kill the innocent. Turner explains that just because it is difficult to honor the principle of distinction, this does not mean that the principle should simply be ignored. Even though he allows for a crude necessity calculus, he reiterates the conventional martialist virtue of restraint. He argues, "in the heat and ardour of Battle [surrender] cannot be done without danger."[92] Laying down one's weapons clearly put the person surrendering at the mercy of his attacker. Soldiers must be clear-headed when fighting. And when lesser evils are to be committed, they must be done so with clear vision. Such decisions should not be driven by one's impulsivity. Killing the innocent in the 'heat of battle' is not a legitimate excuse, but necessity and the lesser evil do provide a certain degree of moral cover.

Others in this tradition would resort to a kind of divine fatalism. In a 1697 Thanksgiving Sermon on God and War, the English preacher, Charles Nicholetts, attempted to reconcile the realities of war with the nature of God as an intervening force in human affairs. He writes, "The design of War is doing the utmost damages, and committing the most Barbarous Cruelties. Hence Towns are not only taken and sack'd, but consum'd in Flames. Standing Corn trod down and destroy'd, yea Innocent Women and Children most horridly Massacred; these and a thousand other Mischiefs *are not only the accidental Effects, but the real Design of War*."[93] For Nicholetts, it is impossible to

---

90 Sir James Turner, *Pallas armata, Military essayes of the ancient Grecian, Roman, and modern art of war* (London: Printed by M.W. for Richard Chiswell ..., 1683), 367.

91 Ibid.

92 Ibid.

93 Charles Nicholetts, *The Great Work of God in this Present Dispensation of Peace* (London: Printed by Hugh Flemman ..., 1698), 9. my emphasis. There was a whole genre of literature in the early seventeenth century which cast war in entirely spiritual language – a Christian martialist literature that described war as a kind of spiritual struggle. This too can be traced back at least to Erasmus. See Erasmus, *Enchiridion Militis Christiani* [1503] (London: Methuen & co., 1905); For more in this tradition, see Daniell Price, *Prælium & Præmium: The Christians War and Rewarde* (Oxford: Printed by Joseph Barnes, 1608); William Gouge, *The Whole Armor of God: or a Christians spiritual furniture to keepe him safe from all the assaults of Satan* (London: Printed by Iohn Beale, 1619); Richard Bernard, *The Bible-battells, Or The sacred art military For the rightly waging of warre according to*

morally rationalize warfare by way of proper intentions or, as Turner suggests, an appeal to the greater good. To engage in war is to accept the reality that certain evils will be committed for certain ends, and what is more there is something disingenuous in quelling our moral apprehensions about the cruelties of war by characterizing them as "accidental" or "unintentional" as the Scholastics had done.

Nicholetts's argument is couched in fatalistic terms: "God may suffer War to continue for a time, but he will not always let it rage."[94] In fact, God may "break this Bow [which signifies "Wrath and Fury"] by curbing and depressing the Wrath that was so predominant in Warriors; by sweetening their Natures, and making them pliable and willing to put up former resentments; or by taking away and cutting off those that were indomitable, and the chiefest incendiaries."[95] In short, War is a tool of God for punishment and an inducement to repent from sin: "peace and War are in his hands, and at his dispose. When the sins of a Nation provoke him, he can bring War (with all the Calamities of it) upon them; and when his Will is accomplish'd, he can restore Peace again unto them."[96] The exact expression of war is entirely up to God; he will either allow men to commit barbarous cruelties or he will curb their natures and cool their resentments. And in this respect, Nicholetts bypasses the entire problem of intentional versus accidental effects. Since war is a mechanism of God's justice, its terrors are what drive us to accept peace graciously:

> we should cast back our eyes upon the effects of this War, and see what lamentable havock has been made, in the one and the other Nation. How many goodly Towns have been laid in Ashes? How many stately Ships of War have been destroy'd? How many Thousand Lives have been lost?

Holy Writ (London: Printed for Edward Blackmore …, 1629); John Downame, *The Christian Warfare Against the Devill, World, and Flesh* (London: Printed by William Stansby, 1634); and William Gurnall, *The Christian in Compleat Armour; Or, A Treatise of the Saints War Against the Devil* (London: Printed for Ralph Smith …, 1655).

94 Nicholetts, *The Great Work of God*, 11.

95 Ibid., 13.

96 Ibid., 19. This view can be found in some authors in the martialist tradition as well. George Marcelline explains that certain cities were languishing "by the means of warres have bin sacked and spoyled, robbed and ruinated, & oftentimes layd wast and desolate, and therefore it is but merely a punishment inflicted by God, upon some place for some notable sinne and heavy affliction, being used of him as a corasive to eate out the dead flesh of security, the onely meanes to humble us for our owne sinnes, and to reduce us from the following of our owne corruptions to the favoring of the Christian religion." George Marcelline, *Vox Militis* (London: By B.A. for Thomas Archer …, 1625), 2.

> And how many poor Families have been ruin'd and undone? For the more we consider this, the more we shall be affected with the Mercy of Peace.[97]

It is striking that Nicholetts refuses to explain away or to subsume the evils of war into a clearly articulated moral framework. Evil may be necessary in the prosecution of a war, but the inability to formally justify the death of innocents leaves the final analysis and accountability to God and his divine and mysterious retributive wisdom. Ultimately, wars are tools of God. He uses them as He wills.

It is not surprising that these novel moral formulations were beginning to be developed just as the technological development of incendiary bombs and the increasing references to the suffering of noncombatants in fortified places was reaching a new level of intensity.[98] For instance, Lewis de Gaya, a French military historian who lived during the reign of Louis XIV, explains that:

> The shivers and pieces of it [the bomb] break and bruise everything they meet with, and the Fire-works that are within it, set fire in all places where it chances to fall. Nothing gives greater terror to the Towns people of a Besieged Town, than Bombes; and the ravage that there they do it so extraordinary, and so gall Men, Women, and Children, that they know not where to be in safety.[99]

Though it was not particularly thematized or developed as an outright strategy, there is a noticeable uptick in references to the suffering that could be inflicted on the inhabitants of fortified places. Another anonymous manual on military discipline provides instructions for preparing "red hot Bullets" which were used during sieges "to fire Houses."[100] As previously mentioned, Monsieur de

97 Nicholetts, *The Great Work of God*, 24.

98 There are exceptions to this. Some authors continued to argue for traditional restraints when taking cities or towns. For instance, Charles Molloy argues "Generals in the measure of killing, look commonly no further than the destruction of those who by force of Arms oppose them; and through Ships or Cities are taken by assault, the which by the Laws of War subjects every individual to the mercy of the Conquerour, yet Children, Women, Old Men, Priests, Scholars, and Husbandmen are to be spared." Charles Molloy, *De Jure Maritimo Et Navali; Or, A Treatise of Affairs Maritime and of Commerce* (London: Printed for John Bellinger ..., 1682), 179.

99 Lewis de Gaya, *A Treatise of the Arms and Engines of War* (London: Printed for Robert Harford ..., 1678), 89–90. For the exact same formulation, see *An Abridgment of the English Military Discipline* (London: Printed for Robert Harford, 1680), 89.

100 *Industrious Souldiers Golden Treasury of Knowledge in the Art of Making War* (London: Printed for Nicholas Boddington, 1690), 146.

Vauban advocated the use of red hot balls "with which the Cannon are charg'd to set Fire upon Houses or Magazines."[101] In describing the different types of artillery one could use to target a besieged place, the Italian strategist Tomaso Moretti explains that *mean* shot can be used to "ruine the Houses within Cities, Castles, &c."[102] And finally, when laying out guidelines for siege warfare, William Freke casually reminds the reader that besieged places are sometimes reduced by "firing the Town."[103] Such references run throughout the literature of the period.

Unlike the sixteenth century, seventeenth-century theorists were clearly much more comfortable directing fire into the residential districts to expedite surrender or to stir up confusion before an assault was to begin. The necessities of self-defense transformed virtually all a city's residents into soldiers which blurred the lines between military and residential districts. Traditional assaults on the battlements were, to be sure, still quite common throughout the seventeenth century;[104] however, this increase in references to the deliberate targeting of towns and residential districts is significant if for no other reason than it required new justification and self-understanding. Nothing could be more calculated than the deliberate bombing of residential districts to drive the enemy to surrender.

To vividly illustrate these trends, take the English conduct during the Nine Years' War (1688–1697). During this time, it seems to have already become an established convention for a besieging force to direct fire into the city center.[105] While this tactic was being used with some effect on land, what is of principle interest in this war is the English use of naval bombardments along the coast of France. Rather than confronting French forces in the open, many English commentators advocated for waging an economic war by disrupting French trade

101 Sébastien le Prestre de Vauban, *The New Method of Fortification* (London: Printed for Abel Swall, 1691), 8.

102 Tomaso Moretti, *A General Treatise of Artillery* (London: Printed by A.G. and J. for Obadiah Blagrave, 1683), 87. See also, Robert Anderson, Of *Granadoes, Carcasses, and Fireballs* (London: Printed for Robert Morden …, 1690); Joseph Moxon, *An Epitome of The Whole Art of Warre* (London: Printed for J. Moxon …, 1692).

103 William Freke, *Art of War; Summary of the Martial Precepts Necessary for an Officer* (London: Printed for Tho. Minors …, 1693), 275.

104 For instance, see Jamel Ostwald, *Vauban under Siege: Engineering Efficiency and Martial Vigor in the War of Spanish Succession* (Boston: Brill, 2007).

105 Take the bombardments of Genoa (1684), Mons (1691), and Brussels (1695). In a description of the siege of Mainz in 1689, the English assessment of the damage was as follows: "None of the Churches or Monasteries were endamaged by the bombs, only one Nunnery; but the greatest part of the private Houses suffered very much." *An Exact Diary of the Siege of the City of Ments* (London: Printed by R. Janeway, 1689), 20.

and production through the bombardment of coastal towns, bringing "Terror and Affrightment" to the local populations.[106]

After the unsuccessful amphibious assault on Brest, which took place 18 June 1694, Roger Coke explains that English ships "made all necessary Preparations to Bomb some of the Maritime Towns of France." Upon arriving at the town of Dieppe a little while later, six men-of-war and nine bomb galliots[107] "threw Eleven hundred Bombs and Carcasses into the City; which set the Houses on Fire in several Places; so that the Townsmen, not knowing whither to run, and despairing to quench it, took a Fright, and began to run away in Disorder."[108] Describing the desolation, Coke continues, "they so ruined it with Bombs, that there was hardly a House left unshattered, nay almost all of it was entirely reduced to Ashes."[109] A very similar attack took place in Saint-Martin-de Ré: "The bombing began about Nine a Clock at Night, and set the Town on Fire in five places. The Enemy quench'd it in three, but the Fire continu'd in the other two with great Violence."[110] From there the English fleet sailed on to bombard the towns of Havre de Grace, Olonne and St. Martin.[111]

The following summer, the English continued their coastal assaults. On 4 July 1695, the English fleet approached the town of St. Malo.[112] A report of the bombardment describes the devastation unleashed there: "We Fired 900 Bombs and Carcasses into the Towne, a great part of which is certainly burnt or beat down, and we believe there is little part of it free from some share of

106 E.L. *A Preservative for our Money* (London: Printed by B. Motte ..., 1696), 6. See also, W. Perron, *Galliæ Flagellum: The Reduction of France Made Easie* (London: [s.n.], 1696).

107 The *man-of-war* was the most formidable warship of the seventeenth century; a *galliot* was typically smaller, a ship with a single mast.

108 Roger Coke, *A Detection of the Court of and State of England, Volume II* (London: Printed for J. Brotherton and W. Meadows ..., 1719), 546.

109 Ibid.

110 *The Present State of Europe: or, The Historical and Political Monthly Mercury, Volume VII* (London: Printed for Henry Rhodes ..., 1696), 268.

111 See *An Account of the Burning of Havre de Grace* (London: Printed for Richard Baldwin ..., 1694); and *Whitehall, July 23, 1696. This evening arrived here on express from my Lord Berkeley with letters dated the 20th of this month, off Ushant* (London: Printed by Edw. Jones ..., 1696). For a complete chronology of the English bombardments along the French coast, see *An Exact and Complete Journal of all the Considerable Actions of the Confederate Fleets and the Armys from the Opening of the Campaigne, for this Present year 1695, until the Conclusion Thereof* (London: Printed for John Whitlock, 1696). For similar accounts in the Flanders campaign, see Edward D'Auvergne, *The History of the Campagne in Flanders, for the Year 1695* (London: Printed for Mat. Wotton ..., 1696).

112 See Sylvanus Urban, "Historical Account of the Town of St. Malo," in *The Gentleman's Magazine, Volume 75* (London: Printed by John Nichols ..., 1794), 105.

Desolation."[113] From there, the fleet proceeded to Granville, "a large and populous city," which was bombarded from nine in the morning to seven at night "with very good success, setting the Town on fire in several places, and leaving it all in Flames, so that we believe it is quite destroyed."[114] The following month, the fleet bombarded Calais, a barrage that lasted from eleven in the morning to five in the afternoon, "during which time we Shot into the Town about 600 Shells, which by the several Fires we saw, we judged to have done very considerable Execution."[115]

The bombing of French coastal towns, in the broader scope of the war, had a limited overall effect on its outcome.[116] What this practice evinces, however, is the paradoxical tension that was continuing to play out regarding the duty to be restrained and disciplined in the treatment of nonfighting residents and the tactical and technical realities of assaults on fortified places. While the English Army had had some success on land in this war, the defeat at Brest had forced them to reconsider their strategy. If towns could not be stormed and occupied, they could be disrupted and generally harassed, putting them, at least temporarily, out of service. The hope was that swift bombardments would lead to the kinds of disruptions that would ultimately provide some breakthrough in the posturing and positioning that was allowing the war to persist.

It is worth pointing out that by the eighteenth century many commentators struggled to grapple with and explain these turn-of-the-century attacks on the French coast. They presented something of a challenge since nothing in the literature on war up to that point came close to providing a justification for these kinds of bombardments, that is, they seemed designed to terrorize noncombatant populations rather than achieve a clear military objective. The novelty of this tactic may explain why historians' assessments vary so wildly. For instance, Paul de Rapin de Thoyras reports, "These bombardings of the French towns soon spread a terror among all that lived near the coast," which "seemed inhuman," but the responsibility actually lie with the French since they "had

---

113 *An Account of the Bombarding of St. Malo, by the Fleet under the Command of the Lord Berkeley, Admiral of the Blue* (London: Printed by Edw. Jones ..., 1695).

114 *An Account of the Bombarding of Granville* (London: Printed by Edw. Jones ..., 1695).

115 *Whitehall, August 19, 1695. This morning came in an express from my Lord Berkeley with an account of the Bombarding of Calais* (London: Printed by Edw. Jones ..., 1695).

116 The French employed very similar tactics in the bombardment of Barcelona (1697). According to one account, "The cruel hostility of the bombardment persisted with stubborn obstinacy for eleven days and nights. The mortars fired from two sloops caused the most ruin. Large parts of the city lie desolate." Serapi Berard, *Manifestacion en que se publican muchos y relevantes* (Barcelona: En Casa de Cormellas ..., 1697), 138.

bombarded Genoa without a previous declaration of war."[117] In other words, the assault on coastal towns was a justified punishment, a reprisal, for a breach in the proper conduct in the lead up to war by the French.

John Campbell argues that these bombardments were justified inasmuch as they were strategic attempts at drawing away French forces from other sites of conflict.[118] James Ralph makes no defense. He simply notes that "this coasting Project was mere Matter of Parade and Terror only ... When, therefore, they had sufficiently gratify'd themselves with the Pannic of the French, they all at once stretch'd over to our own Coast."[119] Walter Harris notes that Lord Berkeley "spread Terror along the Coasts of France" due to the fact the region had become a "Nest of Privateers, which had most infested the Channel, and molested the English Merchants."[120] This appears to recapitulate Sir Cloudsley Shovel's position, who had previously argued that the ostensive purpose of the bombardments was to put a stop to privateers, but he eventually came to the conclusion that the "unexpressible Terror of the Citizens, and the Consternation of all the Maritime Towns" had a more productive effect.[121] He writes,

> This Bombardment was very rationally concerted, and according to the Crisis of Affairs prov'd very beneficial to the English, tho the Execution done upon that nest of Pyrates was not very considerable. Our Navies alarming the Maritime Provinces of France, amus'd the Country-People with a wonderful Consternation, their fears became contagious; and the more Inland Cities began to apprehend a visit from the Forces, which had embark'd, either in reality or Imagination aboard the English Navy.[122]

What had begun, according to many English sources, as a kind of justified punishment transformed into an effective terror campaign, or at least commentators like Shovel believed so. He goes on to explain that these fears forced

117 Paul Rapin de Thoyras, *The History of England, Volume 3* (London: Printed for John and Paul Knapton ..., 1744), 255. This same argument is made by Frederic Hervey, *The Naval History of Great Britain, Volume 3* (London: Printed by William Adlard ..., 1779), 9.

118 John Campbell, *Lives of the Admirals, and other Eminent British Seamen, Volume 3* (London: Printed by John Applebee, 1744), 126.

119 James Ralph, *The History of England, During the Reigns of King William, Queen Anne, and King George the First, Volume 2* (London: Printed by Daniel Browne ..., 1746), 505.

120 Walter Harris, *The History of the Life and Reign of William-Henry* (London: Printed by Edward Bate ..., 1749), 412.

121 Sir Cloudsley Shovel, *Secret Memoirs of the Life of the Honerable Sir Cloudsley Shovel* (London: Printed and Sold by B. Bragge ..., 1708), 58.

122 Ibid.

France to divert regiments to coastal towns to assuage the public.[123] By the late eighteenth century it seems that this view had become quite common.[124] One author notes that the attack on Havre de Grace was so effective because it "struck such a terror into the French ministry, that they became divided in their councils; for they found it almost impossible that they could invade Britain, while her navy did their duty."[125] In other words, this aggressive offense was the best way of keeping the French navy away from the British coast.

Perhaps the most compelling explanation – the most indicative of the shift in norms – comes from Sir John Dalrymple, a minor figure in the Scottish Enlightenment. In his description of the attacks, Dalrymple is much more interested in the principles that underwrite the logic of French and English targeting to begin with. In particular, he pays close attention to King Louis XIV's response to the English coastal bombardments:

> Lewis XIV [sic] retaliated the injury by bombarding the noble city of Brussels [13–15 August 1695]. The English had an excuse; because in bombarding sea towns, they could destroy ships, docks, naval stores, and the timber-works of the harbours; the French had no excuse but revenge, or the hopes of saving themselves by the fear of their retaliating upon others.[126]

Dalrymple points out that in retaliation for the coastal attacks, Louis XIV laid siege and bombarded Brussels. Since the French were retaliating non-strategically, merely with the hope of instilling fear in their enemies, their cause had been vitiated, but since the English were taking aim at legitimate war-time infrastructure, they were perfectly justified. Implicitly the argument here is that terror is only permitted when taking aim at strategically necessary targets.

What is striking about this array of explanations is the variation. That there were so many ways to justify this behavior suggests that the commentators

123 Ibid., 59.

124 At least among English commentators. French historians downplayed the damage. Mérigot, for instance, claimed that the costal bombing was not that extreme considering the cost of preparations. He believed that the English attacked out of revenge for Louis XIV's support of James II to depose and take the throne of King William III. J.G. Mérigot, *Histoire universelle: depuis le commencement du monde* (Paris: libraire sur le quai de augustins … 1788), 353.

125 *A History of All the Invasion of England* (Windsor: Printed by C. Knight …, 1794), 168.

126 Sir John Dalrymple, *Memoirs of Great Britain and Ireland* (Edinburgh: Printed for John Bell and William Creech …, 1788), 47.

were attempting to grapple with something novel. It is compelling that the justifications employed above are nowhere to be found in the martialist literature up to this point. That particular tradition had not yet evolved the moral framework to justify the bombing of towns to disrupt military production, but, as seen in the emerging lesser evil discourses above, such justifications were beginning to emerge. Military manuals in the seventeenth century simultaneously promoted restraint and prudent judgment while at the same time permitting towns to be deliberately bombed for certain necessities of war, to disrupt war-related industry and trade, among other things. These new norms that permitted the bombing of residential districts would slowly begin to take hold in the eighteenth century, indeed, just as the sacking of defeated fortresses was in the early stages of being phased out of siege warfare.[127] These arguments would continue to gain force in the nineteenth century.

There persisted a paradoxical tension between the duty of restraint and certain types of expediency in the eighteenth century. Some commentators, John Gittins, for instance, continued the tradition of requiring troops to setup camp outside of villages and towns so that soldiers would not be tempted to "burn, plunder, or pull down any Houses."[128] He explains that since men join the military for one of two reasons, love of honor or "The hopes of a License to do Evil," strict military discipline is crucial to prevent the latter from dominating the ranks.[129] Soldiers should be restrained and dispassionate. But these same soldiers might now deliberately bomb cities and towns in the process of a siege.

127 As Gavin Daly points out, "The Duke of Marlborough conducted twenty-six sieges in the Low Countries during the War of the Spanish Succession (1701–14), without a single town being taken by storm or subjected to sack. All ended with negotiated capitulation. Likewise, capitulation was the norm of many sieges that took place in the Iberian Peninsula during the same war." Gavin Daly, "'The Sacking of a Town is an Abomination': Siege, Sack, and Violence to Civilians in British Officers' Writings on the Peninsular War – The Case of Badajoz," *Historical Research* 92:255 (2019); 160–182. 164. Of course, sacks did occur up to the nineteenth century, but they were becoming much less common. For more on the duties and obligations of an orderly surrender to avoid a sack, see John Muller, *The Attack and Defence of Fortify'd Places* (London: Printed for J. Millan …, 1757). Furthermore, by the mid eighteenth century, Stephen Payne Adye confidently states that "Towns taken are never now given up to plunder, as was formerly the practice among the ancients." By this late stage sacking the town, while still an occasional occurance, had become an anomaly. Stephen Payne Adye, *A Treatise on Courts Martial* (London: Printed for J. Murray and S. Highley, 1796), 279.

128 John Gittins, *A Complete System of Military Discipline* (London: Printed by J. Humfreys …, 1735), 172.

129 Ibid., Preface. For a prohibition against marauding, see *A System of Camp-Discipline* (London: Printed for J. Millan …, 1757), 40.

That soldiers might now be called upon to launch bombs into towns, however, raised new moral dilemmas. Granville Sharp argues that soldiers had a duty to forgo any "unlawful act" even if "he commits it by the express Command of the highest military Authority in the Kingdom."[130] Military honor (i.e. fidelity to "the king of kings," namely God) should be internalized to such a degree that even when ordered to commit an atrocity the soldier is able "to think for himself."[131] He explains that even though the law rarely prosecutes soldiers for obeying illegal orders "the time will certainly come, when all such temporalizing military Murderers must be responsible for the innocent blood that is shed in an unjust War, if they have rendered themselves accessories to it by an implicit, and therefore, criminal obedience to the promoters of it."[132]

Sharp's recommendation is a remarkable, indeed radical, sentiment, but it clearly sat uncomfortably with the tradition of martial discipline that had arisen up to that point. It was difficult to sustain the tension between the duty to obey lawful commands and the duty to assess whether the commands were lawful. What Sharp is suggesting is that now, rather than being merely the hands and feet of princely commands, soldiers must think for themselves and be actively involved in assessing whether or not the war is just, and the commander's orders are lawful.[133] As indicated above, the standard view in the martialist tradition, even by the eighteenth century, was that certain cruelties could be authorized by a legitimate authority. For instance, Alexander Bruce gives expression to the moral calculus of this period when he writes: "the perfidious and deliberate Murder of Thousands of Men, acted by Command of publick Authority, shall not be a greater Crime, than the Slaughter of one or two by a private Soldier."[134] To Bruce, the greater moral concern was not illegitimate orders but unrestrained soldiers. Within the context of the martialist tradition's standard view, Sharp's position is clearly an outlier, as he himself admits, yet it notably reflects the tensions within the martialist tradition resulting from these changes in warfare.

---

130 Granville Sharp, *Remarks on the Opinions of Some of the Most Celebrated Writers on Crown Law Respecting the Due Distinction between Manslaughter and Murder* (London: B White and R. Horsfield, 1773), 67.

131 Ibid., 68.

132 Ibid., 68–69.

133 Another commentator in this period notes: "Unjust war is murder. All who engage in such warfare, therefore, whether voluntarily or by compulsion, are, to a certain degree, criminal." *The Analytical Review, Volume 10* (London: Printed for J. Johnson …, 1791), 130.

134 Alexander Bruce, *The Institutions of Military Law, Ancient and Modern* (Edinburgh: Printed by the Heirs and Successors of A. Anderson, 1717), 287.

Sharp's vaulted rhetoric notwithstanding, many treatises and technical manuals on fortification continued to make reference to the use of red hot bullets and carcasses to burn residential houses down.[135] For instance, one anonymous author pays particular attention to the cruelties of

> sieges and the sacking of great, wealthy, and populous cities. Besides the houses and most sumptuous buildings frequently set on fire, and the number of people killed by the various instruments and devices of slaughter made use of for that purpose; if the siege continues 'till the provisions are all consumed, what multitudes often perish by the more lingering and painful deaths of famine and pestilence?[136]

Another writer notes: "Red hot bullets, are shot in sieges to fire houses, and do the more mischief in a town."[137] Francis Grose laments: "Falling on the tops of houses they broke through them, and as soon as they had taken fire, they burst, and spread out on every side a flame, which was difficult to extinguish with water. This instrument, which gave origin to grenades, fire-pots, and the like machines contrive for the destruction of the human species."[138]

With the rise of incendiary bombs designed to destroy a fortress from the inside out, new justifications for bombing residential homes started to arise. Antoine Feuquières, for instance, explains that "It is an undoubted Axiom, that the sooner the Body of any Place can be demolished, or its interior Works ruined, or the Garrison fatigued by the Effects of Bombs, the greater is the Expedition with which the besiegers are Masters of that Place."[139] There are multiple ways of engaging a fortified place: attacking the defenses and parapets, harassing the inhabitants and garrison with ricochet balls, or burning the interior works with incendiary bombs. He continues, however, by recommending: "If the Place attacked be large and populous, Bombs are necessary, because they destroy Houses, and continually fatigue the People, who are soon

---

135 *An Introduction to the Art of Fortification* (London: Printed for and Sold by John Brindley …, 1745), See Bullets; and *A Military Dictionary Explaining and Describing the Terms, Phrases, Works, and Machines Used in the Science of War* (London: Printed for G. Robinson …, 1778), See Bullets.

136 *Considerations Upon War, Upon Cruelty in General, and Religious Cruelty* (London: Printed for T. Osborne, 1758), 55–56.

137 *A General History of Sieges and Battles, by Sea and Land, Volume V* (London: Printed for J. Curtis …, 1762), 73.

138 Grose, *Military Antiquities Respecting a History of the English Army, Volume 2*, 327.

139 Antoine de Pas de marquis Feuquières, *Memoirs Historical and Military: Containing a Distinct View of All the Considerable States of Europe* (London: Printed for T. Woodward …, 1735), 265.

harassed with their Vigilance for the Preservation of their Lives, and Effects, without any Probability of evading the Loss they sustain by the Fire from above."[140] Feuquières had observed that it was an effective strategy to set fire to the whole town since this forced the inhabitants to pressure the garrison to capitulate.[141] At a minimum, targeting civilian infrastructure was accepted as one of many valid ways of going about taking certain types of fortified places.

Other commentators were beginning to point out the risks of targeting non-combatants. In explaining the technique of attacking a village, Thomas Simes writes:

> If the war is in an enemy's country, which you do not chuse to spare, it would be an easy matter to set fire to the four corners of the village, and oblige the besieged to surrender themselves presently; but besides the inhumanity of using means that tend to lay waste a whole country, it is likewise very dangerous to throw all the inhabitants of the open country into despair, because then flying into the woods, they form into bodies … and spread about ever where to knock the straggling soldier on the head.[142]

There are prudential reasons for not burning one's way through enemy villages. However, when assaulting a fortress, Simes concludes, it may be necessary to set parts of the place on fire; "This way is indeed something rude and violent, but it is allowed by the rules of war, and an Officer must not scruple to put it in practice, when he finds it necessary for the preservation of the troops under his command."[143] As this passage suggests, the martialist tradition, which had been long governed by the principle of just cause and the proportional use of punitive force was slowly evolving to allow for certain instrumental and 'necessary' cruelties. As Richard Lee explains, certain evils are only permitted "when it has some Relation to the Design of the War; that is, when some Advantage accrues from it to ourselves."[144]

In response to these developments, some authors shifted the moral burden more squarely back onto the political leadership and military planners. Maurice de Saxe, one of the most well-known commanders in the eighteenth century, for instance, redoubled his commitment to martial discipline, infusing

140 Ibid.

141 Ibid., 266.

142 Thomas Simes, *A Treatise on the Military Science* (London: Printed for John Millan, 1780), 274.

143 Ibid., 279.

144 Richard Lee, *A Treatise of Captures in War* (London: Printed for W. Sandby …, 1759), 74.

it more deeply into every aspect of civilian life. He continued to prioritize the character and disposition of commanders: "He should be endued with a capacity prepared for everything; with activity, accompanied by judgment; with skill to make a proper choice upon all occasions; and with an exactness of discernments."[145] But these responsibilities had become much more thorough, and they ultimately provided some exculpatory cover for invading armies. For instance, Saxe extends the logic of "exactness of discernment" to the planning and development of fortifications and the entire infrastructural development of a state. Saxe complains that "few cities have been originally founded for the purpose of sustaining a regular siege; but were indebted to trade for their extent, and to chance for their situation."[146]

The implication to Saxe's thinking is that the cruelties of siege warfare can in part be traced back to some more formative germ of bad planning.[147] Bad location choice, ad hoc design, inadequate preparations, and poor accommodations for the nonfighting residents all gave rise to the cruelties that the early commentators on the tragic necessities of bombardment indicated above. Even in siege warfare, martialists like Saxe were confident that if proper design and preparation were seen to, the innocent would less likely be killed, which, inversely seems to presuppose that those who die from an assault have their leaders and civil engineers to blame, not the firebombs of the enemy.[148]

In a manner similar to Saxe, though with a rather different set of conclusions and recommendations, Comte de Guibert (1743–1790), an influential

---

145 Maurice de Saxe, *Reveries, or Memoirs Concerning the Art of War* (Edinburgh: Printed by Sands, Donaldson, Murray, and Cochran ..., 1759), 221.

146 Ibid., 115.

147 Some seventeenth-century martialists make similar claims. See R.N. *Fortification or Architecture Military* (London: Printed by Tho. Cotes ..., 1639). Jonas Moore argues that "If this Nation ever receive Damage by another, it will be for want of Knowledge and Practice of Military Discipline, and Arts." Jonas Moore, *Modern Fortification; Or, Elements of Military Architecture* (London: Printed for Obadiah Blagrave ..., 1689), Dedicatory. See also John Muller, *Treatise of Artillery* (London: Printed for John Millan ..., 1768).

148 Of course, even Saxe's formulations did not preclude the fact that militaries might legitimately punish the inhabitants of villages who refused to pay contributions to an army in the field. Saxe argues that they could be threatened either with execution or with the promise that their whole village would be burned down. Saxe, *Reveries*, 128–129. But in these cases, violence against villagers was not random or arbitrary; it followed the strict protocol of martial field discipline. In other words, inhabitants were harmed only for crimes that could be directly attributed to them. See also, J.H., *The Complete Gentleman Soldier; Or, a Treatise of Military Discipline* (London: Printed for Thomas Ballard ..., 1702); and Humphrey Bland, *A Treatise of Military Discipline* (London: Printed for Daniel Midwinter ..., 1743).

tactician and general, argues that it would be better if nations "dispense with fortified places."[149] He laments that in current war, fortifications

> are too much multiplied; they are computed at too great a value in the balance of the strength of a state, and in the present system of war. They have rendered wars more ruinous, by compelling armies to be reinforced and augmented: they have rendered them less scientific and decisive, by having occasioned the great Taffies, or the art of fighting battles, to be neglected, and by having in general confined the views and operations of the military.[150]

While fortresses may have prevented certain "incursions and devastations," their immense cost in both money and men has come at the expense of the citizens of a state.[151] What is more, he implies that the proliferation of fortifications has engendered a kind of laziness in warfare. Though he does not directly speak to the emerging tactics of bombing residential districts of fortified cities, one could reasonably infer that this is an example of the new kind of imprecision in war that was being permitted. Armies were losing the art of troop movement and dynamic field engagements, which require a great deal of discipline and training. Guibert argues that a nation would be better defended with a maneuverable army at its border rather than a string of costly fortresses.

5 Conclusion

There are a few key observations to be made about the prudence of leadership and the promise of martial discipline that characterize this period. The first is the degree to which it sought to render warfare scientific and technical.[152] The above theorists assumed that the ravages of war, including harm to the innocent inhabitants of towns and cities, could be controlled and effectively eliminated by minimizing the uncertainty of war. This was aided by the belief that properly stocked, well-compensated armies should encamp away from population centers. Soldiers would be steeped in the kind of moral training

149 Comte de Guibert, *A General Essay on Tactics* (London: Printed for J. Millan … 1781), 129.

150 Ibid.

151 Ibid.

152 As will be seen, these are trends which also carry through the nineteenth and twentieth centuries.

that would restrain them even in those desperate moments when they were storming the breach of a fortified city.

Considering the amount of attention the Scholastics paid to the launching of bombs at fortified places, it is striking that very few commentators in this tradition speak about the moral difficulties surrounding siege warfare. But when they do, martialist authors focus on the moral restraint of soldiers and proper planning and design as the key to eliminating the threat to noncombatants. It is only when cities and town are built ad hoc that the innocent residents are put at risk. In this sense, their deaths can be traced back to imprudent leadership and bad planning. In general, however, the operative, if idealized, vision of war seems to be much like Aquinas's – that the innocent would ordinarily have very little contact with soldiers. In those cases where soldiers needed provisions, military discipline regulated the way such contributions should be collected, which permitted certain limited punishments to those townspeople who resisted payment. While these treatises rarely, if ever, directly speak about the protections extended to those not engaged in combat, the underlying assumption is that any cruelties noncombatants face is the product of errors in judgment, poor training, and/or a breakdown in military discipline.

CHAPTER 3

Humanists and Republicans

1 Introduction

Concurrent with and in contradistinction to the Scholastic and martialist traditions, there emerged a body of humanist thought in the early Modern Period that produced a distinct set of arguments about how and under what conditions it was permissible to kill the innocent.[1] In contrast to the martialist writers of the previous chapter, these theorists were steeped in much more sophisticated intellectual histories.[2] These authors critically engaged with a range of sources: scripture, Scholastic authors, and Greek and Roman writers of antiquity.[3] While developed to a certain extent in opposition to the theological framework embodied in the Scholastics, by the early seventeenth century there was a great deal of interplay between these schools of thought.[4] As will be seen, while humanists tended to reject the emphasis on proper intentions

---

1 Among International Relations scholars there have been various attempts to cluster and classify the major perspectives on law and war and the character of international society more generally since the early modern period. In the English School, Martin Wight and Hedley Bull divide schools of thought along the Machiavellian, Grotian, and Kantian traditions. Though, more recently, Alexander Wendt has divided international society into three cultures: Hobbesian, Lockean, and Kantian. See, Martin Wight, *International Theory: The Three Traditions* (Leicester, U.K.: Leicester University Press, 1994); Hedley Bull, *The Anarchical Society: A Study of Order in World Politics* (New York: Palgrave Macmillan, 2012); and Alexander Wendt, *A Social Theory of International Politics* (New York: Cambridge University Press, 1999).

2 As noted in the previous chapter, some of the transformations in the martialist tradition can be explained by the interplay between republican and humanist scholars, like Machiavelli and Grotius.

3 For more on the debates between humanists and scholastics, see Richard Tuck, *The Rights of War and Peace: Political Thought and the International Order from Grotius to Kant* (New York: Oxford University Press, 1999). Ryan Greenwood argues that the antecedent traditions of canon law and the much more neglected medieval Roman law both contributed to the just war traditions of the modern era. See Ryan Greenwood, "War and Sovereignty in Medieval Roman Law," *Law and History Review* 32:1(2014); 31–63.

4 For instance, Machiavelli famously rejects those authors who discuss "imaginary republics" and who provide unrealistic requirements about "how they ought to behave." Niccolo Machiavelli, *The Prince* ed. Quentin Skinner (New York: Cambridge University Press, 1988), 55. Erasmus rejects the "argument of Thomas," instead preferring "the teaching of Christ." Desiderius Erasmus, "Dulce bellum inexpertis: War is a treat for those who have not tried it" [1515], in *Collected Works of Erasmus, Volume 35, Adages III iv 1 to IV ii 100* trans. Denis Drysdall, ed. John Grant (Toronto: University of Toronto Press, 2005), 427.

 | DOI:10.1163/9789004515482_005

as a moral exculpatory bright line separating permissible and impermissible killing, by the seventeenth century more humanists were directly engaging with these arguments and presenting alternative formulations. Several twenty-first-century scholars readily point out that both Scholastics and humanists were arguing to protect certain classes of people in war; however, the way these traditions formulated their arguments differed significantly. These differences would prove to be consequential in the development of civilian casualty discourses in subsequent centuries.

While humanist discourse was polyvalent and diverse – in fact, many Scholastic commentators have humanist sympathies – authors in this tradition typically made some appeal to the principle of necessity, right, and the common good to explain the death of the innocent. These formulations focus on the delicate balance between the grim practical necessities of enacting a right and the mitigating duties of charity and mercy. The interplay between these variables required a complex set of judgments that mostly elided the problem of intentions. To be sure, a precise understanding of what "right" permits varies to some extent in this tradition. Some give moral priority to self-preservation, while others expand this category to include punitive responsibility, to punish those who have violated the natural law and/or the law of nations. Nevertheless, within this discourse, while the death of the innocent was regrettable, it could be justified instrumentally. While the formulation and intensity of these permissions differ greatly among the commentators indicated below, these arguments evince a proto-consequentialist character in that that they routinely formulate necessity in terms of costs and benefits and moral trade-offs. A wise and prudent leader will consider the true costs of his claim to right and make a prudential calculation based on an intricate mix of disparate variables. Of course, the below arguments cannot be reduced to something as distinct as utilitarianism; most of these authors still adhere to some conception of the natural or divine law. What one finds in this discourse, however, is a complex relationship between the classical just war tradition and the necessity calculations that were beginning to emerge, especially as it pertains to explaining the permissibility of killing of the innocent.

Again, it deserves to be reiterated, this tradition is quite expansive and riddled with deep conflicts and contradictions. The purpose of this chapter is not to cavalierly compress, say, Machiavelli and Erasmus into a singular or unitary voice.[5] Rather, the goal here is to see how the killing of the innocent

5 For a clear contrast of these two authors, see Harry Burke, "Audience and Intention in Machiavelli's *The Prince* and Erasmus' *Education of a Christian Prince*," *Erasmus Studies* 4:1 (1984); 84–93.

is explained by an array of authors who can loosely be classified together, who are working on similar issues, with analogous or roughly comparable assumptions. Though grounded in necessity and right, this tradition incorporates a range of additional variables: fear, risk, and what counts as a reasonable tradeoff. This section will outline some of the main contributors to this school of thinking from the sixteenth to late-eighteenth centuries.

2 Niccolò Machiavelli

In many respects the great Florentine philosopher Niccolò Machiavelli (1469–1527) could be counted among the martialist tradition writers. Many martialists, after all, were writing directly in response to his arguments. What is more, a crucial aspect of Machiavelli's thinking is that military preparedness must be understood holistically. Well protected cities are overseen by capable princes who are skilled in passing good laws and effectively managing their armies.[6] Republican cities go soft and moral virtue begins to breakdown when there are long periods of peace. It turns out that for Machiavelli martial discipline and military preparedness are fundamental to political health. The tragic reality of Machiavellian politics is that domestic political health and wellbeing are predicated on a tense and disquieted international arena. Martial virtue does not lead to peace but fortitude and readiness.

Machiavelli is included here with the humanists because unlike the martialists he provides one of the earliest formulations of justified cruelty based on the principle of necessity, self-defense, and the common good in the early modern era.[7] As noted above, the theoretical framework of necessity was largely absent from the martialist tradition, especially in its earliest forms. Several of those commentators quite expressly rejected the "Italian manner" which presumed "warre to be a necessary evill."[8] One might recall that the ultimate goal of those writing in the martialist tradition was something resembling

6 In the *Prince* he argues: "The principal foundations on which the power of all governments is based (whether they be new, long-established, or mixed) are good laws and good armies." Niccolo Machiavelli, *The Prince* in *Selected Political Writings*, trans. David Wootton (Indianapolis: Hackett, 1994), 38.

7 Tuck argues that "the ideas of Machiavelli about the relationships between cities were not very different from those of many apparently more orthodox humanists." Tuck, *The Rights of War and Peace*, 12. Others believe that Machiavelli fails as a humanist. See Joseph Bien, "Politics of the Present: Machiavellian Humanism," *Philosophy and Phenomenological Research* 42:2 (1981); 197–204. 197.

8 François de la Noue, *The Politicke and Militarie Discoverses of the Lord de la Nouue* trans. E.A. (London: Printed for T.C. and E.A. by Thomas Orwin, 1587), 220.

international peace. Though perhaps idealistic, martialist writers envisioned just commonwealths playing a key role in obtaining that end. For Machiavelli, as with many others in this tradition, the still-emerging international system of states was revealing its own logic, one that rendered the idea of permanent peace elusive if not hopelessly impossible.[9]

Machiavelli's understanding of necessary evil can be summarized by a famous line from the *Prince* (1513): "For anyone who wants to act the part of a good man in all circumstances will bring about his own ruin, for those he has to deal with will not all be good. So it is necessary for a ruler, if he wants to hold on to power, to learn how not to be good, and to know when it is and when it is not necessary to use this knowledge."[10] As this passage suggests, princes must "learn how not to be good," to be willing to commit atrocities for the sake of a greater good, namely the viability of the state. Though the *Discourses* (1517) are written for a very different purpose and audience, he continues this same line of thinking by explaining that it would be better to live the life of a private citizen rather than an absolute prince:

> Such methods are horrific and destructive not merely of a Christian way of life but of a merely human existence. Nobody ought to willingly adopt them. Anyone should prefer to remain a private citizen rather than become a king through the destruction of so many people's lives. Nevertheless, anyone who does not want to choose the option of doing good, if he wants to stay in power, will find himself well-advised to adopt these evil methods. But instead, men pursue policies that are neither good nor bad, and these are extremely dangerous.[11]

Princes will often find themselves in morally compromising situations, and if greater cruelties are to be avoided, one must be ready, without hesitation, to commit evil, which includes the prospect of destroying many innocent people's lives.

The grim realism of this sentiment is exactly what made Machiavelli so challenging, not only for the martialists but for many writers of this period.

---

9 See Robert Sullivan, "Machiavelli's Balance of Power Theory," *Social Science Quarterly* 54:2 (1973); 258–270.

10 Machiavelli, *The Prince*, 48.

11 Niccolo Machiavelli, *The Discourses* in *Selected Political Writings* trans. David Wootton (Indianapolis: Hackett, 1994), 131. Incidentally, this passage may derive from Socrates, who argues: "A man who really fights for justice must lead a private, not a public, life if he is to survive for even a short time." Plato, "Apology," in *Plato: Five Dialogues* trans. G.M.A. Grube (Indianapolis: Hackett, 2002), 32a p. 36.

For instance, the late sixteenth-century Huguenot, Innocent Gentillet (1535–1595), attacks Machiavelli, "this stinking doctor," for "teaching a prince all kinds of evill and wickednesse, which may lead to the highest degree and top of all vices, and of all evill."[12] And in the early seventeenth century, William Sancroft (1617–1693), who would later become the Archbishop of Canterbury, reminds the reader, *pace* Machiavelli, that "We know that a good aime, much lesse a good pretence cannot justifie a bad action, and therefore wee ought to be as Solicitous about the lawfulnesse of the means, as about the goodnesse of the end. It is a Maxim in Morality, that *bonum oritur ex integris*, and in Christianity, that wee must not doe evill that good may come of it."[13] For many centuries Machiavelli was derided as a teacher of evil, a reputation he would not live down until well into the nineteenth century.

To his defenders, however, it was not ultimately clear that Machiavelli intended such a brutal and, indeed, evil message. In Edward Dacres's 1600 edition of the *Prince*, he notes that "this book carryes its poison and malice in it; yet mee think the judicious peruser may honestly make use of it in the actions of his life, with advantage."[14] As this passage suggests, there were dangers to taking the text *literally* – one should not simply imbibe the poison. To his supporters it was far from clear that Machiavelli was arguing that princes should be wicked, that is willing to callously exterminate the innocent. After all, there are reasons to think that passages like the ones indicated above can be read as promoting a covert agenda. As John Wolfe had explained in his 1584 preface to the text, these controversial passages help to show "exactly what difference there was between a prince and a tyrant, between government by many good men and government by a few bad ones, a well-regulated commonwealth and a confused licentious multitude."[15] In other words, the text should be read not as a recommendation but as an esoteric warning. According to this reading, Machiavelli intends no hyperbole; it really is better to be a private citizen than an absolute prince, especially a private citizen with the definitive handbook about how tyrants will scheme to brutalize the innocent to retain power. By understanding how tyrants are compelled to violate the limits of humanity, the innocent public can understand when such schemes are afoot and thwart

12 Innocent Gentillet, *A Discourse upon the Means of Wel Governing and Maintaining in Good Peace, a Kingdome* (London: Printed by Adam Flip, 1608), 312.

13 William Sancroft, *Modern Policies Taken from Machiavel, Borgia, and Other Choice Authors* (London: Printed for Tho. Dring ..., 1655), see Colasterion.

14 Edward Dacres "The Epistle" in Nicholas Machiavel, *Prince* (London: Printed by R. Bishop ... 1600).

15 Quoted in Victoria Kahn, *Machiavellian Rhetoric: From the Counter-Reformation to Milton* (Princeton: Princeton University Press, 1994), 128.

any such attempt at control. Along these same lines, another great Italian legal philosopher, Alberico Gentili (1552–1608), defends Machiavelli by explaining: "The purpose of this shrewdest of men was to instruct the nations under pretext of instructing the prince."[16] Once the public knows and understands the tactics of tyrants, they can be preempted and undermined. In this light, Machiavelli could be read as presenting a guidebook not to cruel princes, but the people tyrannized by them.

It is also possible that Machiavelli was developing a surreptitious plan to trick tyrants into acting *more* morally. He can be seen as driving would-be tyrants to the logical and indeed unsavory ends of their actions. Since most people are unable to be as truly evil as absolutism would in principle demand – after all, princes need to "*learn* to be evil," which implies an often-overlooked presumption of at least moderate goodness – Machiavelli simply needs to sufficiently problematize the requirements of tyranny. For example, he writes: "If you take control of a state, you should make a list of all the crimes you have to commit and do them all at once. That way you will not have to commit new atrocities every day, and you will be able, by not repeating your evil deeds, to reassure your subjects and to win their support by treating them well."[17] This is no small task for rulers who are, as Machiavelli explains, "neither perfectly good nor wholly evil" but something in between. Being moderately good means that would-be tyrants may not be morally capable of murdering all potential rivals at the start. To begin with, just consider the epistemic demands of this approach to politics. How would one truly know when one act of violence would foreclose upon future acts of violence, when killing a determined opposition would work and when it would generate a deep-seated hatred?

To further demonstrate the plausibility of this "moral trick" reading, take Machiavelli's advice for conquering republics. He claims that "in republics there is more vitality, more hatred, more desire for revenge. The memory of their former freedom gives them no rest, no peace ... the best thing to do is to demolish them or to go and live there oneself."[18] A prince who conquers such a territory must be willing to totally wipe out the inhabitants who live there. As a dedicated republican, however, it is difficult to imagine Machiavelli genuinely advocating the destruction of republics in this way. It is much more likely, as Wolfe and Gentili indicated above, that these passages are part of a strategy to either inform the public about the dangers of tyranny, or to make

16 Qtd. in Kahn, *Machiavellian Rhetoric*, 128.

17 Machiavelli, *The Prince*, 30.

18 Ibid., 18.

certain political calculations so morally unsavory for prospective tyrants that they would be dissuaded from taking such drastic actions to begin with. In other words, the moral and psychic burden of "necessity" may undermine the turn to such drastic behavior.

The difficulty with these attempts to soften Machiavelli's justifications for killing the innocent, however, is that he clearly affirms the same instrumentality of violence in the process of political founding. In the *Discourses*, he explains that if a prospective prince "has blood in his veins, he will be appalled at the thought of imitating the evil times and will burn with an immense desire to copy the good. Truly, if a ruler wants to acquire worldly glory, he ought to want to rule over a corrupt city, not in order to destroy it completely, as Caesar destroyed Rome, but to re-establish it, as Romulus did."[19] After all, taking over a corrupt city will require both foundational *and* punitive violence since corrupt men can only be tamed with the threat of brutality. He explains "enemies of the new order" are bound to attack it. This is why a political founder must be willing and able to, as Friar Savonarola had been unable to do, "force the public to believe" in the foundational vision of the state through violent means.[20] He illustrates this point in the *Discourses* when he explains that "Moses was obliged, in order to have his laws accepted and his proposals adopted, to murder vast numbers of men, men who opposed his plans for no other reason but envy."[21] In other words, the moral virtue implicit in the civil law "Thou shalt not kill" is necessarily, and indeed, paradoxically predicated on a prior law of political founding: "Thou shalt kill." Founding *requires* violence.

For Machiavelli there are, of course, limits to this instrumental evil. Violence needs to have a proper object and it needs to work in achieving those ends. He explains, for example, that while Agathocles, the Greek tyrant of Syracuse, had mastered fortune and succeeded in acquiring and maintaining power, he was not a good example of princely virtue. Even though he employed violence effectively – liberating the poor by murdering all the city's oligarchs – he did not do so for an appropriate end. He reveled in the public's adoration rather than striving to make them more virtuous. This is not a critique of Agathocles's use of violence but its poorly conceived instrumentality. Machiavelli understood that tradeoffs are inevitable, and some "well applied" cruelty might be necessary for good ends. He explains that cruelty is:

19 Machiavelli, *The Discourses*, 113.

20 Machiavelli, *The Prince*, 20.

21 Machiavelli, *The Discourses*, 207.

> well applied (if we may, indeed, speak well of what is evil), when it is only once exercised, and that too, when it is dictated by the absolute necessity of self-preservation, and even then converted as much as possible to the benefit of the public. But it is ill applied when, though, practiced with caution and reserve in the commencement, it increases instead of diminishes with time.[22]

Even though Agathocles had clearly mastered fortune and power politics by choosing the proper objects of violence, he did not count as a virtuous prince because he did not achieve glory; he acquired power through all manner of cruelty and betrayal, but he did not establish a virtuous republic.[23] This shows that Machiavelli saw instrumental violence not as an end in itself but dedicated to some conception of the public good.

It is worth considering the innocence of those killed instrumentally through Machiavelli's foundational political violence, e.g. those enemies of Moses who were slain for the sake of there being a "Thou shalt not kill." Those who opposed Moses were not "innocent" but "envious." Here Machiavelli anticipates the Hobbesian state of nature, where prior to the existence of good laws and good armies, the public is greedy, fickle, and corrupt. These enemies of a foundational vision are not immoral, per se; they are more like competitors promoting alternative moral systems. Prior to the foundation of the law, people are waiting to be molded into something noble, virtuous, and good. Indeed, much like Hobbes would come to argue in the seventeenth century, Machiavelli's pre-political state is one of immense insecurity and amorality. There is no framework to justify behavior other than self-preservation and rational self-interest. The structural conditions of an un- or yet-to-be-regulated multitude – i.e. those contexts prior to the rise of the virtuous prince – make it virtually impossible for a people to be consistently good. And in this respect, the thousands of men that Moses killed in the process of founding the Hebrew nation could only be described as guilty in some pre-political or structural sense. They were sacrificed for the prospects of building a moral and political sphere.

---

22 Ibid., 434.

23 For more on the various interpretations of Agathocles, see Victoria Kahn, "*Virtù* and the Example of Agathocles in Machiavelli's *Prince*," *Representations* 13 (1986); 63–83. John McCormick paints a much more paradoxical picture of this historical example. Much of Machiavelli's historical account of Agathocles was either obviously wrong or the same moral failings could be attributed to other princes who he celebrated in the *Prince*. See John McCormick, "The Enduring Ambiguity of Machiavellian Virtue: Cruelty, Crime and Christianity in *The Prince*," *Social Research* 81:1 (2014); 133–164.

For Machiavelli, some of these same considerations governing foundational violence are at play in the unregulated state of war that arises between states in times of conflict. When an enemy threatens a political community from without, it threatens the existence of the morality, virtue, and security of the city state. The community risks being subsumed under a foreign power, which, like the enemies of Moses, means that the moral framework preserved by the prince could be destroyed and supplanted by that of a competitor. To avoid the annihilation of the polity, certain cruelties can justifiably be inflicted on those who threaten a state's security from without. In this respect, the excesses of violence are reserved for enemies outside the city's walls, which, like the pre-political realm is also a space of amorality. Princes must do what is necessary to ensure the protection of their city state and the moral character of its citizens.

Machiavelli does not directly thematize the issue of civilian casualties, but one does catch glimpses of these issues in his *Art of War*, which, though published in 1521, is still in many ways a fifteenth-century document.[24] While he spends some time addressing the emerging realities of siege warfare, in this essay one finds countless references to the burning, occupation, and destruction of towns and villages. His writing is tactically situated in the pre-siege era which emphasized raids and pitched battles. Though he does not explicitly reflect on the plight of the innocent, there is no question that Machiavelli takes it for granted that war will be a highly intrusive affair for the inhabitants of towns and villages who were not directly involved in the fighting and who had no part in the war effort.

In fact, part of what Machiavelli believes best explains the decline of Rome is that the Christian religion "has wrought such a change in the manners and customs of mankind, that they are now no longer under a necessity of defending themselves with such a degree of obstinacy and despair as they did in former times."[25] It was better, Machiavelli laments, when:

24 As Charles Oman explains, "Many of [Machiavelli's] premises were quite sound, and his observation of recent facts was keen. But unfortunately for his reputation as a prophet in the military sphere, all his recommendations of a practical sort bear no relation whatever to the actual development of tactics and organization during the later years of the century. He 'backed the wrong horse' in almost every instance." Charles Oman, *A History of the Art of War in the Sixteenth Century* (London: Methuen & Co., 1937), 93–94.

25 Niccolo Machiavelli, *The Art of War* in *The Works of Nicholas Machiavelli, Volume 2* trans. Ellis Farneworth Printed for Thomas Davie ..., 1762), 56.

> all such as were vanquished in battle, were either put to death, or carried into perpetual slavery in the enemy's country, where they spent the remainder of their lives in labour and misery: if a town was taken, it was either demolished, or the inhabitants were stripped of their goods, dispersed all over the world, and reduced to the last degree of poverty and wretchedness: so that the dread of these evils obliged them to keep up good discipline in their armies, and to honour all those that excelled in the Art of War.[26]

War had become, as Machiavelli's alter ego, Fabrizio, complains, too tame. And this has had a deleterious effect on the ability of republics to cultivate the moral and political virtue that would produce great republics. He writes, it is worthy to imitate "those states, which being jealous of their neighbours, and abhorring the thoughts of slavery … take all proper means to defend their liberties, and therefore continue free and respectable. This, I think, may suffice to shew the causes of our degeneracy, and the present neglect of military discipline amongst us."[27] Nothing focuses one's attention like the threat of annihilation or slavery. To be perfectly secure in one's liberties, one must be willing to expel and exterminate those who pose a threat. As indicated above, this is a tragic moral paradox in Machiavelli's thought: the less cruel war is, the less virtuous and viable republics become. The more pacific the interstate system is, the more republics will begin to falter in their discipline and the more likely they will begin to degrade.

The long-term influence of Machiavelli's thought on Europe and North America has been well-documented.[28] But it has not gone without notice that Machiavelli's appeal to necessity and, in particular, necessary cruelties in the service of the common good, took some time to percolate and enter the formal discourse of military ethics in the early modern period.[29] No matter how nuanced or esoteric his thinking actually was, subsequent commentators were reluctant to affirm what appeared to be mere expediency in the pursuit

---

26 Ibid.

27 Ibid., 57.

28 For analysis on Machiavelli's influence over English and American republicanism, see J.G.A. Pocock, *The Machiavellian Moment: Florentine Political Thought and the Atlantic Republican Tradition* (Princeton: Princeton University Press, 1975). See also Quentin Skinner, *The Foundations of Modern Political Thought: Volume 1, The Renaissance* (New York: Cambridge University Press, 1978).

29 Allen notes, "Machiavelli's influence on the thought of his own land in his own century seems to have been almost negligible" largely due to the active suppression of his printed works by the Church. See J.W. Allen, A *History of Political Thought in the 16th Century* (London: Methuen & Co, 1928), 502.

of justified ends. Many of the emerging international legal theorists of this period were unwilling to abandon some conception of divine and/or natural law and concomitantly some notion of Christian virtue and piety, which is to say a prince must be classically virtuous in his conduct in wartime. To be sure, the Scholastic tradition, with its focus on proper intentions, Christian piety, and the natural law, was still quite influential up until the end of the eighteenth century. Nevertheless, running concurrent with the Scholastics, this Machiavellian tradition of necessity was slowly gaining influence.

3 Desiderius Erasmus

Unlike Machiavelli, Desiderius Erasmus (1466–1536), known as the "Prince of Humanists" due to his immense fame and reputation in this period,[30] argues that even though men may have beastly impulses, it is not inevitable or even necessary that they act on them, especially regarding war:

> But how more justly should this be wondered at, what evil spirit, what pestilence, what mischief, and what madness put first in man's mind a thing so beyond measure beastly, that this most pleasant and reasonable creature Man, the which Nature hath brought forth to peace and benevolence, which one alone she hath brought forth to the help and succour of all other, should with so wild willfulness, with so mad rages, run headlong one to destroy another?[31]

In fact, rather than creating men with horns, claws, tusks, quills, spurs or any other type of natural weaponry, man is born "naked, weak, delicate, unarmed, with very soft flesh and smooth skin. No part of his body seems to be intended for fighting and violence."[32] Our defenselessness at birth is indicative of the fact that God calls us to mutual assistance rather than to war.[33] So instead of being persistently driven to fight, man's rational capacity can resist and suppress such base desires. War is a temptation, and a peculiar one since the tragedy of war "conteineth so many mischiefs, that it would abhor any man's heart to speak thereof."[34]

30 See Jonathan Arnold, "Desiderius Erasmus (1467–1536): Prince of Humanists," in *The Great Humanists: An Introduction* (New York: I.B. Tauris, 2011), 103–122.

31 Desiderius Erasmus, *Against War* (Boston: Merrymount Press, 1907), 5.

32 Erasmus, "Dulce bellum inexpertis," 402.

33 Ibid.

34 Erasmus, *Against War*, 10–11.

By providing a fuller account of self-interest, Erasmus contends that one would come to recognize that war is almost never worth it. War's cruelties are innumerable: "The poor commons pillaged, the nobles overcharged; so many old men of their children bereaved, yea, and slain also in the slaughter of their children; so many old women destitute, whom sorrow more cruelly slayeth than the weapon itself; so many honest wives become widows,"[35] and that "so many thousands of innocents should be brought into extreme calamity and mischief without deserving."[36] What is perhaps most troubling about war is that it "cannot be contained within its original limits, but, having begun in one quarter, it infects not only neighbouring regions like a contagious disease but even, through the effects of trade, of family or political alliances, drags remote ones into the general uproar and turmoil."[37] In this respect, war is better understood as a contagion, a form of madness, a pestilence that Christian men must do all in their power to eradicate.[38] Contagions are difficult to morally regulate, and this will have pronounced implications for what is permissible in war.

Despite these strongly worded denunciations, even Erasmus admits that in certain cases, war might be unavoidably necessary. A Christian prince ultimately may be put in a position where there is no other option. He writes, "A good prince will never start a war at all unless, after everything else has been tried, it cannot by any means be avoided."[39] It is only after he has set aside his emotions and has considered all his options rationally and morally that a prince can proceed to war. A rational calculation, however, means a rigorous assessment of "the anxieties, expense, dangers, and the long and difficult preparations."[40] The "true cost of war," faced honestly, will dispel any hopes of quick territorial or financial gain. He writes, "If we will only do the arithmetic and weigh up in honest terms the cost of war and the cost of peace, we shall soon discover that the latter can be bought for a tenth part of the worry, effort, distress, risk, expense, and bloodshed that war involves."[41] For Erasmus, the decision to go to war ought to be truly painstaking.[42] This is because, *pace* Machiavelli, rather than serving as a basis for republican virtue, he argues

35 Ibid. 11.
36 Ibid. 49.
37 Erasmus, "Dulce bellum inexpertis," 404.
38 For linkages between Erasmus and early Stoic arguments against war, see Doyne Dawson, *The Origins of Western Warfare: Militarism and Morality in the Ancient World* (Boulder, CO: Westview, 1996), 184.
39 Erasmus, *The Education of a Christian Prince* in *The Education of a Christian Prince with the Panegyric for Archduke Philip of Austria* (New York: Cambridge University Press, 1997), 103.
40 Ibid.
41 Erasmus, "Dulce bellum inexpertis," 416.
42 Erasmus, *The Education of a Christian Prince*, 103.

that war *erodes* the moral fabric of society. People become less virtuous the more they fight, and what is more, the cascade of violence makes future wars more likely. Men easily delude themselves into thinking that any war they are fighting is just, when they are simply giving expression to their own greed or ambition.[43] All wars, even so-called just wars, simply produce calamity. Better to seek arbitration when conflicts arise between princes.[44]

It is a striking feature of Erasmus's reluctant militarism that one of the many reasons he believes a Christian prince should avoid war is because "the greatest part of all the great evils which every war entails falls on people unconnected with the war, who least deserve to suffer these calamities."[45] The object of violence in war is never truly contained. The cascade of cruelty will inevitably spill over onto the innocent. Indeed, Erasmus suggests that there is no escaping this since the first premise of war is the pillaging of one's own citizenry; since princes cannot engage in war without "opening hostilities against his own subjects" (i.e. conscription, taxes, quartering, etc.), it is only natural that atrocities inevitably will be directed against foreign citizens.[46] Ultimately, this is why true princely "fame derives from the art of peace."[47]

It should be noted, however, that the premise of Erasmus's preference for the art of peace (over Machiavelli's art of war) is that "the law are silent in the midst of war (as Cicero says), or if they say anything beneficial, they cannot be heard over the din of combat (as Marius put it elegantly)."[48] It should be avoided *because* it is untamable: "arable land is laid waste, villages are deserted, shrines are torched, towns reduced to ruins, houses looted …"[49] War is an uncontrollable calamity, "engulfing everything in disaster and wickedness."[50] In another place he explains that war is such "an evil thing, it is the exclusive

43 This is a direct challenge to Martin Luther who argues that in a just war against a foreign adversary "it is a Christian act and an act of love confidently to kill, rob, and pillage the enemy, and to do everything that can injure him until one has conquered him according to the methods of war. Only, one must beware of sin, not violate wives and virgins, and when victory comes, offer mercy and peace to those who surrender and humble themselves." Martin Luther, *Secular Authority: To What Extent it should be Obeyed* [1523], in *Martin Luther: Selections from His Writings* ed. John Dillenberger (New York: Anchor Books, 1961), 398.

44 Erasmus, *The Education of a Christian Prince*, 105–106.

45 Ibid., 106.

46 Ibid., 107.

47 Erasmus, *Panegyric for Archduke Philip of Austria* in *The Education of a Christian Prince with the Panegyric for Archduke Philip of Austria* (New York: Cambridge University Press, 1997), 133.

48 Erasmus, *Panegyric*, 137.

49 Ibid.

50 Ibid., 138.

responsibility of evil people" – implying that when a state does go to war it should reserve this business for people who are already corrupted.[51] With these characterizations in mind, it is difficult to know how a necessary war could ever be restrained. The implication of this line of thinking is that when war comes, the cruelties inflicted upon the innocent will also necessarily come.

Of course, only a tortured reading of Erasmus could conclude that he *allows* for the death of innocents in war. The moral force of his argument is quite evidently that war must be stopped because of the uncontrollable wickedness it produces, and that every effort should be made to preserve the innocent in war. The world, however, is not made up of good Christian princes, and even when wars are fought by Christian men, he points out that there is always a temptation to sin. This is because war is a temptation to engage in cruelty under the guise of justice. Therefore, it is impossible for even Christian soldiers to always be perfectly restrained. In going to war, one is tacitly approving of some violence against the innocent.

These ideas become more challenging considering Erasmus's support for war against the Turks.[52] Erasmus concludes that though wars against Turks are not always justified, "there are times when failure to resist the Turks simply means the surrender of part of Christendom to these barbaric enemies, and the abandonment of those of our brethren who are already enslaved beneath their foul yoke."[53] Erasmus cannot be described as a holy war crusader. Even in this admission that war against the Turks can be justified, he stipulates that Christendom should strive to set its own house in order before looking to rectify injustice from without, purging the "Turkish vices: avarice, ambition, power-lust, self-satisfaction, impiousness, luxury, love of pleasure" etc.[54] He argues that war is permissible only "when you have left nothing untried and no stone unturned in your search for peace."[55] Much like the Scholastics, who were beginning to theorize about Turkish innocence, he also stipulates that when war is eventually undertaken, "Christian clemency demands that every

---

51 Erasmus, "Dulce bellum inexpertis," 438. Like in Thomas More's *Utopia*, certain jobs, e.g. slaughtering animals, should be performed by people who were already corrupted. Thomas More, *Utopia* [1516], trans. G. Logan and R. Adams (New York: Cambridge University Press, 2003), 71.

52 See Mura Idris, *War for Peace: Genealogies of a Violent Ideal in Western and Islamic Thought* (New York: Oxford University Press, 2018), 135.

53 Erasmus, "On the War Against the Turks," in *The Erasmus Reader* ed. Erika Rummel (Toronto: University of Toronto Press, 1990), 317.

54 Ibid., 324.

55 Erasmus, "Dulce bellum inexpertis," 438.

effort be made to confine the number involved to the minimum and to end the war, with the least possible bloodshed, as quickly as may be."[56] Even if the laws of war are silent, every effort is made to minimize civilian casualties, but this is not to say that such death will be eliminated. As indicated above, Erasmus painstakingly explains how the terrors of war will inevitably spill over onto the innocent. The goal of the Christian soldier is ultimately a losing battle.

The international order Erasmus envisions is quite different than Machiavelli's. Rather than the inchoate system of international states in perpetual competition, where war is inevitable and ineradicable, Erasmus believes that Christian states can eliminate war amongst themselves. Conflict will only be necessary at the edges, in those clash-of-civilizations moments, that is, when a nation outside of Christendom poses an existential threat. When all options have been tried, and a prince is staring at the destruction of Christian nations, he must be willing to resort to war. Since warfare produces innumerable cruelties, it can only be tolerated in those moments of utter desperation. Despite these stringent moral restraints that Erasmus strives to put in place, the fact that war is untamable inadvertently produces a tacit approval for the cruelties that will emerge in a necessary war. Christians should be as merciful as possible, minimizing unnecessary bloodshed, but war remains untamable.

4 Alberico Gentili

In 1598, Alberico Gentili, the Italian legal scholar who served in the court of Queen Elizabeth I, published *De Jure Belli Libri Tres*. This text, one of the founding documents of the nascent field of international relations, gives another early expression to a view of war governed by necessity and the common good.[57] In many respects, Gentili fuses the proceduralism of the Scholastics with the principle of necessity developed in both Machiavelli and Erasmus.[58] And while Gentili (and later Grotius) distances himself from some of the pacifist-oriented conclusions of Erasmus, whom he chides as a "flighty dilettante,"[59] it is important to note that they all draw on similar notions of

56 Erasmus, "On the War Against the Turks," 320.

57 For Gentili's interest in Machiavelli as a moral philosopher, see Erica Benner, *Machiavelli's Ethics* (Princeton: Princeton University Press, 2009).

58 See Ursula Vollenhun, *The Idea of International Society: Erasmus, Vitoria, Gentili and Grotius* (New York: Cambridge University Press, 2017), 108.

59 Alberico Gentili, *De Jure Belli Libri Tres, Volume 2*, trans. John Rolfe (London: Humphry Milford, 1933), 29.

necessity stemming from the dynamics at play in the relationships between nations, or in Erasmus's case, civilizations.[60]

Gentili agrees with the Scholastics that war is the purview of public authorities, and it must be waged for a just cause. He is, however, less interested in the Scholastic problem of right intention or "good motive," which, he says, "is a problem for theologians," and instead focuses on the formal causes of war, which ultimately determine whether the war is just and if restraints can reasonably be imposed on belligerents in war.[61] Unlike the Scholastics, Gentili believes that when driven to war, some degree of justice may be found on both sides. He evinces an early awareness that the dynamics of the state system may, in some instances, incentivize violence. Fear and insecurity can drive states to reasonable, if tragic, conclusions. In this respect, like Machiavelli before him, he gives expression to an early view of the systemic effects produced by a society of nations.[62]

Gentili pays close attention to what he calls "natural causes" of war. While there is no natural enmity between men, war might be said to be natural inasmuch as it concerns the violation of the natural law. Unlike the Scholastics, Gentili argues that the most basic feature of the law of nature is self-preservation. Quoting Cicero, he explains, "Every method of securing safety is honorable. This has been taught to philosophers by reason, to barbarians by necessity, to the nations by custom, to wild animals by Nature herself. The law is not written but inborn."[63] With this in mind, Gentili concludes that "one who is attacked by an armed enemy makes a necessary defence, and his action is that of necessary defence; and so also does one against whom an enemy has been making preparations [i.e. a preemptive war]."[64] Gentili cautions, however, against thinking that all fear-driven preemption is justified. He notes that "a sense of security is the most common cause of disaster" – a false sense of security, after all, is no security. If one is truly paying attention, sometimes preemption is necessary; "we ought not to wait for violence to be offered us, if

60 Idris, *War for Peace*, 182.

61 Gentili, *De Jure Belli Libri Tres*, 35.

62 See Martin Wight, *Systems of States* (Leicester: Leicester University Press, 1977); Ursula Vollerthun, *The Idea of International Society: Erasmus, Vitoria, Gentili, and Grotius* (New York: Cambridge University Press, 2017); and Valentina Vadi, *War and Peace: Alberico Gentili and the Early Modern Law of Nations* (Boston: Brill, 2020).

63 Gentili, *De Jure Belli Libri Tres*, 59.

64 Ibid., 58.

it is safer to meet it half way."[65] However the threat must truly be justified and reasonable.[66]

There are a few interesting features of Gentili's defensive war. He affirms the long-standing just war tradition in that a defensive war can end in an offensive punishment. One is, after all, allowed to redress injuries (though, he cautions against re-litigating past wrongs). A defender can exact an appropriate retribution for an unjust invasion. He also claims that defensive wars can be waged "for the sake of others," based on the "fundamental principle, that nature has established among men kinship, love, kindness, and a bond of fellowship."[67] Even what he calls "offensive wars" have a defensive quality. If a neighbor withholds a necessary resource, or if a people is driven into exile and is in need of making a new home elsewhere, these exigencies, though they might take the form of an ostensibly offensive action, are more basically predicated on the principle of self-preservation. Additionally, certain violations of the laws of nature may warrant offensive military action. Cannibalism, piracy, those engaging in human sacrifice, and other "monsters of lewdness" deserve to be punished.[68] Idolatry when coupled with the murder of the innocent might also be a pretext for a just intervention.[69] Like the Scholastics, however, these otherwise offensive actions can also be characterized as defensive since those who violate the laws of nature in such a fundamental way have become enemies of humankind. The proliferation of such enemies would signal a fundamental breakdown of human society.

There is, however, a moral ambiguity in Gentili's formulation of preemptive self-defense and the status of the innocent. He writes, "wars are just even though so many things which come from them are evil, because their final aim is good … And since its aim is good war itself is also just and good."[70] He continues, "the purpose of war is peace, to which end and to justice … one cannot attain except through bloodshed and those evils arising from strife."[71] This formulation implies that wars may require certain necessary evils in

65 Ibid., 61.

66 He explains that in lieu of a general rule about what counts as a justifiable cause for fear, the most commonly justified case is when one opposes "powerful and ambitious chiefs" (Ibid., 64).

67 Ibid., 67.

68 Ibid., 123, 124.

69 Ibid. 123. This was a common view in this era. For instance, Émeric Crucé also argues that against savages, thieves, and pirates "War will always be a good thing against them, if they cannot be brought to reason." Émeric Crucé, *The New Cyneas* [1623] trans. Thomas Willing Balch (Philadelphia, Allen, Lane, and Scott, 1909), 68.

70 Gentili, *De Jure Belli Libri Tres*, 28–29.

71 Ibid. 29.

the pursuit of just ends. However, to complicate matters, Gentili also stipulates that militaries are governed by "the laws of God which were made for all men, and the one of these with which we are here concerned is, that we should spare women and children."[72] Descriptively, Gentili appears to make certain allowances for the evils stemming for a just war – for instance, "Cities are sacked when taken; they are not sacked when surrendered"[73] – but normatively the laws of God prohibit harm being directed against certain classes of people. Though, since Gentili defends offensive wars to punish those who have violated the laws of nature, he may presume some level of collective guilt. After all, consider those crimes against humanity that he mentions. Entire communities would be implicated in the social practices of cannibalism and human sacrifice. All members of those communities supporting these crimes against the natural law will likely be guilty, perhaps even women and children. Even though he claims that "Children should always be spared, and so should women," it is likely that women and children who participate in cannibalism or human sacrifice do not retain their innocence.[74]

In more conventional military conflicts, Gentili offers little wiggle-room in his prohibition of killing the innocent. He addresses and subsequently rejects two common exceptions to this rule: "If there is an order of God to the contrary" (i.e. if God commands you to slaughter the innocent of an enemy country) and retaliation for when "the enemy wrongs our women and children."[75] Regarding the first, Gentili recommends extreme caution; it is an "extraordinary" command, and "we shall not, as do the hasty and careless, make of them a precedent for our own conduct."[76] In terms of the second, Gentili is skeptical that, as the interpreters of the law he is reading suggest, "the rule must be adhered to."[77] He, in fact, argues that "he is foolish who connects with the laws of war the unlawful acts committed in time of war. In this connection I make no allowance for retaliation."[78] Even if enemies stoop to certain lows of military conduct, justified military action must adhere to certain standards.

While he indicates that evils may be necessary in the prosecution of a just war, Gentili makes it difficult to justify the death of innocents by way of proper motives; however, he does create a small bit of room by way of necessity. Some kinds of killing are entirely off limits. For instance, one cannot use

72 Ibid., 255.
73 Ibid., 315.
74 Ibid., 251.
75 Ibid., 255.
76 Ibid.
77 Ibid.
78 Ibid., 257.

poison in war because the contamination of streams will lead to the "slaughter of innocents."[79] And no matter what necessity demands, Gentili argues, "To me it seems cruel, and therefore unjust, to vent fury upon a weak and innocent throng."[80] Unleashing indiscriminate violence in the heat of passion is unbecoming a just belligerent. Again, having the right to wage a war does not mean one is free to do whatever is perceived to be necessary. But the normative line between necessary and permissible harm is difficult to delineate.

Take for instance Gentili's assumption that in certain circumstances it is permissible to sack or destroy cities. He begins by noting that "moderation should be observed" and that the "actions of the war" need to be brought into harmony with the "reasons for making war."[81] Much like the classical just war tradition, there needs to be a proper application and proportionate use of force. When deciding whether a sack is in order, one must do so in a calm and collected manner. Gentili writes, "we must take care, says Cicero, that nothing be done rashly and nothing cruelly; and he declares that it is the part of a magnanimous man, after examining the circumstances, to punish the guilty but spare the multitude."[82] A magnanimous leader will always be actively looking to spare the innocent, but, of course, he may not always have the luxury to do so. Here Gentili provides some clues as to what those difficult exigencies look like. A particularly recalcitrant enemy, one that has engendered years of bitter struggle may be justifiably destroyed, along with those who violate their treaties and inspire fear in their neighbors.[83] These are, of course, hardly matters of precision. So, while Gentili's otherwise forgiving leaders will look to preserve the innocent multitude, they also must "look to the future" and make decisions that will produce the greatest level of security.[84] Much as Machiavelli had concluded, the international system and the behavior of neighboring states may not always extend a prince the luxury to preserve the innocent in the face of certain kinds of evils.

It is worth reiterating that justified cruelty, much as was the case with Erasmus, represents a painstaking assessment in Gentili's theory of war. The purpose of these delineations is to make the resort to cruelty more restrained and less terrible. In the end, there is a high degree of accountability built into his decision-making process. For instance, he concludes his moral prohibition against wars of retaliation with the following argument:

79 Ibid., 159.
80 Ibid., 256.
81 Ibid., 315.
82 Ibid., 316.
83 Ibid., 318–20.
84 Ibid., 319.

> At some time the enemy will have to render account to them for his wrong; he will surely render an account to God, and he will render it to the rest of the world, if there is no magistrate here to check and punish the injustice of the victor. He will render an account to those sovereigns who wish to observe honourable causes for war and to maintain the common law of nations and of nature.[85]

Just wars may lead to some evils, but these evils cannot be justified by way of proper motive or a strict appeal to right; they are justified by virtue of the end for which the war is prosecuted – a just defense or a righteous prosecution of a violation of the laws of nature – and what is necessary to obtain those ends. As a practical matter, Gentili believes that wrongdoers will be punished by "those sovereigns" who affirm the natural law and law of nations. And if no earthly power is there to right a nation's wrongs, the evildoer will ultimately be required to "render an account to God."

5 Hugo Grotius

Despite the variations in emphasis, the above theorists hold in common that certain evils might be necessary, in certain desperate cases, for justified ends. In effect, the explicit linkage of justified harm with necessity became more formally developed in this period.[86] The Dutch theorist, Hugo Grotius (1583–1645), unquestionably the most influential writer on international politics in this period, gives the clearest expression of these developments. In *The Rights of War and Peace* [1625], for instance, he begins with the general principle that was affirmed by Vitoria and other Scholastic commentators, that certain people – women, children, ministers, farmers, foreigners, and merchants, should be exempted from the "calamities of war."[87] However, these protections are not absolute: "there may be circumstances, in which absolute justice will not condemn the sacrifice of lives in war, yet humanity will require that the greatest precaution should be used against involving the innocent in danger, except in cases of extreme urgency and utility."[88] While the principle of humanity requires a belligerent to strive to take every precaution to protect the innocent,

---

85 Ibid., 257.

86 For Gentili's influence on Grotius see Ursula Vollerthun, *The Idea of International Society: Erasmus, Vitoria, Gentili, and Grotius* (New York: Cambridge University Press, 2017).

87 Hugo Grotius, *The Rights of War and Peace* [1625] trans. A.C. Campbell (New York: M. Walter Dunne, 1901), Bk. 3, Chap. 11, 361–363.

88 Ibid., Bk. 3, Chap. 11, VIII. 361.

in some extremely urgent cases utility may necessitate harm being inflicted on the innocent. And elsewhere he continues: "an individual or belligerent power may, in the prosecution of a lawful object, do many things, which were not in contemplation of its original design, and in themselves would not be lawful to do."[89] For instance, sometimes, he explains, "very important advantages may attend striking a terror, by preventing the same crimes in future from being committed."[90] Under normal conditions terrorizing the innocent is a moral liability, but if such action prevents greater evil in the future, it may be permitted. One must always be cautious, however, to avoid "unnecessary violence." Like Machiavelli above, the effective use of cruelty will foreclose on future violence.

Grotius develops the principle of utility in his work, but moral norms are still undergirded by the principles of natural and divine law. He explains that using "extreme rigour" when taking a town is prohibited because it is not "conducive to the termination of the war" but also since it is "repugnant to every principle of Christianity and justice."[91] There is something like a belt and suspenders approach here; not only is "extreme rigour" typically unnecessary and counterproductive, but God also opposes such excessive cruelty. One should strive to be as charitable as the circumstances allow. Nevertheless, running throughout Grotius's writing is a clear tension: the pursuit of lawful ends may sometimes require militaries to violate the natural law principle which stipulates that the innocent should be protected from the calamities of war.[92] When cruelty is absolutely necessary, it is permitted.

---

89 Ibid., Bk. 3, Chap. 1, IV, 292.

90 Ibid., Bk. 3, Chap. 11, XV, XVI, 363.

91 Ibid., Bk. 3, Chap. 1, XIII, 368.

92 These sentiments appear to be shared by Thomas Hobbes, as well. Hobbes writes, "There is little therefore to be said concerning the Laws that Men are to observe one towards another in Time of War, wherein every Man's Being and Well-being is the Rule of his Actions. Yet thus much the Law of Nature commandeth in war, That Men satiate not the Cruelty of their present Passions, whereby in their own Conscience they foresee no Benefit to come. For that betrayeth not a Necessity, but a Disposition of the Mind to War, which is against the Law of Nature." Cruelty is only permitted under two conditions 1) when necessity demands it, and 2) when a future benefit will come of it. Both of these are contextualized by an authentic reason for going to war in the first place which is fear. He continues, "For nothing but Fear can justify the taking of another's Life. And because Fear can hardly be made manifest, but by the Action dishonorable, that betrayeth the Conscience of one's own Weakness, all Men, in whom the Passion of Courage of Magnanimity have been predominant, have abstained from Cruelty; insomuch, that though there be in War no Law, the Breach whereof is Injury, yet there are in War those Laws of Actions in War, is Honor." Thomas Hobbes, *The Elements of the Law* [1650] in *The Moral and Political Works of Thomas Hobbes* (London: Printed in the Year 1750), 50.

The moral tradeoffs between necessity rooted in a principle of right on the one hand and Christian mercy on the other can best be seen in punitive military action. Like Gentili, Grotius believes that a just war was not merely defensive. A prince could justifiably punish those who engaged in "gross violations of the law of nature and of nations," including "pirates, general robbers, and enemies of the human race."[93] Those who engage in such crimes open themselves up to direct harms. However, Grotius explains, "It is lawful to attack a ship maintained by robbers, although in that ship, or that house there may be many innocent persons whose lives are endangered by such attack."[94] In other words, it is both lawful and permissible to attack pirates due to the threat they pose to humanity, even if there are innocent people on board the ship.

Upon first glance, this appears to be a simple restatement of the traditional Scholastic argument of double effect. One may incidentally or accidentally kill the innocent if the end one pursues is lawful and the harm is not directly intended. But there are a few aspects to this formulation that deserve closer scrutiny. To begin with, the morally significant feature of Grotius's framework is not the intentionality of harm but the tradeoff between right and charity. There is a delicate balance between the overall good that will come from the harm in question and the extent to which one can feasibly be charitable. This tension can be seen in the section immediately following the passage on killing the innocent in the process of destroying a pirate ship. Grotius seeks to temper the contextual lawfulness of unintentional innocent death with an appeal to charity. While one may be technically permitted to pursue a lawful object, say, to kill pirates and robbers, leading to some foreseeable and indirect death of innocents, Grotius argues that "there are many instances, in which the law of charity will not allow us to insist upon our right with the utmost rigour."[95] In other words, strict duty to right and adherence to the law – i.e. the duty to punish those who violate the natural law – is itself not always morally appropriate.[96]

93 Grotius, *The Rights of War and Peace*, Bk. 2, Chap. 20, XL, 247. For subsequent attempts to assess violations of the nature law, see Christian Thomasius, *Foundations of the Law of Nature and Nations* [1705] in *Institutes of Diving Jurisprudence* ed. and trans. Thomas Ahnert (Indianapolis, IN: Liberty Fund, 2011).

94 Grotius, *The Rights of War and Peace*, Bk. 3, Chap. 1, IV, 292.

95 Ibid.

96 Aristotle called this type of justice equity. Sometimes the strict adherence to the law is itself a form of injustice. See Aristotle, *The Art of Rhetoric* trans. Robin Waterfield (New York: Oxford University Press, 2018), 1374, b2–b10, 52.

Secondly, over and above the tension between one's duty to the law and to charity, Grotius frames one's judgment about the justified killing of the innocent in an inchoate utility calculus. As seen above,

> things, which fall not within the original purpose of an action, and the happening of which might be foreseen: unless indeed the action has a tendency to produce advantages, that will far outweigh the consequences of any accidental calamity, and the apprehensions of evil are by no means to be put in competition with the sure hopes of a successful issue.[97]

When deciding to use lethal violence that will fall on both the guilty and the innocent, one must carefully weigh the potential benefit the "accidental calamity" will produce. It is only in those moments when the morally questionable action produces advantages that "far outweigh" the original evil that such cruelty is permitted. Much as was the case with many of the commentators above, there is a painstaking process of moral deliberation that defies simple quantification. One must honestly assess the extent and long-term effect of the evil in question, the extent to which the punishment will bring unnecessary harm onto the innocent, and, finally, the overall good that will come from the use of violence.

There are good reasons to believe that Grotius was drawing on Aquinas's formulation of double effect in this argument. After all, the way he formulates this idea bears some of the conceptual trappings of classical double effect reasoning; for instance, the harms in question "were not in contemplation of its original design" and they "fall not within the original purpose of an action" but were "accidental calamities." What is more, Grotius makes explicit reference to Aquinas and his use of the parable of the tares: "Suffer the tares to grow, says *our divine teacher* [i.e. Aquinas], least in rooting up the tares you root up the wheat also."[98] One might recall from Chapter 1 that this is the argument Aquinas uses to caution against rooting out unrighteous sinners who are settled amongst the innocent.[99] Culling the wicked may disproportionately harm the innocent.

Notice, however, where Grotius places the conceptual weight of his moral argument. The death of innocents is not permissible because it is an

97 Grotius, *The Rights of War and Peace*, Bk. 3, Chap. 1, IV, 292.

98 Ibid.

99 On this theme, Grotius explains that a whole people can be "implicated in the misconduct of the majority, which holds the representative character of the state" (Ibid., 263). The subsequent losses must be endured by the innocent.

unintentional side-effect of an ostensibly good objective. Much like Gentili, Grotius instead emphasizes the principles of lawfulness and right, the degree to which justice can practically and effectively be served, and the degree to which the harm will "produce advantages that will far outweigh the consequences."[100] Killing the innocent is permissible in those moments where one is governed not only by the principle of right, where the enemy has become such a menace to humankind (like pirates and robbers) that punitive action is necessary, but where one is also certain that the deaths of the innocent are worth it in the sense that they will lead to an overall good effect.[101] In this respect, the right to kill the innocent is moderated and softened by both the duty of charity and a version of the lesser evil argument; while one may be permitted to right a wrong through retributive justice, acting strictly on this right may cause more harm. Plus, as Grotius explains, "whenever there is any doubt, it is always the safer way to decide in favor of another's interest, than to follow the bent of our own inclinations."[102] Indeed, since in war "the innocent must often be involved in the sufferings of the guilty … the balance should incline in favor of peace."[103]

---

100 In fact, in Grotius's reframing of Aquinas's original case for double effect, i.e. when a private individual kills a robber in self-defense – he conspicuously omits intention in favor of a lawfulness grounded in necessity: "Thus by the laws of all known and civilized nations, the person is judged innocent, who kills another, forcibly attempting or endangering his life; a conspiring and universal testimony, which proves that in justifiable homicide, there is nothing repugnant to the law of nature" (Ibid., 57). Later, he continues, "In strict justice, it cannot be denied that we have a right to kill a robber, if such a step is inevitably necessary to the preservation of our property" (Ibid., 81). It is not only that the robber excites horror, but that an individual exercises his right to wound or kill an aggressor.

101 Jean Jacques Burlamaqui, one of Grotius's prominent commentators, takes a slightly different angle in assessing the death of women and children in these cases. He writes, "With regard to old men, women, and children, it is certain that the right of war does not of itself require that we should push hostilities so far as to kill them; it is therefore a barbarous cruelty to do so. I say, that the end of war does not require this of itself … It may also be said, that when the heat of action hurries the soldiers, as it were in spite of themselves, and against the order of their superiors, to commit acts of inhumanity, as for example, at the siege of a town, which by an obstinate resistance, has irritated the troops; we ought to look upon those evils rather as misfortunes, and the unavoidable consequences of war, than as crimes that deserve to be punished." Jean Jacques Burlamaqui, *The Principles of Natural and Politic Law, Volumes 1–2* [1751] (Philadelphia: Carey and Lea, Chesnut Street, 1830), 195–196. Killing the innocent inhabitants of a town is simply an "unavoidable consequence" of siege warfare.

102 Ibid., Bk. 3, Chap. 1, IV, 292.

103 Ibid., Bk. 2, Chap. 14, VI, 276. This proto-utilitarian assessment is also affirmed by Burlamaqui, who writes, "We must above all things, exactly weigh the good or evil which we may bring upon ourselves by it. For if in making war there is reason to fear that we

6 Samuel Pufendorf

Appealing to certain aspects of the martialist tradition, the German legal philosopher, Samuel Pufendorf (1632–1694) explains that since the practice of war is seemingly inevitable, it would be prudent to "make War a sort of Trade and Profession, and to form it into an Art; that so at least, the Injuries which the Innocent must suffer may be less insupportable, when the Liberties of the Sword and the Miseries natural to a State of War are more restrained."[104] By treating war as a profession, not only will cruelties be minimized, but when they do occur they will be more justified. Along these lines he explains that the emerging protections of the innocent – of women, children, prisoners, and farmers – must be understood as the product of custom or tacit consent rather than being grounded in the natural law.[105] He goes on to explain, however, that while the protections extended to these people are in some limited sense

shall draw greater evils on ourselves, or to those who belong to us, than the good we can propose from it; it is better to put up with the injury, than to expose ourselves to more considerable evils than that for which we seek redress." Nevertheless, he recapitulates the standard humanist argument: "I say generally speaking; for there may be cases of necessity so pressing that the care of our own preservation obliges us to proceed to extremities, which in any other circumstances would be absolutely criminal." Burlamaqui, *The Principles of Natural and Politic Law*, 170, 196.

104 Samuel Pufendorf, *Of the Law of Nature and Nations* [1672] trans. Basil Kennett (London: Printed for I. Walthoe …, 1729), Bk. 8, Chap. 7, 850.

105 Ibid., Bk. 2, Chap. 3, 151. In stark contrast, Pufendorf's contemporary, John Locke (1632–1704), prefers the language of *right* to that of *consent* when speaking about protection of the innocent. In the *Second Treatise of Government* (1690), Locke, characterizes *right* in war two ways: 1) There exists the *right* of the just invader to seek equitable reparation for a wrong or "to destroy that which threatens me with destruction," (§16), and 2) There is also the *right* of the innocent noncombatant to have her life preserved (§179). Both rights form the parameters of legitimate action in a just war. One can infer from the conditions of a just conquest that the conqueror only has power over those "who have actually assisted, concurred, or consented to that unjust" domestic resistance; and elsewhere, Locke continues, "The right then of conquest extends only to the lives of those who joined in the war," preserving "the right of the innocent wife and children" of the guilty combatant (§§179, 182). He admits, however, that conquerors "seldom trouble themselves to make the distinction, but they willingly permit the confusion of war to sweep all together" (§179). That this is descriptively the case does not make it normatively acceptable. Locke was wary of these "miscarriages of violence" (§182). Though, one must also note that this formulation is ambiguous. The wife and child are innocent inasmuch as they did not "assist in [the war]" (§183). It is not clear what the relevant parameters of "assistance" are that rise to the level of guilt. As it stands, Locke is unique in this tradition in that makes *no* provision for killing the innocent. John Locke, *The Second Treatise of Government* [1690] in *The Works of John Locke, Volume 4* (London: Printed for C. and J. Rivington …, 1824). For more on this Lockean, rights-based approach to immunity, see

binding, one is not strictly required to honor these restrictions when fighting a just war: "Whoever therefore wages War in a just Cause, may slight these Formalities at Pleasure, and govern himself purely by the Law of Nature: Unless he thinks it more for his Interest to observe them, as a Means to render the Enemies less severe towards himself and his men."[106] Like Grotius, Pufendorf argues that while one may have a legitimate claim to the extreme justice of right, there may be prudential reasons to not pursue one's perfect right. Then again, there may also be good reason to violate the custom of extending protection to certain classes of people.

In this respect, like many of the martialist commentators above, Pufendorf places great weight upon prudential leadership and the professionalization of war. The judgment about how much suffering can be inflicted on the innocent depends on the justness of the war and a careful assessment about the best way to prosecute the conflict. In short, wise leaders will know the precise amount of violence to inflict:

> The Law of Humanity would have Men consider not only what an Enemy may suffer without Injury, but what it may be proper for a generous Conqueror to inflict. And therefore they ought to be particularly careful, that as far as possible, and their own necessary Defence and future Security will permit, to proportion the Evils they inflict upon their Enemy, to the Measures and Moderation observed by Civil Courts in punishing Criminals and Offenders.[107]

Like Grotius, one must cautiously weigh what is necessary for one's security – now and in the near future – against the amount of necessary cruelty obtaining that security requires. The ideal goal of this assessment is to determine exactly what measure of force is needed. For Pufendorf, this doesn't mean however that "it is always unjust to return a greater Evil for a less," because, he continues, in war "it is lawful for me to make use of such Means as I th"nk will best prevail against a Person, who, by his injurious Treatment of me, hath put me out of all Capacity of doing him any Injury, till such Time as it shall be agreed between us to forbear Hostilities."[108] To effectively punish an enemy

---

Kai Draper, *War and Individual Rights: Foundations of Just War Theory* (New York: Oxford University Press, 2016).

106 Pufendorf, *Of the Law of Nature and Nations*, Bk. 2, Chap. III, 151.

107 Ibid., Bk. 8, Chap. 6, 838.

108 Ibid., Bk. 8, Chap. 6, 837–38.

nation, sometimes *greater* harm is required than was initially inflicted in order to subdue an enemy that may pose an indeterminate threat in the future.

However, like Grotius, even though one may be technically permitted by the principle of necessity to use overwhelming force, Pufendorf argues that militaries should be temperate for no other reason than the avenues of violence they open at any given moment could be turned back on them at another time. In fact, Pufendorf explains that what is permitted by a legalistic appeal to right may not be the most reasonable or appropriate course of action. In fact, in some cases Pufendorf concludes that it might be better for the guilty to go unpunished if punishment leads to greater social harm. And like Erasmus, he recognizes that "war will, in all Probability, produce greater Evils than Good," so those who forego their right to revenge a wrong are the ones who should be commended.[109]

To make sense of the moral tradeoffs at work in Pufendorf's thought, it is worth looking at his critical reassessment of the Scholastic argument of double effect. Much like the martialists, Pufendorf places a great deal of conceptual weight on the military leadership's responsibility to evaluate competing considerations and to take due care in the process of waging a just war, the assumption being that they will have a reasonably good sense of the potential harm a particular attack will produce. In this respect, the scope of what counts as a truly "accidental consequence" is much narrower in Pufendorf's taxonomy. The moral significance of good intentions entirely depends on the degree to which the actor is or should be aware of the harm he is committing. In this respect, Pufendorf rejects good intentions as meaningfully exculpatory when it comes to killing the innocent.

When attacking a city or town, a sufficiently wise leader should be able to foresee the harm that would be produced by lobbing bombs into the city center. To this end, Pufendorf distinguishes between "voluntary" and "involuntary" ignorance. Voluntary ignorance stems from "idleness and unattention," i.e. when a person "knows not that which he could, and ought to have known."[110] If, for example, military leadership pursues a strategy that kills the innocent out of voluntary ignorance, this would imply that they willfully ignored something they should have known and factored into their prudential calculus. On Pufendorf's account, they could not rationalize their behavior citing good intentions alone. If an injury can, in principle, be foreseen and anticipated, there is a duty to factor this expected harm into one's strategic planning. For Pufendorf, just because a harm is foreseen, however, does not mean that it is

109 Ibid., Bk. 8, Chap. 6, 834.

110 Ibid., Bk. 1, Chap. 3, 31.

prohibited. As seen above, when pursuing a just cause, one may violate mere conventions.

Pufendorf argues that it is only with certain kinds of involuntary ignorance – which he further divides into "antecedent ignorance" and "invincible ignorance" – that good or benign intentions become morally significant. Being drunk, for instance, is a form of antecedent ignorance. It is involuntary in that sense that when individuals put themselves in this condition, their judgment is impaired. Pufendorf argues that while drunk they may not intend the harms they commit, they nonetheless still bear any liability for actions stemming from their state of self-imposed ignorance. Being drunk, of course, does not excuse bad behavior. As Aristotle had explained, the disposition to drunkenness is itself a moral failing. For Pufendorf, it is only "invincible ignorance" – where one "knows not such Things, as he had neither Ability nor Obligation to know"[111] (in other words, genuine accidents) – that a good intention can override a negative consequence.[112] Unlike the Scholastics, when pursuing a legitimate right, only those evils that could not in principle have been known or anticipated in advance could be excused by an appeal to good intentions.

Pufendorf argues that military leadership cannot claim "invincible ignorance" to justify the death of innocents in war. Only genuine accidents can be excused by good intentions. Furthermore, by foreclosing on this more expansive moral loophole, Pufendorf is seeking to subsume the harm inflicted on the innocent into a clear proportionality framework, where evil can be committed as long as it is understood to be both necessary and effective to obtain just ends. As noted above, Pufendorf argues that in certain circumstances greater amounts of harm could even be justified if it led to greater security in the future. The criteria of justified harm inflicted on the innocent turns back to the careful assessment of the implements and strategies of violence, necessity, the moderating effects of charity, and the greater good.

---

111 Ibid.

112 A variation of this position is advanced by Alexander Shields, who explains that "casual killing contrary to intention without knowledge" is no breach of the prohibition against killing the innocent; however, "Tho a man kill an innocent unwittingly & unwillingly, besides his knowledge and against his will; yet he may be guilty of sinful homicide, if he was obliged to know that he was in hazard of it, and neglected to consider, lest a man might be killed by what he was doing." Alexander Shields, *A Hind Let Loose* (Edinburgh?: [s.n.], 1687), 640–641.

7 Jeremy Taylor

Reflecting the turn to casuistic reasoning and the shift to more probabilistic formulations of the greater good, Jeremy Taylor (1613–1667), a well-known cleric under Cromwell, builds on many of the same ideas expressed by Gentili and Grotius in his massive (and now largely neglected) *Ductor Dubitantium* (1660).[113] Part of the difficulty Taylor finds in providing a Christian assessment of war is that the

> Christian Religion hath made no particular provisions for the conduct of warre under a proper title, because it hath so commanded all the actions of men, hath so ordered the religion, so taken care than men shall be just, and doe no wrong, hath given laws so perfect, rules so excellent, threatening to sever, promises so glorious that there can be nothing wanting toward the peace and felicity of mankind, but the wills of men.[114]

He concludes that there is no principle of reason or law of nations that one can appeal to in order to justify war, especially between Christian nations.

Taylor, however, concludes that war is not "measured by laws, but by privileges; that is, things being gone beyond the laws of Nature: he is left to his natural powers and defenses."[115] In other words, since there is no law of God forbidding self-preservation, he concludes that this natural impulse can serve as a sufficient basis for war. Using the language of Gentili, Taylor in fact argues: "The learned are taught by reason, the barbarous nations by necessity, the civil by custom, the very beasts also by Nature, to defend their head, their body, their life from all injury by all means. For God hath nowhere forbidden that a man shall be defended."[116] While God never expressly justifies war by way of the natural impulse to self-defense, he does not proscribe the act of self-defense, and since self-defense necessarily entails acts of violence, violence used for self-defense is permitted. He concludes, "Whatsoever is absolutely necessary is certainly lawful, and since Christ has nowhere forbidden Kings to defend themselves and their people against violence; in this case there is no

---

113 W.J. Brown argues that even though this is Taylor's *magnum opus* it is probably his least well-known work due to its impossible size. W.J. Brown, *Jeremy Taylor* (London: Society for Promoting Christian Knowledge, 1925), 179.

114 Jeremy Taylor, *Ductor Dubitantium* (London: Printed by James Flesher …, 1660), 377.

115 Ibid., 378.

116 Ibid.

law at all to be considered; since there is a right of nature which no law of God hath restrained."[117]

In attempting to assess what the appropriate measure of necessity entails, Taylor concludes that it is safest to analogize public necessity to that which is already legitimate in private. And in this respect, a prince is justified in punishing certain injustices (the breaking of covenants and promises) and the violation of certain rights. However, he insists that the measure of violence must be proportional to the crime. Taylor continues:

> He that is injur'd may drive away the injury, he may fight against invaders, he may divert the warre if necessary; but he may not destroy the innocent with the guilty, the peaceable Countrymen with the fighting Souldiers: and nothing can legitimate that but an absolute necessity, that is, it must not be done at all when it can be understood and when it can be avoided; and there is no direct action of warre but it is to be rul'd by necessity, and justice and charity.[118]

Like Pufendorf, if harms can be foreseen and avoided, they cannot be explained away as "accidental effects," rather, only by necessity. The implication of this formulation is clearly that when strict necessity demands it, foreseen (i.e. "when it can be understood") but unpreventable harm is permitted. What is striking about Taylor's formulation, however, is that, like many of the commentators above, it is justified not by way of proper intentions but rather by the complex mix of justice (or right), necessity, and charity.

8 George Dawson

George Dawson (1637–1700), a prominent jurist of this period, notes that even though commanders should take care "that no innocent People (if it be possible) be put to death in the time of War" – making explicit provision for children, women, clergy, and academics – this sits in tension with the "Law of War and Arms," which gives invading armies:

> Liberty … to lay waste and destroy all the Cities and Towns of the Enemy; all their Castles and Fortifications; all their Ports and Ships; all their Corn and Fruits; yea, all Churches and Temples, and all things dedicated and

117 Ibid.

118 Ibid., 379.

> devoted to the Service and Honor of God; and no wonder, for if that Law permits the killing and slaying of all Men, we need not marvel if it permits also the spoiling of them.[119]

Burning and ravaging were permitted, Dawson explains, in three cases:

> In Case of such Necessity, as we may justly suppose to be excepted at the first Introduction of Dominion or Property, as if I should break (or throw away where it can never be found again) the Sword of Tirius, when I see Caius coming in a mad Fit to kill me with it. 2dly, When I take or waste so much, as will make satisfaction for my just due, unjustly from me detained. Or 3dly, When I take or waste so much, as may be a just punishment for some wicked and flagitious Act against me committed.[120]

To summarize Dawson's criteria, it is only when one faces an imminent threat, where one sees an enemy approaching in a "mad fit" that justifies a preemptive or preventive laying waste. This is justified under the principle of self-preservation, which is "the primary Law of Nature."[121] The second two cases are examples of retributive justice, where no more is to be destroyed "than what is fit and proportionable to the Debt and Injury."[122] For instance, if one had been denied goods or payment, one would be justified in taking or destroying up to the amount not paid; likewise, one would also be justified if the enemy had violated the natural law – say, murdered the innocent – they would be subject to severe punishment in due portion to their crimes.

While Dawson's criteria may seem overly permissive, he also provides four qualifications to the right of lying waste. The first is that one should only ravage and burn if it is certain the punishment will work. For instance, he explains that one should not punitively ravage towns with easy access to the sea, because they will be immune to certain hardships stemming from the damage, i.e. they

---

119 George Dawson, *Origo Legum: Or A Treatise of the Origin of Law, and Their Obliging Power* (London: Printed for Richard Churchill at the Rose and Crown in St. Paul's Church-Yard, 1694), 164. The judge and jurist, Sir Robert Wiseman (1609–1684), also argues limits to the right of laying waste: "That the enemies Country, when it offers to yield, be not laid waste, burnt, or destroyed; That when a Town is to be stormed, women, children, aged, Ecclesiastical person, so far as is possible, may be spared" and when an army is moving through an enemy nation "no offence being given, be not injured, but kept from spoil and rapine." Sir Robert Wiseman, *The Law of Laws: or, The Excellency of the Civil Law, Above all other Humane Laws* (London: Printed for R. Royston …, 1686), To the Reader.

120 Dawson, *Origo Legum*, 165.

121 Ibid., 154.

122 Ibid., 165.

would be able to import the goods they need to survive. In other words, one must consider if the punishment would be efficacious; gratuitous ravaging is unjustified. Secondly, if too many innocents would be harmed the punishment might be inequitable. Thirdly, one need also to consider the cultural value of the buildings being destroyed. The "famous Structures of some cities," like churches, temples, tombs, and sculptures should be preserved if possible.[123] Dawson explains that to destroy such items is to wage war on "the whole race of mankind, and the men of all Ages, past, present, and to come."[124] It is also a sign of an "irreligious spirit," as though one is waging war "against the Saints of God in Effigie."[125] And finally, hanging over all decisions is the ultimate question "*Cui Bono*? To what end, or for what good should these be spoiled? A wise General would be ashamed surely of such vain havock or waste."[126] As Grotius indicates above, while technically justified in pursuing a punishment, one must weigh the many competing harms that may stem from strict justice. Sometimes charity is to be preferred to strict right. The first and fourth considerations seek to assess whether the punishment is effective, and, more broadly if it will produce an overall good. The second and third considerations seek to determine whether the punishment is discriminating and proportional. In other words, one must assess whether the punishment will be properly directed, reaching mostly those who are guilty. Sufficiently severe crimes, however, would warrant the loss of innocent life and the destruction of places of cultural and religious significance.

Not only should one seek a proportional and directed punishment, i.e. making sure that "great Punishments are not to be inflicted for small Offences,"[127] one must have an eye to the end of hostilities, not only to exacting precise justice but to ensure peace. What governs the rather extreme course of action – ravaging and lying waste – is the prudence and wisdom of an authorized leader. Reflecting the position commonly expressed by the martialists of the previous section, Dawson explains, "a due temper is to be used by all Martial Men," and a little later he continues, "A just and due temperament therefore is here also to be used by Commanders and Generals."[128] While the laws of war permit rather rigid punishments, Dawson concludes by saying "the rule of Charity is more extensive than that of strict right: Hundreds may be in a

123 Ibid.
124 Ibid.
125 Ibid.
126 Ibid.
127 Ibid.
128 Ibid., 165, 166.

City, who are altogether innocent as to the Causes of War; and for these to be deprived of what they have, for the fault of others is an hard case."[129]

Dawson's formulation does not give specific expression to the question of intention; much greater weight is placed on the duties engendered through necessity. If a general ultimately judges that a violation of the natural law warrants the ravaging and destruction of towns and cities, the subsequent death of the innocent is permitted. In Dawson's formulation, however, those victims do not lose their status as innocent. And in this respect, when Dawson speaks of "a great and manifest Necessity," the object is always directed at that which is required to punish the guilty.[130]

9 Christian Wolff

The German Enlightenment figure, Christian Wolff (1679–1754), widely believed to be the most significant intellectual figure between Leibnitz and Kant, wrote *The Law of Nations* (1749) as an attempt to establish "international law as an instrument of public order and ultimately moral progress."[131] Though, unlike some of the humanists indicated above, for Instance, Gentili and Grotius, who attempted to eliminate the formal classification of just and unjust belligerent, Wolff reasserts this older just war distinction. Nations that engage in unjust wars are little better than robbers. As such, *everyone* killed by an unjust aggressor is, technically speaking, innocent due to the fact that when the "unjust belligerent kills, he kills without right."[132] Just as was the case with Pufendorf, when a belligerent possess the right to rectify a transgression of the natural law, he is free to do what is necessary to those ends.[133] However,

129 Ibid., 166.

130 Ibid., 165.

131 Pablo Kalmanovitz, "Sovereignty, Pluralism, and Regular War: Wolff and Vattel's Enlightenment Critique of Just War," *Political Theory* 46:2 (2017); 218–241. 220; see also, Nicholas Onuf, "*Civitas Maxima*: Wolff, Vattel and the Fate of Republicanism," *American Journal of International Law* 88:2 (1994); 280–303.

132 Christian Wolff, *The Laws of Nations Treated According to the Scientific Method* [1749], trans. Joseph Drake and Thomas Ahnert (Indianapolis: Liberty Fund, 2017), §778, 578.

133 This idea can be found in John Locke's formulation of just war as well. He explains that the innocent party has the "right … to destroy the other whenever he can, until the aggressor offers peace, and desires reconciliation." Though it is less clear that the belligerent's right overrides the right of the innocent in Locke's thought. He clearly resisted the strict appeal to right and the law when it caused harm, e.g. it would be wrong "to pull down an innocent man's house to stop the fire, when the one next to it is burning." Locke, *Second Treatise*, §§20, 159.

like many in this tradition, Wolff is wary of rashness or the kind of enflamed passions that would lead men to prefer settling the controversy with blood instead of seeking out "meetings and conferences."[134] After all, "Since nothing is more dreadful for the human race than war, we must not think that in bestowing the right of war nature was prodigal. Only extreme necessity makes a place for this violent remedy."[135]

In this respect, Wolff, like Pufendorf, believes that the nonfighting population could be punished if they were killed out of necessity and "with right." For instance, he argues that "the destruction of the property of an enemy is legal, if without it we cannot attain our right."[136] Along these same lines, he continues, "since destruction is legal as a just punishment, the destruction of cities, towns, and villages is legal, if some special offence has before occurred which deserves such a punishment."[137] Wolff clarifies that a "special offence" may be a wrong committed by "those who inhabit the cities, towns and villages," or it could be a violation by the sovereign power himself.[138] Crimes of the sovereign implicate the citizenry, since, "all the property of subjects alike with public property is looked at as property of the nation, and the act of the sovereign power is imputed to the entire nation."[139] Much like the early Scholastics, a nation could bear a kind of collective guilt for the crimes committed on behalf of the nation. Of course, in the same way Grotius and Pufendorf soften the rigors of justice, it is not always prudent for a belligerent to lay claim to strict right. Wolff explains that "the justice of the punishment is to be determined from its purpose," and, he continues, "to decide of course how much is to be inflicted for the terrorizing of the enemy, that provision may be made for security in the future."[140] And, as was the case with Pufendorf, sometimes particularly recalcitrant enemies can be subjected to *greater* levels of harm than they initially inflicted.[141]

134 Wolff, *The Laws of Nations*, §571, 418.
135 Ibid., §630, 463.
136 Ibid., §824, 615.
137 Ibid., §825, 616.
138 Ibid.
139 Ibid.
140 Ibid. There are limits to Wolff's use of terror. One could not, for instance, kill prisoners of war or captives. One can redress wrongs and seek to end war quickly, but one must be cautious not to turn cruelty into "mere licence." To balance these competing considerations, Wolff argues that one should look to "the advantage which is hoped for in the future from terror." Ibid., §802, 598.
141 Ibid., §807, 601.

As has been seen throughout the humanist discourse, in Wolff's approach to innocent death one must carefully weigh the appropriate amount of terror that can justifiably be inflicted on an enemy. In this process, one must consider not only the justness of one's purpose but also the degree to which the cruelty will enhance rather than diminish one's security in the future. This formulation, of course, obviates the appeal to intentions. Wolff is attuned, however, to the kind of sadism and self-deception that could potentially arise when seeking to retaliate for a wrong. He anticipates some of the dangers human psychology may fall victim to in the recourse to violence. First, there is an ever-present temptation to "represent to ourselves as a benefit that which we desire."[142] Princes may be too easily persuaded that violence is justified out of their desire, and as a result, the deceived belligerent will be "filled with a mistaken notion of good."[143] Furthermore, once convinced he is pursuing a just end, the sovereign may begin to derive pleasure from the unjust use of violence, and "consequently the slaughter and mangling of human beings and the destruction of property belonging to innocent people delights them."[144] One must always be wary of these two subversive impulses: "To injure another for the sake of one's own dventtage" and the desire "to injure many innocent people without deriving any 168dventtage from it, because injury of itself gives pleasure, undoubtedly is so atrocious as to combat any idea of humanity."[145] Confusing a just end with "mere advantage" and taking pleasure in retributive violence, even when there is no advantage in doing so, are both indicative of a deeply flawed moral judgment.

Wolff is eager to prevent strict right from devolving into mere license. He is abundantly clear that a flawed and overly permissive conception of right would lead to "promiscuous slaughter."[146] He is especially wary of appeals to utility. "Utility alone," Wolff writes, "is not a just cause of war"; instead, a just war must have "persuasive reasons" beyond mere advantage and utility.[147] When one does not have a legitimate claim to right, *all* killing is unjustified, and as seen above, even the enemy army counts as innocent. When one is engaging in a just war, however, certain instrumental cruelties are permitted, and those not directly engaged in fighting do not need to be spared in every

142 Ibid., §626, 458.
143 Ibid.
144 Ibid.
145 Ibid., §626, 459.
146 Ibid., §792, 591.
147 Ibid., §645, 477.

case. This formulation can particularly be seen in Wolff's account of siege warfare. While he rejects the use of poisons, bombing residential districts is permitted. He writes, "For a besieged city is resisting with hostile force, and by the throwing of the bombs, by which the destruction of buildings particularly is aimed at, we use as it were a means of compelling the enemy to surrender the city."[148] As the language here clearly suggests, Wolff believes it is acceptable to bomb the city itself to compel the enemy to surrender. The same can be said of raiding and pillaging defeated cities. He writes, "In pillaging of cities and houses excesses are avoided with difficulty, because soldiers claim an unbridled license as a right. Therefore, it should hardly be resorted to unless the greatest necessity should demand it."[149] Again, one sees here the delicate balance between "unbridled license" and the demands of necessity in the service of a just end.[150] Furthermore, certain activities, like bombing a city and pillaging, will produce reliably consistent outcomes. As such, they should only be pursued in situations of great necessity and under the sure auspices of right.

10 Emer de Vattel

Principally influenced by Grotius and Wolff, Emer de Vattel (1714–1767) was one of the most influential theorists in the field of international law in the eighteenth century.[151] Like Wolff, Vattel claims that, in principle, only one side can legitimately lay claim to justice and right;[152] although, he admits that there

148 Ibid., §879, 649. It is worth comparing the timing of this justification with the martialist writers of the previous chapter. By the mid eighteenth century, using incendiary bombs to weaken the resolve of a besieged community had become widespread. It is not surprising that both traditions were grappling with this issue and veering toward various conceptions of necessity.

149 Ibid. §846, 632.

150 Wolff's contemporary, Francis Hutcheson (1694–1746), reproduces the limiting framework of necessity for killing the innocent in war. He claims that, in general, the laws of war are humane, and that "departing from them may increase mischiefs further than is necessary for the purpose of war, by destroying women or children, or persons wounded, who however could have been of no service against us." However, there is great difficulty in enforcing these norms since "No person is punished, or counted infamous in his own country, for murders of cold blood, rapes, butchery of women and children, or any cruelties committed against an enemy during war." Francis Hutcheson, *A System of Moral Philosophy, Volume 2* (London: Sold by A. Miller …, 1755), Bk.2, Chap.10, Sect. 5, 353.

151 See Chares Covell, "Wolff and Vattel," in *The Law of Nations in Political Thought: A Critical Survey from Vitoria to Hegel* (London: Palgrave Macmillan, 2009).

152 Emer de Vattel, *The Law of Nations* [1758], eds. Béla Kapossy and Richard Whatmore (Indianapolis, IN: Liberty Fund, 2008), Bk. III, Chap. III, §39, 489.

may be some epistemic uncertainty in the moment as to which side actually possesses right in any given case. When the question of right is ambiguous or uncertain, both sides must be provisionally seen as lawful, that is, until there is greater clarity about the causes and reasons for resorting to war.[153] This, of course, complicates the much more black and white distinctions at work in Wolff's framework which contends that all those fighting in the service of right are innocent.

For Vattel, the recourse to right is shrouded in much greater epistemic uncertainty. This is in part due to the fact that "men are under a necessity of regulating their conduct in most cases by probabilities, those probabilities claim their attention in proportion to the importance of the subject: and … their right to obviate a danger is in a compound ratio of the degree of probability, and the greatness of the evil threatened."[154] The perception of threat and what counts as a violation of state sovereignty or security are not always well defined, especially if potential threat is cast in more probabilistic terms. This implies that a nation's right must be seen as subordinate to a prudential calculus which considers both the likelihood and intensity of the perceived threat or violation, and the degree to which retaliation is both necessary and likely to produce a good end. In this respect, Vattel's framework leads to an international order governed by perceptions. States must form some reasonable assessment of what equilibrium entails as well as what kind of behaviors would threaten international balance, and hence the security of the states within the system.[155] Here one begins to see the early formulations of what twentieth-century International Relations scholars identify as the security dilemma, where one state's defensive posture might be interpreted as offensive action.[156]

Despite these persistent ambiguities, Vattel does affirm that perceptions of right can, in fact, be correct, and as such inform legitimate state action. In other words, like Wolff, there can be objectively valid action in the realm of international politics. Though, again like Wolff, Vattel argues that even when one is in the right there is a duty to be restrained and circumspect in the decision to punish wrongdoers, through war or some other type of punitive

153 Ibid., Bk. III, Chap. III, §40, 489.

154 Ibid., Bk. III, Chap. III, §44, 493.

155 Nearly half a century later, this would be further articulated by Friedrich von Gentz, the German diplomat who would provide some of the first analysis of the impact the French Revolution had on the balance of power in Europe. Friedrich von Gentz, *Fragments Upon the Balance of Power* (London: C&R Baldwin Printers, 1806).

156 Vattel, *The Law of Nations*, Bk. III, Chap. III, §50, 499. See Robert Jarvis, "Cooperation under the Security Dilemma," *World Politics* 30:2 (1978); 167–214.

intervention. In other words, the mere possession of right is not meant to make the recourse to war preferable or easy. He writes:

> Whoever entertains a true idea of war, – whoever considers its terrible effects, its destructive and unhappy consequences, – will readily agree that it should never be undertaken without the most cogent reasons. Humanity revolts against a sovereign, who, without necessity or without very powerful reasons, lavishes the blood of his most faithful subjects, and exposes his people to the calamities of war, when he has it in his power to maintain them in the enjoyment of an honourable and salutary peace.[157]

He is quite clear that in most cases, the risks of going to war far outweigh the benefits: "It is only in extremities that a just and wise nation or a good prince has recourse to it."[158]

Even with some of the ambiguity inherent to states in international society, Vattel does enumerate certain conditions when a nation would clearly have the right to go to war. He explains,

> The right of employing force, or making war, belongs to nations no farther than is necessary for their own defence and for the maintenance of their rights. Now if anyone attacks a nation, or violates her perfect rights, he does her an injury. Then, and not till then, that nation has a right to repel the aggressor, and reduce him to reason. Further, she has a right to prevent the intended injury, when she sees herself threatened with it.[159]

One might also retaliate if one has received direct injury.[160] These represent what he calls a nation's "perfect rights."[161] Preventive, punitive, and defensive wars are all permissible within this framework. However, even in these more clear-cut cases, Vattel argues that it is not always advisable to lay claim to one's perfect right. One must consider each situation in its distinctiveness and determine that the use of force is both expedient and prudent.

157 Vattel, *The Law of Nations*, Bk. III, Chap. III, §24, 482.

158 Ibid., Bk. III, Chap. III, §25, 483.

159 Ibid, Bk. III, Chap. III, §26, 483.

160 Ibid. For instance, "1. To recover what belongs or is due to us. 2. To provide for our future safety by punishing the aggressor or offender. 3. To defend ourselves, or to protect ourselves from injury, by repelling unjust violence." Ibid., Bk. III, Chap. III, §28, 484.

161 Ibid., Bk. III, Chap. III, §26, 484.

Furthermore, one must also be cautious and self-critical enough to not pursue war for vicious ends – i.e. as a mere pretext for increased power and wealth.[162] Such ends are ultimately short-sided and will make a state more vulnerable in the long-run. Much as Wolff had indicated, "Nations that are always ready to take up arms on any prospect of advantage, are lawless robbers: but those who seem to delight in the ravages of war, who spread it on all sides, without reasons or pretexts, and even without any other motive than their own ferocity, are monsters, unworthy the name of men."[163] And elsewhere he writes, "The prince who violates all laws, – who no longer observes any measures, – and who would in his transports of fury take away the life of an innocent person, – divests himself of his character, and is no longer to be considered in any other light than that of an unjust and outrageous enemy."[164] Those states who honor no limits become "enemies to the human race."[165]

The challenge to Vattel's conception of right in war, however, is that there is a much smaller framework to conceptualize how the innocent might properly be treated. One might recall that in Wolff's framework, all those who fought on the side of right were innocent, and all those who fought on the side of the unjust were guilty and could be liable to necessary evils to bring about a just end to the war. Indeed, even members of the unjust enemy nation could be killed. Vattel agrees with Wolff that when a nation engages in an unequivocally unjust war "he is loaded with the guilt of those which he inflicts on an innocent nation."[166] However, since in Vattel's framework both sides may perceive themselves to be in the right, both may make similar appeals to necessity to justify the recourse to certain cruelties. He quite specifically states that "when the end is lawful, he who has a right to pursue that end, has, of course, a right to employ all the means which are necessary for its attainment."[167] This is complicated by the fact that, for Vattel, "it is understood that the whole nation declares war against another nation … Hence, these two nations are enemies, and all the subjects of the one are enemies to all the subjects of the other."[168] If both sides perceive themselves to be operating under the auspices of right, it is unclear where the limits of cruelty actually lie.

Vattel does soften his analysis by clarifying that while women and children are to be counted as enemies, they should not be treated like "men who bear

162 Ibid., Bk. III, Chap. III, §30, 485.
163 Ibid., Bk. III, Chap. III, §34, 487.
164 Ibid., Bk. I, Chap. IV, §54, 110.
165 Ibid., Bk. III, Chap. III, §34, 487.
166 Ibid., Bk. III, Chap. III, §24, 482.
167 Ibid., Bk. III, Chap. VIII, §136, 541–42.
168 Ibid., Bk. III, Chap. V, §70, 509.

arms."[169] Under normal conditions, killing women and children are typically not necessary to attain the ends of a just war.[170] Therefore, whenever a state goes beyond the limits of necessity it "is reprobated by the law of nature."[171] Furthermore, even when certain cruelties are permitted by a nation's appeal to right, and the exigencies of the case at hand allow for extreme measures, Vattel explains that "it becomes a generous prince to refrain from an exertion of his rights at the expense of an innocent individual."[172] What is more, he also contends that there are prudential reasons to not be inordinately violent to the unarmed inhabitants living in enemy lands, especially since an invading army will need to live off of contributions from the locals.[173] In the end, all of these variables must be factored into a prudential calculus that elides rigid quantification – the degree to which punitive military action accords with justice and right, the extent to which specific acts of violence are necessary to those ends, and the duty of a sovereign to at least consider being charitable and generous to the innocent in each case. As Vattel indicates, "it is very difficult, always to form a precise judgment of what the present case requires, and as, moreover, it belongs to each nation to judge of what her own particular situation authorises her to do."[174]

Military leadership must make prudential judgments based on probabilistic considerations as to the most expeditious way to bring an enemy nation to terms. Whatever is necessary to obtain just ends within reason is permitted,

169 Ibid., Bk. III, Chap. V, §72, 510; see also Bk III, Chap. VIII, §145, 549.

170 One must be cautious when assessing these protections. Marcus Schulzke overstates the degree to which Vattel promoted a robust conception of noncombatant immunity, i.e. the shift away from collective guilt to a guilt tied to one's participation in the war. While women and children are not guilty in the same way as soldiers are, their membership in a guilty nation opens them to certain harms. Their distance from war does insulate them from direct harms in most cases; however, they do not possess a right to immunity in any robust sense of term. Whatever immunity they possess is intimately linked to the conception of necessity employed by the invading army, i.e. those actions viewed as necessary to achieve victory (including ravaging and burning) are permitted. See Marcus Schulzke, "The Evolution of Civilian Immunity and the Right to Life," in *Just War Theory and Civilian Casualties: Protecting the Victims of War* (New York: Cambridge University Press, 2017), 29. Mark Grimsley points out that while Vattel did seek to minimize harm to noncombatants, his notion of military necessity created a significant breach in the moral edifice of war. Mark Crimsley, *The Hard Hand of War: Union Military Policy Toward Southern Civilians, 1861–1865* (New York: Cambridge University Press, 1995), 214. For more discussion on this issue, see James Turner Johnson, "Secularized Just War Doctrine: Grotius, Locke, and Vattel," *Ideology, Reason, and the Limitation of War: Religious and Secular Concepts, 1200–1740* (Princeton: Princeton University Press, 1975).

171 Vattel, *The Laws of Nations*, Bk III, Chap. VIII, §137, 542.

172 Ibid., Bk. II, Chap. XVI, §260, 406.

173 Ibid., Bk. III, Chap. VIII, §147, 551.

174 Ibid., Bk. III, Chap. VIII, §137, 542.

including the killing of women and children. Under normal conditions, however, those not fighting should not be treated the same way as soldiers, but since they are, in fact, part of an enemy nation, they can be subjected to punitive violence, again, under certain conditions. While it is impossible to set out all such conditions in advance, Vattel does provide certain examples as to when the killing of the innocent might be required. Not surprisingly, given the period in which he is writing when the incendiary bomb had become commonplace, the assault of fortified cities proves to be the most illustrative case. Vattel explains that the sack of towns might be permitted as a punishment of "savage" nations who "observe no rules" of war.[175] He writes, "On certain occasions ... a country is totally ravaged, towns and villages are sacked, and delivered up a prey to fire and sword. Dreadful extremities, even when we are forced into them! Savage and monstrous excesses, when committed without necessity!"[176] When a besieged place holds out longer than prudence requires – when such resistance is considered reckless or fruitless – those who surrender might be put to death.[177] Poison and assassination are prohibited,[178] but, like Wolff, he affirms the practice of bombing residential districts to expedite surrender: "To destroy a town with bombs and red-hot balls, is an extremity to which we do not proceed without cogent reasons. But it is nevertheless warranted by the laws of war, when we are unable by any other mode to reduce an important post, on which the success of the war may depend, or which enables the enemy to annoy us in a dangerous manner."[179]

For Vattel, a large part of the assessment of a nation's conduct in war has to do with the power differential between the enemy nations. For instance, he explains, "the pillaging of a country, or ravaging it with fire, is not, in a

175 Ibid., Bk. III, Chap. VIII, §141, 544.

176 Ibid., Bk. III, Chap. IX, §166, 570.

177 Ibid., Bk. III, Chap. VIII, §143, 547.

178 Ibid., Bk. III, Chap. VIII, §155, 557.

179 Ibid., Bk. III, Chap. IX, §169, 572. Several near contemporaries of Vattel's make similar arguments. Cornelius Bynkershoek (1673–1743) noted how odd it was that the French Ambassador would send condolences to his English counterpart in 1666 when the great conflagration destroyed London, particularly since the two countries were at war. He writes, "I think it disgusting to trifle with mere words, for what else is it than trifling when you express sorrow for the conflagration of a city to which you would wish to set fire yourself." Cornelius Bynkershoek, *A Treatise on the Law of War* (Philadelphia: Fry and Kammerer, 1810), 19. G.F. von Martens (1756–1821) argues: "All the means necessary to the reduction of a fortress are justified by the laws of war; consequently, there are cases which may authorize the demolition or burning of the suburbs." G.F. von Martens, *The Laws of Nations Being the Science of Natural Law, Covenants, Power, Founded Upon the Treaties and Customs of Modern Nations in Europe* trans. William Cobbett (London: Fleet St, 1829), 302. Again, it is worth noting that these developments had a great deal of influence on the eighteenth-century martialist writers mentioned in the previous chapter.

general view of the matter, a violation of the laws of war"; it could be seen as a violation if, however, this approach was taken by an "enemy of much superior strength."[180] A weaker nation may be pushed to these extremities in the way a more powerful nation would not. Again, for Vattel, though somewhat fuzzy in practice, the necessity framework he was advancing was designed to be a restraint on conduct, not mere license. He writes, "All damage done to the enemy unnecessarily, every act of hostility which does not tend to procure victory and bring the war to a conclusion, is a licentiousness condemned by the law of nature."[181] Of course, since necessity militates against any precise standard of objectivity, it would behoove states to agree upon rules of conduct ahead of time. A "voluntary law of nations" would go a long way in terms of helping determine which actions are "unlawful and odious."[182] Given the subjective and probabilistic means by which right is ascertained, it would be better if nations agreed in advance as to what the laws of war should be.

11 Conclusion

For commentators in the humanist or republican tradition, intentions were rarely thematized, and when they were, the principles of right and necessity carried greater moral heft. In most cases, authors in this tradition sought to bring the killing of the innocent more explicitly into a necessity framework.[183] Of course, necessity was typically framed within certain limits – a complex balance between one's right, what is feasible, and what is charitable. Additionally,

180 Vattel, *The Law of Nations*, Bk. III, Chap. IX, §173, 574.

181 Ibid., Bk. III, Chap. IX, §171, 573.

182 Ibid., Bk. III, Chap. IX, §173, 574.

183 *Pace* Cécil Fabre and Seth Lazar, at *no point* did Grotius, Pufendorf, Wolff, and Vattel argue that "combatants on either side of a conflict are under an equally strict duty not to intentionally kill noncombatants." Cecil Fabre and Seth Lazar, "Introduction," in *The Morality of Defensive War*, eds. Cecil Fabre and Seth Lazar (New York: Oxford University Press, 2014), 1. Lazar also attributes this view to Grotius and Wolff in, Seth Lazar, "Necessity and Non-combatant Immunity," *Review of International Studies* 40:1 (2013); 53–76., 53n2. As has been seen throughout this chapter, these theorists routinely elided the moral exculpatory force of good intentions, instead favoring prudential wisdom and the calibrated use of necessary violence to pursue just ends – i.e. in certain difficult cases one could deliberately kill the innocent. For more on these early lesser evil arguments, see Pablo Kalmanovitz, "Early Modern Sources of the Regular War Tradition," *The Oxford Handbook of Ethics of War*, eds. Seth Lazar and Helen Frowe (New York: Oxford University Press, 2018).

it is also significant the degree to which these moral arguments about the death of innocents were adapted and reconfigured to address the emerging trends in warfare. In fact, the centuries-long progression that gave rise to individual self-defense through intentions and double effect, and then the application of these moral tools to the reality of siege warfare, and so on, is a clear indication of the way ethical language forms a kind of scaffolding for subsequent judgments. The moral frameworks developed in this period provide a kind of toolkit for successive generations to draw from and build with. And as will be seen in the following chapters, none of these ideas go away. In subsequent periods and for different reasons, these arguments are adapted and repurposed to address historically contingent needs. In the following chapters, one will see how the arguments of necessity put forward by these humanist writers will be joined with subsequent utilitarian formulations to form the basis of nineteenth century military ethical norms.

PART 2

The Nineteenth Century

CHAPTER 4

Killing Noncombatants in the Nineteenth Century

1 Introduction

Archibald Forbes, a Scottish war correspondent for the *Morning Advertiser*, was embedded within the Prussian First Army Corps as it marched on Metz in the early weeks of the Franco-Prussian War (1870). Though Metz itself was not bombarded – the fortress was starved into submission – the villages and towns that lie en route had been utterly decimated. In his reporting, Forbes provides a moving account of the profound devastation suffered by the residents of Lauvallière, one such small town that lay in the pathway of the advancing army. As he entered the town, he made note of just how thorough and absolute the destruction had been: "Gaunt and grim stood its houses, burnt to a shell, the graves lying all around, the holes in the fields where the shells had burst."[1] But what caught his attention was a particular scene:

> I came on a little group, the extreme pathos of which made my heart swell. It was a family, and they sat in front of what had once been their home. That home was now roofless. The stones of the walls were all that was left. The garden was a wreck, and the whole scene was concentrated desolation. The husband leant against the wall, his arms folded, and his head on his chest. The wife sat on the wet ground, weeping over the babe at her breast. Two elder children stared around them with wonder and unconcern – too young to realize the misfortune which had overtaken all the family. The silence of the grief-stricken group was more eloquent than any words. No home, no food, a wagon, and a couple of starved horses, a field with four graves in it. This is the brief summary of the belongings and condition of the family. All were alike; although this family was the one I happened to see … some similar piteous sight must soon have been visible by every house in every village all around Metz.[2]

1 Archibald Forbes, *My Experiences of the War between France and Germany, Volume 1* (London: Hurst and Blackett, 1871), 413.

2 Ibid., 414.

 | DOI:10.1163/9789004515482_006

We do not know the fate of this family, but the implication of Forbes's assessment seems to be that many such families would have a very difficult time surviving as their livelihood and sustenance had been completely decimated. Indeed, the town's entire infrastructure was now non-existent. As Forbes indicates, the "concentrated desolation" this poor family faced was indicative of the wider destruction experienced by virtually all the villages and towns on the way to Metz.

The Prussian officers who memorialized the Metz campaign confirm, in a general way, Forbes's account, albeit with much less pathos:

> But what a country there was round Metz! All around were villages and farms burnt down, fields devastated, gardens trampled underfoot, and to crown it all, far and wide, fields of corpses. That was the view offered to the eye. Why, no less than 400,000 warriors had there on this narrow space contended with each other for six days, leaving death and destruction behind.[3]

One of the bitter realities of nineteenth-century war was that villages and towns that stood in the pathway of approaching armies were subject to extreme hardships. 'Death and destruction' routinely followed soldiers on the move.

One does, in fact, get a brief glimpse as to how the domestic population managed to survive – if only just barely – the extreme desolation brought about by this campaign. John Bellows, a Quaker philanthropist who organized aid for displaced populations around Metz, kept notes in a pocketbook about the conditions of the villagers he brought relief to. One note reads: "50 already dead of starvation," about which he later reflected: "This I afterwards found incorrect. They die of privation and trouble mingled – not of sheer starvation. I have found many cases of persons who have died from fright, and from grief at seeing their families reduced to such crushing misery!"[4] In other words, it wasn't merely the hardship and privation, but the utter demoralization of the desolation that was just as destructive. For many, hopelessness and grief itself had simply become too much to bear. While Bellows does not provide concrete or systematic numbers, one can infer that the residual noncombatant casualties from the Metz campaign were non-trivial.

---

3 Generals and Other Officers, *Franco-German War, 1870–71* trans. J.F. Maurice (London: Swan Sonnenschein and Co, 1900), 183.

4 John Bellows, *The Track of the War around Metz and the Fund for the Non-combatant Sufferers* (London: Trübner and Co …, 1871), 24.

Another commentator on the war, Sir Henry Hozier, notes that despite the seeming excess of force insinuated in Forbes's and Bellow's descriptions above, many observers were generally favorable of German conduct at the beginning of the war:

> the conduct of the German troops was excellent; they were not more remarkable for courage and discipline than for honourable treatment of the invaded country. The picture as subsequently darkened; complaints were made that the German leaders acted like Tilly and Wallenstein; and dreadful stories were told of murdered free-shooters, of villages burnt by way of reprisals, of barbarous executions, of innocent citizens made hostages, of devastations carried out on system, as if by the savage hordes of Atilla.[5]

It is worth considering that the destruction of the towns around Metz was undertaken in the so-called 'honorable' phase of the war, when the German soldiers were more constrained by military discipline. After all, nothing in the laws of war prohibited the destruction of towns that were defended or that stood in the way of a legitimate military target.

As Hozier's quote suggests, things got much worse as the war progressed. Of course, from Forbes' description it is perhaps hard to imagine that things could get any worse for the non-fighting population, but as the desperation of the French civilian resistance grew there were increasing encounters between irregular combatants, the *francs-tireurs*, and the Prussian army. As Field Marshal von Moltke explains, eventually the *francs-tireurs*' "gruesome work had to be answered by bloody coercion. Because of this, our conduct of the war finally assumed a harshness that we deplored, but which we could not avoid. The *francs-tireurs* were the terror of all the villages; they brought on their own destruction."[6] Reprisals against individual residents and sometimes whole villages became common.[7] In the end, the orderly, premeditated

5 Sir Henry Montague Hozier, *The Franco-Prussian War: Its Causes, Incidents, and Consequences, Volume 2* (London: William Mackenzie, 22 Paternoster Row, 1872), 282–283.

6 Helmuth von Moltke, *Moltke on the Art of War: Selected Writings* ed. Daniel Hughes (London: Greenhill Books, 1993), 32.

7 As Bastian Scianna argues, violence against civilian resistance was widely practiced and not specifically a Prussian problem. In this sense, there is something problematic to interpreting Prussian behavior during the war as antecedent to the crimes that would be perpetrated in World War II. See Bastian Matteo Scianna, "A Predisposition to Brutality? German Practices against Civilians and *Francs-tireurs* during the Franco-Prussian War 1870–1871 and their Relevance for the German 'Military *Sonderweg*' Debate," *Small Wars and Insurgencies* 30:4–5 (2019); 968–993.

devastation around Metz eventually gave way to a downward spiral of reprisals and murder. One might say that it was actually this orderly and systematic destruction of people, animals, and buildings produced by armies battling in confined spaces that eventually gave way to these targeted reprisals. As conditions grew increasingly desperate, the real (but often-times imagined) threat of *francs-tireurs* grew in intensity.

2 Counting Casualties

Outside of anecdotal accounts like the ones provided by Forbes, Bellows, and the Prussian officers indicated above, we know very little about civilian death in the nineteenth century, and even less from earlier periods. Explaining the record-keeping during the American Civil War, Drew Faust writes, "in an era when military record keeping was itself flawed and incomplete, no one thought to account for civilians. Their losses remain the stuff of anecdote and even legend – largely unacknowledged casualties of a war even more devastating than its official statistics imply."[8] Indeed, accounts of noncombatant death in the nineteenth century war are spotty and unreliable.[9]

Even Peace Societies, which were perhaps the most attuned to the evils inflicted upon residential populations, could not offer reasonable estimates in their literature on the costs of war in the nineteenth century. For instance, in 1869, the London Peace Society translated and reprinted Paul Leroy-Beaulieu's accounting of the casualties and financial losses brought about by the major European and American wars between 1853 and 1866. He admits that almost nothing concrete can be said of the costs borne by the non-fighting populations. He writes,

> And at length, when we have made all these calculations, shall we then have accomplished our task? By no means! All the private losses, the ravage of the lands, the spoiling of crops; incase of siege or maritime war, the ruin of cities and the destruction of shipping; all these losses, impossible

8 Drew Faust, *This Republic of Suffering: Death and the American Civil War* (New York: Alfred Knopf, 2008), 138.

9 For instance, James McPherson estimates that 50,000 civilians died in the Civil War, but this number seems to be a guess. James McPherson, *Battle Cry of Freedom: The Civil War Era* (New York: Oxford University Press, 1988), 619, n53. To begin with, it entirely omits the plight of former slaves.

> to be estimated, must be always kept in view, although they cannot be calculated.[10]

As this passage suggests, there was a sense that the indirect suffering of war was significant but, in some sense, unquantifiable. Even though civilian casualties are entailed in the "private losses" that Leroy-Beaulieu enumerates, it is worth noting that that they are not specifically identified as a countable cost of war.[11]

While counting casualties may seem obvious to the late-twentieth or twenty-first century observer, there were good reasons why nineteenth century governments and militaries would not have considered counting noncombatants killed in war a viable or even necessary task. To begin with, even in the twenty-first century, there remains a great deal of epistemic uncertainty as to who ought to be counted as a noncombatant, which says nothing of the best methodologies to count them.[12] Not only does any such classification rest on normative assumptions as to who counts as a civilian, but there are also spatio-temporal considerations as well. In the end, to be counted as a civilian casualty it matters not only *where* one dies, i.e. one's proximity to a lethal military action, but also *when* one dies. If a person dies years after the end of a war due to complications stemming from a war-related injury, this premature mortality

---

10 Paul Leroy-Beaulieu, *Contemporary Wars: 1853–1866; Statistical Researches Respecting the Loss of Men and Money Involved in Them* (London: London Peace Society, 1869), 5. There were many books which sought to enumerate the costs of war during this period. See John A. Bolles, *Essay on a Congress of Nations for the Adjustment of International Disputes* (Boston: Whipple and Damrell, 1839); Rev. George Cheever, "Number of Slain in Different Battles," in *Religious and Moral Anecdotes* (London: Charles Griffin and Company, 1848); Hollis Read, *Foot-prints of Satan; Or, The Devil in History* (New York: E.B. Treat ..., 1872); *Facts and Illustrations in Reference to War, Peace, and International Arbitration* (London: London Peace Society, 1872).

11 In another nineteenth-century attempt to calculate the costs of the Mexican American War, Abiel Livermore explains, "It is quite as difficult to ascertain accurately the mortality, as it is the cost of the Mexican war. Persons of different views and temperaments will give different estimates. All that we can accomplish in either of these matters, is an approximation to the reality." Abiel Livermore, *The War with Mexico Reviewed* (Boston: The American Peace Society, 1850), 108.

12 For instance, one need only take note of the great range of civilian casualty estimates from the Iraq War. See The Iraq Family Health Survey estimated 151,000 (March 2003–June 2006); Iraq Body Count documented 112,667–123,284 (March 2003–March 2013); the *PLOS Medicine* Survey estimated 500,000 (March 2003–June 2011); the *Lancet* survey calculated 601,027 violent deaths out of 654,965 excess deaths; the "Opinion Research Business" survey estimated 1,033,000 deaths as result of the conflict. Neta Crawford looks at these figures in *Accountability for Killing*, 135–144.

may not be captured in the casualty figures. Then, of course, there remains the potentially much larger category of "indirect death," that is, those deaths that stem from depleted or destroyed infrastructure.[13] All the attendant suffering from disease, famine, and other complications of war were typically more extensive than direct casualties from fighting. In fact, *most* battlefield casualties in war prior to the twentieth century were due to illnesses and infections that simply could not be treated under such difficult conditions.[14] For these reasons, many contemporary scholars forgo civilian casualties in favor of "total battle deaths" or "excess mortality."[15]

Despite some valiant contemporary efforts to give an account of nineteenth century civilian death, these numbers will continue to remain shrouded in epistemic uncertainty, and not only for the normative reasons above.[16] It should also be remembered that the very idea of "the civilian" as we understand it today was still somewhat obscure at that time. Indeed, throughout the nineteenth century there were continuous discussions about the relative merits of standing armies versus ad hoc militias and the degree to which an individual (i.e. a military-aged male) bears a certain liability due to the fact that he could transform into a combatant at any moment.[17] If townspeople could, at any moment, form into a militia, the residents of those towns would seem to lose the character of civilians, especially if those militia members could toggled back and forth between combatants and townspeople, engaging in guerrilla-like tactics. All this militated against the idea of the discrete category of innocent civilians.

The prevalence of town militias helps to explain why there was no conceptual category of "civilian casualties" much less "collateral damage" at that

---

13 For more on indirect death, see Geneva Declaration on Armed Violence and Development, *Global Burden of Armed Violence, 2008* (Geneva: Geneva Declaration Secretariat, 2008).

14 Richard Garfield and Alfred Neugut, "The Human Consequences of War", in *War and Public Health*, eds. Barry S. Levy and Victor W. Sidel (New York: Oxford University Press, 1997), 28.

15 See Bethany Lacina and Nils Petter Gleditsch, "Monitoring Trends in Global Combat: A New Dataset of Battle Deaths," *European Journal of Population* 21 (2005): 145–166; Peter Ekamper, Govert Bijwaard, Frans van Poppel, and L.H. Lumey, "War- and Famine-Related Excess Mortality Among Civilians in the Netherlands, 1944–1945," *Journal of Maps* 16:1 (2020); 124–131.

16 W. Eckhardt, "Civilian Deaths in Wartime," *Bulletin of Peace Proposals* 20:1 (1989); 89–98; Nicholas Jewell, Michael Spagat, and Britta Jewell, "Accounting for Civilian Casualties: From the Past to the Future," *Social Science History* 42 (2018); 379–410.

17 For a discussion of this issue, see William Paley, *The Principles of Moral and Political Philosophy* (Boston: Published by West and Richardson, 1815), 473–76.

time.[18] As Figure 4.1 shows, these words do not appear together with any regularity until World War II. "Collateral damage" emerges even later as part of the sanitizing terminology describing the anticipated harm to civilians and property in a nuclear strike during the Cold War. By the early 2000s, with the US wars in Afghanistan and Iraq, "collateral damage" became something of a synonym to civilian casualties. It is important to note that the great deal of humanitarian effort to protect civilians emerged in conjunction with the terminology that sought to delimit the object under protection. It was only after the "civilian" came into conceptual focus in contradistinction to the "combatant," that a new space for moral deliberation emerged. Since such conceptual categories did not exist in the nineteenth century, it is not surprising that the statistical compilers of nineteenth century war did not expressly look for those casualties. And, of course, the absence of the category of "civilian" was in a large part due to the fact that anyone could transform into a belligerent through the necessities of self-defense.[19]

---

18 For more on the development of the category of the citizen, see Stephen Rockel and Rick Halpern (eds.) *Inventing Collateral Damage: Civilian Casualties, War, and Empire* (Toronto: Between the Lines, 2009). Hugo Slim points out the historical reluctance to affirm the "civilian idea." Hugo Slim, "Why Protect Civilians? Innocence, Immunity, and Enmity in War," *International Affairs* 79:3 (2003); 481–501. Lawrence Freedman argues that "The concept of the civilian as a distinct category from the military can be traced back to the start of the nineteenth century, but it was not in general use until its end." Lawrence Freedman, *The Future of War: A History* (New York: Penguin, 2017), 32. For more on these developments, see also, Richard Shelly Hartigan, *The Forgotten Victim: A History of the Civilian* (Chicago: President Publications, 1982); Amanda Alexander, "The Genesis of the Civilian," *Leiden Journal of International Law* 20:2 (2007); 359–376; Helen Kinsella, *The Image Before the Weapon: A Critical History of the Distinction between Combatant and Civilian* (Ithaca, NY: Cornell University Press, 2011); Claire Garbett, *The Concept of the Civilian: Legal Recognition, Adjudication and the Trials of International Criminal Justice* (New York: Routledge, 2015); Emily Crawford, *Identifying the Enemy: Civilian Participation in Armed Conflict* (New York: Oxford University Press, 2015); and Frederik Rosén, *Collateral Damage: A Candid History of a Peculiar Form of Death* (London: Hurst, 2016). The term "collateral damage" seems to have originated in discussions of nuclear deterrence in the 1960s, well after intention became the guiding moral principle demarcating just and unjust civilian death. See T.C. Schelling, "Dispersal, Deterrence, and Damage," *Operations Research* 9:3 (1961); 363–370. Since the post-9/11 wars in Afghanistan and Iraq, "CIVCAS" has become the way to refer to civilian casualties. While this abbreviation may have been employed internally within the US military for some time, it was popularized around 2009. See Stanley McChrystal, "Tactical Directive," NATO/ISAF UNCLASS (6 July 2009), https://www.nato.int/isaf/docu/official_texts/Tactical_Directive_090706.pdf and Stanley McChrystal, "Commander's Initial Assessment," NATO/ISAF UNCLASS (30 August 2009). https://static.nzz.ch/files/6/5/4/Afghanistan_Assessment_1.3630654.pdf.

19 To see how integral armed militias were to war planning, Clausewitz theorizes their tactical value: "Militia and bands of armed civilians cannot and should not be employed

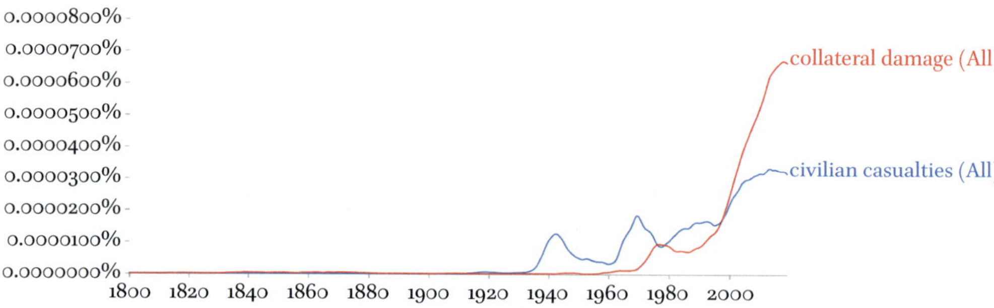

FIGURE 4.1 Ngram search of 'civilian casualties' and 'collateral damage'

That the idea of civilian casualties emerged so late certainly does not mean that those whom we now call civilians were not dying prior to the twentieth century. Scattered throughout the war reporting of the nineteenth century, one catches glimpses of civilian death here and there. Much like with Forbes' pathos above, soldiers, reporters, and noncombatant bystanders to war made notes of the suffering they experienced and witnessed. One can read of the injury and suffering of "women and children," "inhabitants," "residents," "villagers," the slightly more antique "innocents," and the increasingly common "noncombatants." These were all terms used to describe innocent non-fighters who were killed in war.

As the realities of war (especially siege warfare) started to put more innocent people in harm's way, new moral vocabularies began to emerge. Even though arguments for a protective status for certain groups of people find their origins in moral developments that go back to antiquity, it was not until the nineteenth century that something resembling the formal category of the "noncombatant," signifying the local population as a whole, started to emerge. In fact, the term "noncombatant" originally emerged in the eighteenth century to describe the nonfighting units that travelled with armies on the march – medical units, baggage trains, and other personal attendants, like cooks, wives, prostitutes, and servants. As Figure 4.2 shows, at the end of the Napoleonic Wars there was increased mention of "noncombatants," with a steady rise and burst around the US Civil War. There was another gradual rise during The Hague Conventions, and then an even greater uptick around World War I.

against the main enemy force … They are not supposed to pulverize the core but to nibble at the shell and around the edges. They are meant to operate in areas just outside the theatre of war – where the invader will not appear in strength – in order to deny him these areas altogether." Carl von Clausewitz, *On War* ed. Beatrice Heuser (New York: Oxford University Press, 2007), 186.

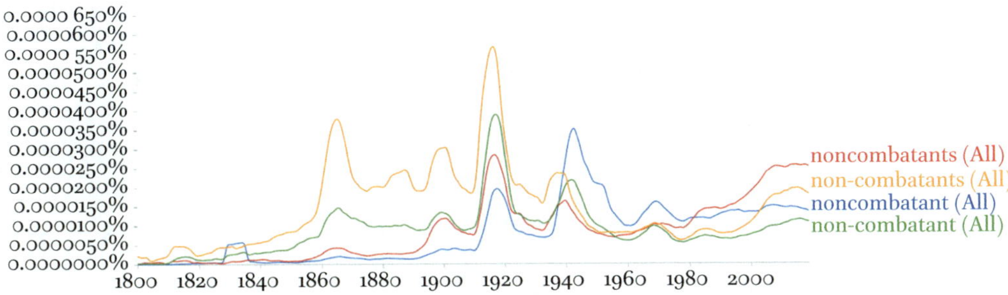

FIGURE 4.2 Ngram search of 'noncombatant(s),' 'non-combatant(s)'

In effect, this shows that the noncombatant was largely a moral and legal construction of the nineteenth century. In fact, much of this terminological development was part of a movement to instantiate an idealized form of warfare that had never existed, i.e. the desire to pit only combatants against opposing combatants. The goal was to transform war into conflict between professional armies. The historical realities of this distinction, however, were much more complicated than the binary combatant-noncombatant implies.

That there was no such conceptual category as "civilian casualties" prior to the twentieth century has contributed to the popular misconception that nineteenth-century war had little impact on nonfighting populations. Looking at the public health effects of war, Richard Garfield and Alfred Neugut dubiously claim that "Prior to the development of the bomber in World War II, countries whose wars were fought on foreign territory were largely spared from civilian casualties. Most wars during the eighteenth and nineteenth centuries are believed to have caused relatively few civilian casualties."[20] Indeed, it is widely assumed that "Post Thirty Years' War – Pre-World War II civilians were generally not targets during warfare."[21] The image of standing armies facing off in isolated, pitched battles has led to the mistaken belief that noncombatants were largely exempt from the most brutal aspects of nineteenth-century war. In the field of Political Science, the rise of Mary Kaldor's New War thesis has also contributed to the wide-spread assumption that disproportionate harm directed at civilian populations has been predominantly a feature of "New War," i.e. war after the 1990s.[22] New War theorists and other proponents of

20 Richard Garfield and Alfred Neugut, "The Human Consequences of War," 30.

21 The Judge Advocate General's Legal Center and School, "Gaps Remain in International Law to Prevent Destruction of Humanity," in *Civilian Casualties in War*, ed. Barbara Krasner (New York: Greenhaven, 2019), 65.

22 See Mary Kaldor, *New and Old Wars: Organized Violence in a Global Era* (Stanford: Stanford University Press, 1998); Mary Kaldor, "'New Wars' Wreak Havoc on Civilian Populations,"

what has become known as the "civilian casualty ratio inversion hypothesis" argue that in early twentieth-century wars, 80–90 percent of war casualties were combatants and 10–20 percent were noncombatants; whereas, by the 1990s this figure had flipped.[23]

Contemporary scholarship in military history has done a great deal to dispel the idea that noncombatants were largely excluded from the cruelties of nineteenth-century war. Stephen Conway, for instance, argues that it has been "demonstrated beyond any doubt that European wars of that time were highly intrusive affairs and could be truly terrible for civilians unlucky enough to be in the vicinity of the competing armies."[24] Likewise, David Bell chronicles how brutal, and indeed total, early nineteenth-century war was. In the Napoleonic Wars, Bell writes, "The line between combatants and noncombatants ... turned blurry, both in regard to enemy populations and to what we would today call the 'home front'... The absolute destruction of the enemy became a moral imperative, raising the prospect of the conflict stretching out indefinitely into the future."[25] One point of clarification is needed here; the distinction did not *become* blurry. It was never as clear as Bell suggests. There was a persistent tension between the idealized conception of how wars should be fought, what norms and the laws of war technically permitted, and what armies actually did. Indeed, the history of warfare in the nineteenth century is one in which nonfighting populations were routinely forced to endure significant hardships and even death.[26] The idealized conceptions of war usually did not align neatly with what the laws allowed and what armies did.

---

in *Civilian Casualties in War* ed. Barbara Krasner (New York: Greenhaven, 2019); 114–121; Martin Shaw, *The New Western Way of War: Risk-Transfer War and Its Crisis in Iraq* (Malden, MA: Polity, 2005). For criticism of this view, see Brian Smith, "Two Dogmas of the New War Thesis," *International Journal of Military History and Historiography* 38 (2018); 92–120.

23 See Christer Ahlström and Kjell-Åke Nordquist, "Casualties of Conflict," *Report for the World Campaign for the Protection of Victims of War* (Uppsala: Department of Peace and Conflict Studies, 1991); Ruth L. Sivard, *World Military and Social Expenditures* (Washington DC: World Priorities, 1991); Emily Crawford, *The Treatment of Combatants and Insurgents Under the Law of Armed Conflict* (New York: Oxford University Press, 2010), 15.

24 Stephen Conway, "British Soldiers at Home: The Civilian Experience in Wartime, 1740–1783", in *Civilians and War in Europe: 1618–1815*, eds. Erica Charters, Eve Rosenhaft, and Hannah Smith (Liverpool: Liverpool University Press, 2012), 130.

25 David Bell, *The First Total War: Napoleon's Europe and the Birth of Warfare as We Know It* (New York: Houghton Mifflin, 2007), 126.

26 See Jeremy Black, *Rethinking Military History* (New York: Routledge, 2004); Alan Forrest, Karen Hagemann, and Jane Rendall (eds.), *Soldiers, Citizens and Civilians: Experiences and Perceptions of the Revolutionary and Napoleonic Wars, 1790–1820* (New York: Palgrave Macmillan, 2009).

3 Noncombatants

One catches glimpses of the cruelties that were permissibly inflicted upon noncombatants in the military manuals and theoretical literature on war in this period. For instance, Carl von Clausewitz, the Prussian general who inspired over a century's worth of war theorizing and strategy with his *On War* (1832), had advised the use of rather stringent measures for noncombatants who were engaged in any form of resistance: "Where the population is concentrated in villages, the most restless communities can be garrisoned, or even looted and burned down as punishment."[27] Noncombatants could be punished rather severely for being, among other things, restless and troublesome.[28]

These recommendations hint at a much broader set of cruelties that might be justifiably inflicted on domestic populations, much like those experienced a half a century after *On War* was published by the residents around Metz during the Franco-Prussian War. As another nineteenth-century commentator on this subject explains, "An army in the strictest state of discipline can hardly pass into a country without great inconvenience to the inhabitants; roads are broken up, the price of provisions is raised; the sick are quartered on individuals, and a general uneasiness and terror are excited."[29] The logistical and infrastructural demands of moving, housing and feeding thousands of young men in hostile territory put incredible strain on local populations. Quartering, demanding contributions, pillaging, not to mention rape, murder, and other terrors were not uncommon.[30] As one frustrated commentator of this period notes:

> The bombarding the civil portion of cities may be a right of war, but it is still inhuman cruelty. The burning of towns and villages whole-sale … may be a military necessity, but it is inhuman cruelty. Plundering citizens by threat of instant death, the placing them on the engines, the massacre in cold blood of irregular troops, and still more of villagers suspected

27 Clausewitz, *On War*, 186.

28 In the mid-eighteenth century, Maurice de Saxe recommended sending "circular letters, threatening the inhabitants with military execution, on pain of their refusal to answer the demand made from them." Maurice de Saxe, *Reveries, or, Memoirs Concerning the Art of War* (Edinburgh: Printed by Sands, Donaldson, Murray, and Cochran …, 1759), 128–129.

29 Richard Wildman, *Institutes of International Law, Volume 1* (London: William Benning & Co., 1849), 67.

30 See Kim Murphy, *I Had Rather Die: Rape in the Civil War* (Batesville, VA: Coachlight Press, 2014).

> of aiding them, may be a mere measure of self-defense; but I call it inhuman cruelty.[31]

The degree to which noncombatants could be subjected *intentionally* to all manner of cruelties, stemming from the demands of necessity, sat uneasily with the wide-spread belief that international law was making war more humane and less cruel. In reality, the contrast between what was permitted by the laws of war and the nascent laws of humanity was rather stark.

Though the concept of civilian casualties had not yet been formalized, nineteenth century commentators on war wrote, to some limited extent, about the status of noncombatants (captured above in Figure 2.1). The idea of the "noncombatant" seems to really have gained a conceptual foothold around the period of the US Civil War (1861–1865) and the Franco-Prussian War (1870–1871). The emergence of the noncombatant in this era is not surprising since it is in this period that we can see a burst of theorizing about the laws of war and international law, more broadly.[32] The belief that war could be both limited and humanized runs through virtually all the legal documents of this period, and it therefore makes sense that attention would be paid to those most vulnerable to the excesses of war.[33]

In many ways, the emergence of the noncombatant was itself the product of institutional changes in warfare. If martialists in ages past privileged military discipline and sought to perfectly command and restrain their own troops, by the nineteenth century there arose a more coordinated effort to demarcate professional soldiers from ad hoc militias and other irregular fighters. If the nascent laws of war were to take hold in this period, it made sense that institutionalized militaries would serve as a regulating agent. What this suggests, however, is that the construction of the noncombatant in this period was part of an idealized, and one might say terminological, attempt to make professional distinctions that did not yet actually exist in practice. In other words, what is truly central about the rise of the noncombatant is the more formal presumption of the *combatant*, which was itself still a deeply contested category in nineteenth-century ethical discourse. In fact, arguably the most central diplomatic concern in this period was not the issue of noncombatants but

---

31 Frederic Harrison, "Bismarkism," *The Fortnightly Review, Volume VIII*, ed. John Morley (London: Chapman and Hall, 1870), 638.

32 See Wilhelm Grewe, *The Epochs of International Law* (2000); Fassbender, B., & Peters, A. "Introduction: Towards A Global History of International Law," in *The Oxford Handbook of the History of International Law*. (New York: Oxford University Press, 2000).

33 See Stephen Neff, *War and the Laws of Nations: A General History* (New York: Cambridge University Press, 2005), 190.

how to account for irregular soldiers.[34] Indeed, civilian militias, like the *francs-tireurs* and ad hoc resistance groups, raiders, and vigilantes, who took up arms to defend their homes in both the US Civil War and the Franco-Prussian War continued to problematize these tidy demarcations.

Part of these trends can also be explained by what Larry Addington identifies as the eighteenth-century shift from dynastic war to nationalist war.[35] Dynastic warfare largely centered on disputes between consolidated monarchies, where armies were composed of feudal and mercenary forces with a smaller professionalized element. In contrast, nationalist war came to be seen as conflicts between peoples. Under nationalist war, an expanded emphasis was placed on public support in the form of conscription along with a professional military force.[36] As a bellwether of these shifts, Clausewitz writes of the late-eighteenth century:

> Suddenly war again became the business of the people – a people of thirty millions, all of whom considered themselves to be citizens. We need not study in detail the circumstances that accompanied this tremendous development; we need only note the effects that are pertinent to our discussion. The people became a participant in war; instead of governments and armies as heretofore, the full weight of the nation was thrown into the balance.[37]

Another early nineteenth-century commentator, John Bolles, explains, "War is now, in effect, declared by the uplifted hands, and by the united voice, of the majority of freemen; it is carried on by the money and the swords of the voters, and has, consequently, become the manifestation of universal feelings and opinions instead of the mere index of individual hostility."[38]

34 The success in constructing the lawful combatant in international law ended up being a pyrrhic victory. At the early twentieth century, when irregular soldiers were mostly phased out, the distinction between solider and non-solider was obliterated by other innovations. Aerial bombing would undercut the spatial dimensions of the battlefield and deeply problematize the combatant-noncombatant distinction.

35 See Larry Addington, *The Patterns of War Since the Eighteenth Century* (Bloomington: Indiana University Press, 1994).

36 These shifts were anticipated in Immanuel Kant's republican peace theory. In his essay on "Perpetual Peace," Kant argued against dynastic war, believing that republican constitutions would lead to less war. See Immanuel Kant, *To Perpetual Peace: A Philosophical Sketch* [1795], trans. Ted Humphrey (Indianapolis: Hackett Publishing Company, 2003).

37 Clausewitz, *On War*, 238.

38 Bolles, *Essay on a Congress of Nations*, 29.

Even with the shift away from dynastic war and the movement towards the professionalization of militaries, by the nineteenth century there remained a long-standing tradition of the public militia, which continued to contribute to the conceptual fuzziness that existed between soldiers and the domestic population. The Dutch diplomat, Jan Helenus Ferguson (1826–1908) noted with frustration:

> It is true, that there are instances in which private citizens have taken an active share in a war waged between their respective Governments, and many may sacrifice their fortune on the altar of the common country and give freely their life-blood in its defence, thus contributing, at their own private risk, to the success of their country's cause, but can these single cases of heroism and devotion be used as argument to upset all principles of Justice and Benevolence, or as motives to arrive at the conclusion that the war must be waged against every individual member of the enemy state?[39]

Local resistance and private self-defense in war risked transforming the entire population into enemy combatants. Indeed, in another account by the war correspondent Archibald Forbes, he describes an incident where residents in a small village between Metz and Nancy attacked and killed Prussian soldiers with pitchforks.[40] Such encounters between private citizens and soldiers were common. To be sure, the permissibility of ad hoc resistance groups and town militias were one of the most contested issues in the late-nineteenth century, largely because it was a clear barrier to any idealized, bright-line distinction between combatant and noncombatant.

By the late nineteenth century these issues had come to a head. The subject of irregular soldiers was hotly debated at the Brussels Peace Conference of 1874. Ad hoc militias had been sanctioned by the French Government but not subsumed under their authority structure in the Franco-Prussian War.[41] More

39 Jan Helenus Ferguson, *Manual of International Law, Volume 2* (London: W.B. Whittingham & Co., 1884), Pt.5., Chap. 32, §198, 310.

40 Forbes, *My Experiences of the War Between France and Germany*, 114.

41 Tracey Dowdeswell argues that this conference helped to crystallize what Mary Kaldoor called 'old war' as a political ideal, i.e. that wars should be fought like a tournament between professional armies. See Tracey L. Dowdeswell, "The Brussels Peace Conference of 1874 and the Modern Laws of Belligerent Qualification," *Osgoode Hall Law Journal* 54:3 (2017); 805–850. See also W. Thomas Mallison and Sally Mallison, "The Juridical Status of Irregular Combatants under the International Humanitarian Law of Armed Conflict," *Case Western Reserve Journal of International Law* 9:1 (1977); 39–78.

than a decade prior, similar concerns about irregular soldiers had factored heavily in the formulation of the Union code of conduct, The General Orders 100, a document written by Francis Lieber to govern Union troop behavior during the US Civil War. The Lieber Code, as it was commonly referred to, was developed at Abraham Lincoln's request specifically to address the problem of how Union soldiers ought to treat spies and irregular guerrilla fighters.[42] In fact, by the end of the nineteenth century the issue of the combatant had hardly been settled, the concept of "small war" and with it irregular local resistance was simply considered an intractable feature of war.[43]

The ability of the civilian population to toggle between combatant and noncombatant status undermined the very idea, not to mention the validity, of something as concrete as noncombatant immunity.[44] Of course, that these issues were debated so persistently and, indeed, so late into the nineteenth century suggests that military practice had not yet caught up to its more idealized expression – that the "combatant" and "noncombatant" were sites of terminological conflict that did not so much express reality as attempt to make it.[45] To make these matters more complicated, even while the issue of irregular soldiers was being debated in numerous peace conferences of this period, the industrialization of the nation state was surreptitiously undermining these efforts. As the reality of industrial warfare was becoming clearer, the inherent

42 See, Sibylie Scheipers, *Unlawful Combatants: A Genealogy of the Irregular Fighter* (New York: Oxford University Press, 2015).

43 Colonel Callwell defines small wars as "all campaigns other than those where both the opposing sides consist of regular troops. It comprises the expeditions against savages and semi-civilised races by disciplined soldiers, it comprises campaigns undertaken to suppress rebellions and guerilla warfare in all parts of the world where organized armies are struggling against opponents who will not meet them in the open field, and it thus obviously covers operations very varying in their scope and in their conditions." Colonel C.E. Callwell, *Small Wars: Their Principles and Practice* [1896] (London: Printed for his Majesty's Stationary Office, 1906), 21.

44 Even the English Canon Law's attempt to settle this issue rested on distinctions that didn't always hold up under closer scrutiny: "Soldiers are irregular if they have been engaged, in a war manifestly unjust, on the unjust or aggressive side; but if they are engaged in a war strictly defensive, they are not irregular even if they kill or mutilate others, because this case is parallel to that of homicide in self-defence." John Brownbill, *Principles of English Canon Law, Part 1* (London: Kegan Paul, Trench & Co., 1883), 94.

45 In Thomas Arnold's lectures on modern history, he explains that one of the great advances in the conduct of war is the rise of regular armies. He writes, "The truth is, that if war, carried on by regular armies under the strictest discipline, is yet a great evil, an irregular partisan warfare is an evil ten times more intolerable; it is in fact no other than to give a license to a whole population to commit all sorts of treachery, rapine, and cruelty without any restraint." Thomas Arnold, *Introductory Lectures on Modern History* (New York: D. Appleton & Co., 1862), 160.

innocence of the non-fighting citizen farmer and industrial laborer were coming under renewed scrutiny. A new and even more problematic relationship was beginning to emerge, not between combatant and irregular soldier, but between combatant and those employed in all manner of war-related industries. As wartime economies were starting to implicate larger swaths of the nonfighting population, it became increasingly difficult to tidily demarcate legitimate from illegitimate targets in a clear-cut way. These difficulties, indeed, persist to this day.

The interest in the combatant-noncombatant distinction is not entirely reducible to the concern with irregular soldiers, though such difficulties clearly predominate the military literature of this period. As these problems were being discussed among war planners and theorists, there emerged a concomitant public discourse that sought to highlight the plight of nonfighting populations in war. Specifically, improvements in war reporting and the activism of peace societies, like that of Bellows indicated above, brought the cruelty of wars home unlike any moment in history. The public witness to war became a powerful force in educating the citizenry, disabusing them of idealized conceptions of heroism and noble sacrifice. Though writing in the early twentieth century (1938), Virginia Wolff aptly describes these sentiments when describing a picture taken of the aftermath of a bombing, where women and children lay mangled and virtually unrecognizable: "Those photographs are not an argument; they are simply a crude statement of fact addressed to the eye. But the eye is connected with the brain; the brain with the nervous system. That system sends its messages in a flash through every past memory and present feeling."[46] The powerfully educational pangs of horror and disgust teach us that war is a barbarity that must be stopped. As Wolff suggests, these descriptive efforts, setting out the grim facts of war, had the effect of priming a receptive public for laws of war that reflected the principles of humanity.

It is not surprising that at the same time humanist institutions and pacifism, more specifically, gained expression as a formal ideological and indeed transnational political movement, various peace societies started springing up all throughout the west.[47] Many of these societies made it a goal to illuminate the cruelties of war with the hope of changing public opinion. Gustave Moynier, one of the co-founders of the "International Committee for Relief to the Wounded" a precursor to the International Committee of the Red Cross, explains that

46 Virginia Wolff, *Three Guineas* in *The Selected Works of Virginia Wolff* (London: Wordsworth, 2007) 781–924. 790.

47 See Chapter 1, Note 1.

> War is evidently destined to become more and more humane under the breath of civilization, and in order to favour this tendency, it will be necessary to brand and repress many abuses. It is, and without doubt will always be, a calamity, but progress will consist in waging war more economically, if we may so express ourselves, in wasted less the human capital, in giving more consideration to the lives of men, which may be styled the money that pays for both defeat and triumph.[48]

Moynier believed that war could be waged more humanely and "economically," that is, in such a way that minimizes the loss of excess "human capital." These sentiments, he suggests, spring largely from "the vast increase in the means of diffusing information." With the rise of "facilities for rapid personal intercommunication" this has led to "the spread of more civilized and humane feelings, and the abatement of national animosities."[49]

William Ladd, the founder of the American Peace Society, makes a similar argument. He explains: "We hope by the force of public opinion alone to increase the number of non-combatants, and to lessen the restraint laid upon them, and to give greater security to their lives and property."[50] By highlighting the plight of noncombatants, peace societies were raising a particular kind of public awareness; indeed, they were helping to construct the very idea of the noncombatant. By appealing to a common humanity, peace activists like Ladd hoped to formally expand the category of the noncombatants to all humankind, with the ultimate hope of eliminating the category of combatant.

There was a symbiotic relationship between these peace societies and the first-hand reports, like those provided by Forbes (noted above), which, even if anecdotal and unsystematic, helped to erode any illusion that wars did not harm the nonfighting residents who lived near zones of conflict. As first-hand observers of war, these correspondents dramatized and humanized all the

48 Gustave Moynier, *Help for Sick and Wounded*. trans. John Furley (London: John Camden Hotten …, 1870), 340–341.

49 Ibid., 365.

50 William Ladd, "Extracts for the Circular Letter of the American Peace Society," in *Thirteenth Annual Report of the Committee of the Society for the Promotion of Permanent and Universal Peace* (London: Richard Clay …, 1829), 25. See also, William Ladd, "Changes in Public Opinion," in *The Essays of Philanthropos on Peace and War* (Portland: Printed by Shirley & Edwards, 1825); in a later work, Ladd argues that "The late improvement in, and increased power of, public opinion furnish another guaranty of peace." William Ladd, *An Essay on a Congress of Nations* (Boston: Whipple and Damrell, 1840), 108. See also, William Jay, *War and Peace: The Evils of the First, and a Plan for Preserving the Last* (London: Thomas Ward and Co, Paternoster Row, 1842).

attendant suffering that flowed from the battlefield.[51] To be sure, these war correspondents did perhaps the most to aid in the construction of the vulnerable noncombatant. For instance, Sir William Howard Russell, an Irish war correspondent during the Crimean War, reported on the horrendous suffering endured by the people in the town of Balaklava (1854). He writes,

> The dead, lay as they died, were side by side with the living. The commonest accessories were wanting; there was not the least attention paid to decency or cleanliness – the stench was appalling – the fetid air could barely struggle out to taint the atmosphere, through the chinks in the walls and roofs. The sick appeared to be tended by the sick, and the dying by the dying.[52]

Orville Victor, reporting on the Quantrill raid of Lawrence, Kansas, which took place during the Civil War in August 1863, writes:

> General Lane was among the first to enter the town. What a sight met his view! Dead and mangled bodies lay along the street, some literally cooking by the blaze of their own dwellings. Everywhere were burning buildings. The hotel was a smoking ruin. Females and children were shrieking in the agony of their bereavement. It was a scene to sadden the heart.[53]

And in a slightly more incendiary and sensationalist tone, Charles Coffin, an embedded journalist who followed the Union during the US Civil War, reported on the horror of numerous Confederate atrocities:

> What act so malignant, bloody, ghastly, and fiendish as the sacking, burning, and massacre at Lawrence! What deed so damning since the barbarities of Scio or Wyoming! What woe so deed! – men, children, murdered, butchered, scalped, the bodies of the dead tossed into the flames! No relenting on the part of the Rebels, but savage, infuriate joy at the sight of

51 One of the great texts highlighting the suffering of war was by one the co-founders of the Red Cross: Henry Durant, *Un Souvenir de Solferino* (Genève: Imprimerie Jules-Guillaume Fick, 1862).

52 Sir William Howard, *The British Expedition to Crimea* (London: George Routledge and Sons, 1877), 192.

53 Orville Victor, *The History, Civil, Political, and Military of the Southern Rebellion, Volume IV* (New York: James Torrey, 1864), 188.

> the war heart's blood of their victims! Woman's prayers and tears availed not to stay their murderous hands or move their brutal heart.[54]

Reports like these increasingly brought the suffering of innocent populations into the public eye.[55]

The growing emphasis on the proper object of violence, and who might be excluded from that violence, can be seen in the attempts by military ethicists and international lawyers in the literature of the nineteenth century to provide some explanation for the death of noncombatants. It is significant that so much energy was spent illuminating the plight of noncombatants by peace advocates and war correspondents. This energy is indicative of how *permissive* the moral climate was. It shows that the laws of war had not yet evolved to exempt certain people from the direct harms of war. In fact, the push for noncombatant protections was always something of an uphill battle. As will be seen, even when supposed protections were set out in international law, the language was oftentimes vague enough to render meaningful restraint illusory. This is in part due to the fact that while the frequency of references to noncombatants increased in the theoretical literature of this period, the scope of this interest remained surprisingly marginal. Many theorists in this period simply did not address the subject.[56] As seen in the Peace Society activism above, that many theorists were silent on this issue cannot be attributed to the fact that noncombatant immunity had already become an established norm – it was not. As Lawrence Freedmen explains, many of those theorizing about war "had little to say to those with no role in fighting. The interest was in where and how the line between combatancy and noncombatancy might be blurred, and what

54 Charles Coffin, *Four Years of Fighting* (Boston: Arno, 1866), 442.

55 For additional Civil War reporting, see Albert Gaius Hills, *A Civil War Correspondent in New Orleans: The Journals and Reports of Albert Gaius Hills of the Boston Journal* ed. Gary Dyson (Jefferson, NC: McFarland and Company, 2013); for Boer War reporting, see Richard Harding Davis, *Notes of a War Correspondent* (Fairfield, IA: First World Library, 2004). For more on war reporting in this period, see Joseph Mathews, *Reporting the Wars* (Minneapolis: University of Minnesota Press, 1957); Philip Knightley, *The First Casualty: The War Correspondent as Hero and Myth-Maker from the Crimea to Iraq* (Baltimore: Johns Hopkins University Press, 2004); Catherine Waters, *Special Correspondence and Newspaper Press in Victorian Culture, 1850–1886* (New York: Palgrave, 2019).

56 Many of the leading texts on international law had nothing to say about the noncombatant. See Richard Wildman, *Institutes of International Law* (London: William Benning and Co., 1849); *Kent's Commentaries on the International Law* (Cambridge: Deighton, Bell, and Co., 1866); Henry Sumner Maine, *International Law* (London: John Murray, Albemarle Street, 1888); and Thomas A. Walker, *The Science of International Law* (London: C.J. Clay and Sons, Cambridge University Press, 1893).

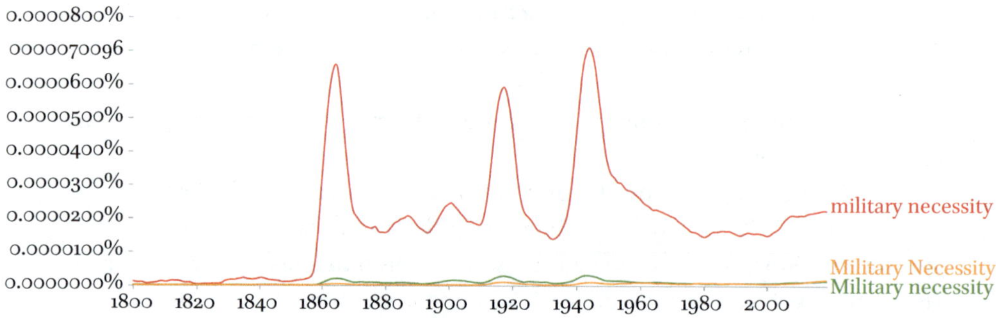

FIGURE 4.3 Ngram search of 'military necessity'

sort of protections might be expected either side of this line."[57] But even this is not quite right. As noted above, the theoretical problem was that the ontological categories were never that clear to begin with.

Despite the apparent tidiness of the combatant-noncombatant distinction, part of what made these formal categories difficult to pin down was the perpetually blurry distinction between the two. In many respects, ideas about what protections noncombatants deserve can be explained by looking at the ascendant ideas about military necessity (See Figure 4.3). While the language of military necessity can be traced throughout the literature on war strategy and military ethics for many centuries, it gained a formal conceptual foothold in the nineteenth century. When the death of noncombatants was addressed in the literature of this period, it was almost exclusively justified by the principle of necessity.

Contrary to another widely held belief in contemporary scholarship, there was no blanket immunity for noncombatants in this period; virtually no nineteenth-century theorist spoke in these terms, and when they do, it is always contextualized by the principle of necessity. For instance, J.S. Risley explains that "The right of violence is limited by the necessities of war, and on this doctrine is founded the immunity of the non-combatant."[58] This is, indeed, one of the few references to noncombatant immunity in this period, and at first blush it appears to indicate some notion of innate protection for certain classes of people. Notice, however, that immunity is founded upon the doctrine of military necessity. In fact, a few passages later, Risley explains how noncombatants residing in or close to fortified cities lose their immunity.[59] What this passage ultimately shows is that "immunity" is a derivative concept.

57 Freedman, *The Future of War: A History*, 33.

58 J.S. Risley, *The Law of War* (London: A.D. Innes and Co., 1897), 107.

59 Ibid., 117.

The argument is effectively tautological: those deemed unnecessary to kill are immune from being killed. Even if the latent assumption here is that, under normal conditions, those not directly engaged in fighting would be immune from being targeted, this status was far from guaranteed. To be sure, this concept has little resemblance to the innate immunity spoken of in the twentieth and twenty-first centuries.

The derivative nature of noncombatant immunity in this period has been obscured in contemporary scholarship by authors who retroject anachronistic assumptions back onto this period. For instance, Igor Primoratz argues "By the eighteenth century, major European powers had accepted civilian immunity as a central tenet of their military practice. In the last decade of the nineteenth century and the first decade of the twentieth, the principle was embedded in the Hague Conventions (1899, 1907) and thus became part and parcel of international law."[60] To begin with, the Hague Conventions had very little to say about noncombatant immunity. These agreements do reiterate and reaffirm the ascendant norms regarding protections for *hors de combat* (prisoners of war, the wounded, and medical personnel), but they say very little about the nonfighting members of a foreign state. Despite some nominal provisions protecting undefended cities, coastal cities could be bombarded by approaching naval forces at will. On land "open towns" could not be bombed, but the residential districts of "defended towns" could be expressly targeted. It is striking that the rules even permitted attacking the detached suburbs of defended cities. Furthermore, there was quite critically no consensus as to how one might differentiate between an open and defended town, which essentially vitiated what little force remained of the so-called immunity provisions.

There are others who retroject this contemporary notion back onto this period. For example, Emily Crawford argues, "Since its earliest incarnations over 150 years ago, the modern law of armed conflict has developed according to a fundamental axiom: only military objects may be lawfully targeted while civilian objects remain immune."[61] Steven Lee writes, "The principle of civilian immunity has taken a beating in the past century. Civilians have become an increasing portion of the increasingly larger number of casualties of war."[62] Again, the implicit assumption in this sentiment is that noncombatant immunity was honored in the nineteenth century unlike wars of the twentieth

60 Igor Primoratz, "Civilian Immunity in War: Its Grounds, Scope, and Weight," in *Civilian Immunity in War* ed. Igor Primoratz (New York: Oxford University Press, 2007), 2.

61 Crawford, *Identifying the Enemy*, 11.

62 Steven Lee, *Ethics in War: An Introduction* (New York: Cambridge University Press, 2012), 165.

century. More recently, Max Abrams has also argued that "The moral ideal of civilian immunity has been the bedrock of international law since the nineteenth century."[63] All of these attempts to ground a robust conception of noncombatant immunity in the nineteenth century suffer from a similar fallacy; while many point to progressive-sounding passages in the legal documents of that era, they fail to appreciate the cruel permissions the laws of war actually extended and what militaries believed they were justified in doing.[64]

While there was, to be sure, greater impetus to clear up the fuzziness between combatants and noncombatants in the nineteenth century, as seen in the idealizing push to subsume all fighting forces into the traditional military structure, this principle of distinction was not designed to extend blanket immunity to civilian populations. Indeed, "noncombatant immunity" is another twentieth century conceptual formulation that is mistakenly attributed to nineteenth-century war. While there were certainly antecedent moral formulations that would eventually give rise to this twentieth-century conception, it is important to recognize that in the nineteenth century, that phrase – "noncombatant immunity" – did not exist (see Figure 4.4).[65] In fact, virtually the only time military manuals mention immunity is to stipulate that certain *property* should be immune from direct harm – works of art, sacred or cultural buildings, monuments, merchant ships, and the property of neutrals more broadly.[66] Similar provisions are *never* explicitly applied to people.

63 Max Abrams, *Rules for Rebels: The Science of Victory in Military History* (New York: Oxford University Press, 2018), 74.

64 For similar arguments, see Robert McElroy, *Morality and American Foreign Policy: The Role of Ethics in International Affairs* (Princeton: Princeton University Press, 1992), 151; Judith Gardam, *Non-combatant Immunity as a Norm of International Humanitarian Law* (Boston: Martinus Nijhoff, 1993), 15; Colin McKeogh, *Innocent Civilians: The Morality of Killing in War* (New York: Palgrave, 2002); and J. Daryl Charles and Timothy Demy, *War, Peace, and Christianity: Questions and Answers from a Just-war Perspective* (Wheaton: Corssway, 2010), 154.

65 While antecedent elements of noncombatant immunity can be traced back quite early in military ethical thinking, it would be a mistake to suppose a singular and unitary conception unproblematically persisted and can be traced through time. There has always been a great deal of epistemic uncertainty as to who counted as a noncombatant and whether immunity was operative in a particular context. In many cases, the perception of noncombatant immunity was a byproduct of a particular understanding of military necessity. See Richard Shelly Hartigan, "Noncombatant Immunity: Reflections on Its Origins and Present Status," *The Review of Politics* 29:2 (1967); 204–220; and Marcus Schulzke, *Just War Theory and Civilian Casualties: Protecting the Victims of War* (New York: Cambridge University Press, 2017).

66 The issue of confiscating private property had been a pointed issue during the Napoleonic Wars (1803–1815). Napoleon accused the British of violating the "right of nations" by unjustly confiscating merchant ships and taking the crew members as prisoners of war.

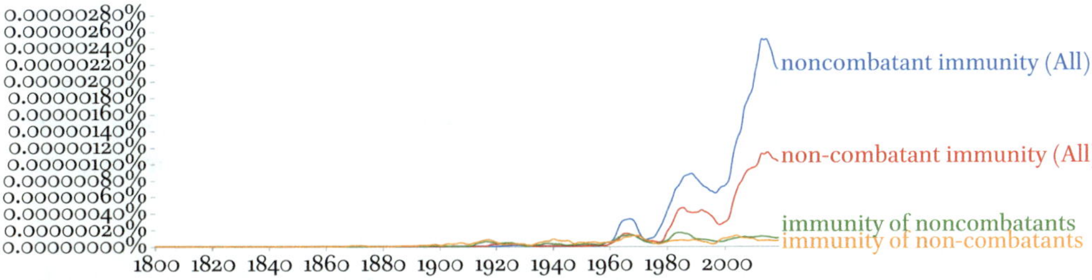

FIGURE 4.4 Ngram search of 'noncombatant immunity,' 'non-combatant immunity' and 'immunity of noncombatants,' 'immunity of non-combatants'

In fact, early twentieth-century commentators on war, those educated in the traditions and norms of the nineteenth century, *knew* noncombatants were never immune from being intentionally targeted. As one interwar theorist, M.W. Royse, descriptively notes: "At no time [...] was [civilian] immunity assured even under ordinary circumstances."[67] As noted above, the so-called protections extended to noncombatants were always subordinate to the principle of military necessity, and that "false notions of immunity" arose over periods of time when the targeting of noncombatants was unnecessary to achieve the aims of a particular military objective.[68] The assumption of immunity began to arise in a context where certain killing was not believed to be necessary to prosecute a successful military campaign. In a similar manner, a post-World War II commentator explains:

> How then is the noncombatant immune from attack? He is legally subject to almost unrestricted artillery and naval bombardment. If he lives in a besieged locality he may legally be starved or bombed. If he lives in a country which does not grow enough food to support its population, a blockade can legally starve him to death. If he lives in an important city, he is subject to bomb and robot attack of the most catastrophic nature. True, in many cases, he may not be the intended subject of attack, but under modern methods of waging war that gives him little protection.[69]

"The Berlin Decree" of 21 November 1806, indicated that British subjects in France would be likewise treated as enemy combatants and taken prisoner.

67 M.W. Royse, *Aerial Bombardment and the International Regulation of Warfare* (New York: H. Vinal, 1928), 155.

68 Ibid., 150.

69 L. Nurick, "The Distinction between Combatant and Noncombatant in the Law of War," *The American Journal of International Law* 39:4 (1945); 680–697. 696.

The methods of death Nurick refers to are not novel twentieth-century inventions; he is referring to norms in warfare that long precede the nineteenth century. Throughout the nineteenth century it was widely assumed that noncombatants could be deliberately targeted and killed under certain conditions, especially if their deaths were believed to expedite victory under a framework of necessity.

Again, part of the challenge to making sense of nineteenth-century beliefs about noncombatant casualties is the binary language it is misleadingly couched within. To distinguish combatants from noncombatants seems indicative of a clear demarcation – that there is a discrete class of people engaged in war-making and another discrete class of people who are not. Commentators throughout the nineteenth century, however, routinely made note of the "hazy distinctions between combatants and non-combatants."[70] As has already to some extent been seen, this binary thinking does not fully capture the ethical formulas of noncombatant protections in this period. There were a range of intermediary positions one could occupy, not to mention the fact that rather than being ontologically fixed, one's exact status was open to a great deal of interpretation. As Gunner Lind explains, "Instead of a single line dividing the fighting men from the rest, we see a hierarchy of arms. Simplified, it had three levels: those whose social role it was to be a warrior, such as noblemen and mercenaries; the rest of the male gender, excluding the very poor and sometimes the clergy; [and] women and children."[71] Historically, rather than being a binary division, the distinction between combatant and noncombatant was almost always indicative of gradations, a sliding scale of targetability. This is especially true of the nineteenth century.

Even when nineteenth century theorists made attempts at drawing a bright line between combatant and noncombatants, the language was still reminiscent of this sliding scale. For instance, in his 1871 treatise on international law, Theodore Woosley argues that "Private persons, remaining quiet, and taking no part in the conflict, are to be unmolested, but if the people of an invaded district take an active part in war, they forfeit their claim to protection. This marked line of separation between soldier and the non-soldier, is of extreme importance for the interests of humanity."[72] At first glance this classification

70 Henry Hozier, *The Seven Weeks' War: Its Antecedents and Its Incidents* (London: Macmillan and Co., 1872), 52.

71 Gunner Lind, "The Genesis of the Civilian in the Western World, 1500–2000", in *Civilians at War: From the Fifteenth Century to the Present*, ed. Gunner Lind (Copenhagen: Museum Tusculanum Press, 2014), 50.

72 Theodore D. Woosley, *To the Study of International Law* (New York: Charles Scribner & Co., 1871), 224. Incidentally, Woosley dedicated this text to his friend, Francis Lieber.

seems to partake of binary categories. Notice, however, the ambiguity of "taking no part" versus "taking an active part." Embedded within these distinctions are a range of intermediate positions that may subject the non-soldier to losing protection. In other words, one does not need to take up arms to lose protected status. There were many ways of taking an active part in war that falls short of strict combatancy. Another commentator in this period, George Wilson, also explains that "Combatant status is extended to those who under government sanction engaged either directly or indirectly in the operations of war."[73] *Indirect* support of an enemy army may also lead to the conferral of some degree of combatancy.[74] It is quite significant that Wilson explains one's status is *conferred* or *extended* rather than recognized or honored. This is because where one ultimately ends up on this scale of targetability is not grounded in some innate or fixed status. As will be shown more clearly below, erstwhile noncombatants could transform into combatants for many reasons, based largely on their proximity to sites deemed to be of military value.

4 Sovereignty and Self-Preservation

There are two significant trends that explain the rise of military necessity and with it the evolving ideas about noncombatant death in the nineteenth century. The first is the rise of the Westphalian nation-state at the end of the Thirty Years' War (1648), and the second is the formulation of philosophical utilitarianism.[75] While the treaties associated with the Westphalian system, signed in Münster and Osnabrück, did not explicitly formalize the notion of the sovereign nation-state, it is widely assumed that the treaties gave conceptual and practical legitimacy to the autonomy of a state over its constituent territory which formalized one of the essential features of nationhood.[76]

73 George Wilson, *Handbook on International Law* (St. Paul: West Publishing Company, 1910), 274.

74 Thomas Walker makes this same point: "The greatest of the triumphs of International Law has consisted in the creation of Non-combatants Classes, and the limitation of strict war-right in respect at once of persons and of property. Such exemption is, however, exactly conterminous with the maintenance of strict neutrality." Walker, *The Science of International Law*, 249. He then proceeds to discuss the many abnormalities of war where individuals assume the character of combatants.

75 Derek Croxton, "The Peace of Westphalia of 1648 and the Origins of Sovereignty," *The International History Review* 21:3 (1999); 569–591.

76 There is, of course, a great deal of debate about the emerging post-Westphalian consensus. See Stephen Krasner, *Sovereignty: Organized Hypocrisy* (Princeton: Princeton University Press, 1999); Benno Teschke, *The Myth of 1648: Class, Geopolitics, and the Making of Modern*

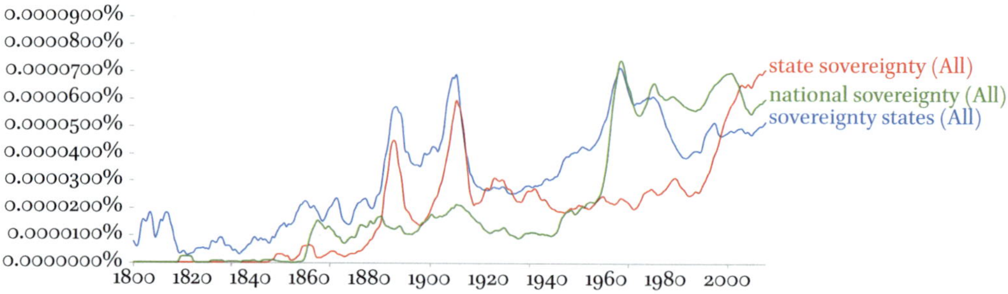

FIGURE 4.5 Ngram search of 'state sovereignty,' 'national sovereignty,' and 'sovereign states'

The fact is that in the wake of Westphalia, states started to become sovereign entities, even if the exact significance of this transformation remained deeply contested. This development of sovereignty as a site of concern in the nineteenth century can partially be seen in Figure 4.5, especially after the Napoleonic wars and the Congress of Vienna (1815). While references clearly trace back to the seventeenth century, there is a noticeable uptick in references to "state sovereignty," "sovereign states," and "national sovereignty" in this period.[77] Clearly, as the nineteenth century progressed, states were engaged in renewed debates about the duties and rights they as sovereign entities possessed.

There is no question that in this period of development sovereignty remained an essentially contested concept. The burst of references in this period attests to these conflicts.[78] Nevertheless, the one feature of the sovereign state that remained uncontroversial and indeed ubiquitous in this period was the duty of self-defense, a concept that was itself deeply entangled with the emerging logic of military necessity. In practice, this meant that states were increasingly being seen as responsible not for punishing religious or ethnic grievances, but for the internal maintenance and defense of the political community contained within the state's borders, i.e. the preservation of the state. As one mid-century scholar, James Reddie, notes: the "right of necessity" that was emerging in this period of time was a moral framework that overrode lesser obligations: "the more imperative right of self-preservation, may

International Relations (London: Verso, 2003); James Turner Johnson, *Sovereignty: Moral and Historical Perspectives* (Washington DC: Georgetown University Press, 2014); David Onnekink (ed), *War and Religion after Westphalia 1648–1713* (New York: Routledge, 2016).

77 "National sovereignty" emerged in the late eighteenth century but appears to have taken hold in the nineteenth century.

78 See Duncan Kelly, "Popular Sovereignty as State Theory in the Nineteenth Century," in *Popular Sovereignty in Historical Perspective*, eds. Richard Bourke and Quentin Skinner (New York: Cambridge University Press, 2016), 270–296.

justify a nation in the nonfulfillment of an inferior legal obligation to another people."[79]

As seen in the humanist tradition, from the sixteenth to eighteenth centuries, there was a growing presumption that sovereign entities (typically sovereign persons rather than states), could do whatever was necessary to defend themselves or to prosecute a right. Self-defense was construed as one right among many. However, this right to self-defense always sat uneasily in the natural law tradition since, according to several theorists covered in the previous chapter, the violation of the natural law could undermine one's perfect claim to the right of self-defense. If the guilty deserve to be punished, it is not clear by what right they could defend themselves. Due to these difficulties, some in the preceding era, like Vattel, presumed that both sides must tentatively be understood as within their right to defend themselves, even if technically one side was guilty and had strictly speaking lost their perfect right.

By the nineteenth century, the presumption that states possess the right of self-defense had become deeply linked to notions of sovereignty. Since political communities were *a priori* justified to exist, the logic of military necessity became intimately linked with the duties of self-preservation. As Thomas Jefferson would explain over a century after Westphalia, indeed, at the very outset of the nineteenth century, "self-preservation is paramount to all law."[80] The desperate struggle to survive gave states much more latitude and license to articulate to themselves what was permissible in self-defense, what was "necessary," and what they could justifiably do when they needed to defend themselves.

Unlike the Scholastic just war tradition, post-Westphalian justifications for war had become more fundamentally tied to the duty of the state to preserve itself. It is worth reiterating the degree to which these formulations chafed against these earlier strands of just war theory, which focused less on the right of self-defense, and more on proper procedures one needed to follow in order to wage wars of *offense* – i.e. for punishing nations that had violated the natural law. Both Scholastic and humanist traditions were much more forthcoming in articulating the conditions when punishment was permissible, where a state had a right if not duty to police those who had violated the natural law. In the post-Westphalian era, the right to justly punish a wrong-doer was increasingly

79 James Reddie, *Inquires in International Law* (Edinburgh: William Blackwood & Sons, 1842), 199.

80 Thomas Jefferson, "Letter LXXI, To Doctor James Brown, October 27, 1808," in *Memoir, Correspondence, and Miscellanies, From the Papers of Thomas Jefferson, Volume 4* (Boston: Gray and Bowen, 1830), 114.

overshadowed by the logic of self-defense; wrongdoers were those who put the autonomy of sovereign states in jeopardy.[81] As Friedrich von Gentz would note at the dawn of the century, "A just and necessary war is at all times a defensive war, whoever be the first to take up arm."[82] As post-Westphalian theorists of the nineteenth century routinely pointed out, "It is manifest therefore that the maintenance of the political association itself is an essential condition for the fulfillment of these reciprocal engagements, and the State or Nation is thus under a primary obligation to preserve itself; in other words, Self-preservation is a primary duty of National Life."[83]

The notion of self-preservation certainly was not invented in the nineteenth century. Indeed, these ideas can be traced back to Machiavelli who himself borrowed these ideas from antiquity; he had argued that "where one deliberates entirely on the safety of his fatherland, there ought not to enter any consideration of either just or unjust, merciful or cruel, praiseworthy or ignominious; indeed every other concerns put aside, one ought to follow entirely the policy that saves its life and maintains its liberty."[84] However, the degree to which this idea had gained ascendance and came to underpin and inform the concept of nineteenth-century conceptions of sovereignty and with it just war thinking did represent a novel shift.[85] As seen in the previous chapter,

---

81 At the very least, intervention that functioned as punishment needed to be articulated in the language of self-defense. See Martha Finnemore, *The Purpose of Intervention: Changing Beliefs about the Use of Force* (Ithaca: Cornell University Press, 2013).

82 Friedrich von Gentz, *Fragments upon the Balance of Power* (London: Printed for M. Peltier …, 1806), x.

83 Sir Travers Twiss, *The Laws of Nations Considered as Independent Political Communities* (Oxford: Clarendon Press, 1863), 3.

84 Niccolo Machiavelli, *Discourses on Livy* trans. Harvey Mansfield and Nathan Tarcov (Chicago: University of Chicago Press, 1996), 301. Though this language has been softened in contemporary military ethical discourse, it survives in Michael Walzer's problem of the Supreme Emergency. See Michael Walzer, *Just and Unjust Wars: A Moral Argument with Historical Illustrations*, Fourth Edition (New York: Basic Books, 2006). See also, Alex Bellamy, "Supreme Emergencies and the Protection of Non-Combatants in War," *International Affairs* 80–5 (2004); 829–850.

85 Leo Strauss argues that modernity originated with Thomas Hobbes's belief that the natural law reduces to one impulse: "the desire for self-preservation." Strauss writes: "If, then, natural law must be deduced from the desire for self-preservation, if, in other words, the desire for self-preservation is the sole root of all justice and morality, the fundamental moral fact is not a duty but a right; all duties are derivative from the fundamental and inalienable right of self-preservation. There are, then, no absolute or unconditional duties; duties are binding only to the extent to which their performance does not endanger our self-preservation. Only the right of self-preservation is unconditional or absolute. By nature, there exists only a perfect right and no perfect duty. The law of nature, which formulates man's natural duties, is not a law, properly speaking. Since the fundamental

Machiavelli was routinely decried as a kind of devil by seventeenth-century commentators, and Hobbes's *Leviathan* was publicly burned in Oxford in 1683.[86] These are hardly ringing endorsements for the ascendance of a strict doctrine of self-preservation. When these ideas did begin to emerge in the humanist tradition of the seventeenth and eighteenth centuries, they were softened and moderated by additional duties of the natural law, like that of charity, mercy, and humanity. Self-preservation was one duty among many. Reductive appeals to this right were seen as far too cruel and, inasmuch as they were unbending, might themselves be unjust. The Scholastic just war tradition also militated against firm notions of self-preservation and self-defense. Since only one of the two belligerents could be right – one was necessarily in the wrong, if not both – a nation that was guilty of violating the natural law lost the right of self-defense and deserved to be punished. Aquinas in fact goes so far to say that "human beings should always observe such precepts in readiness of spirit, namely, that they be ready, if necessary, not to resist and not to defend themselves."[87]

Very much in the Machiavellian and Hobbesian tradition, the post-Westphalian consensus was, in the words of Henry Halleck (one of the great legal minds of the nineteenth century), that the "law of nations make no distinction … between a just and an unjust war," and as such it was the perception of threat and the demands of self-preservation that governed justified military action.[88] Wars of self-preservation and defense, however broadly construed, became *de facto* just wars. No state could be accused of violating its right when it engaged in an act of perceived self-defense. Indeed, the epistemic fog around assigning blame for encroaching on a state's right to self-preservation was aptly summarized by John Bassett Moore in his pre-World War I compendium on international law: "We may think that the attack was not justified by the cause which was assigned. But, as an independent state, we have no right to judge the motives which actuated another state in asserting their rights

and absolute moral fact is a right and not a duty, the function as well as the limits of civil society must be defined in terms of man's natural right and not in terms of his natural duty. The state has the function, not of producing or promoting a virtuous life, but of safeguarding the natural right of each. And the power of the state finds its absolute limit in that natural right and in no other moral fact." Leo Strauss, *Natural Right and History* (Chicago: University of Chicago Press, 1965), 181.

86 See Patrick Springborg, "Hobbes, Heresy, and the *Historia Ecclesiastica*," *Journal of the History of Ideas* 55:4 (1994); 553–571.

87 Thomas Aquinas, *Summa Theologica* Volume 3 (Part II, Second Section), trans. Fathers of the English Dominican Province (New York, 2007). II-II, Q. 40, Art.1, 1354.

88 Henry W. Halleck, *International Law or Rules regulating the intercourse of states in peace and war* [1861] (London: Kegan, Paul, Trench, Trubner, 1893), 15.

and vindicating wrongs which they supposed its citizens or subjects had sustained."[89]

In the nineteenth century, the post-Westphalian conception of sovereignty was largely Machiavellian and Hobbesian in essence; wars of necessity became linked to the perceived existential viability of the state.[90] In this respect, a necessary war began to be seen in those cases in which "without war we cannot exist."[91] Over time, those actions believed to be necessary for self-preservation came to govern *jus in bello* considerations, which would have significant implications for the treatment of noncombatants in war. Some of these trends can be seen in the humanist discourse of the previous chapter, but these would take on a noticeably different connotation within the emerging philosophical tradition of utilitarianism. With the rise of utilitarian thought, the framework of necessity established by the humanist and republican writers of the preceding centuries would become unmoored from natural law restraints and begin to shift more decisively in the direction of a pleasure/pain, cost-benefit utility framework.

5 Utilitarianism

The second major explanation for the shift in justifying civilian death by way of military necessity can not only be traced to the emerging nineteenth-century ideas of state sovereignty but also the rise of philosophical utilitarianism.[92] As seen in Figure 4.6, after 1820 there was a noticeable surge in references to "utilitarian" and "utilitarianism." While its intellectual founder, Jeremy Bentham (1748–1832), had been publishing since the late eighteenth century, his ideas started to gain wider public attention in the early nineteenth century, right around the time James Mill and he founded the *Westminster Review* (1823), a radical journal that sought to publicize utilitarian approaches to a range of public issues.

---

89 John Bassett Moore, *A Digest of International Law, Volume VI* (Washington, D.C.: Government Printing Office, 1906), 939.

90 This, of course, renders the language of self-preservation exceedingly elastic. For instance, within this framework Montesquieu could argue, "The object of war is victory; that of victory is conquest; and that of conquest, preservation." Montesquieu, *Spirit of the Laws* in *The Complete Words of M. de Montesquieu, Volume 1* (London: Printed for T. Evans in the Strand, 1777), 6.

91 Thomas Walker, *A History of the Law of Nations* (London: Cambridge University Press, 1899), 255.

92 For more on how the tradition of utilitarianism intersects with the ethics of war, see William Shaw, *Utilitarianism and the Ethics of War* (New York: Routledge, 2016).

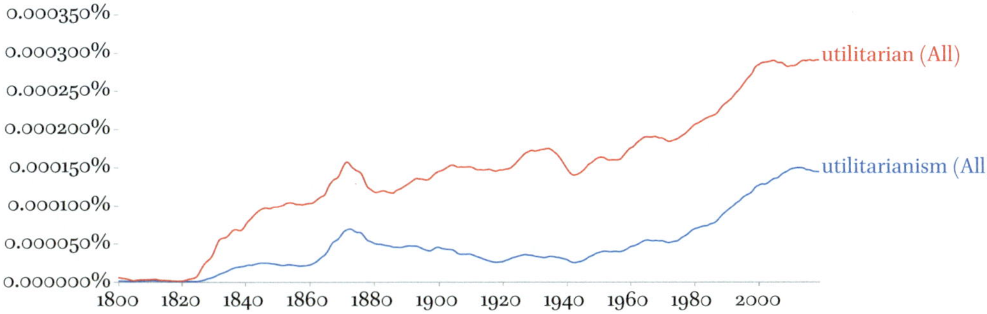

FIGURE 4.6 Ngram search of 'utilitarian' and 'utilitarianism'

Bentham had in fact given expression to an early conception of military necessity that was appropriately suited for an international system that had clarified its moral priorities, i.e. the indubitable *a priori* value of the state and with it the right of self-defense.[93] Bentham writes, "Understand by military necessity, the necessity of giving by law to military functionaries authority to produce, on each occasion, in any shape whatsoever, whatsoever evil may be at the same time sufficient and necessary to the exclusion of greater evil."[94] Military actions are acceptable if they are "preponderant-evil-excluding," which even extends to "the openly committed offenses against person and property."[95]

Under moments of extreme duress, the moral priority of a state shifts from one that honors legality and justice to a position that emphasizes the efficiency of the national forces. Bentham explains:

> regulations which, in a civil case, would be established with a view to justice, are accordingly, in this or that military case, made to give way to others, which are regarded as most conducive to the maximum of efficiency on the part of the national force. This sort of conflict being admitted – follows the observation – that in time of war, the demand for corroboration of power is at its maximum; the demand for justice at its minimum;

93 Nineteenth-century commentators were aware of Bentham's linkage to the Hobbesian tradition of absolute sovereignty. Frederick Pollock points out that, like Hobbes, Bentham argues that "the power of the sovereign must be treated as unlimited. The difference between free and despotic governments is in the constitution of the sovereign authority, not in its power." Frederick Pollock, *An Introduction to the History of the Science of Politics* (London: Macmillan and Co., 1890), 99–100. For more on this, see Mark Francis, "The Nineteenth Century Theory of Sovereignty and Thomas Hobbes," *History of Political Thought* 1:3 (1980); 517–540.

94 Jeremy Bentham, *The Constitutional Code* in *The Works of Jeremy Bentham, Volume 9* (Edinburgh: William Tait, 1843), 381.

95 Ibid.

> in time of peace the demand for justice is at its maximum; the demand for corroboration of power at its minimum.[96]

Under certain conditions, the demands of justice must be subordinated to the corroboration of power necessary to produce maximum efficiency of the national force. Bentham's utilitarian thinking gave conceptual clarity to the way a state might internally prioritize resources and how it might place legitimate, though strenuous, demands on the public during a time of war. But, as indicated above, it also served to frame the logic of acceptable civilian death more broadly. The "openly committed offenses against person and property" might be justified and deliberately pursued, not by an appeal to good intentions, but by determining whether they are "preponderant-evil-excluding."[97]

Bentham's friend and colleague, James Mill (1773–1836), father to the great nineteenth century philosopher John Stuart Mill, would lay out a very similar argument to Bentham's. War should be prosecuted in such a way that it "produce[s] the loss of [men in arms and the property of the government] as rapidly as possible, till the end or purpose of the war is obtained."[98] Suffering should not be gratuitous, and every effort should be made to neutralize an enemy without killing him. However, some necessary minimum amount of death is to be expected. The utilitarian formulation strives to rapidly kill the necessary threshold of (presumably) soldiers needed to attain victory.

Regarding the civilian population, Mill argues that armies had a duty to distinguish combatants from noncombatants and that the protection extended to the civilian population could only "be infringed upon special and justifying circumstances."[99] Of course, what those justifying circumstances are were not immediately clear. Mill explains that "the question, in settling the difficulties of international jurisprudence, is not whether an advantage is gained, but whether the advantage, such as it is, be not gained, at too great a cost."[100]

---

96 Ibid., 393.

97 As Alfred Mele and Steven Sverklik show, Bentham, Austin, and Sidgwick were not familiar with double effect arguments, and, when they formulated intention-based arguments, they believed that foreseeing consequences counted as intention, which undermined many of the late Scholastic formulations. Alfred Mele and Steven Sverklik, "Intention, Intentional Action, and Moral Responsibility," *Philosophical Studies: An International Journal for Philosophy in the Analytic Tradition* 82:3 (1996); 265–287. 267. To further understand the restraint of military necessity, see Jeremy Bentham, "A Plan for an Universal and Perpetual Peace," in *The Works of Jeremy Bentham, Volume VIII* (Edinburgh: William Tait, 1839). In this essay he lays out the parameters of a defensive war.

98 James Mill, "Law of Nations," *Essays* (London: Printed by J. Innes, Oxford Street, 1823), 21.

99 Ibid., 22.

100 Ibid., 23.

Mill's notion of "cost" was expansive; it must consider the harms inflicted upon both the winners and losers. He explains, "If it be certain that the losing party, in consequence of the destruction in question, loses more than the gaining party gains, it is certain that the two parties, taken together, are losers by proceeding."[101] Military necessity must also weigh the total amount of suffering; otherwise, both armies might, in the end, be losers.[102]

The progressive nature of this formulation notwithstanding, nineteenth-century utilitarians noted with frustration the high degree of imprecision in terms of identifying what the "justifying circumstances" of cruelty were that would lead to harm being inflicted on the civilian population. Henry Sidgwick (1838–1900) reiterates a version of James Mill's argument which seeks to calculate advantage not merely in terms of the immediate advantage and need of a belligerent, but in terms of total aggregate suffering. He writes, "It is clear that the aim of a moral combatant must be to disable his opponent, and force him to submission, but not to do him (1) any mischief which does not tend materially to this end, nor (2) any mischief of which the conduciveness to the end is slight in comparison with the amount of the mischief."[103] He continues, however, "Unfortunately, this second limitation is inevitably so vague as to leave room for great differences of opinion as to its proper application; and, moreover, its application must continually vary with variations in the arts of war, and in the circumstances and prevailing sentiments of civilised men."[104] To Sidgwick, the only feasible way to specify appropriate harms was in having the parties agree to limit the scope of war from the outset, such as by banning weapons that "inflict decidedly more suffering that others, without crippling the enemy at all in proportion."[105] While stronger prohibitions against unnecessary harm to civilians would be preferable, Sidgwick recognized the difficulty to reigning in prevailing opinion.

Much of Sidgwick's writing on war builds on arguments against unnecessary harm. For instance, he advances an early version of the "hearts and minds" approach taken by the US military in the twentieth and twenty-first centuries. He explains that "The effect of scaring the enemy into submission by harsh treatment of non-combatants is on the whole too uncertain and

101 Ibid.

102 This argument was also made by the pseudo-utilitarian John Austin. He explains that England would never have gone to war with her American colonies "if their dominant opinions and sentiments had been fashioned on the principle of utility." John Austin, *The Province of Jurisprudence Determined, Volume 1* [1832] (London: John Murray, 1861), 50.

103 Henry Sidgwick, *The Elements of Politics* (London: Macmillan and Co, 1897), 267.

104 Ibid.

105 Ibid., 268.

remote to outweigh (1) the danger of rousing the sympathetic indignation of neutrals, together with (2) the serious inconveniences to which an invading army is exposed in the midst of a population embittered by private injuries."[106] Harming innocent civilians may sour international opinion against one's war efforts, and the domestic population may stir up even greater antipathy among the occupied population. Restraint and placing limits on cruelty in advance would ensure the moral calculus of war would not swing too far off balance. Like Mill, utility must be understood in its most expansive sense.

Unlike Herbert Spencer (1820–1903), another English utilitarian who gave expression to an early form of social Darwinism, Sidgwick rejected the presumption that war could be valued in broad terms, for instance, even as a social evolutionary tool of development, i.e. as a utility in the broader scope of human evolution and socialization. Though Spencer was, like Sidgwick, less optimistic about war among modern industrial powers, he had argued that war in early human societies produced sufficient levels of social cohesion and cooperation that would ultimately foster the traits needed for peace. The necessities of self-defense led to a kind of "self-subordination" where individuals risked their lives for the preservation of the group.[107] Over time, "as war declines – gradually as the compulsory cooperation needful in dealing with external enemies becomes unnecessary, and leaves behind the voluntary cooperation which effectually achieves internal sustentation; there grows increasingly clear the code of conduct which voluntary cooperation implies."[108] To Spencer, while modern war was a great evil that should ultimately be eradicated in the service of continued human development, at some early stage primitive war nevertheless did serve a crucial social and organizational function.[109]

In stark contrast to Spencer, Sidgwick argues that "the morality between man and man is [negatively] affected by war between society and society, they have no tendency to prove that morality between man and man is thereby 'adapted' to the needs of a society in this condition of external enmity."[110]

106 Ibid., 269.

107 Herbert Spencer, *The Principles of Ethics, Volume 2* [1897] (Indianapolis: Liberty Fund, 1978). Pt. IV, Chap. 3, §260.

108 Herbert Spencer, *The Principles of Ethics, Volume 1* [1897] (Indianapolis: Liberty Fund, 1978), Pt. I, Chap. 8, §55.

109 Just to be clear, Spencer argues that war is the "greatest evil," and it is only in "resistance to invasion" when there is a "descent upon our shores" that the recourse to violence is permissible. He even went so far as to argue that the government should not have the right to wage war. See Herbert Spencer, "Letter V" in *The Proper Sphere of Government* (London: W. Brittain …, 1843), 19.

110 Henry Sidgwick, "The Inductions of Ethics," in *Lectures on the Ethics of T.H. Green, Mr. Herbert Spencer, and J. Martineau* (New York: Cambridge University Press, 2012), 230.

He rejects the view that, even in historical terms, war had been a key source of moral development, that the socialization and cooperation demanded by war allowed humans to develop conceptions of the common good, not to mention the individual sacrifices that would be necessary and worth it to achieve that good. To Sidgwick, this would be like saying that individuals could become *more* moral by participating in organized crime.[111] After all, it too requires coordination and high degrees of organization. To Sidgwick all crime, be it at the collective or interpersonal level, affects humans negatively, which is why one should be exceedingly cautious in the recourse to justify any war.

As Sidgwick lamented, when assessing the utility of a particular war, so much depends on the "prevailing sentiments of civilized men" and an estimation of the "mischief" the particular war would cause in relation to those ideas. The values of harm must be weighed against the value of the associated ideas/ideals that are at risk. He, after all, disagreed with John Stuart Mill (1806–1873) on the merit of the US Civil War. Sidgwick writes:

> I think as things are now, it would be just as sensible for us to exterminate Brazilian society as for the North to exterminate Carolinian. If the war were to end now, I should not regret it. The New Englanders have cleansed themselves from the guilt of slavery. Let the crusade cease; let us draw a 'cordon' round the abnormal society, and let the oligarchs try their experiment. It would get the world on faster.[112]

Sidgwick, of course, did not support slavery, but he believed the potential mischief of war was vulnerable to what he took to be the crusading rhetoric of the northern anti-slavery apologists. War is uncertain and it opens the door for greater cruelties. Even for an ostensibly good cause it would be better to prevent war from occurring to begin with.

In contrast, John Stuart Mill expressed vociferous support for the North during the US Civil War. He writes:

> War is an ugly thing, but not the ugliest of things: the decayed and degraded state of moral and patriotic feeling which thinks nothing worth *war*, is worse. When a people are used as mere human instruments for

---

111 However, this is *exactly* how political scientists explain the rise of the state. See Charles Tilly, "War Making and State Making as Organized Crime," in *Bringing the State Back In*. eds. Peter Evans, Dietrich Rueshemeyer, and Theda Skocpol (New York: Cambridge University Press, 1985), 169–191.

112 Henry Sidgwick, *A Memoir* (London: Macmillan, 1906); 102.

> firing cannon or thrusting bayonets, in the service and for selfish purposes of a master, such war degrades a people. A war to protect other human beings against tyrannical injustice; a war to give victory to their own ideas of right and good, and which is their own war carried on for an honest purpose by their free choice – is often the means of their regeneration. A man who has nothing which he is willing to fight for, nothing which he cares more about than he does about his personal safety, is a miserable creature, who has no chance of being free, unless made and kept so by the exertions of better men than himself.[113]

Mill argues that it was difficult to employ the utilitarian logic of greatest possible good to war when soldiers were being sent to kill and die for the benefit of greedy monarchs, for dynastic wars of conquest and domination. Wars of liberation, like the US Civil War, were different. While protecting individuals from tyranny might bring some suffering to the innocent, it may still produce a sufficient overall good, not only in terms of physical liberation, but also deliverance from spiritual and moral apathy.[114] This is, however, suggestive of a painstaking and uncertain moral calculus.[115] When military actions are tied to "right and good" they could be a powerful humanizing force in the world, but it is unlikely that Mill would agree that all perceptions of right are equally valid. Having something to fight for does not mean that one would be fighting for something truly beneficial or good. Eliminating tyranny and slavery are, to be sure, utilitarian goods, but the effects of war remain unpredictable, even if the end is good.

---

113 John Stuart Mill, *The Contest in America* (Boston: Little, Brown and Company, 1862), 31. For a favorable early American review of this document see Evert A. Duyckinck, *National History of the War for the Union, Volume 2* (New York: Johnson, Fry and Company, 1862), 157.

114 This is the position of Francis Lieber, the architect of the Union soldier code of conduct for the Civil War. He writes, "just wars are not demoralizing … The tone of morality of those who engage in a patriotic war is eminently raised." Francis Lieber, *Manual of Political Ethics, Volume 2* (Boston: Charles C. Little and James Brown, 1839), 645.

115 Critical of Mill's thinking on war, Mountague Bernard, one of the most prominent English international lawyers of this period, notes that "It would not be enough, for this purpose, to allege a general probability that good would in many cases be done by intervention. But, in fact, good is hardly ever done by it – good, I mean, in any degree commensurate with the evil. On the contrary, even when it dethrones a tyrant, puts an end to a ruinous anarchy, or stanches the effusion of blood in a civil war, it has a direct tendency to produce mischiefs worse than it removes." Mountague Bernard, *On the Principle of Non-intervention* (Oxford: J.H. and Jas. Parker, 1860), 9.

Despite these concerns, Mill was generally much more optimistic than other utilitarians in this period to justify resorting to war for the right reasons. He was frustrated at Britain's unwillingness to intervene in the US Civil War for humanitarian reasons. The emerging principles of self-defense had made it difficult to justify any war that had no immediate relevance to a state's national defense. Mill contended that the nation should articulate to itself what the criteria of non-defensive intervention would be. He writes,

> To go to war for an idea, if the war is aggressive, not defensive, is as criminal as to go to war for territory or revenue; for it is as little justifiable to force our ideas on other people, as to compel them to submit to our will in any other respect. But there assuredly are cases in which it is allowable to go to war, without having been ourselves attacked, or threatened with attack; and it is very important that nations should make up their minds in time, as to what these cases are.[116]

Intervention for the sake of liberation or other humanitarian causes *could* be justified. Even though Mill admits that war is an ugly thing, he was convinced that slavery and tyranny were worse. Nations could be a source of good in the world and war could also be the source of a nation's moral regeneration.[117]

Utilitarian thinkers differed in their approach to war in general, and they also differed in terms of their support for particular wars. As seen above, Sidgwick and J.S. Mill disagreed about the relative merits of the US Civil War, but they nevertheless provided a powerful conceptual tool – a moral calculus – that would enable leaders and military planners to conceive of the mischief and ugliness of war in a much wider frame. Certain ugly things might be permitted to prevent even uglier things. Of course, "ugly" is an aesthetic concept, one that is difficult to precisely pin down. For Mill, tyranny and slavery were uglier than war. His utility calculus might allow certain hardships to be extended to civilians to morally reinvigorate and redeem the whole population. Sidgwick, however, distrusted the crusading rhetoric of the North and held that the mischief of war would not override the harms committed in the

116 John Stuart Mill, "A Few Words on Non-Intervention" [1859], *New England Review* 27:3 (2006); 252–264. 258.

117 In contrast to Mill, Charles Tennant employs a utilitarian framework to show that modern war will *always* cost too much. He calculates that from 1691 to 1859 England had spent £3,041,471,701 in interest on repaying war loans. He points to the immense waste in financial and human capital that renders all war financially reckless. Charles Tennant, *Utilitarianism Explained and Exemplified in Moral and Political Government* (London: Longman, Green, Longman, Roberts, & Green, 1864), 446.

name of justice. Though they disagreed about particulars, all these utilitarian theorists affirmed the language of moral trade-offs. Lesser evils should be permitted in the service of a greater good.

6 The Link between Philosophical Utilitarianism and Military Necessity

What is so shocking about the rise of utilitarianism in this period is the degree to which it absolutely overshadowed and subsumed the entire military ethical discourse of the period, as though it were tailor made for nineteenth-century war-theorizing. Though, admittedly, in the process, much of the progressive character of utilitarian theory was diminished.[118] Few military planners and theorists who employed utilitarian principles in this period were willing to robustly consider the well-being of their enemy in their understanding of military necessity as the above utilitarians recommended, or, if they did, it was simply assumed that the enemy would benefit from sharp but brief conflicts.

Nevertheless, it is important to point out the almost universal success of the utilitarian-positivist approach to international law and the norms governing war-making at this time.[119] This discourse effectively overlapped with the military *practice* of restraint through the establishment of reasonably effective discipline which had arisen during the seventeenth century and persisted up to the twentieth century. The underlying assumption of this discourse was that it is almost always more effective to concentrate on dangerous adversaries rather than wasting energy on harming noncombatants. As noted above, the central assumption of noncombatant immunity, based on the principle of military necessity, was that those that did not need to be killed were immune from harm. Usually this meant that unarmed populations would not be killed.

In fact, prohibitions against harming noncombatants were typically construed in the language of the potential harm such killing would bring to the fighting force. For instance, Patrick Macdougall explains that the inhabitants of an enemy nation should not be harmed by the invading army, but not because they have immunity; instead, he notes that "If the men are allowed to rob and to pillage, the inhabitants will remove with all they can carry away beyond the reach, and after they proceeds of a few days' robbery shall

118 Much like Scholastic arguments for double effect, utilitarian formulas could be unscrupulously deployed to justify all manner of cruelty.

119 For some of these linkages, see Mónica García-Salmones Rovira, *The Project of Positivism in International Law* (New York: Oxford University Press, 2013).

be expended, difficulty will be found in getting supplies."[120] There were overriding prudential reasons for treating the local inhabitants with fairness and restraint. Indiscriminate violence and murder would have made it harder for the invading army to resupply and to pass through the country without local resistance. As can be seen in this example, this moderation in resorting to violence in many cases worked to the benefit of civilians, with or without formally established laws of war prohibiting the killing of noncombatants. Given these trends, it is not surprising that the utilitarian-grounded international agreements of this period focused primarily on the duty to minimize gratuitous harm, as this had already become a kind of operational logic of war. While it was true that sovereign states, animated by the duty of self-preservation, could pursue those actions which were believed to be militarily necessary to ensure their survival, in most cases it was simply understood that the willy-nilly slaughtering of noncombatants did not further those ends. Better to challenge those who had access to the weapons and nodes of military power. In practice, however, this formulation opened noncombatants up to certain intentional harms if those harms were deemed to be necessary.

It is worth noting what a remarkable shift this represented from the prior three centuries of military ethical theory. The principle of double effect, which had permitted the unintentional killing of innocent people, burst onto the scene in the sixteenth century as a conceptual technology used to justify killing the innocent in war, and it was a mainstay in military ethical discourse right up until the end of the eighteenth century. By the nineteenth century these arguments had almost completely vanished. In the first decade of the nineteenth century, William Duane (1780–1865), who would later become the US Secretary of the Treasury, noticed that contrary to "the general laws of justice," there had arisen "a new law, that of necessity" which was premised on the idea of self-preservation.[121] Duane was attuned to the seismic intellectual shift that was beginning to take place. The emerging legal positivism of John Austin (1790–1859) likewise rejected the appeal to "unmeaning abstractions, or to senseless fictions" like "eternal and immutable justice" preferring instead positive law rooted in the principle of utility.[122] Towards the end of the nineteenth century, Amos Hershey (1867–1933), an American political scientist professor who would later accompany the US delegation of the American

120 Patrick Macdougall, *The Theory of War: Illustrated by Numerous Examples from Military History, Volume 1* (London: Longman, Brown, Green Longmans, & Roberts, 1858), 25.

121 William John Duane, *The Law of Nations, Investigation in a Popular Manner* (Philadelphia: Printed by William Duane, 1809), 28. Duane worried that the law of necessity would be subject to great abuse: "Necessity can never exist contrary to justice." (Ibid., 33).

122 Austin, *The Province of Jurisprudence Determined*, 48.

Commission to Negotiate Peace after World War I, noted that by the outset of the nineteenth century, the "pure law of nature school [had] almost wholly disappeared."[123] The moderating conditions of the natural law, evidenced in both the Scholastic, martialist, and humanist traditions, simply evaporated. Consequently, while some theorists wrote about the incidental death of civilians, almost no prominent ethicist, strategist, or international law jurist justified civilian casualties by appealing to intentions or some variation of double effect in the nineteenth century.[124] These arguments almost entirely fell off the map.[125]

Unlike the humanist discourse, which in many cases did refer to some conception of the natural law, virtually every nineteenth-century commentator on the ethics of killing in war appealed directly to philosophical utilitarianism

---

123 Amos Hershey, "History of International Law Since the Peace of Westphalia," *The American Journal of International Law* 6:1 (1912); 30–69. 64–65. More recently, Gardam argues that unlike the Scholastic just war theory, "the nineteenth century was the era of the growth of positivism" and as such the divine law and the law of nature "had no part to play in the scientific discipline of international law." Judith Gail Gardam, *Non-combatants Immunity as a Norm of International Humanitarian Law* (Dordrecht: Martinus Nijhoff Publishers, 1993), 10.

124 For instance, the early twentieth-century moral theorist, Walter McDonald, writes, "I confess I cannot see in all this any basis of explanation of why it should be wrong to cause directly the death of an innocent man, but right to produce the very same effect indirectly. The action of firing a shell into a besieged town and thereby killing innocent people, is the same whether it be performed by an agent in his character of soldier waging just war, or of a personal enemy seeking to satisfy private an unjust resentment. How is the effect produced more directly in one case than in the other?" Walter McDonald, *The Principles of Moral Science* (Dublin: Browne and Nolan, 1903), 153.

125 An important caveat is needed here. Of course, Catholic writers continued to argue that the intentional killing of innocents was morally wrong; however, these arguments were almost never incorporated into the wider secular literature on the laws of war. This is surprising since Christopher Kaczor traces the idea of double effect to the nineteenth century French Jesuit Jean Pierre Gury (1801–1866); he "is often considered the originator of the modern notion of double-effect reasoning." Christopher Kaczor, "Double-effect Reasoning from Jean Pierre Gury to Peter Knauer," *Theological Studies* 59 (1998); 297–316. 300. See Jean Pierre Gury, *Compendium theologiae moralis* (Ratisbonae: Typis et sumptibus Georgii Josephi Manz, 1868), Art. III, §1, 401, 179. For additional Catholic intention-based arguments against killing the innocent in war in the nineteenth century, see Jacques Paul Migne, *Theologiæ cursus completes* (Paris: Apud Editores, in via Gallice Dicta, 1839), Sec. II, Chap. IX, Dub. VII, 606; M.J. Piercy, *Catechism of the Diocese of Paris* (London: Richardson and Son, 1850), 178; Joseph Rickaby, *Moral Philosophy or Ethics and Natural Law* (London: Longman, Green & Co., 1888), 202–204; John Elmendorf, *Elements of Moral Theology: Based on the Summa Theologiae of St. Thomas Aquinas* (New York: James Pott & Company, 1892); and Edward Génicot, *Theologiæ Moralis Institutiones* (Lovanii: Typis et sumptibus polleunis et ceuterick, 1898), 339.

or some crude utilitarian calculus to underwrite their sense of military necessity. For instance, in his 1873 *Commentaries upon International Law*, Sir Robert Phillimore characterizes the laws of war this way:

> The great principle upon which all these rules are framed, is that of, on the one hand, compelling the enemy to do justice as speedily as possible, and, on the other hand, of abstaining from the infliction of all injuries both upon the subjects of the enemy, and upon the Government and subjects of third Powers which do not, certainly and clearly, tend to the accomplishment of this object. Wanton devastation of the enemy's territory, wanton cruelty exercised towards his subjects, are, therefore, according to the principles and practice of Christian nations, unjustified and illegal.[126]

Though understood as a clear restraint against the cruelty inflicted on subjects, Phillimore concedes that "It is, in fact, in many cases, only through the privations and distresses of the subject that his Government will be induced to put an end to the war by according the justice demanded."[127] Even though military necessity was designed as a conceptual and moral limit to the harm meted out against innocents – utilitarianism, after all, was the vanguard of progressive politics in that period, striving for the greatest good for the greatest number – it was never designed to proscribe *all* harm against noncombatants as noncombatants. As a moral system, utilitarianism is built on the principle of the lesser evil and moral trade-offs. So, inasmuch as it aims for the overall wellbeing of all engaged in hostilities (directly and indirectly), this necessarily entails that "privations and distresses" can be placed on some indeterminate minimum number of subjects.

Sir Edward Creasy also appeals to the "Principle of Modern Utilitarianism," which stipulated that "in order to test the soundness and value of a line or rule of action, you must examine whether its whole tendency is or is not to promote in a reasonable degree the advantage, not merely of the acting party, but of all who are liable to be affected by it, if it were generally acted on as a rule."[128] And in a certain respect, despite what may seem as its vague and overly-permissive quality, when theorists referred to military necessity, it was,

126 Sir Robert Phillimore, *Commentaries upon International Law, Volume 3* (London: Butterworths ..., 1873), 78–79.

127 Ibid., 79.

128 Sir Edward Creasy, *First Platform of International Law* (London: John van Voorst ..., 1876), 416. It is worth noting that this is a strange hybrid of the Kantian categorical imperative and nineteenth century utilitarian formulas.

like with Creasy's remark, infused with this progressive, harm-reduction tendency. An authentic conception of military necessity would benefit all parties, even those on the opposite side of the battlefield.

The logic of utilitarianism can be traced through many definitions of military necessity in this period. In his 1861 treatise on international law, General Henry Halleck explains the function of military necessity this way:

> The law of nature gives to a belligerent nation the right to use such force as may be necessary, in order to obtain the object for which the war was undertaken. Beyond this, the use of force is unlawful: *this necessity forms the limit of hostility between subjects of the Belligerent States*. They, therefore, have no right to take the lives of non-combatants, or of such public enemies as they can subdue by other means, nor to inflict any injuries on them or their property, unless the same should be necessary for the object of the war.[129]

Notice that this formulation does not eliminate hostilities against subjects of belligerent states; it simply helps to form a limit, a threshold. Killing civilians must not be excessive, gratuitous, or vindictive; it must advance a clear strategic military aim, which must ultimately refer back to "the object for which the war was undertaken" – ostensibly, the victory in war and the preservation of the state, which, as an arbiter of rights and protections, made it the supreme utilitarian good.[130]

Much along the same lines, Article 14 of the Lieber Code (1863) states, "Military necessity, as understood by modern civilized nations, consists in the necessity of those measures which are indispensable for securing the ends of the war, and which are lawful according to the modern law and usages of war." And in Article 16, Lieber writes: "Military necessity does not admit of cruelty, that is the infliction of suffering for the sake of suffering or for revenge, nor of maiming or wounding except in fight, nor of torture to extort confessions."[131]

---

129 Halleck, *International Law*, 2.

130 In a similar manner, William Paley writes that "The license of war then acknowledges two limitations: it authorizes no hostilities which have not an apparent tendency to effectuate the object of the war; it respects those positive laws which the custom of nations hath sanctified, and which, whilst they are mutually conformed to, mitigate the calamities of war, without weakening its operations, or diminishing the power or safety of belligerent states." William Paley, *The Principles of Moral and Political Philosophy* (Boston: Published by West and Richardson, 1815), 472.

131 Instructions for the Government of Armies of the United States in the Field, prepared by Francis Lieber, LL.D., Originally Issued as General Orders No. 100, Adjutant General's Office, 1863, Washington 1898: Government Printing Office. Incidentally, Francis Lieber

Later in that century, Thomas Walker further clarifies this point: "Men may condone the cruelty which is dictated by necessity, but the necessity which grounds excuse must be instant, absolute, and overwhelming."[132]

As these quite significant nineteenth-century documents suggest – indeed, many of these documents were precursors to the international laws of war formally established in the Hague Conventions of 1899 and 1907 – the point of justifying civilian casualties by way of necessity was to narrow the scope of legitimate civilian death, to "form a limit" as Halleck explains. It was a progressive feature of the emerging positivist, utilitarian-informed international legal consensus to clearly demarcate necessary from unnecessary civilian death. In this respect, it was widely understood as a conceptual tool of restraint.[133] In this context, what made harm inflicted on noncombatants wrong was not the *intent* to harm, which had been the moral focus of double effect justifications, but whether such harm was gratuitous or unnecessary, that is, whether it was proportionally instrumental in advancing an objectively good military aim or not. By the mid- to late-nineteenth century, the push to humanize warfare through reason and international agreement was thoroughly embodied in the utilitarian logic of military necessity.

Seeing military necessity as a form of utilitarian restraint is also preserved in some of the international legal agreements of this era. Take the Saint Petersburg Declaration of 1868. This document was the outgrowth of a meeting organized by Czar Alexander II to promote the banning of exploding bullets, which were believed to needlessly increase the suffering of soldiers but to have no real impact on battlefield mortality. The document states:

> That the progress of civilization should have the effect of alleviating as much as possible the calamities of war; That the only legitimate object which States should endeavour to accomplish during war is to weaken the military forces of the enemy; That for this purpose it is sufficient to disable the greatest possible number of men; That this object would be exceeded by the employment of arms which uselessly aggravate the sufferings of disabled men, or render their death inevitable; That the

---

and General Halleck engaged in a lengthy correspondence while Lieber was developing the Code. See Brainerd Dyer, "Francis Lieber and the American Civil War," *Huntington Library Quarterly* 2:4 (1939); 449–465.

132 Thomas Henry Walker, *The Science of International Law* (London: Cambridge University Press, 1893), 341.

133 See Burrus Carnahan, "Lincoln, Lieber and the Laws of War: The Origins and Limits of the Principle of Military Necessity," *American Society of International Law* 92:2 (1998); 213–231.

> employment of such arms would, therefore, be contrary to the laws of humanity.[134]

The language of this document gives the impression that military necessity was in some sense quantifiable; for an action to conform to military necessity, it should aim at producing the minimum amount of suffering, indeed, only the minimum amount of death required to assure victory. Emulating these utilitarian sentiments, Henry Maine explains the principle governing the use of violence in war is this:

> that in the mode of carrying on the war no greater harm shall be done to the enemy than necessity requires for the purpose of bringing him to terms. This principle excluded gratuitous barbarities, and every description of cruelty and insult that serves only to exasperate the sufferings or to increase the hatred of the enemy without weakening his strength or tending to produce his submission.[135]

The humanizing trajectory of international law would eventually eliminate those things that would uselessly aggravate or exasperate the suffering of the enemy and would not bring about a clear military advantage.

In this respect, military necessity could inform the laws of humanity inasmuch as it compelled militaries to strive for a sufficient minimum of overall casualties. As paradoxical as it may seem, *intentional* or *deliberate* destruction was perfectly acceptable according to the laws of war, that is, so long as it was not *excessive* or *wanton*. These were the exact arguments made against the exploding bullet. Since they provided no strategic advantage, and because such bullets generated needless suffering, they violated the utility calculus that underwrites the laws of war. And while the Saint Petersburg Declaration of 1868 is largely directed at unnecessary suffering inflicted upon soldiers, the same logic could be applied to the excesses of civilian death. Killing civilians in a gratuitous and haphazard manner, in ways that simply did not advance a clear strategic aim could therefore be prohibited. However, quite importantly, when it was perceived to be instrumental and necessary, those deaths could be intended.

---

134 Declaration Renouncing the Use, in Time of War, of Explosive Projectiles Under 400 Grammes Weight. Saint Petersburg, 29 November/11 December 1868. https://ihl-databases.icrc.org/ihl/full/declaration1868.

135 Henry Maine, "The Modern Laws of War," in *International Law: A Series of Lectures Delivered Before the University of Cambridge, 1887* (New York: Henry Hold and Company, 1888), 145–46.

Throughout the nineteenth century, building on the martialist tradition which can be traced back to the sixteenth century, there was a growing sense that there was an indubitable relationship between the "improvements in the practice of War" and the "growth of those civilized habits which are the only augury of the final establishment of Peace."[136] In other words, effective armies were not only efficient on the battlefield, in terms of military discipline and in the execution of orders, but their efficiency could be seen in their overall humanitarian trajectory.[137] Only restrained and efficient militaries, those professionally and in some sense scientifically seeking the minimal amount of suffering to achieve victory, would be both victorious in battle and in sustaining peace.[138] In this respect, the laws of humanity were understood as an achievement rooted in a progressive utilitarianism.

While some theorists of this period were critical of the potential excesses of military necessity,[139] most theorists who employed this type of reasoning emphasized its implicit limit-forming quality. The priority of necessity could offer some guidance as to what counts as proportional: "'Military necessity,'

---

136 Sheldon Amos, *Political and Legal Remedies for War* (London: Cassell, Petter, Galpin & Co., 1880), 332.

137 For a similar point see George B. Davis, *The Elements of International Law* [1887] (New York: Harper and Brothers, 1900), 301.

138 For more on the development of the martialist tradition, see Chapter 2. Incidentally, literature promoting martial virtues persisted well into the nineteenth century, typically in the form of formalized military law. See William Hough, *The Practice of Courts-martial, and Other Military Courts* (London: Published by Messrs. Parbury, Allen & Co …, 1834); Theodore Thring, *A Treatise on the Criminal Law of the Navy* (London: V.& R. Stevens and Sons, 1861); Great Britain Admiralty, *The Queen's Regulations for the Royal Nave: Revised* (London: Printed by Harrison and Sons, 1862); Henry Steuart Tompson, *Catechism on the Army Discipline and Regulations Acts, 1879. Courts-martial* (London: W. Mitchell & Company, 1880).

139 Amos writes, "The extension of the War, and of its consequences, material and moral, to non-combatants, tends to prolong, to embitter, and to renew War quite as much at sea as on land, where such extension is abandoned." Amos, *Political and Legal Remedies for War*, 307. Thomas Arnold argues, "But take the cases themselves, and examine them in all their circumstances; the degree of suffering inflicted, the innocence and helplessness of the sufferers, the interests at stake, and the possibility of otherwise securing them; and if any man can defend the lawfulness in the abstract of the starvation of the inhabitants of Genoa, I will engage also to establish the lawfulness of the massacres of September." Thomas Arnold, *Introductory Lectures on Modern History* (New York: D. Appleton & Co., 1862), 17. See also James Lorimer, *The Institutes of the Law of Nations; a Treatise of the Jural Relations of Separate Political Communities* (Edinburgh: Blackwood and Sons, 1883). Also, Carnahan shows that Confederate officers in the US Civil War routinely decried the excesses of military necessity stemming from the Lieber Code. See Carnahan, "Lincoln, Lieber, and the Laws of War," 217.

in the proper sense of the term, extends only to what is needed to compel the opposing government to come to terms."[140] But, of course, despite its apparent precision, the exact amount of permissible suffering a military might need to mete out remained shrouded in epistemic uncertainties that were not always fully appreciated in the literature. The precise amount of cruelty that would ultimately be needed at any given moment depended on many difficult to quantify variables: how capable the receiving nation was at bearing up under the hardships of war, how calloused its leadership was to the suffering of their subjects, and some implicit judgment as to whether or not the suffering of innocents was necessary for the ends in question.[141] Far from being precise, beliefs about necessity varied from situation to situation, general to general. In the end, military necessity frequently smuggled in assumptions that vitiated any attempts at precise quantification.

The force of utilitarian reasoning in the military manuals of this period is to demarcate a limit of acceptable suffering. Though, through this formulation there emerges what amounts to explicit permission to commit some direct harm to civilians, that is, if that lesser evil produces an overall desirable effect. The utilitarian calculus allowed for – indeed, expressly permitted and in many cases explicitly sought after – some strategically minimal amount of civilian death. The progressive attempts to humanize war did not entail the eradication of all harm to civilians or to confer them some indubitable status, like blanket noncombatant immunity. In a certain sense, it may be more accurate to say that the utilitarian logic of military necessity expressly informed the degree to which noncombatant protections would be extended in any moment. Despite its seemingly clear limitations, the humanizing element embedded in the nineteenth-century theory of military necessity was founded on the attempt to *limit* not *eliminate* harm inflicted on noncombatants, to designate a scope of

140 Jeremy Rabkin, "Proportionality in Perspective: Historical Light on the Law of Armed Conflict," *San Diego International Law Journal* 16:2 (2015); 272–73.

141 Take for instance the experience of John Mance, who was part of the English besieging force in the assault on a fortress at Campveer during the Walcheren Campaign of 1809. He was especially embittered at the callousness of the town's governor: "Many of the unoffending inhabitants of this town were literally murdered, by the combustibles of war, which scattered death around: the women and children might have been spared, if the iron-hearted commander had acceded to our flag of truce, mercifully offered by the Earl of Chatham, to allow them and the helpless to pass out of the town, under the protection of the British troops, but this governor would not agree." John Mance, *The Soldier Spiritualized; or National, Compared with Spiritual Warfare* (London: Partridge and Co., 1859), 110.

legitimate suffering, that is, to demarcate spaces for killing and then strive for some minimal permissible amount of death. Those that fell within the scope of permissible or necessary harm could be deliberately targeted and killed, a small price to pay for less overall suffering.

7 Conclusion

While it is commonly believed that the principle of noncombatant immunity can be traced back to the nineteenth century, this chapter illustrates some of the challenges to this view. To begin with, the discrete category of the civilian was still in its infancy. The concept of the noncombatant was much more common, but it was not always clear who deserved to be so classified. Rather that reflecting an obvious binary relationship – i.e. between those who are fighting and those who are not – the literature of this period is more suggestive of a sliding scale of targetability. In certain cases, noncombatants would transform into combatants. What is more, the dilemma posed by the public militia and the issue of irregular soldiers also allowed individuals to toggle back and forth between the roles of combatant and noncombatant. This chapter also shows that the formalization of military necessity and self-preservation, bolstered by both the rise of utilitarianism and the legacy of the Westphalian sovereignty system, contributed to a moral environment that permitted militaries to intentionally kill some limited number of noncombatants under certain conditions.

CHAPTER 5

Deliberately Targeting Noncombatants

1 Introduction

The humanist discourse of the seventeenth and eighteenth centuries was transformed and bolstered by the logic of utilitarianism. Both systems of thought placed a great deal of moral weight on necessity, which infused military necessity with a special significance, transforming it into a rough metric for the potentially cruel requirements of an effectively waged war. Under normal conditions, killing innocent bystanders was not necessary to expedite victory. In fact, gratuitous death might *inhibit* the ability of an army to effectively wage war – international or domestic opposition to the war might increase. Killing local inhabitants may also make it harder to collect contributions. For violence to be effectively deployed it must be both sufficiently minimalistic and exceedingly necessary. Normally, this meant that fighting was to be directed at armies rather than noncombatants. The idealized conception of warfare in this period was pitched battle between well-regulated, professional armies. Nevertheless, what the moral formulation of necessity implies, however, is that in some cases and under certain conditions it is permissible to directly target civilians. Under the limiting logic of necessity, military manuals of this period routinely gave expression to exceptional cases where the innocent could be targeted. A deliberate action that caused harm was permissible if it was proportional and instrumental in expediting the war's end. In short, under certain conditions, it was widely assumed that militaries could intentionally kill civilians.

In the extensive commentary on war and international law in this period, there were four domains where civilians could be intentionally targeted and killed based on military necessity: 1) States had the right to lay waste to border lands in order to alleviate proximate threats or to punish enemies. If a neighboring country routinely made incursions (or threats of incursions) into one's homeland, that country could be destroyed for punitive and practical reasons. Laying waste to border lands was permitted to punish unjust aggression and to prevent such attacks from continuing. But beyond this, it was widely assumed that the civilians of a belligerent nation were expected to endure the hardships of war, which included ravaging, burning, and other brutalities, especially when they offered material support to a cruel enemy army.

2) As the scope of international commerce grew, along with the size and capability of navies, it became strategically feasible to blockade entire nations.

 | DOI:10.1163/9789004515482_007

Throughout the nineteenth century, enemy nations could be besieged and blockaded with the intended purpose of starving the civilian population, making life so unbearable for residents that they would pressure their government into seeking an end to the war.

3) Civilians living in or around fortified or defended cities could be directly targeted to expedite surrender. It was not a violation of the laws of war to intentionally direct fire at residential districts within or near a fortress or defended place. Detached suburbs of a fortified place could also be directly targeted, and any villages or towns lying en route to a fortress or defended place could be cleared. It was also permissible to drive residents back into the bombarded fort once bombing was underway. While it was generally believed to be honorable to permit women and children to leave before bombing commenced, commanders were under no obligation to do so.

Finally, 4) if the murder of civilians by a hostile force was seen as deliberate and went unpunished, a country had the right to repay the wrong through reprisal, sometimes also referred to as retaliation or retorsion. The terms meant different things at different times in the history of military ethical theory, but by the late nineteenth century it was still permissible to kill an innocent person for crimes they did not commit, for crimes committed by one of his compatriots. For instance, if an enemy nation or a foreign army engaged in acts of cruelty against the civilian enemy population, this would open their population up to similar cruelties. Historically, reprisal was seen as a type of private, sub-national form of justice; later, it became associated with acts of state that were just short of war. However, over time, they came to be seen as mechanisms of restraint in war. Since militaries would repay violations of the laws of war with in-kind violence, this would encourage military commanders to exercise additional caution so their civilian population would not be treated with similar brutality. If one army refused, for instance, to honor the principle of distinction, the other would retaliate the same way. In practice, however, this meant that under certain conditions, militaries were permitted to intentionally kill people they knew to be innocent as punishment for crimes they did not commit.

Collectively, these norms helped to constitute the spatial quality of the nineteenth-century battlefield. They demarcated places where erstwhile noncombatants would become legitimately subject to direct violence. Those residing in borderlands and in the suburbs of fortified cities could be intentionally targeted; if not outright killed, their very livelihood and material sustenance could be deliberately destroyed. Understanding noncombatant immunity as something closer to a spatially linked concept helps explain the risks associated with the threat of retaliation. Those who violated, or were perceived to violate, the ascendant norms of war risked opening their civilian populations to greater amounts of direct harm. Reprisals threatened to shatter these more tidy, spatially constrained conceptions of the battlefield.

As seen in the previous chapter, many wrongly assume that the concept of noncombatant immunity finds its origins in the international legal theory of the nineteenth century. With one or two exceptions, virtually no commentator or international jurist in this period spoke in these terms, and none spoke unequivocally about noncombatant immunity. Of course, the humanitarian trajectory of the law was designed to narrow the window of permissible harm. Rather than immunity, many nineteenth-century commentators simply believed that given the efficiencies of war excess killing was becoming much less necessary and that war itself might one day be phased out. Nevertheless, it should be remembered that rather than a radical break or pivot in the norms, the Hague Conventions (1899, 1907) largely reproduced the ascendant norms of utility and military necessity of this period. As such, one should be cautious of reading into these documents more expansive definitions of noncombatant immunity. Furthermore, as will be seen, there is very little difference between the moral formulations of the early nineteenth century and those that were published in the interwar period, right up to the outbreak of World War II.

2 Enduring the Hardships of War: Laying Waste

General Philip H. Sheridan's campaign through the Virginia Shenandoah Valley (1864) was so brutal that life for the civilian population was increasingly untenable. Describing Sheridan's march through Millwood, Virginia, one Southern diarist of this period writes, "The whole country was enveloped with smoke and flame. The sky was lurid and but for the green trees one might have decided the shades of Hades had descended suddenly."[1] Accounts like these can be found all throughout the region. Indeed, as John Heatwole argues, "the civilian population of the Valley was affected to a greater extent than was the populace in any other region during the war, including those in the path of General Sherman's infamous march to the sea in Georgia."[2]

1 "Mathella Page Harrison: Diary, August 17, 1864," *The Civil War: The Final Year Told by Those Who Lived It* (New York: The Library of America, 2014), 322.

2 John Heatwole, *The Burning: Sheridan's Devastation of the Shenandoah* (Charlottesville, VA: Rockbridge, 1998) x. More recently, Thomas Carson has questioned the destructiveness of this campaign. Thomas Carson, *Lincoln's Ethics* (New York: Cambridge University Press, 2015), 197.

Many first-hand accounts corroborate the challenges the noncombatant population faced. As one contemporary chronicler of the campaign writes,

> Young girls with flushed cheeks, or pale with tearful or tearless eyes, have pleaded with and cursed the men whom the necessities of war have forced to burn the buildings reared by their fathers, and turn them into paupers in a day. The completeness of desolation is awful. Hundreds of nearly starving people are going North. Our trains are crowded with them. They line the wayside. Hundreds more are coming – not half the inhabitants of the valley can subsist on it in its present condition.[3]

A Confederate sympathizer, Edward Pollard, explained the hardships this way:

> The horror and crime of this devastation was remarkable even in Yankee warfare. They impoverished a whole population; they reduced women and children to beggary and starvation; they left the black monuments of Yankee atrocity all the way from the Blue Ridge to the North Mountain. It is remarkable that the worst of Yankee atrocities were always done in the intoxication of unexpected success, when no longer the fears of previous disasters held in check their cruel cowardice, and intimidated their native ferocity.[4]

Unlike the violence meted out by Sherman's Georgia campaign, the brutality of the valley campaign was "committed systematically, and by order."[5]

As a telegram from Lieutenant General Grant explained to Sheridan, "If the war is to last another year we want the Valley to remain a barren waste."[6] In another letter to Major-General Halleck, he ordered that the pursuit of the Confederates through the Shenandoah should be so rigorous and thorough "that crows flying over it for the balance of this season will have to carry their provender with them."[7] To his Chief of Cavalry, Brigadier General A.T.A. Tobert, Sheridan wrote, "the object is to make the Valley untenable for the raiding

3 Robert Tomes, *The War with the South: A History of The Late Rebellion, Volume 3* (London: Virtue & Yorston, 1867), 492.

4 Edward A. Pollard, *The Last Year of the War* (New York: Charles B. Richardson, 1866), 109.

5 Quoted in Heatwole, *The Burning: Sheridan's Devastation of the Shenandoah*, xi.

6 Philip Sheridan, *Personal Memoirs of H. Sheridan, General, United States Army, Volume 1* (London: Chatto & Windus, Piccadilly, 1888), 486.

7 *The War of the Rebellion: A Compilation of the Official Records of the Union and Confederate Armies, Series 1, Volume 3* (Washington D.C.: Government Printing Office, 1892), 223.

parties of the rebel army."[8] Quite expressly, the Union aim was to make the region uninhabitable.

The grim work of laying waste to the Valley was even hard on the morale of many of the Union soldiers commanded to undertake the task. In his memoirs of the war, Samuel Levis Gracey, the chaplain of the Sixth Pennsylvania Cavalry, writes, "Our regiment avoided the detail for this unpleasant duty, and marched in rear of the destroyers, who, stretched in long line across the Valley, did their work thoroughly."[9] He continues, "The day had been an unpleasant one; the weather was hot and the roads very dusty, and the grief of the inhabitants, as they saw their harvest disappearing in flame and smoke, and their flock being driven off, was a sad sight. It was a phase of warfare we had not seen before, and though we admitted its necessity, we could not but sympathize with the sufferers."[10] In fact, tensions in the region remained high, Gracey explains, because, in retaliation for the burning of homes and livelihoods, the Confederate Commander John Mosby, the "Gray Ghost," reportedly killed a number of Union prisoners. In response, the Union army took hostages from the local population, "to the great distress of their families."[11] The compacted cruelties of ravaging, murder, and kidnapping made life in the region one of desperation and struggle for the inhabitants.

Some of the young Union infantrymen, as Edmund Ollier reports, were overzealous in the punishment they meted out. He writes,

> more was done than had been contemplated by the commanders; for, when once the reins are given to an infuriated soldiery, many individual acts of atrocity will be committed, which the sternest of generals would refuse to sanction. The torch was often applied without the slightest attempt at discrimination between the guilty and the innocent. It was no part of the design to burn houses; yet many houses were burned, even apart from those which Sheridan purposely committed to the flames in revenge for the death of Lieutenant Meigs. Some of the unfortunate people lost their all; families were reduced in a moment from wealth to indigence; hundreds were left with scarcely the necessities of life; and it was affirmed that not half the inhabitants of the valley could subsist on it in the state to which it was then reduced.[12]

8 Samuel Levis Gracey, *Annals of the Sixth Pennsylvania Cavalry* (Philadelphia: E.H. Butler and Co., 1868), 286.

9 Ibid., 286.

10 Ibid., 286–87.

11 Ibid., 287.

12 Edmund Ollier, *Cassell's History of the United States, Volume 3* (New York: Cassell Petter & Galpin, 1874), 542.

The punishments heaped onto the Southern communities in the Shenandoah Valley was not incidental but intentional. Notwithstanding the destructive enthusiasm of the young soldiers, it was a calculated strategy designed to end a bloody war that had gone on far longer than expected.

It should be remembered, that the strategy in the Shenandoah was a sharp divergence from Lincoln's policy of 1861 where he had cautioned against "radical and extreme measures, which may reach the loyal as well as the disloyal."[13] John Pope's General Orders, No. 5 of 1862 had in fact stipulated that "vouchers will be given to the owners [of confiscated property], stating on their face that they will be payable at the conclusion of the war, upon sufficient testimony being furnished that such owners have been loyal citizens of the United States."[14] This is suggestive of an earlier phase of the war where the Union policy was devised to avoid wanton destruction and to, at the very least, discriminate between loyal and disloyal civilians.[15]

As the war slogged on, these early conciliatory policies gave way to the much more indiscriminate "hard hand of war."[16] The logic of necessity, recalibrated for local resistance and the continued support for the Confederacy, demanded sharper measures. Indeed, part of this rationale of shifting the burden of war more directly onto the citizenry can be explained in Sheridan's memoirs. He writes, "Death is popularly considered the maximum of punishment in war, but it is not; reduction to poverty brings prayers for peace more surely and more quickly than does the destruction of human life."[17] As this letter suggests, Sheridan knew the devastation of the Valley would bring immense hardships onto the civilian population, but he believed that targeting civilian infrastructure under these conditions was justified since it was more effective than outright murder. Reducing people to poverty leads to "prayers for peace." But of course, as seen in the fallout from the devastation en route to Metz in the previous chapter, in many cases there was very little difference between outright "destruction of human life" and "reduction to poverty."

---

13 Abraham Lincoln, *The Life and Public Services of Abraham Lincoln* ed. Henry Raymond (New York: Derby and Miller, 1865), 222.

14 John Pope, "General Orders, No 5," in John Codman Ropes, *The Army Under Pope, Volume 4* (New York: Charles Scribner's Sons, 1881), 174.

15 Tooley argues that though Pope did not intend this, his orders were the origin of the "new aggressiveness toward civilians." T. Hunt Tooley, "'All the People Are Now Guerillas': The Warfare of Sherman, Sheridan, and Lincoln, and the Brutality of the Twentieth Century," *The Independent Review* XI:3 (2007); 355–379. 357.

16 Mark Grimsley, *The Hard Hand of War: Union Military Policy Toward Southern Civilians, 1861–1865* (New York: Cambridge University Press, 1995).

17 Sheridan, *Personal Memoirs of H. Sheridan*, 488.

Hard measures were necessary to prevent even greater cruelties. It would be difficult to conclude that Sheridan's moral calculus was that targeting civilian infrastructure was a tragic yet efficient way of ending the war, and that the civilian population was only being indirectly targeted, as the logic of double effect had implied in centuries past. It is clear he assumed that, as members of an enemy nation, the residents of the Shenandoah were not strictly speaking innocent. They were not fully exempt from the cruelties of war. In fact, to a certain extent, they *deserved* to be targeted, at least within the limits of necessity; their houses and food stores could be destroyed, and they could be kidnapped or forced into refugee status.

Part of Sheridan's justification for the destruction of the valley was explained by appealing to the intersection of guilt and necessity. Inasmuch as civilians enabled the enemy to prosecute the war, they more formally entered the military necessity calculus, which eroded whatever protections they might have had as noncombatants. Sheridan explains, the inhabitants of the Valley "were bitterly hostile to the Government; they had promoted as much as they could every Confederate expedition against Washington; indeed, without their assistance it was scarcely possible that such expeditions could be made."[18] Their moral and material support of the Southern cause made them guilty and as such they ceased to resemble noncombatants in the strict sense of the term. Lincoln's Secretary of War, Edwin Stanton, expressed a similar sentiment: "No aphorism is more universally received than that the sole object of a just war is to make the enemy feel the evils of his injustice, and, but his sufferings, amend his ways; he must therefore be attacked in the most accessible quarter."[19] Punishing the nonfighting population, forcing them to feel the hardships of war by attacking them on "the most accessible quarter," was permissible inasmuch as it would lead them to "amend their ways" and cease supporting the rebel cause.

Given this framework, it is not surprising that in General Sherman's March through Georgia he too claimed: "War is cruelty, and you cannot refine it. Those who brought war on our country deserve all the curses and maledictions a people can pour out."[20] In their moral and material support for the Confederacy, the innocent brought suffering onto themselves. Similar sentiment can be seen in a testy exchange with General Hood on the evacuation of citizens from Atlanta (1864) – where Hood accused General Sherman of firing past the

18 Walker, *The Science of International Law*, 552–553.

19 Ibid., 553.

20 Willis Fletcher Johnson, *Life of William Tecumseh Sherman* (Philadelphia: Edgewood Publishing, 1891), 351.

military lines into the city, an "unprecedented measure" that "transcends, in studied and ingenious cruelty, all acts ever before brought to my attention in the dark history of war" – Sherman reminded Hood that it was "you, who, in the midst of peace and prosperity, have plunged a nation into civil war – 'dark and cruel war.'"[21] Whatever cruelties Southern civilians endured could be traced back to the original guilt of the secessionist cause, and, perhaps even more proximately, their direct material support of the Confederacy.

3 Laying Waste and the Laws of War

The devastation of the Shenandoah Valley and Sherman's March to the Sea were not deviations from the laws of war. Under the framework of military necessity, civilians could be starved, have their livelihoods and provisions destroyed, and in some cases be intentionally killed (through reprisals and hostage-taking) if those actions were believed to be necessary to achieve legitimate military aims. As seen in the eighteenth-century humanist discursive tradition in Chapter 3, military necessity permitted invading armies to lay waste to borderlands. The great Swiss philosopher Emer de Vattel explicitly affirms this logic: "On certain occasions, however, matters are carried still farther: a country is totally ravaged, towns and villages are sacked, and delivered up as prey to fire and sword. Dreadful extremities, even when we are forced into them! Savage and monstrous excesses, when committed without necessity!"[22] In fact, Vattel could see two cases where necessity might demand "ravaging and burning": 1) "chastising an unjust and barbarous nation, of checking her brutality, and preserving ourselves from her depredations," and 2) to "render it uninhabitable, in order to make it serve as a barrier, and to cover our frontier against an enemy whose incursions we are unable to check by any other means."[23]

Similar arguments were repeated throughout the nineteenth century. José María de Pando (1787–1840), Spain's former Secretary of State, writes: "Sometimes, it is true, the terrible law of war permits fields to be cut down, towns to be plundered, and sword and flame to be unleashed everywhere; but only to punish an unjust and ferocious nation, or to maintain a barrier to the

21 "Official Correspondence from Major General W.T. Sherman to General J.B. Hood, Sept. 9, 1864," *The Rebellion Record: A Diary of American Events with Documents, Volume 11* ed. Frank Moore (New York: D. Van Nostrand, 1868), 315–316.

22 Vattel, *Law of Nations*, 570.

23 Ibid.

incursions of an enemy that cannot be stopped otherwise."[24] George B. Davis (1847–1914), a veteran of the US Civil War who would later become the Judge Advocate General of the US Army, explains:

> Devastation: Devastation is therefore justified: (1.) When dictated by military necessity, as when property is destroyed in battle by the fire of artillery, or by the movement of troops; or where the vicinity of a besieged place is cleared of houses, crops, forests, and the like in order to facilitate siege operations … (2.) Where a portion of territory is laid waste by way of retaliation, or, in an extreme case, to prevent an incursion of the enemy, or to deprive him of food and forage necessary for the support of his armies.[25]

Militaries were well within their rights to destroy noncombatant infrastructure if such territory was a vital resource to the enemy or if it represented a haven for cross-border attacks.

At least according to the North, the Confederacy had met these criteria. Not only had the South become an unjust nation through the perpetuation of chattel slavery and rebellion against the Union, but the Shenandoah Valley had also been strategically significant in terms of cross border incursions. The mountains had given Southern armies concealed access to the North. The Valley had provided General Lee significant tactical coverage as he marched on northern cities. Specifically, he was able to use the Blue Ridge Mountains as a screen on his march to Gettysburg. As far as the North was concerned, the citizens of the Valley were not only barbarians, but they also inhabited the staging ground for the South's cross-border terror campaigns.

Much of the 1864 brutality in the Shenandoah Valley can be traced back to accusations stemming from the hardships caused by the 1863 Gettysburg campaign, specifically the burning of Chambersburg, Pennsylvania.[26] Nearly three thousand cavalrymen, under the command of Brigadier General John McCausland, demanded contributions from the city. When the citizens failed

---

24 José María de Pando, *Elementos Del Derecho Internacional* (Madrid: Imprenta de Algeria y Charlain …, 1843), 399.

25 Davis, *The Elements of International Law*, 305.

26 William Thomas III provides a compelling account of the evolution of the Shenandoah Valley Campaign and its increased intensity. William G. Thomas, III, "Nothing Ought to Astonish Us: Confederate Civilians in the 1864 Shenandoah Valley Campaign" (2006). *Faculty Publications, Department of History*. 48. https://digitalcommons.unl.edu/history facpub/48.

to comply, the soldiers set fire to hundreds of buildings. George Pound, one of the chroniclers of the Valley campaign, remarks that:

> The burning of Chambersburg, the central event of this incursion [as part of the Gettysburg campaign], was wholly indefensible. This was an open town, of no strategic value, and not amenable to the penalty a peaceful village sometimes incurs through being made the refuge of troops that use its dwelling for firing upon their assailants. Factories and granaries had been destroyed in the Valley by Union commanders; but to lay waste to an enemy's subsistence in a mutual field of operations, whose crops furnish him the means of occupying it, is a different matter from putting the torch to a town with intent to lay it in ashes.[27]

What made the burning of Chambersburg wrong from the North's perspective was not the *intent* to lay it in ashes, but the fact that such destruction was believed to be gratuitous.[28] It was neither part of a broader campaign to render the region uninhabitable, nor was the destruction compatible with occupation. Of course, many in the South hoped that such a campaign of extermination and destruction would take place and that Chambersburg was only the beginning of similar attacks. As one editorial in the *Richmond Express* records:

> Glad we are that retribution has at last put forth its terrible arm and assumed its most terrible shape. We hope it will be pushed to the furthest extremity to which it is capable of going. We should be glad to hear that the whole valley of the Susquehanna was one long, unbroken, irresistible flame, not to subside as long as a house or a tree, or a blade of grass, or a stalk of corn, remained to testify that it had even been inhabited by man. No sight could be more agreeable to our eyes than to behold every part of Yankeedom within reach of our armies, converted into a

27 George E. Pond, *The Shenandoah Valley in 1864, Volume 11* (New York: Scribner, 1883), 109–110.

28 General Jubal Early, however, claimed this assault had been justified by the law of reprisal: "A number of towns in the South, as well as private country houses, had been burned by the Federal troops … I now came to the conclusion that we had stood this mode of warfare long enough, and that it was time to open the eyes of the people of the North to its enormity, by wan example in the way of retaliation." Jubal A. Early, *A Memoir of the Last Year of the War for Independence, in the Confederate States of America* (New Orleans: Blelock & Co., 1867), 56.

> mass of ashes – to see every beast that walked on four feet, and could not be driven off for our own use, slaughtered and left to rot on the ground.[29]

The Southern fantasies of total destruction actually help to confirm, in some small way, how unusual the attack on Chambersburg was. Burning a single town with no strategic value was widely condemned as indefensible. Paradoxically, at least according to the norms governing border lands, a dedicated campaign of laying waste to Pennsylvania's Susquehanna Valley might have been *more* defensible than the burning of a single city of questionable military value.

In any case, given these kinds of Confederate incursions into the North, not to mention the culmination of the Gettysburg campaign and the subsequent assaults that were taking place on the outskirts of Washington D.C., which also stemmed from operations in the Valley, this opened Virginia – border lands *par excellence* – to attack. Rendering the Valley "uninhabitable" was perfectly justifiable according to Vattel's logic and the ascendant norms governing laying waste in the nineteenth century.[30]

Ravaging the land, rendering it uninhabitable, was a hard tactic that remained a live option throughout the nineteenth century. Thinking back to Vattel in 1860, just prior to the outbreak of the Civil War, Archer Polson, a prominent nineteenth-century legal scholar, simply concluded: "In a case of extreme necessity, it is lawful to devastate and lay waste an enemy's territory, and to destroy all buildings, &c., therein, as far as is requisite for the success of military operations, but the lawfulness of such proceedings is limited by that necessity in the view of all communities."[31] Henry Halleck (1815–1872), one of the leading American military theorists of the late-nineteenth century, likewise explains,

> There are numerous instances in military history where whole districts of the country have been totally ravaged and laid waste. Such operations have sometimes been defended on the ground of necessity, or as a means of preventing greater evils ... Thus, the Czar, Peter the Great, laid waste as an extent of fourscore leagues of his own territory, to check the advance of Charles XII, of Sweden ... Such violent remedies are to be sparingly

29 Quoted in, Jacob Hoke, *Historical Reminiscences of the War; or Incidents which Transpired in and About Chambersburg* (Chambersburg, PA: M.A. Foltz, 1884), 191.

30 For more on Vattel's influence in the Civil War, see John Fabian Witt, *Lincoln's Code: The Laws of War in American History* (New York: Simon & Shuster, 2012).

31 Archer Polson, *Principles of the Law of Nations* (Philadelphia: T. & J.W. Johnson & Company, 1860), 45.

> applied; there must be reasons of suitable importance to justify the use of them. A prince who should, without necessity, imitate the Czar's conduct, would be guilty of a crime against his people; and he who does the like in an enemy's country, when impelled by no necessity, or induced by feeble reasons, becomes the scourge of mankind.[32]

What these passages show is that laying waste to borderlands – to punish a barbarous people or to create a buffer zone – was widely believed to be justified by the principle of necessity. Of course, this was not a tactic one could employ lightly. While it remained a live option, many commentators knew that ravaging enemy nations' lands was permissible according to the laws of war, even though it pushed the principle of utility to its very limit.[33]

As the nineteenth century progressed, however, fewer theorists actively mentioned laying waste, much less countenanced laying waste to barbarous nations (some of this is captured in Figure 5.1). Even when it was believed justified, theorists routinely attempted to circumscribe the acceptability of such actions based on necessity. Ravaging districts or communities for "feeble reasons" makes the aggressor "the scourge of mankind," guilty of "monstrous excess."

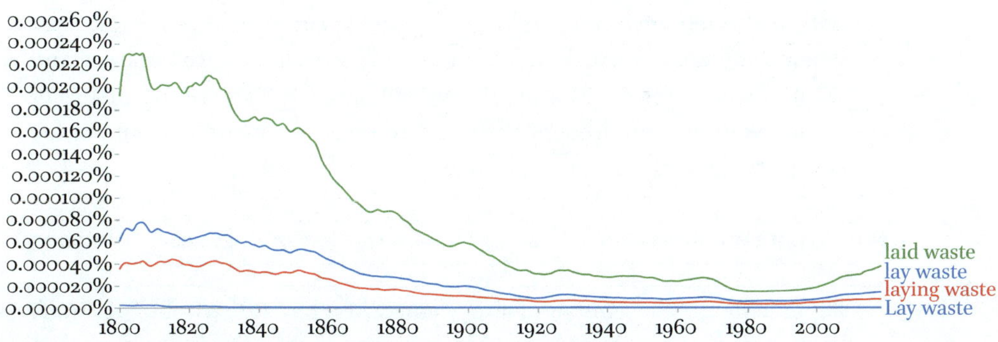

FIGURE 5.1 Ngram search of 'lay waste,' 'laying waste,' and 'laid waste'

---

32 Halleck, *International Law*, 76–77.

33 Reflecting on the troubled moral calculation that would permit an army to lay waste to enemy lands, James Mill writes, "It may not doubt be true, that the plunder and devastation of a province, or other portion of a country, must have an effect in diminishing the resources of the government for carrying on the war. In this point of view it must be allowed that the destruction of private property is of some importance to the invading nation with regard to the result of the war. But the question, in settling the difficulties of international jurisprudence, is not whether an advantage is gained, but whether the advantage, such as it is, be not gained, at too great a cost." James Mill, "Law of Nations," *Supplement to the Encyclopedia Britannica* (London: J. Innes, 1825), 22–23.

The understanding that governs the application of the principle of necessity to punishing barbarians and creating buffer zones, however, was difficult to reconcile with the assumption that states alone have a right to determine what constitutes self-preservation in their individual cases. As Halleck explains, "The necessity of self-preservation, and the right to punish an enemy, and to deprive him of the means of injuring us, by converting those means to our own use against him, lie at the foundations of the rule, and constitute the right of a belligerent to an enemy's property of any kind."[34] It follows that depriving an enemy of their land through ravaging and burning (or occupation) is a right founded in the necessity of self-preservation. And despite there being a strong impetus to be restrained in the use of this extreme tactic, states routinely believed that they were justified in employing such violence. As Sir Walter Scott (1771–1832) reports, these atrocities were common in the Napoleonic campaigns. Soldiers were:

> driven to desperation by rapid marches through roads blockaded with snow, or trodden into swamps, became reckless and pitiless, and straggling from their columns in all directions, committed every species of excess upon the inhabitants. These evils are mentioned in the bulletins of Napoleon, as well as the general orders of Schwarzenberg. The peasants, with their wives and children, fled to the caves, quarries and woods, where they were starved to death by the inclemency of the season, and want of sustenance ... The towns and villages, which were the scenes of actual conflict, were frequently burnt to the ground; and not only in the course of the actions of importance.[35]

As the American philosopher Thomas Upham (1799–1872) likewise chronicles:

> Repeatedly, in the course of European wars, has the whole Palatinate been laid waste; not merely cities, but villages, country-seats, cottages, fields, gardens, everything. The Ukraine during the last century was laid waste, in the same manner, by Catherine of Russia. Almost the whole of La Vendee, thickly peopled as it was with an industrious and rural population, and everywhere bearing the marks of a high rate of culture, was subjected during the French Revolution, to a most horrid and complete

---

34 Halleck, *International Law*, 448.

35 Sir Walter Scott, *The Life of Napoleon Buonaparte, Emperor of the French, Volume 2* (Exeter: J. & B. Williams, 1832), 229.

> devastation. The devastation of the Peninsula by the armies of Napoleon was almost as great.[36]

Similar descriptions of destruction and devastation could be provided for the Crimean War (1853–1856), the Austro-Prussian War (1866), and the Franco-Prussian War (1870–1871). Despite the cautious and one might say strong moral disapprobation against the use of this tactic, there appears to have been very little difficulty in justifying these types of incursions.

In the nineteenth century, noncombatants could accrue various degrees of guilt and liability. As Vattel's quote above indicates, a "barbarous nation" may invite upon itself certain hardship and cruelties. In fact, Article 21 of the Lieber Code makes this point clear: "The citizen or native of a hostile country is thus an enemy, as one of the constituents of the hostile state or nation, and as such is subjected to the hardships of war."[37] But under the logic of necessity, enduring hardships takes on a slightly different meaning. Here one sees a shift away from the much more indirect and incidental cascade of suffering that typically qualifies noncombatant immunity in contemporary ethical discourse, where citizens only accidentally and indirectly experience the hardships that were originally intended only for combatants. When framed by the logic of military necessity that governed norms of laying waste in the nineteenth century, citizens were not excluded from intentional harm to expedite victory. Much like those residents of barbarous nations that Vattel mentions, noncombatants might be knowingly harmed – their houses and crops destroyed – in order to produce a more beneficial outcome overall. It was widely believed that enemy populations *deserved* the intentional hardships that were placed on them.

Theorists of this period never believed it necessary to provide a moral or strategic justification for violating the immunity of a noncombatant in this way. In the literature of this period there were efforts to extend some protections to noncombatants, but these were always conditional and to some degree circumstantial. As seen above, necessity routinely transformed the nonfighting population into legitimate targets. General Sherman puts this idea a bit more forcefully: "one thing is certain, there is a class of people, men, women, and children, who must be killed or banished before you can hope for peace

36 Thomas Upham, *The Manual of Peace* (New York: Leavitt, Lord, and Co., 1836), 58.

37 Instructions for the Government of Armies of the United States in the Field, prepared by Francis Lieber, LL.D., Originally Issued as General Orders No. 100, Adjutant General's Office, 1863, Washington 1898: Government Printing Office.

and order, even as far south as Tennessee."[38] One can read this passage both descriptively and normatively. The hard hand of war will inevitably lead to civilian death, but one also gets the sense that Sherman understood that some brutality is necessary for wars to end. Achieving peace, in the broadest sense of the term, requires some amount of noncombatant death, though, to be precise, the intentionally targeted nonfighting populations were not, strictly speaking, noncombatants. That "class of people" was made of zealots or ideologues who morally and materially supported the Confederate cause. As such, they had become tainted by their direct or indirect support for an enemy regime. As callous and inhumane as this sounds, one should remember that the laws of war at that time presumed some minimum threshold of death, of both combatants and implicitly noncombatants. In this respect, at least according to nineteenth century-norms, Sherman's remark represents the very essence of restraint and humanitarian thinking in military ethics.

The hardship and even death that noncombatants might suffer due to military necessity were well established principles of war throughout the nineteenth century. Even as certain protections for noncombatants were becoming more formalized and widely accepted, the guilt of an enemy population and the demands of necessity made the recourse to deliberately targeting some number of civilians highly reasonable, if not moral. Military necessity created acceptable parameters for non-discrimination. In 1875, William Oke Manning (1809–1878) reiterates the fact that necessity may lead to noncombatant suffering:

> It is clearly a belligerent's right to destroy the enemy's property as far as necessary in making fortifications, or in making approaches to those of a place besieged, or in the indiscriminate destruction that takes place in a bombardment, and similar necessary operations of warfare. It may also be allowable, in extreme cases, to destroy the produce in an enemy's country, so as to render it impossible for his army to subsist there, a measure of which the possible efficacy was shewn by the results of the campaign in Russia in 1812.[39]

38 Letter from Major-General Sherman to Secretary of War Stanton, 21 June 1864, in *The War of the Rebellion: A Compilation of the Official Records of the Union and Confederate Armies* (Washington D.C.: Government Printing Office, 1880), 132.

39 William Oke Manning, *Commentaries on the Law of Nations* (London: Cambridge: Macmillan, 1875), 186.

Not only was "indiscriminate" destruction sometime necessary, but so too was starving the civilian population by destroying their produce and laying waste to their farmlands.[40]

4 Starvation and Blockades

Certain forms of indiscriminate violence had long been permitted in war. As seen above, the destruction of enemy farmlands, infrastructure, and property were permitted in the process of laying waste to enemy lands. These forms of indiscriminate violence were designed to undermine an enemy nation's ability to prosecute the war. If no one in that country could eat, then the war would soon end. As Article 17 of the Lieber Code explains, "War is not carried on by arms alone. It is lawful to starve the hostile belligerent, armed or unarmed so that it leads to the speedier subjection of the enemy." Destroying food supplies and starving the domestic population were accepted practice throughout the nineteenth century, including the US Civil War.[41] As Lance Janda explains:

> The application of force against an enemy's noncombatants and resources, the central tenet of total war, had been used since the dawn of civilization when it suited political and military ends. But Union Army commanders were the first in American history to use these tactics on a widespread scale, and they played a crucial role … in the subjugation of the south.[42]

These tactics would continue throughout the nineteenth century. In the 1898 invasion of Cuba, for instance, T.S. Woolsey notes that "The non-combatant country population was forced to settle within the range of the guns of the Spanish intrenched in the towns, and, there, destitute of food or the means of growing or getting food, it starved."[43]

40 Simone Hutter's recent work on deliberate starvation under current International Humanitarian Law illustrates what a shift has taken place in noncombatant ethical theory since the nineteenth century. See Simone Hutter, *Starvation as a Weapon: Domestic Policies of Deliberate Starvation as a Means to an End under International Law* (Leiden: Brill, 2015).

41 See also, Andrew Smith, *Starving the South: How the North Won the Civil War* (New York: Macmillan, 2011).

42 Lance Janda, "Shutting the Gates of Mercy: The American Origins of Total War, 1860–1880," *The Journal of Military History* 59:1 (1995); 7–27, 8.

43 Quoted in Ellery Stowell, *Intervention in International Law* (Washington D.C.: John Byrne 1921). 121.

The tactic of starving one's enemies is virtually as old as war itself, and it was long seen as permissible by the laws of war.[44] This tactic was specifically defended in the laws governing naval blockades.[45] Francis Upton (1852–1921), a distinguished New York lawyer appointed to the US Counsel for Captors during the US Civil War, explains that while maritime jurisprudence gives nations an undisputable right to blockade enemy nations, "it must, nevertheless, be conceded to be one of the harshest measures in its operation of any know in the code of international law."[46] Its permissibility did nothing to mitigate its brutality. François d'Ivernois (1757–1842), the great Genevan jurist, argues that "Never, surely, was a doctrine better calculated to reduce the civilized world to a state of barbarism."[47] Archer Polson also explains that the blockade "is a severe right, a severe right to be construed strictly, not extended by implication, and a belligerent right not to be exercised for mere profit or convenience, e.g. to obtain a commercial monopoly."[48] Much as with the norms governing laying waste, this was not a measure that could be approached lightly. It was heavily circumscribed by the logic of necessity.

The passages quoted above likely give the impression that jurists were worried about the degree to which starvation blockades indiscriminately targeted the innocent non-fighting residents of an enemy nation. This is not the case. What made blockades so severe to the legal minds of this period was not the indirect harm they inflicted on the most vulnerable of the enemy population, but the degree to which they unfairly disadvantaged and inconvenienced neutral powers and merchants who had no direct part in the conflict. John Macqueen, a Scottish lawyer writing on international law, worried that "This operation is performed and maintained without the slightest regard to the injury which may thereby be occasioned to neutrals."[49] Indeed, the treatment

44 In the eighteenth century, Christian Wolff argued: "Starvation and thirst go along together. Therefore, as it is not censurable to coerce an enemy by starvation to surrender, for example, or desist from war and make peace, so neither can one be censured if he compels an enemy to do the same thing because of thirst. Christian Wolff, *The Law of Nations Treated According to the Scientific Method* [1749] ed. Thomas Ahnert, trans. Joseph Drake (Indiana: Liberty Fund, 2017), 650.

45 For a more detailed treatment of the evolving norms around this subject, see Lance Davis and Stanley Engerman, *Naval Blockades in Peace and War: An Economic History Since 1750* (New York: Oxford University Press, 2006); and Philip Drew, *The Law of Maritime Blockade: Past, Present, and Future* (New York: Oxford University Press, 2017).

46 Francis Henry Upton, *The Law of Nations Affecting Commerce During War* (New York: John S. Voorhies, Law Bookseller and Publisher, 1863), 276.

47 François d'Ivernois, *Effects of the Continental Blockade Upon the Commerce, Finances, Credit, and Prosperity* (London: Printed by Cox, Son, and Baylis …, 1810), 14.

48 Polson, *Principles of the Law of Nations*, 55.

49 John Frasier Macqueen, *Chief Points in the Laws of War and Neutrality, Search and Blockade* (London: William and Robert Chambers, 1862), 27.

of neutrals was a delicate matter that was covered in many law and military manuals in this period.[50] The exclusive moral concern about this tactic was the treatment of neutrals who, for whatever reason, might violate the blockade. In some historical cases the penalties for even inadvertent blockade running had been quite extreme.[51] Indeed, stemming from these concerns some commentators in the nineteenth century argued that a belligerent should only have the right to blockade military ports.[52]

By the nineteenth century, penalties for those nations who violated a blockade did not typically rise to the level of execution, but they were significant enough to warrant moral concern. The military manuals of this period spend a great deal of time laying out the conditions in which enforcing a blockade would be permissible. In fact, many of these rules had been set out in the Treaty of Paris (1856) at the end of the Crimean War. To be valid, it had to be proven that 1) an actual blockade had been declared (rather than being a "paper blockade"), which meant it needed to be enacted and maintained (if the blockade lagged or couldn't be maintain it was seen as *de facto* lifted); 2) the third party knew that the blockade was in effect; and 3) the neutral party was attempting to violate the blockade.[53] Only after these elements had been shown did the blockading power have the right to confiscate the transporting vessel along with contraband goods.[54]

In the extensive commentary on blockades, very few commentators see them as problematic since they disproportionately affected those weaker and sicker residents of an enemy nation. In fact, many argued that blockades were

---

50 William Hazlitt and Philip Roche, *A Manual of the Law of Maritime Warfare* (London: V & R Stevens and G.S. Norton, 1854); Thomas Spinks, *The Ecclesiastical and Admiralty Reports, Volume 1* (London: Printed by A. & G.A. Spottiswoode, 1855); Arthur Young, *A Digest of Maritime Law Cases, 1837–1860* (London: Horace Cox, 1865); William Hall, *The Rights and Duties of Neutrals* (London: Longmans, Green and Co., 1874).

51 William Manning records instances from the fifteenth century where those who attempted to break a blockade were "to be looked upon as criminals, and suffer death by the law of arms." Manning, *Commentaries on the Law of Nations*, 320. Vattel writes approvingly of King Demetrius who "hanged up the master and pilot of a vessel carrying provisions to Athens at a time when he was on the point of reducing that city by famine." Vattel, *The Laws of Nations*, Book III, Chapter, VII, §117, 533. See John Dahlgren, *Maritime International Law* (Boston: B.B. Russell, 1877), 33.

52 David Dudley Field, *Outlines of an International Code* (New York: Baker, Voorhis and Company, 1876), 573. Field was unique as he was one of the few commentators who argued that "Starving the enemy by cutting off supplies of food or water" is only lawful as an act of retaliation (Ibid., 498). This was, by far, a minority opinion in this day.

53 Manning, *Commentaries on the Law of Nations*, 321.

54 Sir James Deane studies eight cases where violations of blockade were punished. Sir James Deane, *The Law of Blockade: As Contained in the Report of Eight Cases Argued and Determined in the High Court of Admiralty* (London: Butterworths …, 1855).

morally preferable to direct invasion. Travers Twiss explains that while blockades target an entire "wrong-doing nation," they are "a more lenient proceeding in the conduct of war than actual assault. The latter involved the necessary sacrifice of human life" whereas "the former gives to the Enemy the option of being spared the effusion of human blood."[55] William de Burgh concurs that the blockade is "a more lenient proceeding in the conduct of a war than actual assault."[56] One gets the impression from this literature that blockade is a courtesy paid by an attacker – rather than killing with bullets and bombs, inhabitants can simply be starved into submission. And even though there were some calls to abolish the practice of blockades, there was little willingness to alter longstanding norms. As another commentator notes, it has

> been suggested that the right of blockade should be abolished entirely. But it seems scarcely possible to advance so far without contradicting practices and maxims which are as old as war itself, and which seem likely to last as long as war, for they are essential to it ... To reduce your enemy by starvation, is as ancient as natural, as efficient and expedient as that of killing him by arms. There is no advantage in a sentimental war; at the best it will be a terrible calamity; and the most really merciful code is that which makes war most sharp, and therefore most short.[57]

Blockades were part of a strategy that, in the targeting of the innocent, by increasing their suffering, war might be ended quickly.

While originally confined to cities or more localized pieces of territory, as the organization and size of navies grew, so did the capacity to execute large-scale blockades of entire nations. In the nineteenth century several such attempts had been made: the British blockade of France in 1806 and the French-led Continental Blockade (the so-called Continental System) that was set up in retaliation (1806–1814); the British blockade of the US from February 1813 to March 1815, which at its peak extended virtually the entire length of the eastern seaboard; the US blockade of Mexico during the Mexican-American War (1846–1848), which covered both Pacific and Atlantic coasts; the Union blockade of the Southern coast (1861–1865) which extended down the Atlantic

55 Twiss, *The Law of Nations*, 191.

56 William de Burgh, *The Elements of Maritime International Law* (London: Longmans, Green, and Co., 1868), 121.

57 "The Wise and Unwise Proposals for Altering the Code of Maritime Warfare," *The Economist* 20:966 March 1, 1862, 226.

seaboard, along the Gulf Coast, and up the Mississippi River; and the US blockade of Cuba during the Spanish-American War (1898).[58]

Rather than isolating single cities, belligerent navies were now capable of closing off entire countries, the effects of which could be rather severe for the civilian populations. Take for instance the effects of the two-month long US blockade of Cuba during the Spanish American War. Lieutenant Jose Muller y Tejeiro describes the horror that unfolded throughout the island:

> The blockade as practiced in Cuba caused a thousand times more victims and more horrors than bursting shells, the burning of cities, the massacre of battles, and all the cruelties of weapons. The blockade makes living expensive, extinguished the means of livelihood, gradually decimates the population, destroys family life, annihilates human beings without distinction – or rather, with one distinction, for it strikes particularly the feeble, the children, the women, and the sick.[59]

He goes on to describe how the terrible effects were felt: "After two months of blockade in Cuba there could be seen in the cities and in the country thousands of human beings looking like ghosts, and men, women, and children dying of hunger in the public roads."[60]

These same concerns were expressed in congressional debates about the validity of the blockade. Republican Senator William Stewart (1827–1909) voiced concern that the US was waging war on the Cubans themselves. He remarks:

---

58 This tactic would culminate in the Blockade of Germany (1914–1919) during World War I which led to between 500,000 and 700,000 civilian casualties. See Leo Grebler, *The Cost of the World War to Germany and Austria-Hungary* (New Haven CT: Yale University Press, 1940); and Suda Bane and Ralph Lutz (eds.), *The Blockade of Germany After the Armistice, 1918–1919* (Stanford: Stanford University Press, 1942); and C. Paul Vincent, *The Politics of Hunger: The Allied Blockade of Germany, 1915–1919* (Athens, OH: Ohio University Press, 1985).

59 Lieutenant José Müller y Tejeiro, *Battles and Capitulation of Santiago de Cuba* (Washington DC: Government Printing Office, 1899), 23.

60 Ibid. This was a cruel irony since the Spanish reconcentration policy, which lead to some 200,000 Cubans dying of starvation, is what helped to generate anti-Spanish sentiment precipitating US involvement in the region. For more, see Nathan Green, *The War with Spain* (Baltimore: International News and Book Co., 1898). For a devastating firsthand account of the starvation throughout Cuba, see the congressional testimony of C.F. Koop "Statement of Mr. C.F. Koop, February 21, 1898," "Condition of the Affairs in Cuba" in *Message of the President of the United States, Communicated to the Two Houses of Congress, on the Relations of the United States to Spain by Reason of Warfare in the Island of Cuba* (Washington, DC: Government Printing Office, 1898).

> The Secretary of the Treasury tells us that we have spent millions in policing the ocean to blockade Cuba and prevent munitions of war and supplies from being landed on the island. When I read that I asked myself were we entirely blameless for the starvation of hundreds of thousands of the noncombatants of Cuba.[61]

He favored a strategy of lifting the blockade and then directly supplying the locals with weapons so they could resist their oppressors: "Let us remove our blockade, and food and munitions of war will find their way to the struggling Cubans, and Cuba will soon be free; but do not for one moment contemplate a war against the patriots for whom the people of the United States have so strong a sympathy."[62]

Despite some minor complaints against American blockade tactics in Cuba, like those put forth by Senator Stewart, nothing in its application violated the laws of war. Such practices were widely accepted as legitimate. Ravaging, burning, starving, and bombing civilians, at least those in proximity to legitimate military targets, and destroying what is now understood as dual-use infrastructure – those industries necessary for the citizenry's survival but which also enable an enemy to prosecute war – had a long pedigree and were virtually unquestioned. Heaping these hardships onto civilians was acceptable and even humane since not only had the enemy accrued some guilt, but those hardships would likely expedite the war's end. As Sheridan had argued, forcing the civilian population into poverty and starvation would lead them to pray for peace and to amend their ways faster than through the outright death of soldiers. This sentiment is put the most succinctly in Article 29 of the Lieber Code, which states: "The more vigorously wars are pursued, the better for humanity. Sharp wars are brief." Targeting noncombatant infrastructure in a deliberate and systematic way, aiming for that minimum threshold of civilian death, could stem from the most humanitarian impulse: a desire to end war.

5 Sieges

Nowhere can the justification of civilian death by the principle of necessity be illustrated more clearly than the long-standing norms around siege warfare.[63]

61 *Congressional Record: Proceedings and Debates, Fifty-Fifth Congress, Second Session, Volume 31, Part 4* (Washington: Government Printing Office, 1898), 3703.

62 Ibid.

63 John Keegan argues that in the fifteenth century "it was already a convention of siege warfare that a refusal to surrender after a breach had been made thereafter absolved the attackers from the obligation to offer quarter or abstain from looting. In the artillery age

As seen in the extensive commentary on siege warfare among the Scholastics, martialists, and humanist writers from the sixteenth to eighteenth centuries, there was broad agreement that if noncombatants resided in or around a fortified place, they lost any protected status they might have had. Cities could be stormed, burned, and bombed and this would inevitably lead to the death of anyone residing therein. The attacking army was not responsible for the hardships and death experienced by the noncombatant population. In fact, many believed that the noncombatant population accrued some culpability for residing near the guilty and that the realities of defense turned them into combatants. By the seventeenth century, many commentators in each of the discursive traditions treated above began to recognize the effectiveness of bombing citizens directly. There was virtually universal agreement that residential districts could be bombed.[64] These trends continued through the nineteenth century and up to World War I.

To illustrate the permissibility of intentionally targeting civilians during a siege, take the bombardment of Vera Cruz during the Mexican American War (1847). As the last remaining obstacle to clear access to Mexico City, US forces, under the command of Major General Winfield Scott, shelled the city for four days. After the city had fallen, the Mexicans estimated that about a thousand people had been killed, a majority of whom were civilians. And those civilians who survived the bombardment faced an uncertain future. As Abiel Abbot Livermore explains in his well-received Peace Society essay reporting on the attack, the survivors were reduced to rather desperate conditions:

> Most of the families, whose houses had been destroyed, had lost everything; all the property remaining to them being the clothes on their backs, because what the flames did not consume was buried under the ruins. Hundreds of persons, as well as fathers of numerous families of children, heretofore relying upon certain incomes, to-day find themselves without a bed to lie upon, without covering or clothing to shelter them, and without any victuals.[65]

We know very little as to what became of these victims of, as Livermore calls, the "legitimate barbarities of war."

---

that convention became absolute." John Keegan, *A History of Warfare* (New York: Random House, 1993), 323.

64 Michael Walzer argues that "Siege is the oldest form of total war … Civilians have been attacked along with soldiers, or in order to get at soldiers, as often in ancient as in modern times." Michael Walzer, *Just and Unjust Wars: A Moral Argument with Historical Illustrations* (New York: Basic Books 1977), 160.

65 Livermore, *The War with Mexico Reviewed*, 136.

Intentionally targeting residential districts did not violate the laws of war. In fact, General Scott was praised for his honorable attitude to the inhabitants of the city, providing a window of time for the foreign consuls, women, children, and other noncombatants "to retire to neutral ships in the harbor, or other places of safety" before the bombardment would begin.[66] A nineteenth-century historian of Mexico, Philip Young, explains:

> We understand that General Scott delayed opening his guns upon the city for eight hours, to give the women and children an opportunity of withdrawing, assuring them of protection for themselves, and whatever property they might choose to carry with them. Not an individual braced this offer, and the subsequent destruction of women and children, during the bombardment, was a source of the most painful regret to the General and the army. Though no accurate return had been made, it was believed that the slain in the city, during the investment, exceeded eight hundred.[67]

Though it was a source of "painful regret" for Scott, bombing the residents became a necessary element to expedite surrender. The European consuls in the city, however, did petition Scott after the bombardment commenced to give the women and children additional time to leave, which he refused, "alleging that he had given due notice of his intention to bombard the city, and that those who remained in defiance of this, had no claim on him to stop the siege in order that they might be removed from peril."[68]

George Furber, a private in a Tennessee cavalry regiment who witnessed the bombardment, invites the reader to picture what such a moment was like:

> though the firing last night and this day had been slow, the mortars have, in these forty-eight hours, thrown over two thousand of the ponderous missiles. These from their high sweeps aloft, have descended with irresistible force on the stone tops of the buildings, passing through them and all the floors, and exploding below; – or else falling into the streets and crowded squares, spreading death and destruction among

66 Ibid., 133.

67 Philip Young, *History of Mexico: Her Civil Wars, and Colonial and Revolutionary Annals* (Cincinnati: J.A. & U.P. James, Walnut Street, 1847), 468.

68 "The Fall of Vera Cruz," *The Ladies' National Magazine, Volume XIII* ed. Ann Stephens (Philadelphia, PA: Charles Peterson, 1848), 147.

> soldiers and inhabitants, men, women, and children, in one common sweep.[69]

While the bombardment was initially directed at the fortress's battlements and magazines, the utter devastation and confusion that stemmed from bombing the city itself was not incidental; it was the climatic build-up of an attack designed to undercut any hope of victory for anyone who remained in the city. At the peak of the attack, the US forces launched Cosgreve rockets – a new type of incendiary missile that had been developed during the Napoleonic Wars – and thousands of other conventional bombs onto the residential district of the city itself.[70] As Karl Bauer explains, the overwhelming force was designed to eviscerate the enemy's will to fight and to make surrender the only way to save the noncombatants who remained in the city.[71]

Not surprisingly, the Mexican defenders were especially critical of Scott's tactics. In their public complaint protesting the assault, they mention nothing of Scott's offer to evacuate women and children prior to the bombardment. Instead, they focus on the unwillingness of the besiegers to approach the walls and the fact that the US Army deliberately targeted noncombatant infrastructure:

---

69 George Furber, *The Twelve Months Volunteer* (Cincinnati: J.A. & U.P. James, Walnut Street, 1848), 528.

70 For the development of this weapon, see Sir William Congreve, *A Concise Account of the Origin and Progress of the Rocket System* (Dublin: Printed by A. O'Neil …, 1817). The Congreve rocket had proven to be devastating but erratic. The British artillery officer, Cavalié Mercer, describes his experience with this weapon during the Waterloo Campaign of 1815: "Meanwhile the rocketeers had placed a little iron triangle in the road with a rocket lying on it. The order to fire is given – portfire applied – the fidgety missile begins to sputter out sparks and wriggle its tail for a second or so, and then darts forth straight up the chaussée. A gun stands right in its way, between the wheels of which the shell in the head of the rocket bursts, the gunners fall right and left, and, those of the other guns taking to their heels, the battery is deserted in an instant." While the guns sometimes proved to be devastatingly effective, Mercer goes on to explain how erratic and unreliable they could be. Some of the rockets "actually turned back upon ourselves – and one of these turned back upon ourselves – and one of these, following me like a squib until its shell exploded, actually put me in more danger than all the fire of the enemy throughout the day." Cavalié Mercer, *Journal of the Waterloo Campaign: Kept Throughout the Campaign of 1815* (London: William Blackwood and Sons, 1870), 279–280.

71 K. Jack Bauer, *The Mexican War, 1846–1848* (Lincoln: University of Nebraska Press, 1974), 250.

> [Scott] was not anxious to measure arms with us – nor venture upon an uncertain deed of arms, selecting as was most agreeable to him, and most in accordance with his character, the barbarous manner of assassinating the unoffending and defenceless citizens, by a barbarous bombardment of the city in the most horrible manner, throwing into it 4100 bombs, and an innumerable number of balls of the largest size, during nights and days, directing his first shots to the powered magazine, to the quarter of hospitals of charity, to the hospitals for wounded … His first victims were women and children, followed by whole families perishing from the effects of the explosion, or under the ruins of their dwellings.[72]

These public criticisms of Scott's actions at Vera Cruz were persistent enough that they warranted some defense.[73] Scott's biographer, Edward Mansfield, perhaps with these criticisms in mind, makes just such a public response:

> Something has been severely said, as to the loss of women and children by the bombardment of the city; but this is unjustly said … it was both unnecessary and impossible that the siege should be delayed, or given up, on account of the inhabitants within, who had long known that the United States army would land there, and who had received from the commander full notice of danger.[74]

Not only did the inhabitants know that the US Army was coming and that the fort represented a legitimate target, but since they refused to evacuate, they had lost their protected status as noncombatants. Indeed, despite the fervent accusations of cowardice and dishonor heaped onto Scott by his critics,

72 "Mexican Narrative of the Events at the Heroic City of Vera Cruz, While Besieged by the American Army," *Niles' National Register*. 8 May 1847. 149.

73 Despite wide-spread acceptance of the tactic, many commentators presented the bombing in a critical light. For instance, according to one eyewitness interviewed by William Jay: "The city, or at least the northern part of it, has been torn all to pieces – the destruction is dreadful. It is impossible to get at the loss of the Mexicans by the bombardment; yet it is certain that *women, children*, and *non-combatants*, have suffered the most." William Jay, *A Review of the Causes and Consequences of the Mexican War* (Boston: Benjamin B. Mussey & Co., 1849), 226. Nassau Senior writes, "The brutality with which the war was carried on equaled its injustice … The ambition of a despot is violent and unscrupulous, but far less so than that of a republic." Nassau William Senior, *Historical and Philosophical Essays* (London: Longman, Green, Longman, Roberts, & Green, 1865), 104. Loring Moody also provides a critical account of Scott's conduct at Vera Cruz. See Loring Moody, *A History of the Mexican War* (Boston: Bela Marsh, 1848), 71–77.

74 Edward Mansfield, *Life of General Winfield Scott* (New York: Published by A.S. Barnes & Co, 1852), 140.

nothing in the laws of war at that time could be used to censure his actions. He had acted perfectly honorably in the killing of hundreds of civilians.

6 Siege and the Laws of War

General Scott's actions were not especially unique. Thomas Walker explains that historically, "The utmost severity was shown by besiegers to fortified places ... To rashly resist the fire of cannon was, according to the general view of the 'law of arms,' to expose all the inhabitants of the town to death."[75] As Walker suggests, it was commonly assumed that all residents of besieged places, both combatants and noncombatants alike, were linked. Civilians living in and among fortified places were subject to the hardships of war. Writing about the 1813 Siege of Acapulco in the War of Mexican Independence, Hubert Bancroft (1832–1918) writes, "I do not remember any instance where a fortress was spared out of consideration for the non-combatants in it, either in the late wars of Europe or in any other late wars."[76] The hardships inflicted on the noncombatant inhabitants were not only permitted, but they were in many cases instrumental in expediting surrender.[77] Sir Howard Douglas (1776–1861) explains

> The bombardment of a fortified place is more destructive to the non-combatant inhabitants, generally very numerous, than to the defenders of the works, who, when not on duty, are lodged in shell-proof barracks, and produces less effect upon the military works than upon the dwellings of the citizens. In fact, the object of the bombardment is to compel the

75 Walker, *A History of the Law of Nations*, 191. Jose Maria de Pando argues "When bombarding a city, it is difficult not to do much damage to public buildings and to the houses of private individuals. Hence, one should not proceed to such an extremity except when it is impossible to take an important place in any other way, especially when the occupation may influence the event of war." Pando, *Elementos del Derecho Internacional*, 400. See also, Andres Bello, *Principios de Derecho Internacional* (Caracas: Almacen de J.M. de Rojas ..., 1847), 157.

76 Hubert Bancroft, *History of Mexico, Volume 4* (San Francisco: A.L. Bancroft & Co, Publishers, 1885), 549.

77 For this reason, Hector Straith argues in favor of building detached citadels: "Citadels are generally in positions that command the interior of the place, and are therefore useful in overawing a population that might otherwise strive to shorten the length of a siege, during which the inhabitants are always great sufferers." Hector Straith, *A Treatise on Fortification* (Croydon: Printed and Published by William Annan, High Street, 1833), 10. As will be seen, however, these advances in engineering did nothing to ameliorate the tensions around bombing noncombatants.

> garrison to surrender, not by the injury which it may sustain, but by the slaughter and misery inflicted on the unoffending inhabitants.[78]

As seen above with the Siege of Vera Cruz, by the nineteenth century, the norms of bombardment universally permitted the deliberate bombing of civilians in order to expedite victory. As James Farrer explains, "Regular and simple bombardment, that is, of a town indiscriminately and not merely its fortresses, has now become the established practice."[79] The ascendance of military necessity and the utility-based approach to suffering had transformed the norms around siege warfare. Laying waste, or sacking a defeated city was widely condemned as gratuitous, but prior to capitulation residential districts outside a fortification could be razed and homes inside the defended city could be intentionally and "indiscriminately" targeted to expedite victory.[80] While some generals, like Scott, permitted the citizenry to evacuate the city; others did not. Surprise or other strategic considerations permitted besiegers to bombard without first warning the inhabitants.

As one nineteenth-century commentator notes, "the fortune of non-combatants is inextricably mixed up with that of the combatants; thus, when a town is besieged, the inhabitants necessarily suffer by the attempts which the besiegers make to overpower the garrison."[81] There was something about the spatial dimension and internal relationship of the residents that rehabilitated the collective targetability of those residing in and around legitimate military targets. The guilt of these residents was not that they were part of an enemy nation. Unlike civilians who were killed in the laying waste of barbarous lands, they were guilty of situating themselves too close to a legitimate military target. As will be seen, the nineteenth-century theory of warfare conceived of targetability spatially, that is, how close one was to a legitimate target. In fact, the only time military manuals problematize intentions in this period is regarding religious and cultural spaces. Not people but sacred and artistically significant buildings should be off limits from intentional harm. The accepted rule regarding fortified or defended cities was that residents should

78 Sir Howard Douglas, *A Treatise on Naval Gunnery* (London: John Murray …, 1855), 174.

79 James Farrer, *Military Manners and Customs* (London: Chatto & Windus, Piccadilly, 1885), 15.

80 R.C.H. Lesaffer argues that sacking was a much more common feature of early modern siege warfare, but that it had been largely phased out by the nineteenth century. R.C.H. Lesaffer "Siege warfare in the early Modern Age: A study on the customary laws of war," in *The nature of customary law: Legal, historical and philosophical perspectives*. A. Perreau-Saussine, J.B. Murphy eds. (New York: Cambridge University Press. 2007), 176–202.

81 William Whewell, *The Elements of Morality: Including Polity, Volume 2* (London: John W. Parker, West Strand, 1845), 373–374.

not put themselves in harm's way. And since all the residents of those places were bound by a kind of collective domesticity and economy, the function of one part was directly linked to the wellbeing of the other. The reality of fortified enclosure bound all the inhabitants together.

What is striking about the permissibility of targeting residential districts in the nineteenth century, is that, unlike the variability and uncertainty of bombardments of prior centuries, reducing fortresses had obtained virtually a scientific degree of precision. For instance, the French general Marshal August de Marmont (1774–1852) was able to precisely quantify the typical duration and probable expense of various styles of fortification assault.[82] Siege warfare had become, quite literally, something of an exact science. Besieging armies, in most cases, were perfectly capable of distinguishing between combatants and noncombatants.[83] For instance, in an entry titled "Bombardment," Major-General Sir John T. Jones explains the difference between "bombarding towns" and "reducing a place by regular siege." In the case of the latter mode of attack, he writes that it "requires that the Engineers should be provided with considerable assistance," whereas the former "is an operation of no Engineer science, and might be carried into effect by the Artillery Officers without Engineers, nearly as well as with them."[84] George Ripley and Charles Dana distinguish between "desultory" and "regular" bombardments. The former occurs when "ships, field batteries, or a proportionally small number of siege batteries, throw shells into a place in order to intimidate the inhabitants and garrison into a hasty surrender"; whereas the latter are described as a "regular attack by direct fire against the defences."[85]

82 Marshal Marmont, *The Spirit of Military Institutions* trans. Frank Schaller (Columbia, SC: Evans and Cogswell, 1864), 110.

83 This is true even if public opinion about bombardments varied. As one war correspondent during the Franco-Prussian War writes, "We should remember, however, that some confusion exists as to what 'bombardment' means. The Germans who call for vigour, and the English public which deprecates severe measures, have in mind a burning up of dwellings to make non-combatants scream for surrender. But the bombarding which I understand to be probably in regard to the Parisian outworks is an artillery attack upon French artillerists, with a view to be in a position to threaten harsher measures." *The War Correspondence of the* Daily News (London: Macmillan and Co. 1871), 90.

84 Sir John T. Jones, "Bombardment," in *Aide-mémoire to the Military Sciences, Volume 1* (London: John Weale, High Holborn, 1846), 165.

85 George Ripley and Charles Dana (eds.), "Bombardment," in *The New American Cyclopædia, Volume III* (New York: Appleton and Company, 1873), 453. Certain munitions, like "red hot balls" were "designed to set fire to a city." William Duane, *The American Library; or Compendium of the Modern Tactics, Volume 2* (Philadelphia: Printed by and for the Author at no 98, Market Street, 1809), 337. James Cutbush describes other types of incendiary bombs built for similar purposes. See James Cutbush, *A System of Pyrotechny: Comprehending the Theory and Practice, with the Application of Chemistry; Designed for*

In short, certain types of artillery assaults were designed to be indiscriminate and promiscuous within the confines of the militarized space. As noted above, at the beginning of the nineteenth century, Sir Walter Congreve had developed a new kind of incendiary rocket that could be fired into a besieged placed with great effect. As one commentator notes, Congreve's "new rockets carry a parachute, which unfolding itself majestically at the very top of the curve which the rocket describes, supports sometimes a bomb, destined to carry destruction on some ill-fated city, (provided it meets with a favorable wind)."[86] Much like the trends in bombing cities in the lead up to the nineteenth century, it was quite common for artilleries to target the cities themselves. This is why, depending on the degree of resistance, bombardments could be

> calculated to set civilian housing on fire, panic the population, and pressure the governor to surrender. Even when there were instructions to limit civilian damage, for example at Copenhagen in 1807, such collateral damage could not be avoided, especially when notoriously inaccurate Congreve rockets were used. After three days of bombardments much of Copenhagen was in flames with several thousand civilian casualties.[87]

These principles of bombardment were not outliers, they were entailed in the laws of war. In Article 18 of the Lieber Code (1863), Francis Lieber explains that "When a commander of a besieged place expels the noncombatants, in order to lessen the number of those who consume his stock of provisions, it is lawful, though an extreme measure, to drive them back, so as to hasten on the surrender."[88] And in Article 19 he concludes that "it is no infraction of the common law of war to omit thus to inform the enemy [prior to removing women and children]. Surprise may be a necessity."[89] Approaching armies could both

Exhibition and for War (Philadelphia: Published by Clara F. Cutbush, 1825). For more on the development of these weapons, see Henry Wilkinson, *Engines of War* (London: Longman, Orme, Brown, Green, and Longmans, 1841); John Scoffern, *Projectile Weapons of War and Explosive Compounds* (London: Cooke & Whitley ..., 1852); and Major F.A. Griffiths, *Artillerist's Manual and British Soldier's Compendium* (London: W. Clowes & Sons ..., 1868).

86 William Theobald Wolfe Tone, *Essay on the Necessity of Improving Our National Forces* (New York: Published by Kirk and Mercein, 1819), 103.

87 Gunther Rothenberg, "The Age of Napoleon", in *The Laws of War: Constraints on Warfare in the Western World*, eds. Michael Howard, George Andreopoulos, and Mark R. Shulman (New Haven, CT, 1994), 92.

88 See also F. Boutros, *Principles of the Law of Nations* (Serampore: Ostell and Lepage ..., 1845), 98.

89 It is easy to overlook the fact that the kindness of permitting women and children to leave a besieged place presented another kind of torment. As Charles Sumner explains:

force noncombatants back into defended cities and they could commence bombardment without advanced notification.

This had, in fact, been Sherman's strategy in the bombardment of Atlanta. Confederate General Hood complained:

> I feel no other emotion than pain in reading that portion of your letter which attempts to justify your shelling of Atlanta without notice, under the pretence that I defended Atlanta upon a line so close to town that every cannon-shot, and many musket-balls from your line of investment, that overshot their mark, went into the habitations of women and children. I made no complaint of your firing into Atlanta in any way you thought proper. I make none now, but there are a hundred thousand living witnesses that you fired into the habitations of women and children for weeks, firing far above and miles beyond my lines of defence. I have too good an opinion, founded both upon observation and experience, of the skill of your artillerists, to credit the assertion that they, for several weeks, unintentionally fired too high for my modest field-works, and slaughtered women and children by accident and want to skill.[90]

Notice that Hood "made no complaint" of the tactic of bombarding Atlanta without warning; instead, he argues that there was a wide-spread perception that he was intentionally bombing women and children. Since Hood could not appeal to the laws of war, he appealed to humanity and the court of public opinion – that Sherman risked being perceived as one who deliberately targeted women and children by the public at large. To this General Sherman coolly reaffirmed the principle of military necessity: "The use of Atlanta for warlike purposes is not consistent with its character as a home for families. There will be no manufactures, commerce or agriculture here for the maintenance of families, and sooner or later want will compel the inhabitants to go."[91] As was repeated throughout this period, it was up to the noncombatants themselves to avoid remaining close to legitimate targets. By choosing to reside in such places, they opened themselves up to direct attack.

It was a striking feature of the logic of siege warfare throughout the nineteenth century that the civilians bore the responsibility to avoid legitimate

---

"They may be spared in this way the languishing death by starvation; but they must desert their firesides, their pursuits, all that makes life dear, and become homeless exiles; a fate little better than the former." Charles Sumner, *The True Grandeur of Nations* (Boston: J.H. Eastburn, City Printer, 1845), 17.

90 *The Rebellion Record: A Diary of American Events, with Documents*, 316.

91 Ibid., 318.

targets. If they did not, they lost their status as noncombatants. Henry Halleck, for instance, goes so far as to argue that "every besieged place is, for the time, a military post; for even when it is not defended by a military garrison, *its inhabitants are converted into soldiers* by the necessities of self-defense."[92] Much like the residents of a barbarous nation who provide material support to an enemy, erstwhile noncombatants could transform into combatants. J.S. Risley also noted that "the inhabitants of fortified towns are so closely associated with the garrison that they are considered to have lost their non-combatant character."[93] George Davis argues, "The very act of war is a sufficient notice to the non-combatant inhabitants of such places that an attack is at least a probable contingency. If they continue their residence, it is presumed that they do so with full knowledge that the place may become the centre of active military operations."[94] Freeman Snow agreed that civilians in fortifications lose their protected status. He writes: "In case of war the very fact of a place being fortified is evidence that at any time it is liable to attack and the noncombatants residing within its limits must be prepared for a contingency of this kind."[95] Snow goes on to explain that

> During the Franco-Prussian war in 22 sieges made by the Germans not a single assault was attempted. It was found to be easier, more effective, and it was claimed more humane to invest the places, seal all approaches, to prevent the supply of provisions or reinforcements and they by long-range siege guns keep up a constant fire beyond the range of the guns of the besieged. This fire was not concentrated upon the fortifications, but was directed upon the town, and causes such ruin and death that the place was forced to surrender through famine and excessive suffering.[96]

Snow explains that Otto von Bismarck defended the bombardment of Paris during the Franco-Prussian War, which commenced without warning:[97] "No doubt the siege of a fortress containing nearly 3,000,000 inhabitants creates much hardship, but the responsibility rests with those who made the capital a fortress and a battle ground. People who take up their residence in such a place must be prepared for such hardships."[98]

---

92 Halleck, *International Law*, 185. (My emphasis).

93 J.S. Risley, *The Law of War* (London: A.D. Innes and Co., 1897), 117.

94 Davis, *Outlines of International Law*, 302.

95 Freeman Snow, *International Law: A Manual* (Washington, D.C. Government Printing Office, 1898). 89.

96 Ibid., 89–90.

97 The siege of Paris lasted from September 1870 to January 1871.

98 Quoted in Snow, *International Law*, 90.

Since military necessity was understood in axiomatic terms when it came to the targetability of specific places – defended places were inherently of military interest – the burden of responsibility could be shifted onto those individuals who continued to reside in close proximity to these places of objective military interest. It was up to the civilians themselves to avoid legitimate targets.[99] Of course, much as it is today, the realistic possibility that civilians could completely avoid military targets always proved to be something of a challenge. To begin with, noncombatants were routinely conscripted to help with siege preparations. In Garnet Wolseley's 1871 *Soldier's Pocket Book*, he explains that in defending a place, a responsible commander should take account of all the civilian population who reside there and employ them to ready the place for assault up to the last minute before expelling them. "Get as much work as possible done by the non-combatants," he recommends.[100] The noncombatant population residing in a city or town may not have been at liberty to flee. And, of course, even if noncombatants could flee the onslaught of the approaching army, it is not clear they had any place to go. As seen above in the status of those refugees living in the wake of the Prussian assault on Metz, life as a refugee was desperate and cruel.

There were, of course, countervailing arguments that sought to minimize the excesses of military bombardments. When reflecting on the necessity of General Lord Cathcart's and Lord Gambier's bombardment of Copenhagen in 1807, Lord Wellington concluded that:

> Bombardments should, in these days, by a compact among civilized nations, be forever abolished. We shudder as we read of women and children, old men and infants, slain by the sword, and exclaim loudly against the barbarities of ancient warfare. The allowed practice of bombardment realizes the same cruelties; for, though the soldier does not exactly see his

---

99 In 1922, Charles Hyde argued that "Close proximity to an object subject to destruction, such as a fortress or an arsenal, must weaken the complaint of him who suffers in consequence of an attack upon it." Charles Cheney Hyde, *International Law: Chiefly as Interpreted and Applied by the United States, Vol. 2* (Boston: Little, Brown, and Co., 1922), 663.

100 Garnet Wolseley, *The Soldier's Pocket-Book for Field Service* (London: Macmillan and Co., 1871), 265. What is more, the leveling of the suburbs and houses outside of the fortification might also be necessary. Samuel Pufendorf explains that a sovereign has the right to demolish the suburbs of his own town: "For Example, when Houses built in the Suburbs, are pulled down in the Time of War. For the Proprietors of these can demand no Compensation for their Losses; because they knew before that they could not build in those Places upon other Terms." Samuel Pufendorf, *Of the Law of Nature and Nations* trans. Basil Kennett (London: Printed for I. Walthoe, R. Wilkin, J. and J. Bonwickle, S. Birt, T. Ward, and T. Osborne, 1729), Bk VII, 947.

> victims, and flesh his sword, yet, as through the long and wakeful night he serves in the batteries, which throw shells among human habitations, he knows well what a scene of blood and lamentation lies beyond the wall, lofty to hide, but vain to protect the miserable sufferers.[101]

Though the bombardment of habitations was permissible by the laws of war, many planners and commentators believed it violated the push to humanize warfare. Reflecting a similar moral concern, William Hall argues that "bombardment really amounts to an attempt to obtain an earlier surrender than would be militarily necessary, though the pressure of misery inflicted on the inhabitants, is an act which, though permissible by custom, is a glaring violation of the principle by which custom professes to be governed."[102] While theorists like Hall sought to provide additional protections to noncombatants, they reluctantly concurred that even though bombardment may contradict the principle of humanity that ought to govern war, targeting residential districts was still acceptable under certain conditions. At least the laws of war continued to permit it.

Indeed, it is only in very few dedicated military treatises of this period that one can find strong prohibitions against bombing residential districts. James Lorimer, the Professor of Public Law at the University of Edinburgh, for example, was one of the few theorists of international law in this era who grounded his arguments of the laws of war in natural law in contrast to the ascendant legal positivism.[103] In his 1872 work on the *Laws of Nations*, Lorimer writes: "The sack of a town, the laying waste of a district for strategic purposes, a bombardment directed against private dwellings, and similar acts, though by no means unknown, I fear, to the practice of modern warfare, are unquestionably antijural, and are strictly forbidden by the laws of war as professed by all civilized States."[104] Later, he continues that "the commander of every civilized army … will always consider it a sacred duty to take every step in his power, in the case of the siege of a fortified town, to cause private property belonging to inoffensive citizens to be respected, as far as local circumstances and the necessities of war will admit."[105] In Lorimer's estimation, besieging armies had a duty to

---

101 Duke of Wellington, *A Memoir of Field-Marshal The Duke of Wellington, Volume 1* (London: A. Fullarton and Co. 1814), 180.

102 William Hall, *International Law*, 335. See also, Edwin Glenn, *Hand-book of International Law* (St. Paul: West Publishing, 1895), 228.

103 John E. Noyes, "Christianity and Theories of International Law in Nineteenth-Century Britain," in *Religion and International Law eds.* Mark Janis and Carolyn Evans (Boston: Martinus Nijhoff, 1999), 238–39.

104 Lorimer, *The Institutes of the Law of Nations, Volume II*, 83–84.

105 Ibid., 361.

protect the property and lives of individuals living within fortified cities. In a similar manner, Thomas Lawrence, another rare natural law theorist of this period,[106] argues that the targeting of noncombatants residing in fortified cities is regrettable but no violation of the laws of war if women and children are incidentally killed in a bombardment; however, "had the guns of the besiegers been deliberately turned upon the dwelling-houses of the bombarded town, or had an open and undefended village been fired into, the persons responsible for such proceedings would have been justly accused of barbarity forbidden by modern usage."[107]

While many commentators on the laws of war made passing references to the opinions held by Lorimer and Lawrence, these were clearly in the minority. For instance, in his 1845 book addressing morality and war, Whewell refers to the view held by some that the general who refuses to allow noncombatants to leave a blockaded town should be "regarded in the same light as one who should murder his prisoners, or should be in the habit of butchering women and children."[108] He explains that such criticisms were, by far, the outliers.[109] They almost never made their way into the strategic and theoretical literature on war-making.

To be sure, even though the discourse among generals and military strategists was feeling some pressure of public opinion in terms of the permissibility of targeting civilians, as seen in the bombardment of Vera Cruz above, it is not entirely clear what impact public disapprobation had in the arena of international law. The pacifist Rufus Stebbins, in his 1857 address to the American

---

106 For a eulogy of Lawrence, see James Brown Scott, "In Memoriam: Thomas Joseph Lawrence, 1849–1920," *The American Journal of International Law* 1920 14:1/2; 223–229.

107 Thomas Lawrence, *The Principles of International Law* (Boston: D.C. Heath & Co., 1895), 344. In a similarly cautious tone, John Thomas writes, "No buildings but the military works should be demolished, unless essential to victory." John Thomas, *A Treatise of Universal Jurisprudence* (London: Published by Longman …, 1828), 417. While Thomas's language is much more restrained, he still leaves room for the principle of necessity. Sir Charles Pasley was similarly restrained; he writes, "In a regular siege, by which I imply, that the houses of the inhabitants shall not be bombarded, by that the fire or the besiegers' mortars shall be directed only against the fortifications, and the troops defending them." Sir Charles Pasley, *Rules Chiefly Deduced from Experiment, for Conducting the Practical Operations of a Siege, Part II* (London: John Weale, 1842), 150. See also Ernest Nys, *Les Origines du Droit International* (Harlem: Erven F. Bohn, 1894).

108 Whewell, *The Elements of Morality*, 374.

109 Ibid. The moral norms around sacking a defeated city, i.e. pillaging and ravaging a city that had already capitulated, were much more fully developed. For some insight into how British officers involved in the sack of Badajoz during the Peninsula Wars morally grappled with their crime, and how public outrage anticipated the moral developments of bombardment in the nineteenth century, see Daly, "'The Sacking of a Town is an Abomination,'" 160–182.

Peace Society, points out some reasons to be optimistic that a shift in moral discourse might be imminent. He remarks: "What a cry of shame rang through Christendom, when, in that storm of iron at Sebastopol [Crimean War, 1854], the British threw their shells into the hospital; and they were compelled, by the indignation of outraged Christendom, to attempt an apology."[110] He goes on to explain,

> in the terrific bombardment of Sweaborg [the Battle of Suomenlinna, 1855], the guns were aimed only at the national works. Not a hundred years since, and the murderous missiles would not have been directed against those adamantine walls of rock only or chiefly, but the messengers of torment and death would have been sent chiefly, if not solely, to the living flesh and exposed homes of men, and women, and children, till the heavens were rent with their shrieks, and the gutters ran with blood, and revenge was drunk with the crimson torrent.[111]

For Stebbins, it was significant that armies now feel the sting of moral disapprobation that would lead them to issue an apology for bombing a hospital. These examples, while they may indicate the inchoate rise of countervailing norms in siege warfare, they in fact highlight how *limited* the restraint of militaries tended to be and how much farther those conventions would need to go before noncombatants would cease to be targeted.

There is plenty of reason to doubt the moral progress implied in the optimism coming out of various peace societies. In a slightly more cynical late-nineteenth-century dialogue on the culture of war, one interlocutor confidently asserts "Christianity has made a stout resistance to the evil passions of mankind, and has even had some effect upon the conduct of nations toward one another, – at any rate, has had some effect upon their mode of conducting warfare. Captured garrisons are no longer put to the sword; women have been respected in this war [Franco-Prussian]"; to which his respondent replies, "The innocent are only starved, not put to death; and the excuse of 'the necessities of war' is always put forward when a village or town is burnt."[112] As seen above,

---

110 Rufus Stebbins, "Address," *The Advocate of Peace, July and August, v. 15–18, 1854–1858* (Washington DC: American Peace Society, 1859), 298.

111 Ibid.

112 Sir Arthur Helps, *Conversations on War and General Culture* (London: Smith, Elder, and Co, 1871), 258. These arguments concerning the humanizing effects of Christianity bring to mind Voltaire who had long ago quipped, "On countless occasions natural religion has stopped citizens from committing crimes. A noble soul has no wish to commit them; a

some viewed the arguments for necessity to be far too elastic. While it was true that after a breach, residents of fortified places were no longer slaughtered, it was still perfectly acceptable to bomb the residential districts to expedite surrender.

As these examples suggest, there was growing public interest in mitigating the cruelty of war. Public sentiments – the 'outrage of Christendom' – were part of a much broader push to humanize war through legal and moral restraint, especially regarding the bombardment of cities. But what these reflections ultimately reveal is the degree to which the ascendant moral norms remained firmly entrenched. It was, after all, a remarkable and praiseworthy event when a belligerent permitted women and children to leave a besieged city (like General Scott in the bombardment of Vera Cruz), when the besieging force only targeted the battlements instead of the residential districts. This is because it was much more common, indeed standard operating procedure in this period, for besieging armies to do what they could to expedite surrender, which included directly bombing the residential districts. Despite the optimism of these advocates of peace, these norms would persist well into the twentieth century.

The degree to which norms around siege warfare remained entrenched can also be seen in the many international conventions which aimed at humanizing war. In virtually every international convention or council of this period, it remained non-controversial that towns, houses, and villages nearby fortified places could be deliberately targeted. Only those noncombatants living in "open" and "undefended" towns were immune from direct harm. Of course, as one will see, this was a much more tenuous distinction than the binary classification seems to imply. Everything hinged on what one perceived open and undefended to be, as these terms were notoriously ambiguous, and there were many heated discussions as to whether such and such town had in fact been defended.

Frequently the license to bomb residential districts was enshrined in international legal conventions. For instance, the 1874 Brussels Declaration explicitly permits the bombing of residential districts:

---

tender soul is horrified by them, bearing in mind a just God of vengeance. But artificial religion encourages all those cruelties that people commit together: plots, sedition, robbery, ambushes, surprise attacks on towns, looting, murder. Each man marches cheerfully off to crime under his own particular saint's banner." Voltaire, "Guerre, War [1764]," in *Voltaire: Political Writings* trans. David Williams (New York: Cambridge University Press, 1994), 9.

> Article 15. Fortified places are alone liable to be besieged. Towns, agglomerations of houses or villages, which are open and undefended, cannot be attacked or bombarded.
> Article 16. But if a town or fortress, agglomeration of houses, or village be defended, the commander of the attacking force should, before commencing a bombardment, and except in the case of surprise, do all in his power to warn the authorities.[113]

While international law had started to make provisions for noncombatants residing in non-fortified or 'open' towns (towns ostensibly of no military value), there remained wide-spread acceptance of the indiscriminate bombardment of noncombatants in or near fortified cities.

Just to see how persistent these norms were, one need only note that neither the 1899 or 1907 Hague Conventions prohibited the indiscriminate bombing of civilians who lived in close proximity to legitimate military targets.[114] In his 1908 analysis of these provisions, Edward Westermarck explains that while certain aspects of land war had softened, there remained a pronounced exception: "during the siege of a fortified town custom still permits the houses of the town itself to be bombarded, with a view to inducing the commandant to surrender on account of the misery suffered by the inhabitants."[115] In his 1910 analysis of these conventions, John Westlake makes the same observation: "Thus firing on the houses of a fortified town is not forbidden, but when it can be avoided it is cruel, it is generally useless, and it ought to be forbidden unless there is reason to suspect that the houses are occupied by troops of the garrison or are used as magazines."[116] Again, notice that the moral language is couched in necessity; "when it can be avoided," bombing civilians

---

113 Project of an International Declaration concerning the Laws and Customs of War (Brussels, 27 August 1874). https://ihl-databases.icrc.org/ihl/INTRO/135 These same provisions are also included in The Laws of War on Land. Oxford 1880. Articles 32–34.

114 See Tami Davis Biddle, "Strategic Bombardment: Expectation, Theory, and Practice in the Early Twentieth Century," in *The American Way of Bombing: Changing Ethical and Legal Norms, from B-17s to Drones* eds. Matthew Evangelista and Henry Shue (Ithaca: Cornell University Press, 2014), 30.

115 Edward Westermarck, *The Origin and Development of the Moral Ideas, Volume 2* (London: Macmillan and Co., 1908), 26.

116 John Westlake, *International Law, Part 1* (New York: Cambridge University Press, 1910), 89. Article 26 of Hague 1899 reiterates Article 16 of the 1874 Brussels Declaration, though, instead of expressing the freedom to attack towns and villages, Article 26 simply emphasized the necessity to warn of an impending bombardment. See Convention (II) with Respect to the Laws and Customs of War on Land and its annex: Regulations concerning the Laws and Customs of War on Land. The Hague, 29 July 1899.

is unnecessary. What makes the bombing of residential districts repugnant is that it is "generally useless." And finally, a few years later Pearce Higgins argues, "The Hague Regulations are silent on [this] point [of bombing residential districts]."[117] They make no explicit provision against such bombing. Their silence is a form of tacit permission.

What is more, in the Convention Concerning Bombardment by Naval Forces in Time of War (Hague 1907), Article 1 does state that "The bombardment by naval forces of undefended ports, towns, villages, dwellings, or buildings is forbidden." But the rest of the document lays out multiple caveats to this prohibition. Article 2 explains that a commander "incurs no responsibility for any unavoidable damage which may be caused by" bombarding "Military works, military or naval establishments, depots of arms or war material, workshops or plants which could be utilized for the needs of the hostile fleet or army, and the ships of war in the harbor." This suggests that property or people nearby these works could be destroyed with impunity. Article 3 stipulates that "After due notice has been given, the bombardment of undefended ports, towns, villages, dwellings, or buildings may be commenced, if the local authorities, after a formal summons has been made to them, decline to comply with requisitions for provisions or supplies necessary for the immediate use of the naval force before the place in question."[118] Refusing a formal summons or declining to comply with requisitions could open up even an undefended town to attack. Some commentators noted that these caveats were so extensive that they eliminated virtually all limitations on the bombing of coastal towns.

As the passages above indicate, the bulk of the responsibility for the nonfighting population rested with the noncombatants themselves. They should avoid places of military interest and comply with summons and requisitions as best they could. The political leadership of a besieged place might also bear the responsibility of not evacuating vulnerable populations well in advance. Despite some rare apologies, military leadership was understood to bear very little if any responsibility for the harm that stemmed from bombing residential districts. Since fortified places represent a clear military objective, choosing to reside in or near a fortification opens one up to attack at any time. And while there were efforts in international law to formally demarcate legitimate places of attack, protecting "open" or "undefended" places from assault, this

117 Pearce Higgins, *War and the Private Citizen: Studies in International Law* (London: S. King & Son, 1912), 48.

118 Laws of War: Bombardment by Naval Forces in Time of War (Hague IX); October 18, 1907. http://avalon.law.yale.edu/20th_century/hague09.asp.

only served to confirm the received opinion that citizens in fortified places could be intentionally targeted.

7 Reprisal, Retaliation, Retorsion

In June of 1864, Major General John G. Foster sent a terse letter to Major General Samuel Jones, the Commander of Confederate Forces at Charleston, concerning the bombardment of the city. Foster indicated that even though advance notice of the attack had been given, women and children had not yet been evacuated. They had been "for many months exposed to fire," a matter that Foster objected to but believed was decided by Jones's "own sense of humanity." The "indefensible act of cruelty" that Foster did rebuke Jones for, however, was "placing defenseless prisoners of war in a position exposed to constant bombardment," a tactic "designed only to prevent the continuance of our fire upon Charleston." To this Foster explains:

> That city is a depot for military supplies. It contains not merely arsenals but also foundries and factories for the manufacture of munitions of war. In its shipyards several armed ironclads have already been completed while others are still upon the stocks in course of construction. Its wharves and the banks of the rivers on both sides of the city are lined with batteries. To destroy these means of continuing the war is therefore our object and duty. You seek to defeat this effort not by means known to honorable warfare but by placing unarmed and helpless prisoners under our fire. I have forwarded your communication to the President with request that he will place in my custody an equal number of prisoners of the like grades to be kept by me in positions exposed to the fire of your guns so long as you continue the course stated in your communication.[119]

There are several compelling aspects to this exchange. The first is Foster's seeming indifference to civilian human shields. Any civilian deaths that resulted from the bombardment were the moral responsibility of Jones who did not make adequate effort to get them out of the city. This sentiment partakes some of what was explained in the previous section on sieges. It was up to the military leadership and the civilians themselves to avoid legitimate military targets, especially after advanced notice of bombardment had been given.

119 "General Foster's Reply," in *The Rebellion Record: A Diary of American Events, Volume 11*, 592.

The civilians who continued to reside in Charleston had lost their noncombatant character due to where they were located, which was intimately linked with the strategic necessity of the site being bombed. Foster, after all, lays out a strikingly clear case for the necessity of the bombardment of the city – "To destroy these means of continuing the war is therefore our object and duty."

What does give Foster pause is the fact that noncombatants, more specifically, *hors de combat*, those soldiers who had laid down their arms and had surrendered to the enemy, were being employed as a deterrent to the bombardment. This was not only dishonorable, Foster notes, but demanded reprisal. Confederate prisoners would suffer the same fate. Though, as Karen Stokes explains, the Union had already started to use Confederate prisoners as human shields in 1863, itself a response to the poor treatment of Union captives. Over 600 Confederate officers were being held in the Charleston harbor, on Morris Island, subjected to barrage from their own forces.[120] As the War Between the States progressed and the breakdown of the Dix-Hill Cartel continued, prisoners of war increasingly became subject to reprisals.[121]

While in the twentieth and twenty-first centuries, "the prohibition against armed reprisals remains one of the most well-established *jus in bello* benchmarks,"[122] throughout the eighteenth and nineteenth centuries, the laws of war made certain allowances for *reprisal, retaliation*, and *retorsion* (sometimes spelled "retortion"), which in certain cases permitted intentional harm to noncombatants.[123] As James Farrer notes, "The worst feature in reprisals is that they are indiscriminate and more often directed against the innocent than the guilty. To murder women and children, old men, or anyone else, on the ground of their connection with an enemy who has committed an action calling for retribution, can be justified by no theory that would not equally apply

120 Karen Stokes, *The Immortal 600: Surviving War Charleston and Savannah* (Charleston, SC: The History Press, 2013). See also, Lonnie Speer, *War of Vengeance: Acts of Retaliation against Civil War POWs* (Mechanicsburg, PA: Stackpole Books, 2002) and Burrus Carnahan, *Lincoln on Trial: Southern Civilians and the Law of War* (Lexington, KY: The University of Kentucky Press, 2010).

121 The Dix-Hill Cartel was the agreement between the Confederacy and the Union governing prisoner exchanges. See "The 'Dix-Hill Cartel' for the General Exchange of Prisoners of War Entered into Between the Union and Confederate Armies," in *The War of the Rebellion: A Compilation of the Official Records of the Union and Confederate Armies, Series 2, Volume 4* (Washington D.C.: Government Printing Office, 1899), 266.

122 Michael Newton, "Reconsidering Reprisals," *Duke Journal of Comparative and International Law* 20 (2010); 361–388. 361.

123 Retorsion, a term that has gone out of fashion, is, as explained by Freeman Snow, "retaliation in kind." Freeman Snow, *International Law: A Manual* (Washington, D.C.: Government Printing Office, 1898), 72.

to a similar parody of justice in civil life."[124] As regrettably frequent as this "parody of justice" was, and notwithstanding Farrer's desire to see such a standard be revised, direct harm inflicted on the innocent was common.

While conceptually related, reprisal, retaliation, and retorsion were not strictly speaking synonyms.[125] In fact, usage of these terms was never formally standardized, which has bred some amount of confusion and contradiction in the literature. The broad and often inconsistent usage of these terms within the international discourse on the laws of war frequently generated some problems in terms of analysis. For instance, in his 1913 compendium on *International Law*, John Westlake complains that "In the literature of international law there are for modes of self-help several terms which are not always carefully distinguished," including retaliation and retorsion.[126] Geoffrey Butler and Simon Maccoby likewise note that Vattel:

> discoursed eloquently on the subject of reprisals, but did not draw attention to what was of cardinal importance. In different paragraphs he mentioned state-authorized reprisals under letters of marquee executed by private citizens and reprisals undertaken directly by the state; but he did not remark that the one was a dying practice while the other was growing in importance.[127]

Not only were these phenomena undergoing significant institutional transformations from the late eighteenth to mid nineteenth centuries, but it was not always clear what the phenomenon under discussion was.[128]

As Butler and Maccoby indicate, much earlier in the history of war, reprisals were more closely affiliated with private war and privateering (state-sanctioned piracy and what can be described as vigilante justice). An individual or corporation, aggrieved by the subject of a foreign power, could petition his own sovereign for "letters of reprisal" or "letters of marque," which authorized him

124 Farrer, *Military Manners and Customs*, 115.

125 See Vattel, *The Law of Nations*, Book II, §§ 339, 341, 342.

126 John Westlake, *International Law, Part II* (London: Cambridge University Press, 1913), 6.

127 Geoffrey Butler and Simon Maccoby, *The Development of International Law* (New York: Longmans Green & Co, 1928), 179.

128 For instance, George Atkinson lumps reprisals, retaliation, and retortion together under the category of "modified warfare, or hostilities, short of open war." George Atkinson, *International Morality; or the Touchstone of the Law of Nations* (London: G. Woodfall and Son, 1851), 106.

as a private citizen to seek recompense and redress.[129] William Blackstone, the great legal mind of the eighteenth century, explains that "These letters are grantable by the law of nations, whenever the subjects of one state are oppressed and injured by those of another: and justice is denied by that state to which the oppressor belongs."[130] Letters of marque permitted privateers to effectively wage private war on the merchants of hostile foreign nations, while letters of reprisal allowed individuals to raise private armies.

In the case of privateering, letters of marque could permit the forced confiscation of goods, a requirement to pay additional customs duties, and the withholding of payments. As Gallaudet explains, "Thus it was for a long time the rule, that when an individual suffered abroad in his rights or his property, he was justified in avenging himself on those who had done the wrong – not only on the offender himself, but on any other person, private or public, belonging to the same nation."[131] In many cases, the letters of marque were based on financial disputes, but sometimes letters of reprisal were issued to communities to avenge murder and other cruelties.[132]

Letters of reprisal had a slightly different history. Cornelius Bynkershoek, an eighteenth-century Dutch legal theorist, explains that these letters of reprisal were only granted if it was clear that the offending individual's state engaged in "an open denial of justice."[133] He writes,

129 These practices can be traced back to *androlepsy*, in ancient Greek law, where "The relations of the deceased were empowered to sieve three men in the city or house, whither the malefactor had fled, till he was either surrendered, or satisfaction made some other way for the murder." In the Roman tradition, *clarigation* were demands for repaying injuries just short of war. They could take the form of hostage taking, that is, until the original injury was repaid. See "Androlepsy" and "Clarigation," in *A New and Complete Dictionary of Arts and Science, Volume 1* (London: Printed for W. Owen, at Homer's Head ..., 1763), 142, 621.

130 William Blackstone, *Commentaries on the Laws of England, Volume 1* (New York: Published by E. Duyckingck ..., 1827), 193.

131 Edward Gallaudet, *A Manual of International Law* (New York: A.S. Barnes & Co, 1879), 198.

132 The history of privateering (i.e. state-sanctioned pirating) goes far beyond this analysis of reprisal. Quite frequently states would authorize private vessels to harass and pillage enemy ships. As one early nineteenth-century opponent of the institution writes, "These ruthless bravadoes, by the habitual practice of rapine and murder, become so hardened in sin and crime, that they seemed to riot and rejoice over the sufferings of their innocent victims." George Coggeshall, *History of the American Privateers and Letters-of-marque* (New York: Geo. Putnam, 1861), xlv.

133 Cornelius van Bynkershoek, *A Treatise on the Law of War* trans. Peter Stephen Du Ponceau (Philadelphia: Published by Farrand & Nicholas, 1810), 184.

> They are an authorization, granted by a sovereign, to take the persons and goods of the subjects of another prince; in order to obtain satisfaction for an injury committed upon his own subjects, for which justice has been denied by the sovereign of the offending party. Thus, an injury committed by force and violence, and not repressed by the competent magistrate, is redressed by the same means and in the same manner.[134]

The point is that these licenses were not meant to set in motion the downward spiral of retribution and counter-retribution. They were designed to publicize and document the grievances of private citizens. Most importantly, they would give the foreign state a chance to resolve the complaint before things got out of hand. In fact, part of the impact of proceduralizing what would have otherwise been willy-nilly revenge-seeking on the part of the citizenry is that it generated the institutional structures that brought states into more formal contact with each other. As Richard Lillich explains, "This practice of granting letters of reprisal was clearly less than an ideal method of fostering international harmony and good will. Nevertheless, it had one important consequence, namely, it entailed the thesis that a dispute between an alien and a national of a host State was also, in some measure at least, a dispute between their States."[135] Though clearly an imperfect vehicle of justice, it played a significant part in bringing states into more extensive diplomatic relations.

To illustrate what a pacifying and indeed stabilizing effect these letters of reprisal had on international politics, the eighteenth-century English barrister, Robert Ward, provides a useful back story to their development. He recounts an incident in 1292 where a physical altercation broke out between two men; an Englishman stabbed and killed his Norman opponent. Friends of the Norman victim petitioned their king to intervene in the affair, but he refused, insisting that they "take their own revenge."[136] What followed was a string of reprisals that culminated in outright war. A band of Normans captured the first English ship they could and "hung up several of the crew, and some dogs at the same time, at the mast head."[137] Private English citizens then retaliated by making informal alliances with the Irish and Dutch, which culminated in a number of atrocities. The Normans in turn set out two hundred ships "hanging all the seamen they could find," to which the English who "fitted out a strong fleet,

134 Ibid., 184–85.

135 Richard Lillich, *The Human Rights of Aliens in Contemporary International Law* (Manchester: Manchester University Press, 1984), 8.

136 Robert Ward, *An Enquiry into the Foundation and History of the Laws of Nations in Europe, Volume 1* (London: Printed by A. Strahan and W. Woodfall …, 1795), 295.

137 Ibid.

destroyed or took the greater part of the Normans, and giving no quarter, massacred them, to the amount of fifteen thousand."[138] Ward explains that it was at this point that the "affair then became too big for private hands, and the Governments interposing in form, it terminated in that unfortunate war."[139] At this point, the conflict transformed from a private to public war. To this, Ward concludes, "while such laxity of discipline and of true subordination prevailed, we can little wonder if the passions of private individuals were allowed to enter into, and mingle with the public administration of the laws of war."[140] When left to private individuals who are not constrained by the chain of command or military discipline, things were bound to get out of hand and all manner of atrocity would arise.

As Ward's account suggests, letters of reprisal were designed to subordinate violence to sovereign control while at the same time providing an outlet for private passions through authorized channels. Before one could start simply raiding merchant ships or murdering foreign nationals, one needed to petition the political leadership for permission to seek justice. As the eighteenth-century British diplomat, Wyndham Beawes, explains, the aggrieved party, in order to obtain the right to take matters into his own hands, had to meet certain requirements:

1) The oath of the party injured, or other sufficient proof, touching the pretended injury, and the certain loss and damage thereby sustained,
2) A proof of the due prosecution for the obtaining satisfaction in a legal way,
3) Of the delay or denial of justice,
4) A complaint to his own prince or state,
5) Requisition of justice, by him, or them, made to the supreme head or state, where justice in the ordinary course was denied,
6) Persistence still in the denial of justice.[141]

Later in the nineteenth century, John Risley also emphasized that for these letters to be justified, an individual harmed by a foreign state must "first appeal to its Courts, and not until justice has been there denied to him has his own Government any claim to interfere on his behalf."[142]

138 Ibid. 296.

139 Ibid.

140 Ibid.

141 Wyndham Beawes, *Lex Mercatoria; Or a Complete Code of Commercial Law, Volume 1* (London: Printed for F.C. and J. Rivington …, 1813), 312. Beawes derived this formulation from Charles Molloy, *De Jure Maritimo et Navali; Or, A Treatise of Affairs Maritime and of Commerce* (London: Printed for John Bellinger …, 1682).

142 Risley, *The Law of War*, 58.

After pursuing all the appropriate legal channels, and only in extraordinary cases, would the letters of reprisal be granted. What is more, even after letters were issued, they typically outlined both the timeframe in which private judgment could be pursued and the response the offended government would take if justice were not given to the individual.[143] These circumscribed acts of hostility were designed to do several things at once. They extended the sovereign's juridical control over private international conflict while at the same time outsourcing the cost of meting out that violence. By limiting the scope of violence private individuals could take, this would, in turn ensure that states would not end up at war with each other. The burden of proof for reprisals to be granted were quite high. Not only did private individuals have to provide demonstrable evidence that they had suffered a wrong, but they also had to show that the foreign government was unable or unwilling to do anything about it. And even when this high demand of evidence had been achieved, these letters typically had a clear timeframe for the aggrieved to exact justice. Letters of reprisal were not indefinitely open-ended.[144]

Though widely accepted as an international norm in the nineteenth century, many theorists were aware of the challenges letters of reprisal posed to international security. They, of course, did not fully eliminate the possibility of exacerbating the threat environment between states. As Sir Walter Scott mentions, the turn to letters of reprisal were sometimes used as an "impolitic expedient" to control "turbulent subjects, committed to a rival sovereign."[145] For instance, in the late sixteenth century, the Scottish sovereign had extended permissions to "the banditti of Liddesdale" to punish English residents who lived nearby, presumably to gain firmer control of his own border. In response, King James V issued letters of reprisal to his own injured subjects, permitting them the "'power to invade the said inhabitants of Liddesdale, to their slaughters, burning, heirships, robbing, reifing, despoiling and destruction, and so to continue the same at his grace's pleasure,' till the attempts of the inhabitants were fully atoned for."[146]

143 For an example of letters granted for the massacre at Amboyna given by Charles II to William Courten, Esq. see Beawes, *Lex Mercatoria*, 314–317.

144 For more on how letters of reprisals could expire or be revoked, see Joseph Chitty, *A Practical Treatise on the Law of Nations* (Boston: Published by Bradford and Read, 1812), 73–77.

145 Sir Walter Scott, *Minstrelsy of the Scottish Border, Volume 1* (Edinburgh: Printed by James Ballantyne, for Longman and Rees ..., 1803), xxii.

146 Ibid.

Scott goes on to explain that this plan worked, but perhaps too well: "For the inhabitants, finding that the sword of revenge was substituted for that of justice, were loosened from their attachment to Scotland, and boldly threatened to carry on their depredations, in spite of the efforts of both kingdoms."[147] As this incident shows, in some cases letters of reprisal were used expediently, as tools to police border disputes and to engage in proxy wars. Private grudges and vicarious disputes could be surreptitiously appropriated, politicized, and then legitimized through the formal/informal mechanisms of the reprisal system.

Beyond the obvious problems that arise with armed conflict, letters of reprisal were also often seen as a questionable mechanism of justice, specifically since they permitted what amounted to the killing of innocent people for someone else's crime. Reflecting on the moral challenge of the reprisal system, the great German jurist and diplomat, G.F. von Martens writes, "the life of an innocent person cannot be taken, unless in extraordinary cases, where there is no other means of obtaining the satisfaction due, and of preventing future violations."[148] As morally unsatisfying as it was, Martens concludes that the death of an innocent person through reprisal killing is justified if it leads to less harm in the future, and this was especially so if the harm being prevented is outright war.

Martens is supported in his analysis by Jeremy Bentham, who in the legal positivist tradition also argues that reprisal should not be thought of as "strictly speaking, vicarious punishment"; he explains that this is because rather than being inflicted on the individual qua individual, "the reprisals inflicted on his subjects operate upon the sovereign himself, either by the compassion felt for their suffering, or by the fear, if patiently submitted to, of alienating the affections of his people."[149] In other words, the person receiving the punishment is not an "innocent victim," per se. The individual is harmed in his capacity as a member of the offending body politic. Since the sovereign did not effectively adjudicate the initial wrong to begin with, he bears the injury since he abdicated his moral responsibility to punish those who had committed the original crime. This assumes that the most effective way a sovereign can be reached is by way of his subjects; punishing the body to get at the head.

147 Ibid., xxiii.

148 G.F. von Martens, *The Law of Nations* (London: Published by William Cobbett, 1829), 272.

149 Jeremy Bentham, *Principles of Penal Law* in *Introduction to the Study of the Works of Jeremy Bentham, Volume 1*, ed. John Bowring (Edinburgh: William Tait, 107 Prince's Street, 1843), 480.

There are, of course, limits to this for Bentham. He explains that "In these cases, what humanity dictates is, that the sufferings inflicted on the innocent should be the least possible, consistent with the production of the desired effect."[150] Bentham was willing to accept harm to and even the death of ostensibly innocent people "as means of preventing the infliction of injuries not warranted by the rules of war."[151] One cannot rely entirely on the honor of combatants to faithfully respect the laws of war, which is why reprisals are a "very necessary coadjutor."[152] Nevertheless, while such cruelties might be necessary to mitigate the greater brutalities of war, Bentham cautions that it is an imperfect form of justice. He concludes his thoughts on the matter by asking: "What satisfaction did the offending person reap from this sacrifice? The degradation and shame of belonging to it. The glory of the sufferer was the disgrace of the judge."[153] Reprisals might be necessity, but they are not an unqualified good.

7.1 *The Beagle Affair*

By the mid-nineteenth century, the norms around letters of reprisal started to change. Rather than a mechanism to authorize private war, it was transforming into a violent act that fell short of war, or, as Bentham had argued, a coadjutor of war itself, a mechanism that functioned to ensure enemy combatants adhered to the laws of war. In 1860, for instance, Archer Polson argued that reprisals were "the first step taken at the commencement of a war, and considered equivalent to a declaration of hostilities, unless an immediate satisfaction is made by the other State."[154] They are an attempt to demand satisfaction for a violation, either before open hostilities or after war has begun. This logic not only served conventional wars between European nations, but also small wars between great powers and indigenous populations. For example, a late-nineteenth-century case dramatically illustrates how the justification for killing innocent people through the right of reprisal functioned. This can be seen in the *Beagle* Affair of 1877.[155]

In April of that year, an Englishman named William Easterbrook was shot and killed by a native on the South Pacific island of Tanna. In response to what appeared to be a crime against a British subject,

150 Ibid.
151 Ibid.
152 Ibid.
153 Ibid., 481.
154 Polson, *Principles of the Law of Nations*, 39.
155 This is not the same *Beagle* that Darwin voyaged on.

> Commodore Hoskins ordered Lieutenant Caffin, of the *Beagle*, to proceed to Tanna to inquire into the murder; and the instructions thus continue: 'Should you be fully convinced that it was not his (Easterbrook's) own misconduct which led to the commission of the deed, you are, should you deem your force sufficient, to endeavour to obtain possession of the murderer; and, if successful, you are to cause him to be executed in the most public and judicial way possible.'[156]

However, as Katherine Cawsey explains, the man responsible for shooting Easterbrook, a native by the name of Yahmunga had, immediately following the murder, "quickly vanished into the darkness and managed to get away on a Queensland recruiting vessel."[157] Yahmunga's escape meant that his brother, Nokwai, and another man falsely accused by an island priest, Nakapok, would be left to face the consequences of his actions.

Upon arrival, Lieutenant Caffin took several Ni-Tanna natives as prisoners, including Nokwai, and indicated that he would continue to hold them as hostages until the islanders gave up the murderer, who, of course, having already absconded, was nowhere to be found. After several violent altercations with the natives due to their perceived intransigence (skirmishes which, incidentally, led to the deaths of eleven natives, including Nakapok), "Caffin decided to hang Nokwai from the yard-arm as an accessory to the crime."[158]

These actions did not sit well with one observer, Sir John Eldon Gorst, who petitioned the crown to offer a ruling on whether or not ship captains had the authority to execute innocent non-subjects in this manner.[159] He was concerned that this encouraged a system where "naval officers were turned into judges of assize, and Her Majesty's ships into perambulating gallows."[160] But after some deliberation the British government ruled that Caffin was perfectly within his right to execute Nokwai, even if he was not responsible for the murder:

156 "The Beagle Affair," *The Law Journal, Volume XIII* (London: E.B. Ince, 1879), 583.

157 Katherine Cawsey, *The Making of a Rebel: Captain Donald Macleod of the New Hebrides* (Suva, Fiji: Institute of Pacific Studies, 1998), 144.

158 Ibid., 146.

159 Archie Hunter, *A Life of Sir John Eldon Gorst: Disraeli's Awkward Disciple* (New York: Routledge, 2013); 104–105.

160 "The Beagle Affair," 583.

> when a British subject is murdered in a civilized country, the representative of the British Government would deem it within his duty, in case of need, to direct the attention of the foreign Government to the crime; and if that Government refused to do justice, if it allowed impunity to the murderer or murderers, a question would arise between the two Governments, and if there was a persistent refusal to punish the murderer or murderers of the British subject, the refusal might be fairly treated as a *casus belli*. But it does not follow that the offended Government is bound to declare war, for it might, if it chose, take a reprisal – that is, meet one act of war by another act of war. The murder of British subjects by savages is, when the murders are unpunished, a *casus belli*, and justifies either war or a reprisal – that is, a single act of war.[161]

The Attorney-General, continued in his ruling by claiming,

> Many British subjects had been murdered at that place, and it was the right of the commander of one of Her Majesty's vessels to levy war upon those savages and shoot them down. He might have destroyed the whole of the village, but he chose to act more mercifully. The chief men of the place refused to comply with the demand of Lieutenant Caffin to surrender the murderer, and then the lieutenant took reprisals by keeping some of the natives as hostages. The trial and execution of the native was as much an act of war as if a party had been landed, had seized the murderer, and shot him on the foreign soil.[162]

Indeed, reflecting some of Westlake's frustration around the variance of word usage, the British government legitimized Caffin's action in the language of reprisal, though here defined as a single act of war (rather than an act short of war). The captain was justified on the grounds that, given the *casus belli*, an actual invasion would have led to the same if not worse, result. Overlooking Nokwai's specific guilt, the ruling concludes that the whole native village had incurred some guilt, either for the murder of Easterbrook or for past murders. Because of this, collective guilt had transferred onto the entire nation; anyone could be killed to satisfy the demand for justice. In fact, as the ruling concludes, killing one prisoner, even if he was technically innocent, was a "mercy," a lesser evil when juxtaposed against the alternative of a retaliatory invasion.

161 Ibid.

162 Ibid., 583–584.

Even though the specifics of Nokwai's guilt remain somewhat obscure, the British government in effect held that if there were no other means of securing justice, executing hostages for the crimes of their fellow tribesmen was acceptable by way of the norms governing the use of reprisals, and this was especially true when dealing with "savages." But upon closer analysis, this ruling was also meant to extend to "civilized nations." While the procedures may differ a little – with "civilized nations" there was an additional requirement to "direct the attention of the foreign Government to the crime" – the result is the same. If a political community were unable or unwilling to redress a wrong, a government could take matters into its own hands, which included killing ostensibly innocent people for the crimes of others.

8 Change in Norms

As Butler and Maccoby indicate, throughout the nineteenth century there was a decisive shift away from seeing reprisals as the purview of aggrieved individuals equipped with licenses to take matters into their own hands. While there were emerging legal mechanisms to address the crimes of foreigners, diplomatic channels, and emerging tools, like extradition, the punishment of foreign nationals whose governments were unable or unwilling to bring them to justice increasingly came to be seen as the responsibility of states and not authorized private citizens. Indeed, as letters of reprisal started to fall into disuse and the linkage of reprisals and private war dissipated, reprisals became much more closely linked with punishments levied by governments just short of war.[163] That is, a reprisal was a formal mechanism of diplomacy and intervention designed to put pressure on another state to act in a particular way. Due to these emerging pressures, many states in the nineteenth century had begun to formalize extradition procedures in order to avoid unnecessary reprisal from foreign states.[164]

The legal standardization of extradition explains some of the shift in usage and the evolving understanding of reprisals. But this transformation was also precipitated by the fact that letters of marque were explicitly prohibited by the Treaty of Paris (1856), the peace settlement of the Crimean War (see Figures 5.2 and 5.3). While letters of reprisal, though technically distinct from letters of

163 Risley, *The Law of War*, 59.

164 For how these norms had been formalized by nation states by the late nineteenth century, see Edward Clarke, *A Treatise upon the Law of Extradition* (London: Stevens and Haynes, 1874).

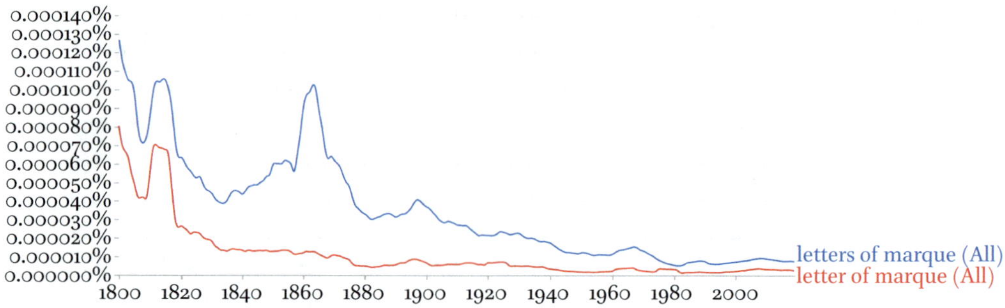

FIGURE 5.2 Ngram search of 'letter(s) of marque'

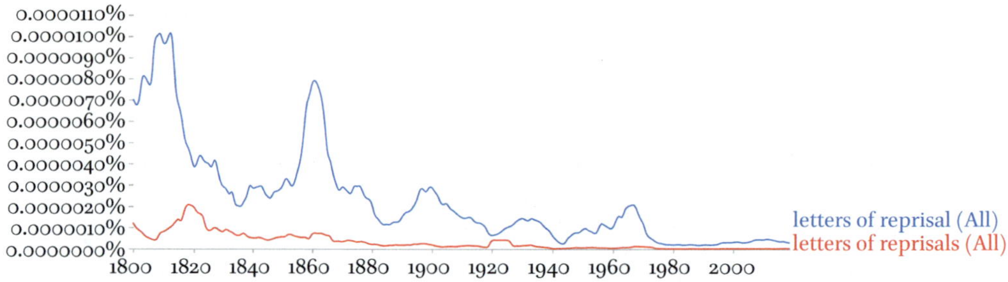

FIGURE 5.3 Ngram search of 'letters of reprisal' and 'letters of reprisals'

marque, remained a valid mechanism of adjudication, it ceased to possess the same degree of prominence as it had in previous centuries.

By the end of the nineteenth century "reprisal" started to resemble its modern form; that is, it came to be seen as a mechanism of deterrence and restraint in armed conflict.[165] Nevertheless it also retained the character of measures short of war.[166] The threat of reprisal was increasingly being used as a tool to deter a belligerent from violating widely agreed upon laws of war. For instance,

165 Halleck calls this the "rule of reciprocity." Halleck, *International Law*, 34.

166 Which is much closer to its use in contemporary Customary International Humanitarian Law, where reprisals are justified only under five conditions under rule 145: (*i*) *Purpose of reprisals.* Reprisals may only be taken in reaction to a prior serious violation of international humanitarian law, and only for the purpose of inducing the adversary to comply with the law; (*ii*) *Measure of last resort.* Reprisals may only be carried out as a measure of last resort, when no other lawful measures are available to induce the adversary to respect the law; (*iii*) *Proportionality.* Reprisal action must be proportionate to the violation it aims to stop; (*iv*) *Decision at the highest level of government.* The decision to resort to reprisals must be taken at the highest level of government; and (*v*) *Termination.* Reprisal action must cease as soon as the adversary complies with the law. See Jean-Marie Henckaerts and Louis Doswald-Beck, *Customary International Humanitarian Law, Volume I: Rules* (New York: Cambridge University Press, 2005); 513–518.

Article 84 of the Laws of Land and War (Oxford, 1880) expressly treats reprisals as a mode of deterrence in war. They are defined as "an exception to the general rule of equity, that an innocent person ought not to suffer for the guilty," even though they are "at variance with the rule that each belligerent should conform to the rules of war, without reciprocity on the part of the enemy."[167] Article 86 explains that "In grave cases in which reprisals appear to be absolutely necessary, their nature and scope shall never exceed the measure of the infraction of the laws of war committed by the enemy."[168]

In William Hall's 1880 compendium on international law, he likewise explains,

> reprisals, or the punishment of one man for the acts of another, is a measure in itself so repugnant to justice, and when hasty or excessive is so apt to increase rather than abate the irregularities of a war, that belligerents are universally considered to be bound not to resort to reprisals except under the pressure of absolute necessity, and then not by way of revenge, but only in cases and to the extent by which an enemy may be deterred from a repetition of his offence.[169]

It is clear that the Oxford Laws of War and commentators like Hall were starting to use the term reprisals in a way that is perhaps more recognizable to twenty-first-century readers. Though extreme and "repugnant to justice," they argue that reprisals were permissible because overall they created a disincentive to violate the laws of war. And yet, Hall goes on to explain reprisals could also still be seen as acts short of war that limit the outbreak of conflict. He writes, "The justification of reprisals being that they are the means of avoiding the graver alternative of war, it must in principle be conceded that anything short of complete war is permissible for sufficient cause."[170] As Hall suggests, reprisals are justified by way of necessity as a lesser evil; they serve as a deterrent in war and a safety valve in international politics that give states a recourse to ameliorating perceived injustices in a way that doesn't commit them to all out hostilities.

While reprisal, retaliation, and retorsion were distinct concepts, defined widely in nineteenth-century international legal literature, over time their use became so conflated and tangled that it is currently difficult to distinguish

167 The Laws of Land and War, Oxford, 9 September 1880. https://ihl-databases.icrc.org/ihl/INTRO/140?OpenDocument.

168 Ibid.

169 William Hall, *International Law* (Oxford: Clarendon Press, 1880), 352. See also Bordwell, *The Law of War Between Belligerents*, 17.

170 Hall, *International Law*, 311.

them. In the early part of the nineteenth century, letters of reprisal quickly fell into disuse and in its place, there arose the concept of *retorsion*. What contemporary theorists today call reprisal, are what eighteenth- and nineteenth-century theorists called retaliation or retorsion (or in-kind retaliation). Retaliation and retorsion were the resort to violence in war where noncombatants (including *hors de combat*) could be intentionally harmed – or intentionally put in harm's way – for the crimes of others. As Risley explains, "*Retorsio facti*, or vindictive retaliation, on the other hand, implies the infliction of the same amount of evil on an aggressive state that it has inflicted on the state aggrieved. It is literally an eye for an eye."[171]

By the end of the nineteenth century, reprisals were becoming associated with what Francis Lieber calls "retaliation." In Article 27 of the General's Orders 100, Lieber explains that "The law of war can no more wholly dispense with retaliation than can the law of nations, of which it is a branch. Yet civilized nations acknowledge retaliation as the sternest feature of war. A reckless enemy often leaves to his opponent no other means of securing himself against the repetition of barbarous outrage." And in Article 28, he goes on to explain that retaliation cannot be "mere revenge" but that it is "a means of protective retribution." One could only resort to such cruelties "after careful inquiry into the real occurrence, and the character of the misdeeds," and even then, the purpose was to impress upon the enemy the risks of continuing such a reckless and unjust tactic or policy. Resorting to cruelty – i.e. intentionally harming noncombatants – was permitted not simply to punish an unjust enemy, but it was designed to be "protective," to deter an enemy from continuing in injustice. And in this respect, the logic of reprisals or retaliations sat squarely in the framework of military necessity – the only way to keep an enemy accountable for how the war is waged is to check cruelty with cruelty.

Military ethicists and commentators were aware of the potential permissibility of harm embedded in the concept of retaliation or reprisal, and many regretted its persistence. For as much as this tit-for-tat approach to injustice might prevent certain cruelties from occurring, it could also precipitously lead to a downward spiral of injustice being heaped on injustice, as was the case in the Franco-Prussian War. When King Wilhelm I addressed the residents of Gitschin, he indicated,

171 Risley, *The Law of War*, 57. See also, Gallaudet, *A Manual of International Law*, 197.

> I carry on no war against your nation, but only against the armies opposed to me. If, however, the inhabitants will commit acts of hostility against my troops without any cause, I shall be forced to make reprisals. My troops are not savage hordes, and require simply the supplies necessary for subsistence … Tell the inhabitants that I have not come to make war upon peaceable citizens, but to defend the honour of Prussia against insult.[172]

Not long after this speech, things quickly deteriorated. In October of 1870, French militia (*francs-tireurs*) ambushed a Prussian unit on their way to Ablis. In response to this, Prussian cavalry torched the village and executed seventeen noncombatant Frenchmen.[173] Instances like this were so common in this war that many commentators repeatedly lamented the tendency of reprisals to facilitate what amounts to a race-to-the-bottom in terms of military restraint.[174] One commentator on the war notes: "Reprisals of the utmost cruelty have been carried on with relentless ferocity, and we begin to wonder how long human nature is to be disgraced by such barbarity. Two civilized nations have become worse than barbarians. Pillage, slaughter of innocent non-combatants, every enormity is daily enacted."[175]

Even though retaliation and reprisal were routinely decried during the Franco-Prussian War, and commentators on the war vociferously protested their abuse, the international community was never able to directly address this issue. Sheldon Amos assumes that reprisals had not been expressly included in the discussion at the 1874 Brussels Conference due to "the strong feelings, not yet abated, which had been aroused in the late Franco-German War."[176] Though it may have not been included for discussion due to the fact that, as Baron Antoine-Henri Jomini claimed, reprisals were "one of the most bitter necessities of war," and no one was willing to give them up.[177]

More recently Tracey Dowdeswell has shown that many of the proposed articles of the Brussels Conference clearly had in mind the issue of reprisal from the recent Franco-Prussian War. Several proposals were, in fact, designed

---

172 W.H. Davenport Adams, *The Franco-Prussian War: Its Causes, Incidents, and Consequences* (London: William Mackenzie, 1871), 83.

173 "About the War and Wounded," *Medical Press & Circular: A Weekly Journal of Medicine and Medical Affairs, From January to June 1871* (London: 20 King William St, Strand, 1871), 12.

174 See George Alfred-Henty, *The Young Francs-Tireurs, and their Adventures in the Franco-Prussian War* (London: Griffith and Farran, 1873).

175 "About the War and Wounded," 12.

176 Amos, *Political and Legal Remedies for War*, 323.

177 Ibid., 324.

to structurally *prevent* the resort to reprisal by addressing what was believed to be the source of the problem, i.e. the proliferation of non-lawful soldiers, specifically *francs-tireurs*. The *francs-tireurs* were civilian militias that had been authorized by the government to fight on behalf of France against Prussia, but which received no material support from the French government.[178] Many of the German atrocities of the Franco-Prussian War had ostensibly been committed in response to the operations of the *francs-tireurs*.

One of the humanizing trajectories of the laws of war in this period was to bring all combatants under a central authority. Irregular combatants, those who had not formally accepted the laws of war, might resort to the kinds of actions that would precipitate reprisal. Therefore, the Brussels Declaration spent so much time attempting to demarcate: the scope of military authority (Articles 1–8), the requirements for being formally recognized as legitimate combatants (Articles 9–11), and the restrictions on the kinds of violence legitimate belligerents could reasonably employ (Articles 12–14). While not expressly addressing the issue of reprisals by name, each of these provisions speaks directly to the issues that had produced much of the perceived injustices in the Franco-Prussian War.

Take, for instance, Article 10, which states: "The population of a territory which has not been occupied, who, on the approach of the enemy, spontaneously take up arms to resist the invading troops without having had time to organize themselves in accordance with Article 9, shall be regarded as belligerents if they respect the laws and customs of war."[179] For irregular troops to respect the laws and customs of war, Article 9 requires 1. That they be commanded by a person responsible for his subordinates; 2. That they have a fixed distinctive emblem recognizable at a distance; 3. That they carry arms openly; and 4. That they conduct their operations in accordance with the laws and customs of war.

The implicit approach to reprisals embedded in the Brussels Conference is that the best way to prevent the downward spiral of cruelty is to bring irregular or non-lawful combatants more formally under the control of a central authority, not to eliminate reprisals themselves. In other words, the operative assumption at the conference was that reprisals work, that is, only if soldiers are brought together under a stricter command structure where the entire body feels the pain of punishment, unlike disjointed bands of guerilla fighters.

---

178 Tracey Leigh Dowdeswell, "The Brussels Peace Conference of 1874 and the Modern Laws of Belligerent Qualification," *Osgoode Hall Law Journal* 54:3 (2017); 805–850.

179 Brussels Declaration 1874. https://web.ics.purdue.edu/~wggray/Teaching/His300/Handouts/Brussels-1874.html.

As it stood, militias were too isolated to feel the pressure of centralized control. By linking military units more formally together, the threat of reprisal as a penal sanction could be more effectively distributed. Furthermore, limiting the excesses of independent militias, specifically the confusion around their exact status, would lead to a humanizing element in war, or so it was believed. These kinds of provisions reflect the tendency of nineteenth-century military ethical theory to limit the scope of justified violence, to demarcate more clearly acceptable from unacceptable killing. As we have seen, under the principle of necessity, reprisal killing was justified only by virtue of its "protective retribution" – it was a way to indicate to the enemy that a moral line had been crossed and to deter further moral excess. In a certain sense, the subtext of the emerging international conventions, like the Brussels Declaration, were to formally demarcate the lines and pressure points that would make clear when reprisals were necessary and justified.

9 The Spatial Dimension of Noncombatant Death

It is worth contrasting for a moment, the spatial features of the nineteenth-century battlefield, as illustrated by the tactics of laying waste, blockade, and siege, and the much more tenuous moral line formed by the threat of retaliation, retorsion, and reprisal. Throughout the eighteenth and nineteenth centuries, the laws of war allowed the intentional killing of noncombatants within clearly demarcated spaces. As women, children, the aged, and the sick and injured got closer to these prohibited spaces, the more they lost the character of noncombatants. Borderlands and fortified cities, in particular, were unique physical and moral spaces that invited different ethical considerations. The idea was not that in those spaces "anything goes"; it was not that anyone could be killed at will (though at some moments in history this was the case regarding siege warfare). The operative moral consideration of these zones of hostilities was whether the harm inflicted would expedite the end of the conflict and prevent greater harms. The key moral consideration was not intention but the metric of necessity in conjunction with superfluous or gratuitous injury. Within this framework, putting violent pressure on noncombatants was believed to be legitimate in order to expedite an end to hostilities, which is how belligerents could justify starving the enemy population through blockades. In this respect, the early humanizing trends in the laws of war were built on their ability not to theorize strict protections for noncombatants; rather it was the ability to spatially demarcate and circumscribe those places where noncombatants could be deliberately harmed. In other words, it was not that noncombatants had immunity but that militaries did not typically see the utility in diverting their energy away from the enemy's fighting force or the direct sites of military operations. Having mapped this topography of permissible violence, it was left to the citizenry to avoid those dangerous spaces, if they could. It is noteworthy that many ethicists in this period clearly placed the burden of responsibility onto civilians or civilian leadership for their misfortune.

In contrast, threats of reprisal, retaliation, and retorsion conspicuously lack this spatial dimension. The precarious threat of in-kind cruelty hangs over and is embedded within every internationally agreed upon limit to war. Perhaps it is better to say that reprisals threaten to invert or turn inside out all the spatial relationships that had been used to demarcate acceptable harm to noncombatants. Lingering behind each prohibition in the international laws of war, say, to not bomb undefended towns, is an implicit justification for the intentional bombing of undefended towns if done in retaliation for a similar harm. William Paley makes this point explicit with regard to violations of the laws of war: "The breach of the rule can only be punished by the subversion of the rule itself: on which account, the whole mischief that ensues from the loss of those salutary restriction which such rules prescribe, is justly chargeable upon the first aggressor."[180] He continues that "the first recourse to them would be followed by instant retaliation; that the mutual license which such attempts must introduce, would fill both sides with the misery of continual dread and suspicion, without adding to the strength or success of either."[181]

Throughout the nineteenth century these norms were firmly established in international law. For instance, in his 1908 commentary on the Hague Convention, Percy Bordwell makes this point explicit. He explains that the prohibition against bombing undefended towns "does not deny the use of bombardment as a penal measure where a legitimate demand for contributions or requisitions has been refused or where the population has been guilty of violating the laws of war, or as a measure of retorsion; but its use even in these cases would be extreme and only justified by imperative necessity."[182] The logic of reprisal, retaliation, and retorsion embeds within each law of war the implicit permission to intentionally violate it.

While they were widely lamented throughout the nineteenth century, reprisals were not only permitted, but they were also believed to be a source of humanizing restraint on war. At their best they precariously guarded the spatial

---

180 Paley, *The Principles of Moral and Political Philosophy*, 471.

181 Ibid. 472.

182 Percy Bordwell, *The Law of War Between Belligerents: A History and Commentary* (Chicago: Callaghan & Co., 1908), 286.

parameters of justified harm as established through norms and the emerging conventions of international law. However, within this principle was the risk of a cascade effect that might entirely invert the spatial quality of eighteenth- and nineteenth-century battlefields – embedded within every limit on war was the threat of an unrestrained war. Military strategists and ethicists believed the intentional killing of noncombatants was justified if it was used to punish a crime and to deter them from further extremities, by way of the principle of necessity. Even though this tit-for-tat meant that war might digress into a race-to-the-bottom, it was one of the only tools that made the resort to cruelty much more painstaking.

10 Noncombatant Death in the Early Twentieth Century

It is important to note that these norms governing justified civilian death, enumerated above, held up through World War I and to the interwar period. Until the outbreak of World War II, commentators repeatedly affirmed the norms indicated above; civilians could be intentionally targeted under certain condition, residential districts could be bombarded, and borderlands could be rendered uninhabitable. In Lassa Oppenheim's 1905 treatise on *International Law*, for instance, he reaffirms a belligerent's right to destroy enemy infrastructure: "The district around an enemy fortress held by a belligerent may be razed, and, therefore, all private and public buildings, all vegetation may be destroyed, and all bridges blown up."[183] And in explaining the norms around siege warfare, he explains,

> it must be specially observed that no rule of law exists which obliges a besieging force to allow all non-combatants, or only women, children, the aged, the sick and wounded, or subjects of neutral Powers, to leave the besieged locality unmolested. Although such permission is sometimes granted, it is in most cases refused, because the fact that non-combatants are besieged together with the combatants, and that they have to endure the same hardships, may, and very often does, exercise a pressure upon the authorities to surrender.[184]

---

183 Lassa Oppenheim, *International Law, a treatise* (London: Longman, Green, and Co., 1905), 151.

184 Ibid., 156–57.

And a bit later, he continues, "It must be specifically observed that no legal duty exists for the attacking forces to restrict bombardment to fortifications only. On the contrary, destruction of private and public building through bombardment has always been and is still considered lawful, as it is one of the means to impress upon the authorities the advisability of surrender."[185]

The belief that applying pressure to the civilian population was an effective way to expedite surrender can be seen in the early twentieth-century reflections on civic morale (commonly spelled 'moral'[186]). In 1903, the French military theorist, Marshal Foch, had spent a great deal of time reflecting on the psychological phenomenon of "will" and "moral" in battle: "A battle won, is a battle in which one will not confess oneself beaten," and "Victory means will."[187] Quoting Von der Goltz, Foch concurs that "It is not so necessary to annihilate the enemy combatants as to annihilate their courage. Victory is ours as soon as the enemy has been brought to believe that his cause is lost."[188] Indeed, the best way to win is to destroy the enemy's hope in victory. And in this respect, the pressure placed on civilians might be the best way to "annihilate their courage."[189] These were principles that could easily be traced back to the terror bombardment of residential districts throughout the nineteenth century.

The degree to which a military was seen as justified in bombing residential districts to break the will of the enemy remained remarkably consistent. In his *War Rights on Land* (1910), J.M. Spaight writes:

> A commander who has forced a garrison to capitulate, does not usually punish the inhabitants for joining in the defence, though he is not bound by any Convention to show such leniency. Indeed … the civilian inhabitants of a beleaguered town are now given certain privileges, such as notification of a bombardment (except in a surprise attack), and in some cases permission to leave the fortress; and it is also usual for humane commanders to avoid deliberately shelling the residential parts of a bombarded town.[190]

---

185 Ibid., 158.

186 See Tami Davis Biddle, *Rhetoric and Reality of Air Warfare: The Evolution of British and American Ideas About Strategic Bombing* (Princeton: Princeton University Press, 2002), 14.

187 Marshal Foch, *The Principles of War* [1903] trans. Hilaire Belloc (New York: Henry Holt and Company, 1920), 286.

188 Ibid., 287.

189 In his 1896 *Manual of Military Ballooning*, Captain B.R. Ward had in much the same spirit noted that "The moral effect produced by the employment of balloons should not be lost sight of in estimating their value for war purposes." Captain B.R. Ward, *Manual of Military Ballooning* [1896] (East Sussex: Rediscovery Books, 2006), 62.

190 J.M. Spaight, *War Rights on Land* (London: Macmillan and Co., 1911), 110.

In a footnote to this passage, he goes on to explain that the first privilege (notification before bombardment) has been enshrined in international law, but "the other two are merely usage which has not become so general as to be universally binding."[191] In other words, it was entirely up to the discretion of the besieging commander as to whether or not the residential district should be bombarded, as international law had not ruled on this issue.[192]

In his 1912 study *War and the Private Citizen*, Pearce Higgins likewise notes:

> The siege and bombardment of a town is an operation of war which bear most cruelly on the ordinary civilian population; the private citizens who are living in their own homes and who generally are not allowed to leave ... are subject to all the dangers of falling shot and shell, and not infrequently their houses are directly bombarded by the assailant in order to bring pressure to bear on the commander of the besieged town so that he may be induced, by the suffering of the inhabitants to surrender.[193]

In his 1914 treatise on *Military Government and Martial Law*, William Birkhimer also notes:

> In the bombardment of places it is difficult to save any particular structure. Every siege gives evidence of this. To destroy a city with all it contains is indeed an extreme measure, not to be resorted to except for cogent reasons, yet it is perfectly justifiable when no other method suffices to reduce the place and this reduction becomes essential to the successful prosecution of the war.[194]

This principle is also affirmed in much more stark language in George Aston's 1914 *Sea, Land, and Air Strategy*. He explains that in bygone eras it was common to "bend the will of another nation" through "slaughter, pillage, and enslavement."[195] He continues, "Fortunately for the noncombatant population, it is only in the case of a beleaguered town that the full effect can be produced;

191 Ibid., 110n5.

192 For a similar point, see also William Birkhimer, *Military Government and Martial Law* (London: Kegan Paul, Trench, Trubner & Co., 1914), 258.

193 Higgins, *War and the Private Citizen*, 44–45. The same case is made in Percy Bordwell, *The Law of Law Between Belligerents: A History and Commentary* (Chicago: Callaghan & Co., 1908), 287; John Westlake, *International Law, Part II* (London: Cambridge University Press, 1913), 89; and Charles Stockton, *Outlines of International Law* (New York: Charles Scribner & Sons, 1914), 302.

194 Birkhimer, *Military Government and Martial Law*, 258.

195 Sir George Aston, *Sea, Land, and Air Strategy* (London: John Murray, 1914), 30.

in such a case the besieger does not hesitate to cut off all supplies of food, and even of water, if he can get at its source of supply, and he then waits for starvation, disease, and thirst to force the inhabitants and garrison to surrender."[196]

Pearce Higgins, however, goes on to point to an ambiguity in the international law that will prove to be significant with the rise of aerial bombing explored below: "It must be noticed that it is only *undefended* towns which may not be bombarded."[197] He writes this to show that the much more tenuous distinction between *defended* and *undefended* had arisen to replace that of *fortified* and *unfortified*. Higgins explains that the distinction is no longer "between fortified and unfortified places. Modern engineering skill has shown the futility of endeavouring to draw such a distinction."[198] Likewise, in his 1908 analysis of the Hague Conventions, William Hull also notes that even though the delegates voted unanimously that "the bombardment of undefended ports, towns, villages, dwellings, or buildings was prohibited," there was no agreement "as to the precise meaning of the term 'undefended.'"[199] Along these same lines George Davis explains that part of the difficulty with The Hague Convention is that

> if any defense is attempted or if a town is occupied or held by the armed forces of the enemy, it ceases to be undefended and, for that reason, may be attacked or fired upon. The inhabitants of such a place, so soon as a garrison is established or military defense is attempted, become charged with the knowledge that the town is defended and, as such, liable to attack, and if they desire to secure immunity from acts of war, should remove their families and belongings from the zone of active military operations.[200]

Indeed, the advances in engineering and the fluidity of troop movements made it quite challenging to formally demarcate defended from undefended – another indication that the so-called immunity from bombardment was far

196 Ibid.

197 Higgins, *War and the Private Citizen*, 45.

198 Ibid.

199 William Hull, *The Two Hague Conferences and Their Contributions to International Law* (Boston: Ginn & Company, 1908), 101. See also James Brown Scott, *The Hague Conventions and Declarations of 1899 and 1907* (New York: Oxford University Press, 1915).

200 George Davis, *The Second Peace Conference, The Rules of War on Land, Working Memoranda* (Washington, D.C.: Government Printing Office, 1907), 28. This passage is another striking example of the claim that nineteenth-century protections for noncombatants were delimited spatially. The closer one was to a legitimate target, the less of a noncombatant one was.

from a clear-cut issue. This led some commentators, like Charles Hyde, to note that "It must be clear that a town does not necessarily cease to be defended because it lacks fortifications, and that conversely, an unfortified place possessing a military force to a population rising in armed resistance must be regarded as defended."[201] He continues that "The price of immunity from bombardment is that the place shall be left open for the enemy to enter."[202] The only guarantee that a city was, in fact, open was its willingness to endure occupation.

As will be seen in the following chapters, the rise of aerial bombing will exploit some of these tensions and limitations within the accepted norms around bombarding civilians. Indeed, as Christopher Duffy suggests, the norms around siege warfare perhaps did the most to determine the character and development of aerial bombing in the twentieth century.[203] In fact, many early proponents of aerial bombing simply assumed that the rules governing warfare in the nineteenth century could easily be applied to new weapons. In a 1909 article on the prospects of aerial bombing, Colonel F.G. Stone is sanguine about the rise of these technologies due to longstanding adherence to the rules governing the bombardment of defended cities. He writes, "It would appear, therefore, that in the absence of any proved derelictions from the international code of honour in the past, we have no justification in assuming … that any nation will cynically cut itself adrift in the future from all moral obligations, on the plea that during war time all means are fair."[204] He was confident in the distinctions enshrined in international law: "The bombardment of defended towns has always been recognized as a legitimate method of bringing about, or helping to bring about, their capitulation, and has been governed by the laws and customs of war in such a way as to minimize its horrors."[205]

---

201 Charles Cheney Hyde, *Land Warfare* (Washington, D.C.: Government Printing Office, 1918), 37. Hyde likewise confirms the principle that "In the case of bombardment, the attacking force is not required by The Hague regulations to confine its operations to fortifications." Ibid., 39.

202 Ibid. James Tryon likewise explains that although The Hague Convention "prevents the bombardment of an undefended city, it does not prevent the enemy from landing its army and making that city a base of military operations or from proceeding from thence to the interior." James Tryon, "The Regulation of War," *The Yale Law Journal* 20:7 (1911); 535–547, 544. See also, Thomas Baty and John Morgan, *War: Its Conduct and Legal Results* (London: John Murray, Albemarle Street, W., 1915), 178–79.

203 Christopher Duffy, *Siege Warfare: The Fortress in the Early Modern World, 1494–1660* (New York: Routledge, 1996), 250.

204 Col. F.G. Stone, "The Limitations of Aerial Bombardment by International Law," *Flight* 11 December 1909. 796.

205 Ibid. Winfield makes the same claim: "If accuracy of attack by aircraft is or becomes possible, then there is every reason to suppose that the rules with respect to them will mutatis

Unlike Higgins, Hull, and Davis, as noted above, Stone argues that the criteria distinguishing defended from undefended towns are hardly controversial at all. He explains,

> it would be legitimate to bombard Portsmouth from aerial vessels in conjunction with an attack by sea or land, or both, against the defences; but it would not be legitimate to bombard Southampton. It would be legitimate to bombard Aldershot Camp if troops were lying there, and the town might suffer incidentally from its proximity to the camp. Again, it would be legitimate to bombard Woolwich Arsenal, and in the case also, the town might suffer incidentally. But when we come to consider London, we cannot admit that it is a 'defended' town in the sense that Portsmouth is.[206]

Though, he admits London is a much harder case. He writes,

> True, there are forts at the mouth of the Thames to prevent access of an enemy's warships … but it would be a strained interpretation of the intentions of the Powers to admit that if warships were attacking Chatham and Sheerness with a view to facilitating their undisturbed progress up the Thames, it would be justifiable at the same time for airships to bombard Cheapside with a view to creating a panic, on the plea that London was 'defended' by the guns at Chatham and Sheerness, and therefore liable to bombardment in conjunction with an attack on those places.[207]

Despite these clear pronouncements, it was far from certain that such easy delineations could be made or widely agreed upon.

Indeed, Erle Richards argued in a 1915 lecture series – in fact, as the Great War was in progress – that the Germans were not justified in bombing the residential districts of London because he believed it was an undefended town. He writes,

> Then again, it is said that the bombardment of defended towns is permitted, and that London is a defended town. It is true that the Hague Convention exempts from bombardment undefended towns only. No

mutandis be those as to naval bombardment." Percy Winfield, "Aircraft Attacks," *The Law Magazine and Review* (London: Butterworths, 1915), 257–271. 271.

206 Stone, "The Limitations of Aerial Bombardment by International Law," 796.

207 Ibid. See also G.G. Phillimore, "Bombardments," *Problems of the War* 15 (1915); 59–65.

definition is given of the meaning of a defended town, and it is argued that their mere existence of defences of any kind make a town liable to bombardment.[208]

On Richard's view, London was not a defended town because he had "never seen any military defences or any trace of fortification."[209] The sentiments expressed by Richards are understandable coming from a resident of London who was living through aerial attacks. Though, throughout history, residents of towns subjected to bombardment routinely considered their towns 'open' or 'undefended' and the attack unjustified. In a much more exasperated tone, Joseph Parker explains in his 1919 *The Laws of Land Warfare*, "No one can tell today if a town is 'seriously fortified' or not."[210] Despite Stone's and Richard's strong sentiments to the contrary, there was no internationally agreed upon criteria to distinguish between a fortified and non-fortified, a defended or non-defended town.[211]

11 Military Manuals

These early debates about defended and non-defended places were not trivial; not only did they seek to form limits in land approaches to cities in the nineteenth century, but they were also crucial to the emerging attempts to integrate aerial war into the traditional military ethical framework. Indeed, just prior to World War I, many of the great powers had made their own formal determinations as to what the acceptable differences were between defended and undefended. The British *Manual of Military Law* (1914) determined that they were not justified in bombing a town just because it "contains supplies of value to the enemy, or railway establishments, telegraphs, or bridges."[212] A fortification was "*prima facie* considered to be defended," but also those places where military forces were stationed or marching through could also be bombed.[213]

208 Erle Richards, *International Law: Some Problems of the War* (Oxford: Clarendon Press, 1915), 33.

209 Ibid.

210 Joseph Baker, *The Laws of Land Warfare* (Washington, D.C.: Government Printing Office, 1919), 199.

211 See also F.N. Maude, "The Bombardment of Open Towns," *The Contemporary Review* 114:275 (1918); 271–78.

212 War Office, *Manual of Military Law* (London: Published by His Majesty's Stationery Office, 1914), 253.

213 Ibid., 253.

The *Manual* also rules on the conventional wisdom regarding the bombing of residential districts; Section 122 states: "No legal duty exists for the attacking force to limit bombardment to the fortifications or defended border only. On the contrary, destruction of private and public buildings by bombardment has always been, and still is, considered lawful, as it is one of the means to impress upon the local authorities the advisability of surrender."[214] And the following section continues: "A town which is defended by detached forts, though they are at a distance for it, is liable to bombardment, for the town and forts form an indivisible whole. The town may, perhaps, contain workshops and provide supplies which are invaluable to the defence, and may serve to shelter a portion of its garrison when not on duty."[215]

The US Army's 1914 *Rules on Land Warfare* tracks very closely with the British attempt to define a "defended place" more formally; however, the definition the US settled on was much more expansive, so much so that almost no major city would be exempt from bombing. Fortresses are always considered defended, but so too are places where troops are garrisoned or through which a force is passing. Towns "surrounded by detached forts" are considered an "indivisible whole, as a defended place."[216] And while the manual prohibits bombarding or dropping bombs from balloons on undefended places, defended places – which included the towns just indicated – could be bombed. In fact, the manual specifically affirms the right of the US Army to "[throw] projectiles from balloons on forts and fortified places."[217]

To reiterate this point, in a 1914 lecture series on International Law given to The Army Service Schools, J.B. Porter addresses not just the legality, but the

214 Ibid.

215 Ibid. See also, *Manual of Military Law* (London: Printed for Her Majesty's Stationary Office …, 1899); and Clinton C. Anderson, *The War Manual, Volumes 1–2* (London: T.F. Unwin, 1916).

216 War Department: Office of the Chief of Staff, *Rules of Land Warfare* (Washington D.C.: Washington Government Printing Office, 1914), 67.

217 Ibid. According to the French *Manual* (1913), when targeting a city, one had to consider it not only from a military point of view, but of a political and economic view as well. If the target constitutes "an essential part of the means of war" it can justifiably be targeted: "We have to distinguish: a) fortresses, strongholds and fortified places; b) towns, villages, constructions, etc., open, but occupied militarily. Fortresses and strongholds are important central points of defense of the country, not only from a military point of view, but from a political and economic point of view. They constitute an essential part of the enemy's means of war, and can consequently be just as bombarded as the enemy army itself." *Les Lois De La Guerre Continentale* (Paris: Librairie Payot, 1916), 44. Incidentally, the French *Manual* may have been the first attempt to align a nation's internal regulations with the formalized principles of the Hague Convention. See George B. Davis, "Les Lois de la Guerre Continentale," in *The American Journal of International Law* 7:4 (1913); 920–923.

morality of applying pressure to the civilian population during a bombardment. He writes:

> Is this pressure immoral or condemnable? It would seem not. It would be going very far to say that projectiles may not be thrown to the interior of a city, since to so hold would serve to establish a place of repose and of shelter for troops. Besides, in modern war it is not of the troops alone that we must demand courage; it is also of the civilian population, which should be firmly persuaded that it is taking part in the war and that its attitude has an influence on the resistance.[218]

Passages like these affirm the nineteenth-century norms which said that residential districts could be deliberately targeted and that civilians must bear up under the hardships of war.

The US manual also rules on the question of allowing noncombatants to leave fortified places. Described as the "American Rule," it reads:

> 218. *No rule compelling belligerent to authorize population to leave besieged place.* – There is no rule of law which compels the commander of an investing force to authorize the population, including women, children, aged, sick, wounded, subjects of neutral powers, or temporary residents to leave the besieged locality, even when a bombardment is about to commence. It is entirely within the discretion of the besieging commander whether he will permit them to leave or not and under what conditions.[219]

Noncombatants could be forced back to suffer the hardships of bombardment in order to expedite surrender. Building on this, sections 220–223 offer additional recommendations for treating noncombatants during the bombing of a fortified place. Section 220 indicates that noncombatants "dwelling in the zone between the opposing forces in the first stages of a siege are treated as inhabitants of the invaded locality."[220] Sections 221 and 222 treat noncombatants who either attempt to leave without permission or who are expelled: "When a commander of a besieged place expels the noncombatants, in order

218 J.B. Porter, *International Law: Having Particular Reference to the Laws of War on Land* (Fort Leavenworth, KS: Press of the Army Service Schools, 1914), 95.

219 *Rules of Land Warfare*, 67.

220 Ibid., 69. Also affirmed in section 127 of British *Manual* (254) and the French *Manual* (44–45).

to lessen the number of those who consume his stock of provisions, it is lawful, though an extreme measure, to drive them back, so as to hasten surrender."[221]

Perhaps the most striking feature of this rule book is section 223, which deals with the case of women and children being expelled or sent out of the besieged army's lines "to get them to a place of safety."[222] Building specifically on sections 218 and 222, which argue that a military is within its right to drive noncombatants back to the besieged place, the rule stipulates that in such a case "It is not necessary to cease or relax fire … or to employ compassion, but fire must not be *intentionally* opened in their direction."[223] Women and children can be driven back into the besieged place, but they cannot be deliberately targeted after having been expelled. This same exact provision – indeed, word for word – is made in the British *Manual*: "130: It is not necessary to cease or relax fire because the enemy sends women and children out of his lines in order to get them to a place of safety, or to implore compassion, but fire must not be intentionally opened in their direction."[224]

These passages anticipate some of the arguments that will arise in the following chapters, but it is worth pointing out that this section is indicative of a moral confusion that will precipitate a subtle transformation in twentieth-century military ethical discourse. Civilians can be forced to suffer the hardships of siege warfare. Towns in close proximity to forts can be bombed, if deemed necessary. In fact, there is no prohibition against bombing residential districts to expedite surrender. Noncombatants can even be forced back into a besieged place, where they could be knowingly and intentionally exposed to fire. But in this vulnerable moment, when women and children are in a kind of no-man's land – they are neither inhabitants of the besieged fort, nor are they free from incidental harm – the manual indicates that they can be driven back but that they should not be intentionally targeted, a subtle and indeed paradoxical moral formulation.

As will be seen, a growing number of theorists writing around the time of World War I – those who saw the laws of war as a restraint to the cruelty meted out to civilians – increasingly began to appeal to intention (i.e. lacking

221 *Rules of Land Warfare*, 69.

222 Ibid.

223 Ibid.

224 *Manual of Military Law*, 254. Though, in both manuals this prohibition is somewhat convoluted. In section 128 of the British *Manual* individuals who leave the fortified place "are liable to be fired on, and may be sent back into the besieged place" (254); the same provision is made in section 221 of the US *Rules* (67). It is acceptable to fire on inhabitants in order to drive them back, but it is not acceptable to deliberately target them. Likewise, in the French *Manual*, 46.

intention) to explain and excuse types of civilian death. And this was especially true as civilians were seen more and more falling into harm's way as airships expanded the range of bombing. For instance, in laying out his interpretation of the Rules of Bombardment stemming from the Hague Convention, Coleman Phillipson points to the role intentions might play. He writes,

> We may reasonably infer from this provision that if any of those protected buildings are destroyed or damaged, or if civilians are killed or injured in the efforts to strike defensive works or military establishments, the assailants will not be chargeable with unlawful conduct if they did not aim intentionally at such buildings or civilians, and did not shoot at random without caring what was struck.[225]

Many commentators on civilian death around the time of World War I were beginning to think of the problem of civilian death in terms of intentional rather than gratuitous harm.

This evolution in military ethical discourse around targeting was uneven. Not every country was showing signs of similar transformations. *The German War Book* (1915), for instance, had an even more expansive view of legitimate military targets than the American and British manuals; it states,

> War is waged not merely with the hostile combatants but also with the inanimate military resources of the enemy. This includes not only the fortresses but also every town and every village which is an obstacle to military progress. All can be besieged and bombarded, stormed and destroyed, if they are defended by the enemy, and in some cases even if they are only occupied.[226]

Along these lines, the book goes on to explain that "fortresses and strong places are important centres of defence, not merely in a military sense, but also in a political and economic sense. They furnish a principal resource to the enemy and can therefore be bombarded just like the hostile army itself."[227] In fact, the criteria laid out for legitimate bombardment led the compilers of the manual to conclude that the Hague Convention's provisions for protecting open or

225 Colman Phillipson, *International Law and the Great War* (London: T. Fisher Unwin, 1915), 161.

226 *The German War Book* trans. J.H. Morgan (London: John Murray, Albemarle Street, 1915), 78.

227 Ibid.

undefended town were unnecessary; they could not think of a case in history where an open town had been bombarded – "modern military history knows of hardly any such case."[228] This is because, given the parameters of legitimacy indicated above, "The only criterion is the value which the place possesses for the enemy in the existing situation."[229] It was up to the invading military to determine if the city was defended or not.[230]

Despite professing a slightly more expansive view of legitimate target for bombing, the Germans shared the virtually unanimous international opinion that bombardments need "not limit itself to the actual fortification, but it will and must extend over the whole town."[231] Quite explicitly, the manual states that "All the inhabitants, whether natives or foreigners, whether permanent or temporary residents, are to be treated alike."[232] Like virtually all other war manuals of this period, the *German War Book* stipulates that women and children could be forced back into a besieged place because "the very presence of such persons may accelerate the surrender of the place in certain circumstances."[233] However, the German manual brought into sharp relief the extreme limits of military necessity. The Germans had adopted the well-known position of *Kriegraison* – the "reason of war" takes precedence over the "laws of war."[234] On this view, a belligerent is permitted "to adopt any measures and employ any means which will contribute to the attainment of the object of the war, even though they are forbidden by the customs or usages of war."[235]

228 Ibid., 81.

229 Ibid., 82.

230 In his scathing critique of German war practices in World War I, James Garner writes "In short, according to the view of the German manual, practically every town within the lines of the enemy is today a 'defended' place and may therefore be bombarded. This extraordinary contention in effect reduces the prohibitions of The Hague Convention in respect to bombardment to a nullity and it is directly contrary to the view of practically all writers on international law as to what constitutes 'defense.'" James Garner, *The German War Code* (Urbana, IL: University of Illinois, 1918), 27. See also George Scott and James Garner, *The German War Code* (Washington D.C.: Published by the US Committee on Public Information, 1918).

231 *The German War Book*, 79.

232 Ibid.

233 Ibid., 80.

234 For a vigorous defense of German militarism, see General Friedrich von Bernhardi, *Germany and the Next War* [1911], trans. Allen Powles (New York: Longmans, Green, and Co., 1914). For the British response, see Arthur Conan Doyle, *Great Britain and the Next War* (Boston: Small, Maynard & Co., 1914).

235 Garner, *The German War Code*, 11.

12 Conclusion

This section has shown that throughout the nineteenth century, right up to World War I, it was common to justify the deliberate killing of the innocent. Borderlands could be invaded to punish "barbarous nations," and hardship, like starvation blockades could be inflicted on the innocent residents of an enemy nation. If citizens resided in or close to a fortified place, they could be subject to legitimate attack. According to many theorists, citizens even became combatants by virtue of where they resided. It has also been shown that, in certain cases, it was acceptable for an invading army to destroy the property and infrastructure that would allow the enemy to prosecute the war, even if this infrastructure were also necessary to keep the citizenry alive. And finally, innocent people could be killed as a form of retaliation or reprisal in the process of enforcing the war convention. To be sure, all these hardships were part of the suffering an enemy population must endure. It is important to reiterate that these norms held right up to World War I. The killing of noncombatants was justified principally based on necessity. Even while armies were increasingly being called upon to discriminate between combatants and noncombatants, the exigencies of war permitted extreme measures.

PART 3

The Twentieth Century: Aerial Bombing and a Shift in Norms

CHAPTER 6

New Possibilities and Problems: Aeronauts, Inventors, and Future-War Fiction on Aerial Bombing

1 Early Experiments

On 17 December 1908, the aeronaut adventurer Roy Knabenshue performed a rather unusual experiment.[1] He had secured permission from the Los Angeles Police Department to "bomb" vulnerable parts of the city: the Courthouse, City Hall, and perhaps some hotels.[2] Knabenshue had originally planned to use fireworks, but, due to safety concerns, he agreed to drop bags full of confetti instead. The purpose of this test was to see how easily a city could be destroyed by a surprise aerial attack, something that had been suspected but not definitively proven. Indeed, experimenters and balloon enthusiasts had predicted that aerial bombing would be the next major breakthrough in war technology, and yet its promise had not yet been fully demonstrated. In the late nineteenth century, Sir Hiram Maxim, the great military inventor, had proclaimed that "when the flying machine succeeds its first great use will be for military purposes ... It will at once become an engine of war and not only to reconnoiter the enemy's positions ... but also of carrying and dropping into the enemy's lines and country large bombs charged with high explosives."[3] He continues that "It does not require a prophet to foresee that successful machines of this character would at once make it possible for a nation possessing them to paralyse completely an enemy by destroying in a few hours" all the crucial industries needed to sustain the war effort.[4] For almost two decades the international community of inventors and aerial enthusiasts knew that bombing would become a reality.

1 For more on Knabenshue's background, see James ap Roberts, "Roy Knabenshue, 1876–1960," *Southern California Quarterly* 54:1 (1972); 31–53.

2 "'Blow-up' By Airship: Sky Pilot Knabenshue Gets Permission to Make an 'Attack' on the City Tonight," *Los Angeles Times* December 17, 1908, 1; See also "Knabenshue Makes Great Flight in Night," *Los Angeles Examiner*, December 18, 1908, 1, 5.

3 Quoted in *Report of the Secretary of War, Volume 1* (Washington D.C.: Government Printing Office, 1892), 598.

4 Ibid.

 | DOI:10.1163/9789004515482_008

Despite this mounting optimism that airships would be used for bombing, it was by no means a foregone conclusion that the pilots would be able to carry out such difficult tasks with the existing technology.[5] Just a few years prior to Knabenshue's experiment, another great nineteenth-century aeronaut, John Mackenzie Bacon, had argued that the technology was simply not well enough developed to carry out such an attack. In 1902, he had declared that "no form of dirigible air ship exists which could be relied on to carry out any required maneuver in such atmospheric conditions as generally prevail."[6] Even in 1908, reliable maneuverability remained questionable.

Many were outright skeptical that bombing technology would ever be practicable.[7] In 1904, Captain Émile Mathieu, who years later would go on to become an officer of the Belgian air force, argued that dirigibles "would never furnish an industrial energy sufficient for the carriage of passengers and merchandise; that it would always remain a means of transport for excursions, and especially for military explorations; and that it would never be possible to utilize it as an aerial torpedo machine."[8] An instructor in Germany's early and experimental air-ship battalion, Captain Georg Paul Neumann, likewise cautioned against "excessively optimistic expectations, fantastic conclusions, and impossible schemes."[9] Indeed, even as late as 1909, balloonists questioned the

5 In the 1890s, there were sometimes heated debates as to whether it would be possible to *navigate* balloons. See Arthur de Bausset, *Aerial Navigation* (Chicago: Fergus Printing Press, 1887); Octave Chanute, *Aerial Navigation* (New York: The Railroad and Engineering Journal, 1891); Barnet Botts, *The Problem of Aerial Navigation* (Paso Robles, CA: Botts, 1894); J.G.W. Fijnje van Salverda, *Aerial Navigation* trans. George Waring (New York: D. Appleton and Company, 1894); and Daniel Caulkins, *Aerial Navigation: The Best Method* (Toledo, OH: The Blade Printing & Paper, 1895). These disagreements were on display at the International Conference on Aerial Navigation (Chicago, 1893). See *Proceedings of the International Conference on Aerial Navigation* (New York: The American Engineer and Railroad Journal, 1894), and M.N. Forney (ed.), *Aeronautics* (New York: The American Engineer and Railroad Journal, 1893).

6 John Mackenzie Bacon, *The Dominion of the Air: The Story of Aerial Navigation* (London: Cassell and Co., 1902), 316. Eric Stuart Bruce, "The Use of Balloons in War," *The Aeronautical Journal* 6:23 (1902); 43–44. 44. For the same argument, see John Alexander, *The Conquest of the Air: The Romance of Aerial Navigation* (New York: A. Wessels Company, 1902), 130–31. See also, Gertrude Bacon, *Balloons Airships and Flying Machines* (New York: Dodd, Mead & Company, 1905).

7 See E. Seton Valentine and F.L. Tomlinson, *Travels in Space: A History of Aerial Navigation* (London: Hurst and Blackett, 1902).

8 Quoted in Emile Guarini, "Free and Dirigible Balloons," *Scientific American* 90:23 (1904); 436.

9 Captain Georg Paul Neumann, "The Possibility of Making Use of Balloons and Motor Air-ships in the Navy," *United Services Institution Journal* 52:369 (1908); 1502–1517, 1502. Neumann would serve in Germany's Air Force during World War I. See Georg Paul Neumann, *The Air Force I Knew, 1914–1918: Memoirs of the Imperial Air Force in the Great War* [1921] trans. J.E. Gurdon, ed. Bob Carruthers (Barnsley: Pen and Sword Aviation, 2014).

viability of bomb-dropping in war. A. Lawrence Rotch concludes: "Undoubtedly scouting is actually the chief role of the present military dirigible, but granting the possibility of launching explosives with effect from larger and faster balloons, capable of a greater radius of action, the airship may become a powerful engine of war"; however, "Whether it is possible to launch explosives with sufficient accuracy, using gravity as a motive force, is still doubtful."[10]

Despite these pervasive doubts, in the first decade of the twentieth century, the technology would rapidly, indeed, exponentially progress, especially on the continent. France and Germany were the most technologically advanced in their experimentation with military airships.[11] Their nascent air programs had limited success with military maneuvering and bomb dropping starting from 1905.[12] In contrast, the US Army had only just started to formally experiment with dropping bombs from balloons in 1907.[13] To be sure, despite the wild optimism that such technology would eventually prove to be successful, it is important to recognize that from the perspective of 1908 it was not a foregone conclusion that Knabenshue would be able to pull off the remarkable stunt.

---

10 A. Lawrence Rotch, *The Conquest of the Air; or, The Advent of Aerial Navigation* (New York: Moffat Yard and Company, 1909), 171–72.

11 The Brazilian aeronaut, Alberto Santos-Dumont, impressed French military observers with his flying ability, see Alberto Santos-Dumont, *My Airships* (London: Grant Richards, 1904).

12 It is often forgotten that an international air race was taking place in the shadow of the Second Hague Convention. Even though aerial bombing was not yet viable, strategists were unwilling to ban an effective technology that might prove decisive in war. Joseph Choate explains that while the Convention of 1907 was taking place "news came from Germany that a dirigible balloon, with a speed of thirty miles an hour, had made a successful ascent, and that in France a cigar-shaped airship had made thirty-one leagues as hour with the wind and eighteen leagues against the wind, and had maneuvered successfully in the environs of Paris. Predictions were freely indulged in that four or five years the air would be as full of airships as the streets then were of automobiles." Joseph Choate, *The Two Hague Conferences* (Princeton: Princeton University Press, 1913), 14.

13 "Balloons for Army: Experiments to be made by Uncle Sam at Fort Omaha," *The Allenton Leader*, 6 March 1907, 5. The US would not start experimenting with dropping bombs from airplanes until 1910. The *United Service Gazette* reports that "Lieutenant [Paul Ward] Beck, in Mr. Curtiss's bi-plane, recently experimented in dropping bombs from a height of 250 feet, at a target 15 feet by 25 feet, on behalf of the United States Army at Los Angeles; and although he was not very successful the possibility of using bombs from these vessels of the air has been proved by the experiment." "Bomb-dropping from Airships," *Antiaircraft Journal, Volume 33* (Fort Monroe, VA: Coast Artillery School Press, 1910), 218. For a more detailed account of these early attempts at bombing, see Juliette Hennessy, *The United States Army Air Arm, April 1861 to April 1917* (Washington, D.C.: Office of Air Force History, United States Air Force, 1985), 45. For a detailed history on the development of US aeronautics, see Henry Woodhouse, "History of United States Army Aeronautics," in *Textbook of Military Aeronautics* (New York: The Century Co., 1918).

In fact, up till that point, all the early attempts to bomb cities with the help of balloons had been more of a spectacle than an effective means of engaging the enemy. The best example of aerial bombing until then had taken place in July 1849. At that time, Austrian forces attempted to attach bombs to kite balloons, which, with a favorable wind, would float over the walls of the besieged city of Venice. However, rather than striking terror in the besieged inhabitants, the "comical-shaped balloons" elicited laughter. As one account records:

> when a small black cloud followed by a feeble detonation indicated that the aerial projectile had harmlessly burst, all Venice broke out into a simultaneous shout of derision and delight. And still more tumultuous, spontaneous, simultaneous, and rapturous was the ironical applause, when the unlucky balloons burst over the heads, and dropped their explosive burdens into the very midst of their yet more unlucky inventors; while the clapping of hands of the imprisoned and starving Venetians, accompanied by shouts of *Viva! Bravo! Buon Appetito!* became perfectly uproarious with enthusiasm.[14]

The following year, the English aeronaut, Henry Tracey Coxwell, "discharged shells, grenades, &c, from the car of a balloon" at a demonstration in Berlin.[15] Forty feet below the balloon's passenger carriage he had hung a wicker basket. Coxwell descended to this basket by rope latter "in order not to risk the gas exploding when the shells were lighted and discharged."[16] To the delight of the crowd below, he was able to drop several shells which exploded midair. While this demonstration proved that balloons could, in principle, be used to drop bombs, neither the German nor the English military observers were interested in building on these experiments. The technology was far too limited. Undaunted, Coxwell would later conclude: "I have no doubt it would be possible to drop with tolerable nicety a host of aerial shells charged with agents calculated to produce stupefaction, if not fatal effects."[17]

---

14 Edmund Flagg, *Venice: The City of the Sea, Volume 2* (New York: Charles Scribner, 1853), 406. For more on the early experiments with bomb-dropping from balloons, see also, F. Stansbury Haydon, *Military Ballooning During the Early Civil War* (Baltimore: Johns Hopkins University Press, 1941).

15 Gerard Manley Hopkins, "Balloons and Aeronauts: In Two Parts," *Once a Week: An Illustrated Miscellany, Volume 13, June to December 1865* (London: Bradbury & Evans, 1865), 530.

16 Victor Plarr, "Henry Tracy Coxwell," in *Men and Women of the Time: A Dictionary of Contemporaries* (London: George Routledge and Sons, 1899), 244.

17 Henry Coxwell, *My Life and Balloon Experiences* (London: W.H. Allen & Company, 1889), 152. In fact, Coxwell would dramatize these experiments in a fictionalized account: Henry

In 1857, a Scottish aeronaut, Malcolm Gillespie, performed another bombing demonstration for the British Government. For this experiment, Gillespie raised "by means of a balloon, a square wooden platform, large enough to contain a dozen men and an unlimited number of loaded shells."[18] From the height of two miles up, Gillespie claimed that he could "effectually destroy the town or works beneath by dropping their shells upon them."[19] However, this experiment left much to be desired. A military observer, Lieutenant Locock, explained that aerial bombing "held out no peculiar advantages, even allowing its feasibility. The mere power of pouring shells promiscuously into a place is no gain, and they can be already thrown in with as much certainty and a great deal faster from shell guns and mortar."[20] He continues, "It would be impossible therefore to judge precisely how to correct the aim, i.e. the position of the platform, and the shell-dropping would be perfectly promiscuous."[21]

In 1859, W.H. James made a bold, indeed, fantastical proposal to the British Government – a "peace-inducing system of aggressive warfare" which consists in the deployment of "hundreds of thousands if necessary, of cheaply constructed balloons" which could be calibrated and timed to drop "phosphoric, self-igniting fire grenades."[22] James envisioned this line of devastation passing over large swaths of the enemy's territory, dropping bombs which would "penetrate and pass through the roofs of all public and private buildings."[23] He continues, such a "well-directed attack … when properly understood and appreciated by mankind, must have the effect of preventing nations from engaging in war, and of so ultimately producing universal peace throughout the world."[24] It is unclear if James actually demonstrated the viability of this new system, but, in any case, the War Department concluded that such an attack would be "considered inapplicable to military service."[25]

In 1898, the US had developed a special balloon corps, as part of the US Army's Signal Corps, for the invasion of Cuba. There is some indication that

Coxwell, *A Knight of the Air; Or, The Aerial Rivals* (London: Digby, Long, and Company, 1895).

18 Lieutenant G.E. Grover, "On the Uses of Balloons in Military Operations," in *Papers on Subjects Connected with the Duties of the Corps of Royal Engineers, Volume XII* (Woolwich: Printed by W. Jackson, 1863), 73.

19 Ibid.

20 Ibid., 74.

21 Ibid.

22 W.H. James, "Balloon Warfare," *Mechanics' Magazine, Volume 4, July to December 1860* (London: Robertson, Brooman, & Co., 1861), 23.

23 Ibid.

24 Ibid.

25 Ibid.

they were experimenting with bombing. As one commentator writes, "Closely allied with the balloon idea is that of the bomb-dropping airship. One of these, lately invented, is said to be under government consideration. The idea is to carry nitro-glycerine bombs in the airship and drop them into the enemy's fortifications, first getting the range by a telescope and by dropping small metal balls."[26] However, as a critic noted, "The practicability of this scheme remains to be determined."[27]

While balloons had been widely employed in war for scouting and directing artillery fire, weaponizing them in a live operational context was a seemingly insurmountable hurdle. The US military was actively investigating the feasibility of a variety of bomb-dropping mechanisms in 1897, but it seems as though they abandoned the idea largely due to the issues around navigation.[28] So, despite the longstanding use of balloons in war, it is easy to overlook just how uncertain this technology was in 1908 and what a challenge Knabenshue's experiment posed to extant beliefs about what was technologically possible. He proposed to engage in a controlled nighttime flight, to single out and "confetti bomb" key buildings in the city, and to do all this undetected.[29] This was nothing less than the elusive Holy Grail of war ballooning.

At midnight, under the cover of darkness, Knabenshue ascended in his airship over the city. To add to the challenge of an undetected bombing raid, the event had been widely publicized in all the local newspapers. The streets were teeming with eager spectators.[30] But even with the wide-spread public enthusiasm, and the crowds of spectators all eager to catch a glimpse of

---

26 "Science in War," *Gunton's Magazine, Volume 16, January–June 1898* (New York: The Gunton Company, 1898), 410.

27 Ibid.

28 See *Annual Reports of the War Department: Report of the Secretary of War* (Washington D.C.: Government Printing Office, 1897). For additional accounts of early ballooning experiments, see Alfred Hilderbrandt, *Airships Past and Present* trans. W.H. Story (New York: D. Van Nostrand Company, 1908).

29 A similar though less ambitious confetti-bombing experiment was carried out by Mark O. Antony who was working with remote controlled dirigibles. Antony's remote-controlled ships were not built to scale. They were flown in an auditorium over an audience onto which confetti was dropped. See "Wireless Control of Airships," *Flight* 6 February 1909, 81.

30 This foreshadows some of the spectacle of aerial bombing in World War I. Rather than taking cover, many London residents took to the street to catch a glimpse of the bombers as the city was being attacked. William Claxton reports that civilian casualties from bombing raids decreased as the war progressed, after "an effective and comprehensive system of warnings was devised, and when people had acquired the habit of taking shelter, instead of rushing out onto the street to see the aerial combats." William Claxton, *The Mastery of the Air* (Glasgow: Blackie and Son, Limited, 1915), 17.

the attack, no one could spot the ship. Some reported that they were able to hear the humming of its motor, but no one knew where the confetti bombs would fall next. In fact, the only time Knabenshue's presence was detected was when he turned a miniature searchlight onto the crowd below. As one report indicates:

> About twenty minutes after the start was made the first bomb was dropped near the city hall. Pedestrians on Broadway were amazed and looked to see whence came the confetti. They could see no sign of the airship and in a few minutes another bomb was dropped. After 'blowing up' the city hall and the greater part of the business section of the city, Knabenshue made a trip about the residential district.[31]

In short, the experiment was a terrifying success. The *Los Angeles Herald* reported that "Aeronauts consider it one of the most remarkable flights yet made, for it proves the feasibility of aerial trips at night. More than that, it demonstrates clearly what has often been a matter of conjecture – that an airship might fly over a city and work terrible havoc before its presence was perceived."[32] Put much more bluntly, the February 1909 edition of *Popular Mechanics* reported: "For 1 hour and 30 minutes, Knabenshue sailed over Los Angeles at night and dropped confetti 'bombs' on the helpless citizens."[33] The article concludes that the experiment definitively shows that "a fleet of airships could easily destroy a city without danger to the aerial fighters themselves," and, what is more, that "Airships can, and will, in the next great war, play a great part in engagements on the battlefield and in bombarding fortifications."[34]

Though the bombing exercise portended a rather foreboding if not apocalyptic future, Knabenshue himself was quite optimistic, indeed, enthusiastic, about the results of the trial. He believed that the rise of this revolutionary technology would make war either less likely or of a much shorter duration: "I have always maintained that a fleet of airships would end a war as quickly as gun fire could start it. Aerial craft must play a great part in any future war, and the side that has the largest and most efficient aerial equipment will win."[35]

31 John H. Wigmore, *Select Cases on the Law of Torts, with Notes, and a Summary of Principle* (Boston: Little, Brown, & Co., 1912), 919.

32 "Airship Safely Bombards City, Knabenshue Drops Missile on City Hall," *Los Angeles Herald* December 18, 1908. 1.

33 "Los Angeles Attacked from Airship," *Popular Mechanics* (February 1909), 152.

34 Ibid., 151.

35 Ibid., 152.

The destructive potential of weaponized balloons "made life so uncertain that nations will be forced to adopt some code of arbitration for settling their differences."[36] By making war so unremittingly terrible, civilized nations would be forced to find other ways of solving their disagreements.

2 Deterrence, Annihilation, or Nonfactor?

In the wake of these experiments, tacticians and politicians varied widely on how this technology might be integrated into the existing militarized geostrategic landscape. Despite the optimism that this technology would end all war, there was a growing unease that these weapons might actually bring about the end of humankind. There was a growing sense of apocalyptic foreboding, as if humans were now unleashing a level of destruction that could not be contained. These extreme thoughts made for an odd juxtaposition. After all, many continued to believe that the technological difficulties of bombing could not be overcome, and these new machines would never be used in future wars. Still others believed that even if these weapons did materialize, they would have no major impact. They would be easily subsumed under the existing laws of war.

The idea that aerial bombing would make war more terrible and therefore less likely can be traced back to the late nineteenth century.[37] In 1894, the American aeronaut pioneer, Octave Chanute, admitted that war balloons were so far limited in use to reconnoitering, but "there is little question, however, that they may add greatly to the horrors of battle by the promiscuous dropping of explosives from overhead, although their limited capacity to carry weight will not enable them to take up a large quantity, nor to employ any heavy guns

36 "To Make Night Trip in Airship, Knabenshue to Ascend at Midnight, Will Demonstrate How City Can be Destroyed," *Los Angeles Herald* 17 December 1908. 8.

37 Though, one eighteenth-century treatise on air balloons predicted that air ships would lead to peace. Using balloons for observation during war will "contribute to the preservation of our existence"; the author argued that the cultivation of this knowledge will "ease the ravages of war." *The Air Balloon: Or, a Treatise on the Aerostatic Globe* (London: Printed for G. Kearsley ..., 1783), 28–29. Also reflecting later debates, another commentator from this period noted: "But like every other destructive discovery, this would avail but little, for friends and foes would have the secret in common, and balloons would be sent to oppose balloons, and we should then have Milton's war in the clouds realized." "An Account of the First Aerial Voyage in England," in *The Gentleman's Magazine, Volume 54* (London: Printed by John Nichols ..., 1784), 773.

with which to secure better aim."[38] He predicted that when man discovers the means to employ aerial bombing "the ultimate effect will be to diminish greatly the frequency of wars and to substitute some more rational methods of settling international misunderstandings."[39] Specifically, he argued it will be due to the fact that in such future war "no part of the field will be safe, no matter how distant from the actual scene of conflict," and this uncertainty will "render nations and authorities still more unwilling to enter into contests than they are now."[40] That same year, an anonymous editorial prayed for the day when Hiram Maxim's flying machine would be ready. Such an invention would certainly "usher in an era of perpetual peace. For what protection can a hostile power have against a foe who can sail through the air and drop a bomb into the heart of its capital city."[41]

In 1895, the doctor-aeronaut-inventor Daniel Caulkins optimistically noted that "The great conflict of our age is much talked of and yearly expected, and every lover of humanity will hail the light of the mighty fleet of air-ships on the clouds and hovering over the war-clouds as a great peacemaker."[42] Likewise, in 1897, another great American aviation pioneer, Samuel Langley, argued that the advent of airships would "change the whole conditions of warfare, when each of two opposing hosts will keep out the foe, and when the difficulties of defending a country against an attacking enemy in the air will be such that we may hope that this will hasten rather than retard the coming of the day when war shall cease."[43] The belief that aerial bombing was a weapon of peace runs throughout late nineteenth-century assessments of the technology.

Many of these same arguments were recapitulated by other great European aeronauts and inventors in the early twentieth century. In 1902, Frederick Walker predicted that if the secrets of aerial navigation were "under the control of a peaceful and highly civilized Power ... war and its concomitant horrors would be a story of the past."[44] In 1909, the highly regarded French aeronaut, Clément Ader, predicted that the dirty work of aerial bombardment would unavoidably lead to the bombing of cities: "The great bombing planes will become veritable terrors! I am convinced that their awesome power and the

38 Octave Chanute, *Progress in Flying Machines* (New York: The American Engineer and Railroad Journal, 1894), 268.

39 Ibid.

40 Ibid.

41 "When War Shall Be No More," *The Argosy, Volume 18, April to September 1894* (New York: Frank A. Munsey & Company, 1894), 446.

42 Caulkins, *Aerial Navigation*, 70.

43 Samuel Langley, "The Flying Machine," *McClure's Magazine* 9:2 (1897); 647–660. 660.

44 Frederick Walker, *Aerial Navigation* (London: Crosby, Lockwood and Son, 1902), ix.

fear of seeing them appear will provoke salutary reflections among the statesmen and diplomats who are the real dispensers of peace and war, and that in the final analysis these airplanes will serve the cause of humanity."[45] In a forecast of the state of Europe in 1918, written from the vantage point of 1909, H.J. van der Leeuw speculated that at the 1914 Third Peace Conference (which, of course, never happened due to the outbreak of World War I), all states would agree to at least a partial disarmament. What is more, by that point in time all the nations of Europe would form into a "League of Mutual Interests" due largely to the dangers of aerial war, which compelled states "to build an ever increasing number of steadily improving air ships."[46] In this optimistic forecast, air power would force states to work through their differences through international cooperation.

The well-known British motoring journalist R. Hearn made similar observations. Writing in 1909, he claimed that the eventual rise of airships would change the "character of war" making it "more costly and more terrible."[47] He optimistically concluded that "even though airships will make war more terrible … its very terror will delay an outbreak."[48] In a later work, Hearne predicted that one day "There will be veritable battleships of the air, and, fitted with pneumatic or other guns, aerial torpedo dischargers, and a good stock of explosives."[49] These developments will "offer unique possibilities for the discharge of explosive, fire-producing, or poisonous gas bombs over a large area unreachable by land or sea forces. Enormous destruction could be wrought by these, and the poisonous gas bombs could render a whole district practically untenable by troops."[50] There was reason to be optimistic about these developments he writes, because "war will be deemed more uncertain, more costly, more terrible; and the enterprise will be too rash to commend itself to any business nation until something more definite is known about aerial navigation. Therefore, the general adoption of airships will defer war, until the new arm can be given its correct value."[51] And when these weapons are fully integrated into the militaries of nations, they will "render warfare more localized in its destruction (that is to say, more humane), more decisive, and

45 Clément Ader, *Military Aviation* [1909] trans. Lee Kennett (Maxwell Air Force Base, AL: Air University Press, 2003), 27.

46 H.J. van der Leeuw, *A Discourse on the Results of Aerial Navigation in the Year 1918*, trans. E. Halliday (Rotterdam: D.J. Storm Lotz, 1909), 14–15.

47 R. Hearne, *Airships in Peace and War* [1908] (London: John Lane, 1910), xxx.

48 Ibid., xxxv.

49 R. Hearne, *Aerial Warfare* (London: John Lane, 1910), 94.

50 Ibid., 113.

51 Ibid., xxvii.

more rapid."[52] Hearne suspected that airships would enable modern militaries to acquire more control over the battlefield. After all, the history of the nineteenth century had shown that over time modern armies no longer engage in "cruel and wanton destruction of children, women, and non-combatants."[53] If these trends were to continue, "airships will have the effect of reducing this accidental destruction."[54] But ultimately, it is only when enemy nations are perfectly balanced in the possession of this destructive power will one have the "surest guarantee of peace."[55] Specifically, "the development of the airship may be looked upon as beneficial to the cause of peace, for in our modern materialistic times, the strongest deterrents of war are those practical arguments which touch the security or the wealth of nations."[56]

In these early days of experimentation, the anticipated horror of aerial bombing was widely viewed as a form of deterrence.[57] The pioneers of aviation quite reasonably thought of themselves as peacemakers. Knabenshue's confidence that the cruelty of aerial bombing would make war more unlikely was reaffirmed several years later by Ivan Bloch who, in his foreboding analysis, *The Future of War* (1914), argues:

> war has become impossible alike from a military, economic, and political point of view. The very development that has taken place in the mechanism of war has rendered war an impracticable operation. The dimensions of modern armaments and the organisation of society have rendered its prosecution an economic impossibility, and, finally, if any attempt were made to demonstrate the inaccuracy of my assertions by putting the matter to a test on a great scale, we should find the inevitable

52 Ibid., xxix.

53 Ibid., 130.

54 Ibid.

55 Ibid., xxxi.

56 Ibid., 132. Richard Ferris makes a similar argument. He notes that both France and Germany have been making rapid advances in aerial bombing but ultimately concludes: "All things considered, however, it is to be expected that the experimenting by army and navy officers all over the world will lead to such improvement and invention in the art of navigating the air as will develop its benevolent, rather than its malevolent, possibilities – a 'consummation devoutly to be wished.'" Richard Ferris, *How to Fly; Or, The Conquest of the Air* (London: T. Nelson, 1910), 378.

57 This anticipates Prime Minister Harold Macmillan's statement from 1966: "We thought of air warfare in 1938 rather as people think of nuclear warfare today." Quoted in Brett Holman, *The Next War in the Air: Britain's Fear of the Bomber, 1908–1941* (Surrey: Ashgate Publishing Limited, 2014), 253. "Deterrence" can be found in the way theorists were conceiving air war prior to World War I.

> result in a catastrophe which would destroy all existing political organisations. Thus, the great war cannot be made, and any attempt to make it would result in suicide.[58]

And while he does not specify aerial warfare, it was not a great leap for inventors and experimenters to make the conceptual link between the destructiveness of air war and peace. By making war more all-encompassing, more unrelentingly total, this would make it increasingly difficult for states to sensibly resort to an aggressive foreign policy.

There were, of course, other species of optimism. Among those who remained skeptical about the technology, many simply assumed that if it continued to improve, the extreme potentialities of aerial bombing would be reined in by the civilizing tendencies of nations.[59] For instance, in a 1909 speech to the British Parliament, Sir Alfred Mond explained:

> I do not think that nations in the future are going to conduct their battles by scattering explosives over houses. That is very unlikely to take place. It would be the very reversal of the rules of war which have now existed for a long time. Nobody expects an enemy to bombard a seaside place like Brighton. With civilised nations warfare is not conducted by simply destroying property and killing civilians, or by dropping dynamite about London, Paris, or Berlin. Such a proceeding would have no effect at all on the ending of the war. No nation would make peace because the enemy was killing civilians.[60]

Mond took comfort in the primitive nature of the technology. He concludes:

> The idea that there could easily be dropped from such machines explosive materials with any chance of hitting any particular object has been entirely disproved by all the experiments that have been made in that direction. No doubt an explosive dropped from an aerial machine would hit somewhere, but that would not be a very useful or scientific method of bombardment. It is a curious fact that the hitting from a moving object in the sky any object on the ground is very much more difficult than an

58 Ivan Bloch, *The Future of War* (Boston: The World Peace Foundation, 1914), xi.

59 It was not unusual for commentators to argue aerial weapons would play a minor role in future war. For an example of this approach, see Major H.L. Hawthorne, "Balloons and Dirigibles in War," *Antiaircraft Journal* 32:2 (1909); 107–116.

60 "Naval and Military Aeronautics," House of Commons Debate 02 August 1909 vol. 8 cc1564–617. Brighton was bombed repeatedly during World War II.

> untrained person would suppose. I do not think that a fleet of airships moving over a fleet of ships on the water would ever hit them at all except by a very lucky accident.[61]

This position can in part be explained by Britain's reluctance to invest in unproven aerial technologies. The strength of their navy and natural island defense provided them with enough security that would allow them to keep this technology at a distance, at least for the time being.[62] Nonetheless, Mond was certain that not only would nations be morally reluctant to bomb cities filled with civilians, but that there was no indication that bombing civilians in the interior of the nation would work, even with a fleet of airships.

Despite Mond's assurances that the horrors of aerial bombing were something of a romantic fantasy, a latent anxiety about its destructive potential lingered. As Knabenshue's ominous flight over Los Angeles's residential district suggested, aerial warfare posed a unique and unprecedented threat to noncombatants living in the interior of a state at war. Indeed, as John Alexander (1902) had forebodingly noted, "what military men are after is an airship from which it will be possible to drop down deadly explosives into 'thick peopled cities.'"[63] Despite the assurances to the contrary, it was easy to foresee that airships would soon possess the capability to destroy city centers many times over and that civilians might be forced to face the brunt of this destruction.[64]

Another early twentieth-century commentator, T.G. Tulloch – a munitions specialist and member of the Chilworth Gunpowder Company – like Mond, believed the tactical effectiveness of employing airships in war was greatly exaggerated, but he also appreciated the much greater threat war-time aviation

---

61 Ibid.

62 It was by no means clear that this security would last. In 1908, Herr Rudolf Martin published a suggestive pamphlet titled "Is a World-War Imminent?" in which he explained that "England was losing her insular character by the development of airships and aeroplanes." "Germany's Air Madness," in *European War Pamphlets, Volume 55* (New York, 1921), 55. Britain did not have a working airship until after 1909. They were much further behind the continental powers in aerial development. See M.W. Royse, *Aerial Bombing and the International Regulation of Warfare* (New York: Harold Vinal, 1928), 87.

63 Alexander, *The Conquest of the Air*, 130.

64 The early twentieth-century anarchist, Emma Goldman, saw the rise of aerial bombing as an inevitable extension of the growing militarism of industrial states. She rebuffed claims that anarchists were the purveyors of violence when those same critical "peace pretenders" routinely expressed delight "over the possibility of the American nation soon being able to hurl dynamite bombs upon defenseless enemies from flying machines." Emma Goldman, "This I Believe" [1908], in *Red Emma Speaks: The Emma Goldman Reader* ed. Alix Kates Shulman (New York: Humanity Books, 1998), 55.

posed to the interior of the nation. If such weapons emerged, there would be no guarantee that states would abide by the laws of humanity. He writes, "But supposed they came via the air in the aerial machines, what then? This whole fifty miles of concentrated essence of Empire lies at the absolute mercy of even a single airship or aerial machine which could plant a dozen incendiary missiles in certain pre-selected spots."[65] He goes on to explain that "Once a certain number of selected centres were alight at about the same time, not all the power of the London Fire Brigade, nor any number of fire brigades, could deal with such a conflagration."[66] To this Tulloch ominously concludes: "It is needless further to press the point I wish to make by piling up horrors. This point is that *by destroying the heart the other members will cease to function*, for it is inconceivable that, with such a chaotic state as would follow the destruction of London, any coherent direction of affairs, either official or commercial, could continue."[67]

Like Tulloch, Major B. Baden-Powell was cautious about over-exaggerating the importance of aerial warfare. He resisted alarmists' calls for the rapid, indeed, haphazard build-up of military airships, and he questioned the reliability of experiments which supposedly "prove the practicability of discharging bombs from airships capable of creating the most awful havoc."[68] Despite the current limitations of these technologies, he was nonetheless eager to see

---

65 T.G. Tulloch, "The Aerial Peril," *The Nineteenth Century and After, Volume 65, January to June 1909* (New York: Leonard Scott Publication Co., 1909), 800–810, 806.

66 Ibid.

67 Ibid., 807. Emphasis in original. Much of the British aerial warfare literature leading up to World War I is a call to arms, evincing a fear that they were falling behind rivals on the continent. In a depiction of a mock battle between a force with airships and one without, Hearne writes: "'Airships!' – What an awful meaning the word would convey to the crowd suddenly struck dumb in their martial rejoicings! Who can describe the horror that would seize their hearts as they helplessly grazed at the strange monsters in the sky, and then realized for the first time that, through their conservatism, their skepticism, their stupidity, they had neglected to build an aerial fleet?" Hearne, *Aerial Warfare*, 137. Claude Graham-White and Harry Harper make a similar point in their assessment of aerial bombing. Though airships are mostly used for reconnoitering, in the future "it will be necessary to reckon air-fleets not in hundreds, but in thousands." and with this in mind "no country can afford to ignore what the future promises in this respect." Claude Graham-White and Harry Harper, *The Aeroplane in War* (Philadelphia: Lippincott, 1912), 211.

68 B. Baden-Powell, "How Airships are Likely to Affect War," *Royal United Services Institution Journal* 54:387 (1910); 555–581, 557. Given the date, he likely has Glen Curtiss's bombing experiment in mind (indicated above). In the summer of 1910, Curtiss dropped "orange bombs" onto a circular mark on a beach to show how easily a bombing attack could be used to strike entrenched soldiers. While military onlookers were optimistic, they were quick to point out that the demonstration "lacked the realism that would accompany the

Britain taking part in the development. This is because, with the help of these new weapons, in future wars "Operations will be quickened, and wars more rapidly lost or won. Raids into the enemy's country, which it seems impossible to entirely prevent, will, on the other hand, tend to hamper and delay his actions, and spread the zone of operations over the whole country."[69] He reminds his audience that such preparation is necessary due to the fact that machines are now actually in existence that can come over, without warning, from the Continent, and it is more than possible that they might be the means of causing considerable damage to us, even risking their own loss thereby."[70] Much like many of the above commentators, William Ennis advises the continued development of weaponized planes. Current experiments had proven promising enough for him to conclude that "There is no doubt but that aircraft may be made efficient agents of destruction on a colossal scale."[71] His hope in pursuing such implements of devastation, however, was ultimately humanitarian: perhaps "Destructive war [would become] so superlatively destructive as to destroy itself."[72]

Throughout this period, speculations about aerial bombing were wildly divergent. Despite the excitement, indeed in some cases alarmist conjecture, many continued to believe that aerial bombing simply would not be reliable enough to factor into future war. Wilbur Wright, one of the great American aviation pioneers, argued that dropping bombs from airplanes would be too impractical since "the machines must rise 1,000 or 1,500 feet above the ground to escape shell fire. At that height accuracy would be impossible in dropping explosives when moving at 40 or 50 miles an hour."[73] Hiram Maxim believed that "wars will be won by the side with the strongest flying-machines and

---

use of real weapons of destruction." Quoted in Jay H. Mowbray, *Thrilling Achievement of 'Bird Men' with Flying Machines* (Philadelphia: National Publishing Company, 1911), 188.

69 Baden-Powell, "How Airships are Likely to Affect War," 569.

70 Ibid. See also Jay H. Mowbray who advocates more experimentation since the "effectiveness of an airship is only conjectural" as an instrument of offensive war. Jay H. Mowbray, *Conquest of the Air by Airship and other Flying Machines* (Philadelphia: National Publishing Company, 1910), 280.

71 William Ennis, *Flying Machines Today* (New York: D. Van Nostrand Company, 1911), 198.

72 Ibid., 203. Victor Lougheed likewise argues that the method of dropping high explosives "into the weakest and most vulnerable points in the enemy's military and social organization" was "sufficiently horrifying to appall the most skeptical tactician or hardened soldier." The rise of this technology would necessarily lead states to work together to develop an international system of disarmament and arbitration. Victor Lougheed, *Vehicles of the Air: A Popular Exposition of Modern Aeronautics with Working Drawings* (Chicago, IL: Reilly and Britton, 1910), 430, 431.

73 "Aeroplane as Scouts; Not Fighters," *Popular Mechanics*, July 1901, 5.

the most of them," but that airplanes would never be strong enough "to work wide destruction by dropping explosives from the air."[74] Still another commentator argued that "As a means of attack, everything indicates that the aeroplane will have no value. Writers and aviators little versed in the military are constantly talking of dropping bombs and carrying destruction to ships and forts by means of new aerial machines. There seems nothing to justify these expectations."[75] General Homer Lea, the celebrated author of *The Valor of Ignorance* (1909), a book which predicted eventual geostrategic conflict between Japan and the US, argued that the science of aviation was "swaddled in fetishism," and that popular beliefs about the prospect of using airships in warfare are "due primarily to prevailing ignorance concerning the laws and forces that govern the conduct of war."[76] In the unlikely event that airships did develop into devastating implements of war, he simply reaffirmed the prevailing sentiment that "the greater the destructive potential of the aeroplane are, the less destructive will it prove to men in his wars."[77] And finally, the American legal scholar Blewett Lee "ridicule[d] the idea of an aerial invasion commenced by the transportation in several hours of an army corps by means of aeroplanes as an episode worthy only of a fantastic romance."[78]

To be sure, in the first decade of the twentieth century, it was difficult to distinguish between the wildly imaginative speculations of future war fantasy from the prudent and cautious planning against immediate threats. Perhaps the most sensible posture was taken by Captain E. Lascelles, who writes:

---

74 "Flying-Machines to Win Wars," *Popular Mechanics* (December 1909), 724. This view was common among British officers and planners. For example, Colonel J.E. Capper suspected that airships, at their best, would lead to the "minor destruction of supplies." Colonel J.E. Capper, "The Military Aspect of Dirigible and Aeroplanes," *Antiaircraft Journal* 33 (1910); 147–168. 159.

75 "The Seeming Triumph of the Mono-plane," *Journal of United States Artillery*, Volume 34 (Fort Monroe, VA: Coast Artillery School Press, 1910), 214. Captain Townsend makes a similar claim: "there is still considerable doubt as to the result in case a small amount of explosive could be dropped with accuracy on a particular target with most of the weight on the side of ineffectiveness, so far as absolute destruction is concerned." Captain G.L. Townsend, "The Use and Effect of Flying Machines on Military Operations," *Infantry Journal* 7 (1910); 244–258. 251.

76 General Homer Lea, "The Aeroplane in War: Some Observations on a Military Delusion," *Harper's Weekly* 54 (1910); 8–9.

77 Ibid. In a much more optimistic overall assessment, the Canadian pilot, Charles Mitchell, nevertheless concludes that aerial bombing "is really easier said than done." Charles Mitchell, *Aerial Navigation in Warfare* (Toronto: [s.n.], 1911), 18.

78 Blewett Lee, "Sovereignty of the Air," *The American Journal of International Law* 7:3 (1913); 440–496. 485.

> While in the popular imagination the fear of destruction as a result of the precipitation of high explosives from the air on to the earth beneath looms large, there are so many questions yet unsolved in regard to the possibility of accurate projection of projectiles from aerial craft, and in regard to the material effect likely to be produced, that the soldier is forced to the conclusion that for some time to come the principle use to which the new engine of war will be put is reconnaissance.[79]

Similarly, W.J. Jackman and Thos. Russell conclude, "The question as to whether it would be practicable to bombard an enemy on land or sea with explosive bombs dropped or discharged from flying machines or airships, is one which is much discussed but hardly yet determined."[80] From the vantage point just prior to World War I, it was by no means clear which role aerial bombing would take and how it would affect future war, particularly regarding noncombatant casualties.[81] Indeed, as John Ledeboer explains in the Preface to Hermann Hoernes's *A Compendium of Aviation* (1911), "Nothing should be regarded as final in the domain of aeronautics. Every one of its aspects should be approached with an open and unbiased mind."[82]

This sensible open-endedness, however, continually provided fruitful ground for wide-ranging fantastical speculation. To understand the range of approaches to this inchoate technology by inventors and experimenters, it is worth taking a slight detour down the path of "popular imagination" that Lascelles speaks of. In many respects, along with the seeming endless progress made in aerial experimentation, future-war fiction writers were setting the tone of these early debates about what aerial bombing would look like and what kind of threat it would pose to civilians.

79 Captain E. Lascelles, "The Airship and Flying Machines in War," *The Calvary Journal* 5 (1910); 208–212. 211.

80 W.J. Jackman and Thos Russell, *Flying Machines: Construction and Operation* (Chicago: Charles C. Thompson, 1912), 252.

81 Despite Italy's early bombing of Libya in 1912, one observer bluntly noted: "the aeroplane does not yet satisfy military requirements." "Military Aeroplanes," *Antiaircraft Journal* 37: (1912); 101. Evan David explains the bombs "fell far from their objectives, and served no military purpose further than to frighten the horses." Evan David, *Aircraft: Its Development in War and Peace and Its Commercial Future* (New York: Charles Scribner's Sons, 1919), 70.

82 See Preface in Hermann Hoernes, *A Compendium of Aviation and Aereostation, Balloons, Dirigibles, and Flying Machines* (London: Charles Griffin & Company …, 1911), ix.

3 Future-War Fiction

The interplay between inventors, aeronauts, military planners, and future-war fiction writers in the decades around the turn of the century was quite extensive. It was not uncommon for those experimenting with balloons and airships, indeed, for planners and tacticians, to begin to imagine how they might be incorporated into real life by way of future-war fiction. Publishers quickly began to capitalize on the popularity of the emerging genre. There was a ready supply of militarily minded authors, in some cases officers or war correspondents, who wanted not only to produce entertaining stories, but to convey a range of ideas about how technology might interact with geopolitical futures. While entertaining, this literature cannot be reduced to mere entertainment. The authors were psychically priming the public for what would come, as if the authors were prefiguring and tapping into the trauma of twentieth-century war.[83] This, of course, is not meant to imply that this genre spoke with a singular voice. One actually finds the same extreme vacillation between ecstatic hope and deep foreboding noted in the non-fiction literature – that airships would make war obsolete, or they will bring about total annihilation. The authors writing in this burgeoning genre routinely fueled the public's near schizophrenic imagination as to what the future of this technology held.[84]

Fantasy literature about flight can be traced back quite a long way. In *Prodromo Dell Arte Maestra* (1670), the Jesuit priest, Francis Lana, had anxiously predicted that if airships were invented "that no walls nor fortifications could then protect cities, which might be completely subdued or destroyed,

83 Paul Saint-Amour notes that "a symptomatic effect can precede its traumatic cause" and that "a *hysteron proteron* occurs with a marked frequency in fictional treatments of historical trauma. Such anachronisms and time travels might powerfully suggest, among other things, how the singularity of a traumatic event can manifest itself in a perceived leveling of past and future into identical orientations of the symptom: time, for the survivor, is spatialized, such that the traumatic event can seem to scatter the shrapnel of its symptoms evenly across the past and the future." This leads to the paradoxical conclusion that symptoms of trauma may "precede traumatic causes." Paul Saint-Amour, "Bombing and the Symptom: Traumatic Earliness and the Nuclear Uncanny," *diacritics* 30:4 (2000); 59–82. 62–63.

84 For more on this tradition, see I.F. Clarke, *Voices Prophesying War: Future Wars, 1763–3749* (New York: Oxford University Press, 1992); Michael Paris, *Winged Warfare: The Literature and Theory of Aerial Warfare in Britain, 1859–1917* (New York: Manchester University Press, 1992); John Buckley, *Air Power in the Age of Total War* (London: Taylor & Francis Group, 1999); and Adam Roberts, *The History of Science Fiction* (London: Palgrave Macmillan, 2016).

without having the power to make any sort of resistance, by a mere handful of daring assailants, who should rain down fire and conflagration from the region of the clouds."[85] In a poem titled *The Air Balloon* (1784), Mary Alcock anticipates the god-like power of those who are free to escape into the sky: "No more of judge or jury will I hear/ The laws of land extend not to the air."[86] Alcock rightly anticipates the degree to which this new technology would challenge not only the spatial conception of the world (what military planners would later call the "third dimension" of war: land, sea, and *air*) but its legal dimension, too.

Most eighteenth-century science fiction novels involving balloons or airships were adventure stories,[87] though some began to toy with the destructive potential of weaponized balloons. In Frederick Pilon's play, *Aerostation* (1785), the balloon owner begins to scheme of ways to contract his balloon out to governments. He calculates: "Let me see, let me see; fourteen first-rate balloons, each carrying twelve cannonades, ten bombs, and four and twenty-two and forty pounders. Ay! fourteen would do it."[88] And in at least one of his numerous adventures, Rudolf Raspe's fictionalized German nobleman Baron Munchausen drops bombs onto a fortress from his balloon.[89]

By the early nineteenth century, several novelists envisioned air skirmishes.[90] Others were just beginning to glimpse the terrible significance this technology would pose to those on the ground. For instance, Alfred Lord Tennyson anticipates what resembles aerial bombing in his poem "Locksley Hall":

85 Quoted in "A Chapter on Balloons," 34–35. John Wilkins imagines the possibility of an "aerial chariot." See John Wilkins, *Mathematical Magick; Or the Wonders That May Be Performed by Mechanichal Geometry* (London: Printed for Edw. Gellibrand at the Golden Ball …, 1680), 209.

86 Mary Alcock, *The Air Balloon; Or, Flying Mortal. A Poem* (London: Printed for E. Mackley …, 1784), 6.

87 *The Modern Atlantis; Or, The Devil in an Air Balloon* (London: Printed for G. Kearsley …, 1784); *The Air-balloon; Or, the Sages Adventures in a Flight to the Moon* (London: Printed for G. Bremner, 1784); and Elizabeth Inchbald, *The Mogul Tale: Or, the Descent of the Balloon. A Farce* (Dublin: Printed for the Booksellers, 1788).

88 Frederick Pilon, *Aerostation; Or the Templar's Stratagem, A Farce* (Dublin: Printed by J. Exshaw, for the Company of Booksellers, 1785), 18.

89 Rudolf Raspe, *Surprising Adventures of Baron Munchausen* [1786] (London: Plummer, 1809).

90 See Julian von Voss, *Ini: Ein Roman aus dem ein und zwanzigsten Jahrhundert* (Berlin: Karl Freidrich Amelang, 1810); Hermann Lang, *The Air Battle: A Vision of the Future* (London: William Penny, 1859); and Robert Easterley and John Wilbraham, *The Germ Growers: An Australian Story of Adventure and Mystery* (Melbourne: Melville, Mullen, & Slade, 1892).

> For I dipt into the future, far as human eye could see,
> Saw the Vision of the world, and all the wonder that would be;
> Saw the heavens fill with commerce, argosies of magic sails,
> Pilots of the purple twilight dropping down with costly bales;
> Heard the heavens fill with shouting, and there rain'd a ghastly dew
> From the nations' airy navies grappling in the central blue.[91]

These sporadic references notwithstanding, as the nineteenth century progressed the predictions of future aerial war became much more pronounced, and its transformative effects were more expressly thematized.

Though not about aerial bombing, the genre of future-war fiction was blown open by George Tomkyns Chesney's *The Battle of Dorking* (1871), a novella written in the wake of Germany's rapid defeat of the French in the Franco-Prussian War. This book inspired a string of authors to try to reproduce Chesney's success. Many of the books in this newly popularized genre started to grapple with the emerging, even if fantastical, technologies.

One of the undisputed masters of this new genre was Jules Verne. In his *Clipper of the Cloud* (1887), the airship *Albatross* originally poses no threat to the military affairs in continental Europe. When the airship glides above a skirmish between Russia and England in Herat (present-day Afghanistan) the narrator explains: "There were heard also the roar of the cannon and the crackling of musketry. But the engineer, never meddled with the affairs of other where his honour or humanity was not concerned. He passed above them … Terrestrial interests were nothing to him who had made the air his domain."[92] According to Verne's vision, one gets the impression that this novel invention might blissfully soar above continental conflicts and petty differences that govern the affairs of men in Europe.

When the *Albatross* flies over Africa, however, Mr. Robur and his crew are much less ambivalent about "terrestrial interests."[93] In one particular instance, Verne explains that the crew members had a very different response when the flight path took the airship over Dahomey, a pre-colonial kingdom in West Africa (present-day Benin) that was "celebrated for the frightful cruelties which signalize its annual festivals, and by its human sacrifices."[94] Flying over-

91 Alfred Lord Tennyson, *The Works of Alfred Tennyson* (Philadelphia: Gebbie & Company, 1881), 53.

92 Jules Verne *The Clipper of the Cloud* (London: Low, Marston, Searle & Rivington, 1887), 119.

93 Robur would take on an even more sinister quality in the sequel with his supremely advanced airship the *Terror*: Jules Verne, *Master of the World* in *The Works of Jules Verne* ed. Charles Horne (New York: Vincent Parke and Company, 1911).

94 Verne *The Clipper of the Cloud*, 150.

head, the airship's crew saw that "thousands of prisoners taken during recent razzias were to be immolated" as part of a ceremony to honor the new king.[95] In this instance, the engineer-pilot could not remain neutral.[96] He passed out dynamite to the crew members so that they could lob bombs onto the people below. Verne doesn't report any casualties from the policing mission, but as a diversionary tactic, the bombing worked: "The king and his court and army and people were stricken with fear at the turn things had taken. They fled under the trees, while the prisoners ran off without anybody thinking of pursuing them."[97]

Frank Stockton, in *The Great War Syndicate* (1889), imagines war being fought with invincible underwater crab machines and instantaneous motor-bombs (which resemble proto-nuclear missiles), both of which anticipate some of this "weapon-to-end-all-war" optimism and anxiety.[98] In the fictional war between the US and Great Britain, a war started by British aggression, the US had outsourced all its fighting to a corporate Syndicate, an entity made up of scientists, inventors, and a cadre of wealthy financiers. The company had dedicated its energy to developing terrifyingly powerful weapons with the singular purpose of ending war. After quickly expelling the British invaders from their assault on the US, the Syndicate's fleet of ships set sail to retaliate against Great Britain, armed with those fantastic weapons of mass destruction

95 Ibid., 153.

96 This is like a scene in Verne's *Five Weeks in a Balloon*, where the passengers onboard the *Victoria* witness a terrible battle being fought between two African tribes. Dick Kennedy, one of the airship passengers, clutches his riffle and proclaims, "I have the greatest mind to interfere in their fight," to which Dr. Fergusson responds: "No, no! Why should we mix ourselves up in this quarrel? How do you know who is right or who is wrong, that you should play the part of Providence? Let us fly from this repulsive exhibition." But after witnessing the terrible ferocity of one of the tribal chiefs in battle, Dick cries: "Ah, the horrible beast! I can't stand this any longer." He shoots the warrior in the forehead. Jules Verne, *Five Weeks in a Balloon: A Voyage of Exploration and Discovery in Central Africa* (London: Sampson Low, Marston, Low & Searle, 1874), 140–41.

97 Verne, *Clipper of the Clouds*, 154. As will be seen, it was widely accepted that aerial bombing might be effectively used to police "savages" from the air. For similar accounts of policing natives from fantastical airships see Robert Paltock, *The Life and Adventures of Peter Wilkins, among the Flying Nations in the South Seas* (London: Printed and Published by J. Limbird ..., 1833); and William Joseph Lancaster [Harry Collingwood], *The Log of the* Flying Fish: *A Story of Aerial and Submarine Peril and Adventure* (London: Blackie & Sone ..., 1887). In contrast, in Tom Greer's *A Modern Dædalus*, a young Irish inventor develops a one-person flying devise. Though inclined to peace, he eventually puts his flying apparatus in the service of the anti-colonial, Irish revolutionaries, which includes bomb-dropping. Tom Greer, *A Modern Dædalus* (London: Griffith, Farran, Okeden & Welsh, 1885).

98 Frank Stockton, *The Great War Syndicate* (New York: Dodd, Mead, & Company, 1889).

in order to ensure a quick end to hostilities: "There was nothing, in the opinion of the ablest naval officers, to prevent the Syndicate's fleet from coming up the Thames. Instantaneous motor-bombs could sweep away all forts and citadels, and explode and destroy all torpedo defences, and London might lie under the guns of the repeller."[99] London was forced to surrender with virtually no loss of life.

Others were not confident that this technology would remain so benign in European conflicts. In *The Great War of 189-* (1893), a book co-written by Philip Colomb and a group of officers and military experts, the authors imagine the Russian army dropping bombs from balloons to terrorize civilians into surrendering.[100] In the chapter describing the fictional "Bombardment of Varna,"[101] the authors paint a rather grim picture:

> A black mass of some kind was seen to drop from the balloon; as it about reached the level of the tallest buildings in the place it suddenly burst into a lurid glare which lighted up the minarets and pinnacles of the old Turkish town. Its course was marked by crashing buildings and falling ruins. It was evidently a dynamite shell of vast proportions, which had been deliberately dropped from the balloon.[102]

The authors continue that "The object could only have been to terrorise the inhabitants by a cruel and wanton destruction of property."[103] Rather than producing a strategic advantage, the narrators of the attack believe that such a tactic suggested Russian weakness: "It indicated to the Englishmen who watched it, rather the parting attempt of Russian hopelessness than a serious effort of war."[104] Through their narration, Colomb et al. reaffirm the commitment to the idea that, as an act of desperation, bombing civilians was ineffectual. In fact, even though the aerial attack had caused an immense amount of destruction

99 Ibid., 162.

100 In another novel that same year, Moffat and White had envisioned a world where aerial bombing had been decisive in a fictional Great War, where French airships had been able to annihilate their enemies. See W. Graham Moffat and John White, *What's the World Coming To?* (London: Elliot Stock, 1893).

101 Notice the historical serendipity: Varna was bombed by Russia in 1915. See "Varna Riddled by Shells," *New York Times*, 2 November 1915. 2.

102 Colomb, J.F. Maurice, F.N. Maude, Archibald Forbes, Charles Lowe, D. Christie Murray, and F. Scudamore, *The War of 189- : A Forecast* (London: William Heinemann, 1893), 228.

103 Ibid., 229.

104 Ibid.

"the inhabitants, did not seem likely, as far as we could judge, to enforce any surrender."[105]

In *The Great War in England in 1897* (1894), William Le Queux depicts the fictitious invasion of England by allied French and Russian forces. In particular, he recounts an incident where the Russians were "about to test the efficiency of a new gigantic engine of war," which elicited a great deal of anxiety among the British officers.[106] He writes,

> It was a huge and curiously-shaped air-ship, and was to be used for dropping great charges of mélinite and steel bombs filled with picric acid into the handsome historic city of Edinburgh! Some of the shells were filled with sulphurous acid, carbon dioxide, and other deadly compounds, the intent being to cause suffocation over wide areas by the volatilisation of liquid gases![107]

With such terrible weapons, the Russian airship, "which might totally wreck a city and kill every inhabitant in half an hour while at an altitude of 6 ½ miles, had been named … 'The Demon of War.'"[108] As the scene unfolds, a Scottish artillery unit – armed with a novel "pneumatic dynamite gun" which could fire fifty shells a minute – desperately tries to hit the balloon before it could ascend to attack Edinburgh. After several near misses, the balloon escapes into the clouds. In utter hopelessness, "Officers and men stood aghast, with blanched faces, scarce knowing how to act. The destructive forces in that controllable balloon were more than sufficient to lay the whole of Edinburgh in ruin."[109] Believing all was lost, the soldiers eventually catch a lucky break. Spotting the balloon through the clouds, they fired one last time. This time they hit the balloon's undercarriage which detonated all the explosives it carried. Unlike the bombing of Varna, which was a sign of weakness, desperation, and dishonor, Queux's narrative suggests that, despite the destructive potential this new aerial technology poses, advances in artillery would be able to keep such threats at bay, albeit, with some difficulty.

105 Ibid., 230.

106 William Le Queux, *The Great War in England in 1897* (London: Tower Publishing, 1894), 248. Queux also depicts artillery units firing on military balloons in William Le Queux, *The Invasion of 1910: With a Full Account of the Siege of London* (Toronto: Macmillan Company, 1906).

107 Queux, *The Great War in England in 1897*, 249.

108 Ibid., 249–250.

109 Ibid., 252.

Edward Douglas Fawcett depicts a much more chaotic and, indeed, devastating picture of aerial bombing. In *Hartmann the Anarchist* (1893), a prospective anarchist conscript, Arthur Stanley, befriends Rudolph Hartmann who, with a crew of anarchist revolutionaries and his airship *Attila*, plans to bomb London. Of course, Stanley was not aware of these sinister designs when he agreed to take a ride on the wondrous air ship. As the ominous plan starts to unfold, Stanley becomes increasingly uneasy. Unable to talk Hartmann out of his terrorist plot, he could only helplessly watch as the destruction unfolds beneath him:

> I looked over, to start back with a shriek. Horror of horrors, the great tower had fallen on the crowd, bruising into jelly a legion of buried wretches, and beating into ruins the whole mass of buildings opposite. Every outlet from the neighborhood was being furiously fought for, hordes of screaming, shrieking madmen were falling, crushing and stamping their victims into heaps, and with the growth of each writhing heap the ghastly confusion grew also. Of the Houses of Parliament pinnacles were collapsing and walls were being riven asunder as the shells burst within them.[110]

The devastation was unrelenting and continuous: "I heard the roar of Schwartz's bombs, and the roar of the burning and falling houses. Huge circles of flame raved beneath us, and shot up their feverish and scorching breath."[111] In the end, after Westminster had been destroyed, they proceeded to bomb the rest of the city; the Bank of England, the Royal Exchange, the Stock Exchange, credit companies, insurance offices, St. Paul's Cathedral and its adjoining neighborhood were all subjected to aerial bombing.[112] Destruction was quite nearly total: "the City was blazing, and a roaring tumult of flames was undulating in every direction from this centre."[113] A disgusted and dejected Stanley flees the *Attila* by parachuting into the apocalyptic terror below.

Like Fawcett, George Griffith further develops the threat of aerial anarchist terror in *The Outlaws of the Air* (1895).[114] In his account, anarchist bandits

---

110 E. Douglas Fawcett, *Hartmann the Anarchist: Or, The Doom of the Great City* (London: Edward Arnold …, 1893), 144–147.

111 Ibid., 147.

112 Ibid., 152.

113 Ibid., 159.

114 Several of Griffith's books deal with terrorists or revolutionaries getting hold of flying machines. See George Griffith, *The Angel of the Revolution: A Tale of the Coming Terror* (London: Tower Publishing, 1894); George Griffith, *Olga Romanoff: Or, the Syren of the*

fortuitously find themselves "in possession of the only air-ship in existence," which they call the *Venguer* (the Avenger).[115] With such a weapon, they realize that "There could be no pursuit, and so no capture. The most inaccessible parts of the world would be open to them, and they could take their choice of refuges in a thousand places where no man could follow them."[116] Since there was no power that could rival them, the outlaws make short work of any military opposition that tried to stop them, disabling battleships that were foolish enough to fire on them. However, when one of the bandits, Max Renault, was captured and sentenced to death, the anarchists planned a daring rescue. This entailed flying their machine over the walls of Newgate Prison. They were easily able to do this with little resistance from the ground.

Having succeeded in their rescue plan, the anarchists then started revenge-bombing large parts of London. First, they bombed a large throng of innocent spectators that had amassed around Newgate: "Two great ragged gaps were torn in the densely packed masses amidst which the shells fell, as though the earth had burst open beneath their feet, and bodies of men and women were hurled up into the air, mingled with limbs and torn fragments of other bodies, and fell back upon the heads of those who thronged the pavements."[117] The attack was unrelenting and indiscriminate. He continues:

> Some made wild rushes into the shops and houses, and as soon as he saw this, Max ordered the guns to play upon them, loaded with fire-shell, and a few minutes later unquenchable fires were blazing furiously in a score of houses and shops all round the area of slaughter and destruction. Then, like the incarnation of an evil spirit brooding above the scene of indescribable horror, the *Vengeur* passed slowly over the streets dropping bombs charged with fire-mixture as she went, and when she had completed the circuit, a ring of flame and smoke encircled the black frowning walls of the prison in which everyone had believed by this time Max Renault would have been hanging by the neck.[118]

Skies (London: Tower Publishing, 1894); George Griffith, *The Lord of Labour* (London: F.V. White, 1911). For an account of wanton destruction by a terrorist with an aircraft loaded with chemical weapons, see T. Mullett Ellis, *Zalma* (London: Tower Pub. Co., 1895). For an account of post-apocalyptic aerial destruction, see M. Shiel, *The Purple Cloud* (London: Chatto and Windus, 1901). For a tale of a criminal investor who uses an airship to bomb targets in Russia to improve the return on his investments, see James Blyth, *The Aerial Burglars* (London: Ward, Lock & Co., 1906).

115 George Griffith, *The Outlaws of the Air* (London: Tower Publishing, 1895), 97.

116 Ibid.

117 Ibid., 204.

118 Ibid.

From Northgate, the aerial attack continued, hitting most of the major landmarks in London: St. Paul's Cathedral, the railway stations, the Bank of England, the General Post Office, Westminster, Scotland Yard, The Royal Courts of Justice, and Parliament itself. Griffith describes the devastation left in the anarchists' wake:

> The blackened and smoking ruins of the most important buildings in London, of the very seat of Government itself, the darkened and wreck-strewn streets, the thousands of human bodies, torn and mangled almost beyond recognition, and the mutilated remains of men, women, and children who had perished in the burning houses or been crushed to death by the panic-stricken throngs in the streets – all these were evidence, too fearfully unmistakable, of the fact that the enemy of society had, for the time being at least, armed himself with a power superior to anything that the forces of law and order could bring against him.[119]

In the wrong hands, such a weapon would be positively apocalyptic. Rather than a deterrent, if airships were acquired by society's enemies, i.e. revolutionaries and anarchists, they would be able to bring most nation-states to their knees. In fact, in gleeful delight with the success of their bombing mission, the anarchists fantasize about reproducing similar results in every other major European city: "What had happened to London might happen to Paris the next day, to Berlin the next, to St. Petersburg the next, and so on."[120]

In Stanley Waterloo's *Armageddon* (1898), the mathematician-inventor David Appleton produces an air machine which, he explains, is "the most destructive force in the world today" – so destructive, in fact, that it will leave "the fate of armies and navies in their hands."[121] But, unlike Fawcett's and Griffith's visions, in this case, the destructive power of this war machine made it a source of peace. As a harbinger of destruction, enemies would be forced to surrender after a few strategic strikes on military targets. The first time the air ship – the *Wild Goose* – was used to bomb the Russian flagship in a massive naval engagement, all the helpless enemy fleet could do was surrender. The "dreadful creature of the skies" – after sinking two ships – had forced the enemy to raise their white flags.[122]

---

119 Ibid., 212.

120 Ibid., 213.

121 Stanley Waterloo, *Armageddon: A Tale of Love, War, and Invention* (Chicago: Rand McNally & Co. 1898), 150, 152.

122 Ibid., 221. Waterloo's account comes quite close to the result of one of the first air-naval bombardments in history. During the Mexican Civil War (1913), American mercenaries,

Also in 1898, Samuel Odell published *The Last War, or the Triumph of the English Tongue*. In this dystopian future, set in the year 2565, Odell envisions a world that has been almost entirely racially segregated.[123] What is more, large portions of the world have been annexed by America, which had grown to include all North and South America. In fact, the US now consists of 185 states. United Europe, the African Union, most of Asia, have all become English speaking and are America's allies. They represent the "Allied Anglo-American Nations." Those living outside of America's influence are Czarist Russia, the Mohammedans of Turkey and Persia, and the Asian countries of China, Japan, and Mongolia.

The religious influences in those countries outside of America's sphere had mutated into some amalgam of Catholic, Buddhist, and Islamic authoritarianism. They are ruled by a Czar-pope, "with one iron hand holding a sword and the other holding the cross and crescent."[124] What culminates is a massive clash of civilizations, the Anglo-American Alliance, progenitors of democracy, progress, and freedom, against the new Holy Empire, small-minded, fanatical, and autocratic. The armies of both sides are made up of vast air fleets, automobile war chariots, and millions of soldiers outfitted with bullet-proof clothing.

When it seems as though the Czar-pope's army might overtake large parts of Europe, the European Allies finally receive relief from a division of American airships, all of which "were supplied with fireballs and grenades with which to destroy fortifications or men over which they should pass."[125] The narrator explains the great relief felt by those living in European cities and how their hope in the war effort had been restored: "the great fleet, amid the thunder of saluting guns and the roar of cheering from the populace, arose with whirring propellers and outstretched wings. Forming in extended order, in two long

---

Dider Masson and Joaquin Alcalade, flew sorties for Pancho Villa. In one bombing run, the airship dropped four home-made bombs onto Mexican gunboats in the Guaymas Harbor. Even though the bombs missed their targets, the ships realized they were totally defenseless and fled. Several crew members jumped overboard in panic. See Alan Axelrod, *Mercenaries: A Guide to Private Armies and Private Military Companies* (London: Sage, 2014), 131.

123 For another racist-imperialist future-war fiction depicting aerial bombing and the subjugation of black and brown peoples, see William Delisle Hay, *Three Hundred Years Hence; Or, a Voice from Posterity* (London: Newman and Co, 1881). This book depicts the world from the vantage point of 2180.

124 Samuel Odell, *The Last War, or the Triumph of the English Tongue* (Chicago: Charles H. Kerr & Co., 1898), 18.

125 Ibid., 103.

lines, the sharp beaks were turned toward Saxony, and with the speed of the wind the fleet sped away upon its deadly errand."[126]

With the help of the American airships, the tides of war quickly turned. In their first engagement at Saxony, the Allies' airpower overwhelmed the enemy: "their surprise turned to terror when grenades fell in showers around them, causing ruin and death."[127] The Russian and Turkish armies were pushed all the way back to Antioch, where the decisive battle of the war would take place. Just prior to the final battle, one of the Allied generals, General Wellington,[128]

> looked out across the plains where twinkled lights, as the illuminators kept watch so that no surprise should happen on either side. Antioch, doomed to destruction under a fire such as never had yet smitten a city, lay silent, except for the rumble of war machines whose noise could not reach his ear; its inhabitants had fled southward, and were still fleeing in order to escape the hail that would come on the morrow. He looked up to the heavens, wondering if the celestial armies were not gazing down upon this field where death would reign shortly, and he felt how infinitely small was he himself in the midst of all these mighty environments.[129]

The next morning, after a relatively brief engagement, largely due to their superior air power, the Allies easily defeated the Russian and Turkish forces. With the enemy army defeated, and after thirty-five years of further pacification, in the year 2600, the world-state, the United States of the World, was formed.

Garrett Serviss's *Edison's Conquest of Mars* (1898) envisions a future where Martians are terrorizing the earth with their technologically advanced air ships: "These flying machines in themselves had given them an enormous advantage in the contest. High above the desolation that they had caused to reign on the surface of the earth, and, out of the range of our guns, they had hung safe in the upper air. From the clouds they had dropped death upon the earth."[130] In fact, the bombs they dropped were of a magnitude of destruction never before imagined by humankind. In the devastating aerial attack on New York, for instance, "all the buildings yet standing in the surrounding towns

126 Ibid., 104–105.

127 Ibid., 128.

128 As a side note, this novel if full of onomastic play. Many of the future war leaders have names of historical significance.

129 Ibid., 144.

130 Garett Serviss, *Edison's Conquest of Mars* [1898] (Auckland, New Zealand: The Floating Press, 2012), 9.

and cities fell in one far-circling ruin."[131] The impact of this bomb strike was so powerful that it sent a shockwave around the globe which "was recorded by seismographic pendulums in England and on the Continent of Europe."[132]

Fortunately, the great inventor Thomas Edison had been developing a flying machine that could match those of the Martians, an airship capable of space travel. What is more, he had developed a weapon – a "disintegrator" – which was able to manipulate the "vibrations of substances" producing such an effect that it could send any object into "violent undulation, sufficient to burst it asunder and to scatter its molecules broadcast."[133] With monetary contributions from all the nations of the world, Edison builds "a hundred electric ships and three thousand disintegrators," in order to bring the fight to Mars.[134] After a series of fierce battles, the Terrestrials are victorious over the Martians.

Much like Odell's *The Last War*, in Robert Cole's, *The Struggle for Empire* (1900), he too envisions a world in which the Anglo-Saxon race had taken over the globe. By the year 2236, London had become the capitol "not only of England, but of the world, the Solar System, and the stars."[135] Since humanity has abandoned frivolous subjects, like Theology, Literature, and Philosophy, dedicating more energy to the practical sciences, they had been able to solve the mysteries of gravity and to invent flying machines. In fact, these air ships "soon became almost the sole means of locomotion," which eventually extended to interstellar ships.[136] In their conquest of the universe, greedy English colonists came up against a formidable rival, in the Sirian System on the planet of Kairet. In response to this perceived threat, the Anglo-Saxons build "Hundreds of interstellar war-vessels [...] with the most terrible destroying weapons, after the manner of the ironclads of olden time."[137] In response the people of Kairet made similar preparations and were evenly matched against the threat they faced. The initial optimism that the Anglo-Saxons could defeat their enemies soon gave way to panic when the massive fleet of airships they had sent was quickly annihilated. Facing such a devastating loss, it seemed as though Earth itself would soon be under threat, and indeed it was. When the Kairet fleet began assaulting the Earth's outer defenses on the moon, Odell writes:

131 Ibid., 4.
132 Ibid., 5.
133 Ibid., 15.
134 Ibid., 33.
135 Robert Cole, *The Struggle for Empire: A Story of the Year 2236* (London: Elliot Stock ..., 1900), 1.
136 Ibid., 8, 10.
137 Ibid., 18.

> Stray shot and shells came whistling down through the atmosphere, often exploding high up, but sometimes dropping on to buildings, crowded streets, or the solitary mountains and valleys of the country. Often, too, entire wrecks and huge fragments of metal came rushing down without the least warning, destroying great buildings and killing hundreds of people.[138]

As the enemy advanced, the vulnerability of London became more apparent. It was subjected to a direct assault: "The air shook with the reverberation. A moment afterwards the metal shower came crashing down on to the city. Torpedoes, shell, and shot fell down in an increasing stream. Houses were dashed to pieces, towers staggered and were overturned, domes were pierced and battered down … After two hours there was scarcely a building left untouched."[139] The annihilation of London would have been complete if not for the last-minute ingenuity of the savior-inventor, Dr. James Tarrant. His "Electro-Ednogen machine" was capable of throwing "out waves of force that will stop the working of the antigravitation apparatus and cause the warshops on to which it is directed to fall to the ground."[140] Equipped with this new weapon, the remaining Anglo-Saxon ships were able to demolish "in a few minutes whole fleets that were bombarding cities, hurling them down on to the ruins they had made."[141]

What is so striking about this compilation of nineteenth-century, future-war fiction is the high degree of variability in terms of how the authors saw destructive aerial weapons integrating into the existing paradigm of war-making. The co-authored account of the bombing of Varna implies that while some unscrupulous countries would resort to such a dishonorable low of terror-bombing residents of cities from the air, such a method of attack ultimately revealed their moral and tactical weakness. Honorable nations would not resort to bombing tactics that violated the established laws of war.[142] However, in Cole's account of Anglo-Saxon colonial aggression, aerial bombing proved to be a liability; they were nearly destroyed by their expansive greed, only to be saved by some more pacific technology. Others, like Queux, suggest that while city-destroying technologies might very well emerge, novel innovations

138 Ibid., 174–75.

139 Ibid., 181–82.

140 Ibid., 186.

141 Ibid., 196.

142 Though, as was demonstrated in the previous chapter, the bombing of residential districts of fortified cities to expedite surrender was an accepted practice throughout the nineteenth century.

in artillery would prove effective at keeping the destructive potential of airships at bay. And finally, many of the above writers focused on the benevolence or cruelty of those who possessed the destructive technology. Stockton's corporate Syndicate, Waterloo's humanitarian-minded inventor, Odell's benevolent American Alliance, and Serviss's inventor-hero, Thomas Edison, all used the superior technology for good. They curtailed injustice and staved off annihilation, which seems to suggest that if the motives of the belligerents are noble, wars might be curtailed, and lives saved. In the right hands, such weapons would end wars before they start, and perhaps even put an end to war itself. On the other hand, Fawcett and Griffith, with their marauding bands of anarchist outlaws, show that if this technology falls into the wrong hands, it may lead to the end of civilization.

Perhaps the most well-known example of aerial bombing anxiety of this period is captured in H.G. Wells's 1907 novel *The War in the Air*.[143] Published the same years as the second Hague Convention and just a year prior to Knabenshue's confetti-bombing run over Los Angeles, Wells envisions a world in which the great cities of the world – and, indeed, civilization itself – have been destroyed by aerial bombing.[144] In the same way Fawcett and Griffith had described the destruction of London a decade prior,[145] Wells vividly depicts the bombing of New York City by a fleet of German warships:

143 Griffith and Wells inspired early silent film adaptations of aerial bombing. See Edwin Porter's, *The Twentieth Century Tramp; or Happy Hooligan and His Airship* (1902); Walter Booth's trilogy: *The Airship Destroyer* (1909); *The Pirates of 1920* (1911); *The Aerial Anarchists* (1911).

144 Many of these post-apocalyptic themes had been explored in Wells's *The Time Machine* (1895) and *The War of the Worlds* (1897). Wells had experimented with bombing themes in *When the Sleeper Awakes* (1899). The post-apocalyptic thematic became a mainstay in the pre-World War II dystopia literature. For instance, Aldous Huxley's *Brave New World* is set at some point after "The Nine Years' War of A.F. 141" where "fourteen thousand aeroplanes" dropped anthrax bombs on civilian populations. He writes, "An enormous hole in the ground, a pile of masonry, some bits of flesh and mucus, a foot, with the boot still on it, flying through the air and landing, flop, in the middle of the geraniums – the scarlet ones; such a splendid show that summer!" Aldous Huxley, *Brave New World* (London: Chatto & Windus, 1932), 50–51.

145 Griffith's influence on Wells is debated. Sam Moskowitz argues, Griffith was "the most popular science fiction writer in England between 1893 and 1895," which implies that Griffith may have been one of the first of Well's exposures to aerial bombing narratives. Sam Moskowitz, *Science Fiction by Gaslight: A History and Anthology of Science Fiction in the Popular Magazines, 1891–1911* (Westport, CT: Hyperion Press, 1974), 39. More recently Harry Wood has shown their relationship was much more hostile than initially believed. See Harry Wood, "Competing Prophets: H.G. Wells, George Griffith, and Visions of Future War, 1893–1914," *The Wellsian: The Journal of the H.G. Wells Society* 38 (2015); 5–23.

> Something had dropped from the aeroplane, something that looked small and flimsy … It hit the pavement near a big archway … Then blinding flames squirted out in all directions from the point of impact, and the little man who had jumped became, for an instant, a flash of fire and vanished – vanished absolutely. The people running out into the road took preposterous clumsy leaps, then flopped down and lay still, with their torn clothes smouldering into flame. Then pieces of the archway began to drop, and the lower masonry of the building to fall in with the rumbling sound of coals being shot into a cellar … Dust and black smoke came pouring into the street, and were presently shot with red flame.[146]

The German bombers had "left ruins and blazing conflagrations and heaped and scattered dead; men, women, and children, mixed together as though they had been no more than Moors, or Zulus, or Chinese. Lower New York was soon a furnace of crimson flames, from which there was no escape."[147] These rather stunning and, indeed, awfully prescient depictions of urban city centers being destroyed are some of what lay behind the deterrence-oriented optimism of pre-World War I thinkers, like Knabenshue and Mond. Since the grotesque horror of urban bombing had already been illustrated for them in pristine literary detail, it was obvious that reasonable states would seek alternative ways to resolve conflict.

But it is interesting to note Wells's rather astute analysis of why he believes militaries would not be restrained and why they would resort to these kinds of terrible actions. For him, the turn to aerial bombing would be the inevitable outcome of the logic of military necessity, which had been the operative ethical framework governing war-making throughout the nineteenth century. Wells perceptively explains that New York had been:

146 H.G. Wells, *The War in the Air* (London: George Bell and Sons, 1908), 206.

147 Ibid., 207. As seen in Verne's account above, Wells's anti-primitivism, namely his willingness to tolerate the death of Moors, Zulus, and Chinese anticipates some of the early twentieth-century debates about employing aerial bombing to police occupied territories. In *Winged Defense*, Billy Mitchell explains: "In Mesopotamia, Irak, as it is called, the air force handles the military occupation of the country in a manner similar to that in which armies have in the past. The result of this occupation has been very satisfactory. The airplanes fly over the country at will, are able to put down uprisings quickly." William Mitchell, *Winged Defense: The Development and Possibilities of Modern Air Power* (New York: Putnam Publishing, 1925), 23. This was codified in the Conference for the Reduction and Limitation of Armaments (Geneva, 1933) attempted to ban bombing civilians "except for police purposes in certain outlying regions." See also, Thomas Hippler, *Governing from the Skies: A Global History of Aerial Bombing* trans. David Fernbach (London: Verso, 2017).

> wrecked as in the previous century endless barbaric cities had been bombarded, because she was at once too strong to be occupied and too undisciplined and proud to surrender in order to escape destruction. Given the circumstances, the thing had to be done. It was impossible for the [German] Prince to desist, and own himself defeated, and it was impossible to subdue the city except by largely destroying it. The catastrophe was the logical outcome of the situation, created by the application of science to warfare. It was unavoidable that great cities should be destroyed. In spite of his intense exasperation with his dilemma, the Prince sought to be moderate even in massacre. He tried to give a memorable lesson with the minimum waste of life and the minimum expenditure of explosives.[148]

If military necessity is determined by the object of the war, then, as Wells suggests, it is unclear how militaries would respond to the extensive war economies of rival nations. If those nations would not surrender or agree to be occupied, states would have very few options to gain the assurance that a threat no longer persisted. Under these conditions, the destruction of cities would become unavoidable; it would be the only way to ensure that the means of prosecuting the war could be curtailed. As Wells predicts, an assaulting army would be forced, by necessity, to make increasingly desperate types of moral calculations. Destroying whole cities may be the only way to eliminate perceived threats.

In this respect, Wells uniquely anticipates some of the internal tensions embedded within military necessity stemming from the tactic of aerial bombing. In the previous century, a besieging force would bombard a city, in many cases intentionally targeting civilians, to expedite surrender. When considering the "immense power of destruction" airships could unleash and the "relative inability to occupy or police or guard or garrison a surrendered position," Wells argues that it is unclear how an attacking army could feasibly accept surrender, much less occupy the surrendered town once bombing commenced.[149] The very nature of the airship – at least as Wells imagines it, that it floats above cities at great heights, more or less out of harm's way – made it immune from retaliatory strikes. The logical progression of utility, Wells concludes, is that cities themselves would need to be destroyed.

The impracticalities inherent to Wells's version of the new technology would lead reasonable states to employ aerial bombing only as a means of

148 Wells, *The War in the Air*, 207.

149 Ibid.

counterattack, as a form of deterrence, since even a direct assault would invite devastating retaliation. All the belligerents would be floating helpless over their enemies while *their* cities are being bombed.[150] On this view, once bombing commences, the world would quickly lie in ruins. Without the ability to occupy and control their defeated enemy, all one could do is bomb cities into oblivion. Wells writes,

> Consequently it was impossible to end a war by any of the established methods. A having out-numbered and overwhelmed B, hovers, a thousand airships strong, over his capital, threatening to bombard it unless B submits. B replies by wireless telegraphy that he is now in the act of bombarding the chief manufacturing city of A by means of three raider airships, while B in a state of passionate emotion and heroic unconquerableness, sets to work amidst his ruins, making fresh airships and explosives for the benefit of A. The war became perforce a universal guerilla war, a war inextricably involving civilians and homes and all the apparatus of social life.[151]

With enough ships in the air, even relatively small powers could threaten greater powers. The threat of mutual destruction made it difficult to conceptualize an offensive use for such technology. This is because, according to Wells's vision, strikes and counterstrikes would leave the world uninhabitable.

What is remarkable about this array of future-war fiction is the high degree of creative energy authors were employing in their attempts to integrate this anticipated technology into the existing paradigm of war-making. As Clarke explains,

> All these tales of war-to-come advanced along the contour lines of contemporary expectations … Most of their tales were admonitory essays in preparedness – arguments for a bigger army, or for more ships. Since most of these authors were responding to some danger or menace represented by the enemy of the day, they were usually careful to present their accounts of the war-to-come as the next stage in the nation's history.[152]

150 One interwar commentator makes this same point: "Apparently the only way to keep from being bombed is to have your men off bombing someone else." Major General James Harbord, "Review: *The Great Delusion. A Study of Aircraft in Peace and War* by 'Neon,'" *Naval Institute Proceedings*, 53:293 (1927), 906.

151 Ibid., 248.

152 Clarke, "Future-War Fiction," 392–393.

It is, therefore, not surprising that the subsequent history of bombing so closely resembles some aspects of the imaginary futures laid out in some of these texts.

In another respect, however, these texts can also be seen as attempts, as Slavoj Žižek argues, "to neutralize [trauma], to integrate it into the symbolic order."[153] As he explains,

> the sinking of the *Titanic* had a traumatic effect, it was a shock, "the impossible happened," the unsinkable ship had sunk, but the point is that precisely as a shock, this sinking arrived at its proper time – "the time was waiting for it": even before it actually happened, there was already a placed opened, reserved for it in fantasy-space. It had such a terrific impact on "social imaginary" by virtue of the fact that it was expected. It was foretold in amazing detail.[154]

Žižek is here referring to the fact that, in 1898, a book by Morgan Robertson, called *Futility*, had already recounted the story of an unsinkable ocean liner called the *Titan* being hit an iceberg and sinking. Civilization had already seen fit to dramatize the hubris of naively positing the superiority of human technology over nature. What made the sinking of the *Titanic* such a powerful historical event was its "symbolic overdetermination" – it was already anticipated and symbolized in the public imaginary prior to the historical occurrence. The historical event reified, and one might even say accentuated, in sublime perfection, the anticipated trauma. The *Titanic*, Žižek writes, "is a Thing in the Lacanian sense: the material leftover, the materialization of the terrifying, impossible *jouissance*. By looking at the wreck we gain an insight into the forbidden domain, into a space that should be left unseen: visible fragments are a kind of coagulated remnant of the liquid flux of *jouissance* a kind of petrified forest of enjoyment."[155] The sinking of the *Titanic* was the historical calcification of latent civilizational anxieties. Prefigured and anticipated, the event became the ecstatic performance of our psychic fear.

In a similar manner, nineteenth- and early twentieth-century future-war fiction writers were coping with their own forbidden anxieties, seeking to anticipate and neutralize the trauma and annihilation that was haunting civilization's collective dreams. As will be seen in the subsequent chapters, when politicians and planners speak about the terror of aerial bombing as the

---

153 Slavoj Žižek, *The Sublime Object of Ideology* (London: Version, 1989), 69.

154 Ibid.

155 Ibid., 71.

twentieth century unfolds, they are remembering not an event, but the near obsessive prefiguration of the event in future-war fiction. In some peculiar sense, the trauma of aerial bombing was already being processed in the late nineteenth and early twentieth centuries. Ironically, in the process of trying to stave off potential harm by alarming the public imaginary as to the impending doom of bombing, the authors of these fictional texts ultimately circumscribed and reserved a space for "the event" to materialize. In many respects, indeed, the twentieth century seemed to have been waiting for it.

4 Conclusion

At the dawn of the twentieth century, it was becoming increasingly clear that airships would have a profound effect on how wars were fought. While there were widely divergent expectations as to how aerial bombing would be integrated into military practice, it was clear that, in principle, they would pose a significant challenge to noncombatant populations. To many early aeronaut adventures, the devastation meted out to city centers would be what makes war so terrible and therefore unlikely. If the technical issues around navigation and load-bearing capacity could be addressed, no one residing in an enemy nation would be exempt from the hardships of bombing with the ultimate effect of making war much less likely. While these technologies were limited and untested, early science fiction writers imaginatively speculated about the future of war. Some believed these new weapons could be integrated into existing practices; others, however, presciently forecasted the terror with which airships would threaten civilian populations and how war would be irrevocably transformed. These early writings on air war would have a profound impact on the articulation of the moral language of aerial bombing in the interwar period, particularly in terms of the harm this technology posed to civilian populations.

CHAPTER 7

Interwar Approaches to Bombing: Two Discursive Traditions

1 Introduction

In the first decade of the twentieth century, aeronaut adventurers, inventors, futurists, future-war fiction writers, and other theorists interested in the emerging technologies of war expressed a range of predictions about what aerial bombing would look like. Some suggested that the technology would never be available – the technological limitations would prevent humans from using airship in war for bombing. Others suspected that when the technology was developed, it would be subjected to the same constraints as conventional weapons. Honorable nations would simply incorporate this technology into the existing paradigm of nineteenth-century international legal norms. There were also those who believed that the ascendant norms would not constrain this technology; the kind of destructive power now available to militaries, especially given the elusive prospect of decisive victory, would be too tantalizing to refuse. Due to the potential strategic advantage bombing offered, some countries would simply refuse to play by the rules. A final group of commentators optimistically believed that the annihilative power of bombing was a moral good. The technology would be so devastating that sensible states would find a way to bring about the end of war. The fear of mutual destruction would quite reasonably lead states to find other ways to resolve their differences. Aerial bombing would lead to peace.

The reality of aerial bombing in World War I did nothing to settle these disagreements. In fact, it exacerbated them. This is largely due to the technology proving to be immensely underwhelming. Of course, towards the beginning of the conflict, very likely with the future-war fiction in mind, many expected something much more apocalyptic. For instance, in January of 1915, Winston Churchill gave expression to a worry that stemmed from the intelligence reports he and his military command were receiving, namely that London would be imminently overwhelmed and bombed by a fleet of German airships. He writes:

> The Director of the Air Department reports that there are approximately twenty German airships which can reach London now from the Rhine,

 | DOI:10.1163/9789004515482_009

> carrying each a ton of high explosives. They could traverse the English part of the journey, coming and going, in the dark hours. The weather hazards are considerable, but there is no known means of preventing the airships coming, and not much chance of punishing them on their return. The unavenged destruction of non-combatant life may therefore be very considerable. Having given most careful consideration to this subject, and taken every measure in their power, the Air Department of the Admiralty must make it plain that they are quite powerless to prevent such an attack if it is launched with good fortune and in favorable weather condition.[1]

Likely recalling Wells's description of German aerial bombing in *The War in the Air*, Churchill's anticipation of the coming aerial war was positively dystopic.[2]

The reality of bombing was nothing like what those fiction writers envisioned. During World War I cities were bombed and civilians were killed, but in the scope of the total destruction the damage from aerial attacks was minor.[3] Mocking these early anxieties, the US inventor, Hudson Maxim, quipped: "It has been predicted that Germany would send across the Channel a large fleet of airships and blow up British towns with the bombs that her great gas-bags might drop out of the heavens."[4] But the result up to that point had left much to be desired: "Now, at last, the much-vaunted and long-anticipated Zeppelin invasion has come, and what is the result? Four peaceful citizens killed, and about ten thousand dollars' worth of property damage."[5] While these deaths

1 Winston Churchill, *The World Crisis, 1915* (London: Thornton Butterworth Limited, 1927), 62–63. This reflects the same fear following Napoleon's use of balloons for reconnaissance during the Peninsular Wars: "The use of balloons by the French in war soon created a panic among English alarmists. It was reported that England would be invaded by clouds of aerial monsters, which would burn the cities and destroy the crops." "A Chapter on Balloons," *Fraser's Magazine for Town and Country, Vol. XL July to December 1849* (London: John W. Parker, 1849), 39.

2 This is highly plausible since Churchill was friends with Wells and regularly borrowed lines from his books. See Sarah Cassidy, "Churchill 'borrowed' famous lines from books by HG Wells," *The Independent* 27 November 2006. https://www.independent.co.uk/news/uk/politics/churchill-borrowed-famous-lines-from-books-by-hg-wells-6229767.html (Accessed 15 September 2020).

3 Raymond Fredette tabulates air raid casualties from 1915–1918; roughly 1,414 people were killed and 3,416 were injured. Raymond Fredette, *The Sky on Fire: The First Battle of Britain, 1917–1918* (Tuscaloosa: The University of Alabama Press, 2007), 262.

4 Hudson Maxim, *Defenseless America* (New York: Hearst's International Library Co., 1915), 210.

5 Ibid. For a similar account, see "The War in the Air," in *Practical Warfare: Chapters on Armies and Navies in Action* (London: E. Nash, 1915).

were certainly unsettling and indeed enraging, they hardly represented the much-touted devastation predicted by futurists and future war fiction writers.

Given the pre-war hype surrounding aerial war, many commentators noted the degree to which the technology remained limited in application, as if the public had been let down by the descriptions in the many fantasy-fiction novels of the prior two decades. As Logan Marshall notes, "so far as bomb-dropping is concerned, the reality has so far fallen short of anticipations."[6] And in his 1919 report on the effectiveness of bombing in World War I, US Secretary of War, Newton Baker, explains:

> The actual loss of life caused by these bombardments was relatively small and the destruction of property, while large, had no appreciable effect upon the war-making power of either nation. Indeed, it may rather be said that the willingness of the enemy casually to slaughter women and children, and to destroy property of no military value or use, demonstrated to England and France the necessity of beating so brutal of foe, and it is more likely that history will record these manifestations of inhumanity as the most powerful aids to recruitment in the nations against which they were made.[7]

Rather than expediting a quick end to war, Baker argues that these weapons were strategically ineffective, and, as such, they remain something of a moral liability. Since the bombardments produced only a negligible effect overall, their military value was limited. In fact, early commentary on the aerial bombing of cities resembles the Union's criticism of the Confederate Army's burning of Chambersburg in the Gettysburg Campaign: it served no strategic military purpose as it was neither part of a campaign of occupation nor an attempt to render the territory uninhabitable. Furthermore, as it was currently employed, aerial bombing was far too weak to terrorize populations to the degree necessary to expedite surrender. Indeed, the underwhelming performance of aerial bombing was so pronounced that even as late as 1932 some commentators could still question whether airplanes would play a major role in future war at all.[8]

6 Logan Marshall, *Horrors and Atrocities of the Great War* (Philadelphia, PA: W. Ziegler, 1915), 267. See also Francis Greene, "Recent Changes in the Art of War," *Proceedings of the New York State Historical Association* 15 (1916); 156–175. 170.

7 "Annual Report of the Secretary of War," *War Department Annual Reports, 1919, Volume I* (Washington D.C.: Government Printing Office, 1920), 68.

8 Frank Thone, "No Flying Devastation for Cities," *Science Newsletter* 21:584 (1932); 386–387.

Some worried that the failure of arial bombing, in particular the degree to which it instigated a kind of ambivalence to air warfare, would prove fatal. H.G. Wells lamented that during a debate on the Air Estimates in the House of Commons in 1927, Commander Kenworthy raised the question of "teaching the use of gas masks to children in the infant schools," in response to which, the hall erupted in laughter.[9] Wells concludes, "Nothing could better illustrate the happy carelessness with which we move towards the next catastrophe. The air manœuvers over London this past summer have demonstrated clearly that it will be almost impossible to prevent the copious gassing of that great warren within a few hours of the opening of any new European conflict of first-class rank."[10]

Though the destruction from aerial attack had been far less severe than previously imagined, many writers still held steadfast to the belief that bombing would be much worse in the next war. For instance, G. Blanchon imagines the future of air warfare from the vantage point of 1917; he writes:

> Air squadrons will advance in cubic formations; they will blend in furious charges, dropping on to the plains below a deluge of bleeding and battered debris. The horror of these fights, which will darken the very sun itself, will surpass everything that man has ever known. The victorious fleet, quickly followed at a height of a few hundred feet by the heavy fleet of bombing and troop-transporting planes, will invade, like a huge flock of birds of prey, the territory of the vanquished foe, hurling all around fire destruction and death.[11]

Robert Reece also forebodingly surmises that since "eighty per cent of the German steel industry was within bombing range of the Allies" a coordinated bombing campaign of industrial centers would have quickly ended World War I; he continues, "If this group had been bombed on a large scale the source of supply of German guns and munitions could have been destroyed."[12] The message was clear – for the next war, the Allies would need more planes if they wish to disable Germany's war industry.

---

9 H.G. Wells, "The Possibility of War between Britain and America," in *The Way the World is Going* (London: Ernest Benn Limited, 1928), 160.

10 Ibid., 160–161.

11 G. Blanchon, *The New Warfare* (London: George Harrap & Co, 1917), 145.

12 Robert Reece, *Night Bombing with the Bedouins* (Boston: Houghton Mifflin, 1919), 10.

In his cool assessment of the development of aerial bombing technologies in World War I, W.O'D. Pierce cautions against overemphasizing the limitations of bombing during the war. He explains that while England had become more effective in its defense against aerial attacks as the war progressed, this should be viewed in a heavily qualified light. He writes:

> The success gained was only a success against an enemy who regarded aerial attack as a sideshow. If a serious and prolonged aerial bombardment on a definite industrial objective had been carried out, then the defense might well have failed in its object. Against an enemy fighting the last stage of war in which its major military plans had proved unsuccessful, the defense was sufficient to make badly organized raids on indeterminate objectives unprofitable.[13]

In a future war, where aerial bombing is not merely a sideshow – that is, where it is properly organized, well-directed, and prolonged – the results would be much more effective and indeed devastating. According to Pierce, no amount of aerial defense could defend against such an attack.[14]

This sentiment is perfectly encapsulated by Prime Minister Stanley Baldwin's famous speech "The bomber will always get through" (1932). Baldwin argued in front of the British Parliament that "there is no power on earth that can protect him from being bombed. Whatever people may tell him, the bomber will always get through."[15] Since defending a territory against a coordinated and persistent aerial attack is impossible – even if most of the bombers are struck down, some will get through – the only hope of survival is an aggressive offense, "which means that you have to kill more women and children more quickly than the enemy if you want to save yourselves."[16] Putting the enemy on the defense, destroying their industry and capacity for war-making, is the only

13 W.O'D. Pierce, *Air War: Its Technical and Social Aspects* (London: Watts & Co., 1937), 106.

14 This had been the standard view of military planners ever since H.G. Wells made the argument in 1907. Brigadier-General Groves, Director of Air Operation of the British Air Forces concluded in 1918 that: "The consensus of opinion in aviation circles is that local defence is of negligible value; that no adequate means of protection against aircraft attack are yet in view; that the best defence against such an attack is the aerial counter-offensive; and that the only effective deterrent to aerial aggression is the threat of reprisals in kind." Quoted in Florence Boeckel, *Between War and Peace: A Handbook for Peace Workers* (New York: Macmillan, 1928), 394.

15 Stanley Baldwin, "The Bomber will Always Get Through," *The Parliamentary Debates, House of Commons, Volume 270* (London: H.M. Stationery Office, 1932), 631.

16 Ibid.

way to ensure victory. Of course, Baldwin was not expressly advocating the bombing of women and children. Like many commentators of that period, his hope was that the coming generations would understand that it is in their best interest to build but *not* employ such weapons: "There are some instruments so terrible that mankind has resolved not to use them."[17] By making the resort to war so costly, the hope was that nations would not be foolish enough to take such a risk.

2 Interwar Period Debates on Aerial Bombing

The contours of the interwar period debates about the role aerial bombing would play in future wars largely followed the logical pathways set out in the last decade of the nineteenth century and the first decade of the twentieth century. Commentary by futurists, inventors, and future-war fiction writers all contributed to the urgency of the issue. It was as though the collective sentiment was that the future was teetering on a scale between apocalypse and world peace. After all, in this period the world was beginning to witness some of the most robust and indeed aspirational formulations of international law but also the most fantastical experimentation with technologies that would prove to be the most devastating in the history of war. Even given the anticlimactic application of bombing during World War I, it was becoming increasingly difficult to deny the looming destructive potential aerial bombing held, even if some outliers tried. W.E. Ellis begins his influential 1914 essay on aerial law and war by saying: "Now that the conquest of the air has become an accomplished fact and the adaptability of aircraft to hostile operations has been fully demonstrated, the question of the necessary modifications of, or additions to the laws of warfare becomes a matter of present interest and importance."[18] Most commentators on twentieth-century war simply assumed that aerial bombing would exist in some form and that these new realities needed to be confronted by the international community.

Broadly speaking, while there was still a great deal of variation in terms of how to understand the specific role bombing would take, there were two discourses that dominated the period, incidentally both of which were tinged with their own degree of optimism and fantastical thinking. On one side of

17 Ibid. For similar arguments see Sir Norman Angell, *The Menace to Our National Defense* (London: H. Hamilton, 1934).

18 W.E. Ellis, "Aerial-Land and Aerial-Maritime Warfare," *The American Journal of International Law* 8:2 (1914); 256–273, 256.

the debate, a growing crowd of theorists and international jurists argued that bombing needed to be aggressively regulated if not entirely banned by the international community.[19] Some in this tradition believed that aerial bombing could be limited by conventional agreements. The laws of war could be adapted to accommodate this destructive technology. In its most expansive form, peace activists and those pushing for total disarmament sought to structurally undermine the possibility of war through international agreements and institutions. Driving this liberal internationalist movement was the assumption that the existing norms governing land and sea bombardments, when applied to the air, would prove catastrophic. For instance, Christopher Thomson writes,

> For every combatant killed in action, ten civilians will quite probably be slaughtered in their workshops or their homes. A horrible prospect is thus opened up, which should make the most callous shrink. To appeal to no higher feelings, the futility of such methods, as a means of settling international disputes, should be apparent. Both victors and vanquished would be left with ruined cities, wide spread distress among the masses of the people, hospitals filled with the maimed and mutilated of all ages and both sexes, asylums crowded with unfortunate human beings whom terror had made insane.[20]

With these potentialities in mind, it would be in states' interests to come together and self-limit the destructive potential inherent to aerial bombing technologies.

Quite in contrast to the international liberals who believed the horrors of aerial bombing could be regulated by positive agreement, bombing realists held to another species of idealism. They continued to argue that military necessity, and the existence of a threat environment where all states had aerial bombers at the ready, would produce its own natural restraints. Some in this tradition believed that bombing might even still be used as a tool of decisive victory. By making wars quicker and more terrible they would ultimately make

---

19 Of course, at this early stage it was not clear what needed to be regulated. Ellis, himself, concludes that in air war the struggle between nations is for air supremacy. And once a nation obtains the upper hand in the air "Much of the 'fog of war' will be dispelled. If one belligerent loses his aerial fleet, his adversary, if well provided with aircraft, will attain such a decided advantage that he can expeditiously push his operations to a decisive conclusion, 'a consummation devoutly to be wished; from both a military and humanitarian point of view." Ibid., 261.

20 Christopher Thomson, *Air Facts and Air Problems* (London: John Murray, 1927), 26–27.

them less cruel. After all, from their perspective, it had been the more-or-less conventional land weapons of World War I that had led to "devastated zones … from 50 to 100 miles in width," not bombs dropped from airplanes.[21] Theorists in this tradition held out hope that aerial bombing would be, overall, far less destructive than conventional land war had been. In fact, the destructive potential of aerial bombing may prevent war altogether, making it a profound tool of peace.

While both traditions ostensibly aimed at peace, they could not have been at greater cross-purposes. The former sought to capitalize on the post-World War I attitude of internationalism and cooperation, which assumed that through international collaboration and coordination, conflict could be eradicated.[22] The formal development of transnational liberal institutionalism, in the form of the League of Nations, first became manifest in this period. In contrast, the legacy of military necessity that had been ascendant throughout the nineteenth century was adapted and updated to include the new reality of aerial bombing. Theorists in this tradition contended that peace and security were only possible through preparation. While tragic, states could limit the horror of war by targeting some sufficient minimum number of civilians. If wars would occur at all, they would be sharp and brief; but more than likely, the threat of aerial bombing would prevent the outbreak of war in the first place.

3 International Liberals: Regulation and Disarmament

While the delivery system was novel – rather than naval or artillery bombardments, bombs would now be dropped from airplanes – initially there was a sense that the logic of bombing could be subsumed under existing norms, especially if, as it was widely suspected, bombing would be employed tactically, as a form of support for land or sea operations. Just prior to World War I, theorists from the US Naval War College suspected that, "In the actual effect of a projectile there may be little difference when it is fired from a gun several

21 This is a prediction from F.W. Lancaster 1916 analysis in which he concludes that: "We are at present only on the threshold of a revolution which aircraft will ultimately bring about in the conduct of warfare." F.W. Lancaster, *Aircraft in Warfare: The Dawn of the Fourth Arm* (London: Constable and Company, 1916). 188.

22 For an account of the utopianism of this period, see E.H. Carr, *The Twenty Years' Crisis, 1919–1939* [1939] (London: Palgrave Macmillan, 2016). For a more recent account, see Zara Steiner, *The Lights that Failed: European International History, 1919–1933* (New York: Oxford University Press, 2005).

miles distant so as to fall within a certain area or dropped from an air craft a few hundred feet above the area."[23] But the authors conclude that despite the degree to which aerial bombing recapitulated conventional forms, it remains open the extent to which rules governing aerial warfare could be subsumed under existing law. The authors much less sanguinely predict, "These new conditions of possible warfare show that the rules for warfare on land and on sea may not be adequate for the regulation of conduct when the extended use of the air is involved in hostilities."[24] This, of course, was the lingering question – as they were framed did the laws of war adequately account for aerial bombing, or did they need to be entirely reformulated? It is often overlooked that from the perspective of World War I there was not a clear answer to this question.

From just prior to World War I throughout the interwar period, however, there were growing concerns that aerial bombing would fundamentally challenge the laws of war as they were practiced. At the Hague Convention of 1899, the contracting states had agreed to a provision put forward by the US envoy, Captain William Crozier, which effectively prohibited "for a term of five years, the launching of projectiles and explosives from balloons, or by other new methods of a similar nature."[25] Crozier was worried that bombs dropped from balloons "would fall, like useless hailstones, on both combatants and non-combatants alike."[26] At that point in time, almost a decade prior to Knabenshue's confetti-bombing experiment, the technology was still so formative and uncertain that Crozier's concern was not yet grounded in a technological reality. Since no state had airships, much less command of the air, there was no reason to oppose banning such weapons. As an act of goodwill, it was easy for states to ban something that they did not yet possess.

When the matter of aerial bombing came up for reconsideration at the Hague Convention in 1907, things had significantly changed, so much so that many states were no longer willing to take such a hard line in banning the prospective technology. In fact, France and Germany were in a minor arms race, with both nations building and testing airships. The ability to fly at increasingly higher altitudes and speeds, not to mention bombing capability, were all rapidly improving. With the limitations of the technology being rapidly overcome,

23 Naval War College, *International Law Situations with Solutions and Notes* (Washington DC: Government Printing Office, 1912), 61.

24 Ibid.

25 Declaration (XIV) Prohibiting the Discharge of Projectiles and Explosives from Balloons. The Hague, 18 October 1907. https://ihl-databases.icrc.org/ihl/INTRO/245?OpenDocument.

26 Quoted in William Hull, *The Two Hague Conferences and Their Contributions to International Law* (Boston: Ginn and Co., 1908), 78.

this led many to optimistically assume that as states gained more control over aircraft, aerial bombing could be integrated into the conventional laws of war. Joseph Baker summarizes this optimism when he explains that "there is no conventional prohibition of the use of aircraft for the bombardment or attack upon fortified or defended places."[27] As the language suggests, the assumption was that airships would be employed much like traditional siege bombardments had been, against defended places. Unlike the 1899 convention, which had expressly banned dropping bombs from balloons for five years, the 1907 agreement simply added a small clause to Article 25: "The attack of bombardment *by any means whatever* of towns, villages, habitations, or buildings which are not defended is forbidden." The "by any means whatever" was thought to be sufficient to cover bombs dropped from airships.[28] Incidentally, this formulation did not restrict the aerial bombing of *defended* villages and habitations – a loophole that would be exploited to full effect in later wars.

Very much in the tradition of the Hague Conventions, Sir Thomas Barclay, one of the most pronounced authorities on international law, in his 1919 handbook on the laws of war, rather curtly declares that "Military aircraft and airmen are bound by the rules that govern belligerents generally."[29] He engages in very little discussion of the difficulties aerial bombing technology would pose and merely insists on restrictions, such as requiring airships to warn in case of an impending bombardment.[30] These views were not uncommon. Despite some lingering anxieties, many commentators almost reflexively presumed that bombing could simply be incorporated into long-standing norms governing land and sea warfare.

The possibility of regulating aerial bombing was to be reassessed at the Third Hague Convention, which was slated to take place in 1914. Of course, with the outbreak of World War I, this meeting never took place. Nevertheless, the reality of mechanized flight created several pressing concerns that, according to many international jurists, could not wait. From the end of the Hague Convention of 1907 to the outbreak of World War I, there were half a dozen

27 Joseph Baker and Louis McKernan, *Selected Topics Connected with the Laws of Warfare as of August 1, 1914* (Washington D.C.: Government Printing Office, 1919), 590.

28 As Lee Kennett explains, "by 1914 only Great Britain had ratified the ban." Lee Kennett, *The First Air War, 1914–1918* (Free Press, 1991), 5. Thomas Holland argues that by adding this provision the Convention had inadvertently made the bombing norms *more* restrictive as Naval Bombardment had been given much greater latitude in terms of bombing non-defended coastal towns. See Thomas Holland, *Letters to* The Times *upon War and Neutrality* (New York: Longmans, Green, and Co., 1921), 62.

29 Sir Thomas Barclay, *Law and Usage of War: A Practical Handbook of the Law and Usage of Land and Naval Warfare and Price* (Boston: Houghton Mifflin Company, 1919), 1.

30 Ibid.

conferences held by various international societies all over the world which sought to regulate some aspect of flight and the problems it posed for armed conflict: the First International Conference upon Aerial Navigation (Paris, 1910),[31] the First International Juridical Conference for the Regulation of Aerial Locomotion (Verona, 1910),[32] the Comité Juridique International de l'Aviation (Paris, 1911),[33] and the Madrid Session of the Institute of International Law (Madrid, 1911). These meetings even continued as the Great War was in progress; for instance, the Pan-American Aeronautic Federation held back-to-back conferences (Santiago, Chile, 1916, 1917).[34] This list does not begin to account for the proliferation of domestic law around aerial navigation that was unfolding at this time.[35]

In these early conferences there were internal divisions between those who believed bombing could be regulated by the existing laws of war and those who were pressing for much stricter regulation. Take, for instance, the conclusions set out at the Institute of International Law's Madrid meeting in 1911. In their final statement, the participants decided that "aerial war is permitted, but on the condition of not presenting greater dangers than land or sea war for the persons or properties of the peaceful population."[36] In other words, states were free to use aerial bombing so long as they reproduced accepted norms.[37]

---

31 See John Cobb Cooper, "The International Air Navigation Conference, Paris, 1910," *Journal of Air Law and Commerce* 19:2 (1952); 127–143.

32 See Arnaldo de Valles, *Premier Congrès Juridique Internacional de La Réglementation de la Locomotion Aérienne* (Paris: L. Larose et L. Tenin, 1911).

33 *Premier congrès du Comité juridique international de l'aviation* (Paris: A. Pedone, 1912).

34 James Brown, "Pan American Cooperation in Aeronautics," *Journal of Air Law and Commerce* 9:3 (1938); 468–502.

35 For more on these developments, see Arthur Kuhn, "International Aerial Navigation and the Peace Conference," *The American Journal of International Law* 14:3 (1920); 369–381; R.F. Williams, "Developments in Aerial Law," *University of Pennsylvania Law Review* 75 (1926); 139–153. 142–44; Peter Sand, Jorge de Sousa Freitas, and Geoffrey Pratt, "An Historical Survey of International Air Law Before the Second World War," *McGill Law Journal* 7:1 (1944); 24–42.

36 Quoted in John Westlake, *International Law, Part II* (Cambridge, The University Press, 1913), 77. Westlake goes on to balk at this formulation: "But who shall measure the proportion of the dangers? And the true question is not whether the danger is greater than those to which mankind has become accustomed, but whether it is not a new danger to which there is no reason that mankind should be exposed." (Ibid.)

37 The assumption was that there were legitimate and illegitimate offensive operational choices to be made. Eric Stuart Bruce writes, "To use the advantage of travelling in the air at altitudes for the purpose of wanton destruction of harmless citizens, and, further, to destroy cities the amasses wealth of art that only centuries, not years, produce, is an unrighteous use of the science of aerial navigation." Eric Stuart Bruce, *Aircraft in War* (London: Hodder and Staughton, 1914), 154.

Not all were satisfied with these conclusions. The German international jurist, Ludwig von Bar, at that same conference, had proposed a much more aggressive plan to limit aircraft in war, which ultimately sought to reassert the much more stringent ban of the 1899 Hague Convention. His proposal states: "Article 1: In general, it is forbidden to use aerostats, balloons or airplanes as a means of destruction in combat. Article 2: However, if enemy aerostats, balloons or airplanes are fired at (by cannons placed on the ground or on board a vessel), they can defend themselves."[38] While some were sympathetic to Bar's suggestions, they ultimately gained little traction at the conference. In a synopsis of the proceedings, one commentator noted that "the opinion of most writers is to extend so far as possible the principles embodied in the rules for war on land and sea to the conduct of war in which aerial domain is involved."[39] At least in these early phases, overwhelming preference was given to the assumption that air war would function much like land war.

The central issue to these debates about incorporating airships into the conventional laws of war was range. The vast distances airships could now cover offered tantalizing prospects in terms of application. It certainly expanded the operational zone of conflict, and there was no guarantee that militaries would use airships simply as a new type of long-range artillery. Inasmuch as aerial bombing recapitulated the traditional modalities of bombardment, it could adequately be subsumed into existing norms, but as W.E. Ellis points out, there was little reason to think this would be the case:

> Both land and naval bombardments are continuations of immediate hostilities before certain places, whereas an aerial bombardment may be an isolated operation. Let us take a concrete case. Suppose the United States is at war with state "X." Land and naval forces are operating against Philadelphia, but as yet have not appeared before any other city. New York is technically defended, which it would be if its coast defenses were manned, even if no troops actually occupied the city. The "X" commander notifies the New York authorities that their city will be bombarded on or after noon the next day by an aerial fleet. Airships appear at noon and drop bombs on the Woolworth Building and the Stock Exchange. Is this legitimate warfare? Yes, according to the letter; but no, according to the spirit of The Hague regulations. This ambiguity suggests the advisability

38 Quoted in *International Law Situations with Solutions and Notes* (Washington D.C.: Naval War College, 1912), 72.

39 Ibid., 75.

> of formulating a rule to the effect that the aerial bombardment of non-military plants or buildings should be permitted only when "defended" places are actually attacked or invested by land or maritime forces.[40]

In other words, even granting the ambiguous classification of the places being bombarded (whether they are defended or not), what makes those assaults permissible is that they could be traced to a kind of intelligible sequence or progression. When a military makes an approach, there is a string of actions that make the advance to some extent intelligible and predictable. One piece of territory must be taken before the next can be assaulted. This makes the resort to bombardment in some sense part of a logical sequence. One could anticipate the way in which assaults would unfold and when, say, noncombatants should seek safety elsewhere (even if militaries increasingly angled for surprise and decisive attacks). In contrast, aerial bombing vitiated the logic of war. Attacks could now be entirely random; bombs could be dropped here and there, with no intelligible sequence in mind.

To be fair, these difficulties were not due exclusively to the advent of airships. Since the early twentieth century militaries had been experimenting with long-range artillery, which produced similar moral conundrums in military ethical theory. In a *Popular Mechanics* article from 1909 on the development of "aerial torpedoes" (primitive motorized bombs), the author optimistically concludes that "the aerial torpedo will outrange the biggest gun now in use, and it is possible that it will be of use against attacking fleets and coast fortifications, and will be even more effective than the aeroplanes itself in holding up rich cities for ransom."[41] Leading up to World War I, there were also multiple reports about experiments with remote-controlled dirigibles that could be used to bomb enemy cities.[42] Concurrent to the development of both-lighter-than-air and heavier-than-air vessels, governments were experimenting with other long range delivery systems which would ultimately culminate in Hitler's *Vergeltungswaffen* (revenge) weapons, the V1 and V2 "flying bombs" used by Germany in World War II.[43]

---

40 Ellis, "Aerial-Land and Aerial-Maritime Warfare," 263–64.

41 "Development in Aerial Torpedoes: The Latest Horror of Modern Warfare is Being Rushed by Armies of the World," *Popular Mechanics* (December 1909), 728.

42 See "Aerial Torpedo Boat," *Popular Mechanics* (May 1907), 523; "'Zeppelin' Armed with Aerial Torpedoes," *Popular Mechanics* (October 1909), 452; "Torpedo Airship Controlled by Wireless is the Latest Invention," *New York Times* May 22, 1910; "Torpedoes of the Air: Bomb-Droppers Directed by Wireless," *Illustrated London News*, September 6, 1913.

43 See Nigel Walpole, *Hitler's Revenge Weapons: The Final Blitz of London* (South Yorkshire: Pen & Sword, 2018).

At the outset of World War I, artillery could already reach distances of about 25–35 kilometers (15–22 miles). Near the end of the war, the Germans had developed siege guns, the Paris Gun, that could hit targets up to 130 kilometers (81 miles) away. The concerns produced by long-range siege gun's strike capability would arise again within the context of aerial bombing. In fact, when the bombardment of Paris commenced on 23 March 1918, many Parisians assumed that the city was being bombed by German airships. As Lieutenant Colonel Henry Miller writes, "Until that time Paris had never been bombarded except from airplanes and Zeppelins, and the first thought of the people was that they were being bombarded from some new type of aircraft that was being operated at such a height that it was practically invisible."[44]

These long-range guns were themselves putting pressure on the norms governing the distinction between what might be called "intentionally indiscriminate" bombardment and "unintentionally indiscriminate" bombardment. While militaries were generally willing to see the intentional bombing of residential districts as necessary and in many cases justified, there was some ambiguity as to whether long-range bombing was covered by these traditional norms. As M.W. Royse explains, "The question of indiscriminate bombardment, however, raises another question, whether such shelling is intentionally indiscriminate or whether the long range [merely] results in such an end."[45] Of course, in practice, there was little difference between being intentionally indiscriminate or just seeming so. As Royse continues,

> Devastation in range and direction, even in modern times under the most improved methods and ideal conditions, results in a scattering of shots or dispersion, which is always broad. A fair estimate of the dispersion in a twenty-five-mile trajectory would be anywhere from 1000 to 4000 yards. In other words, artillery fire at that range, once the target was "straddled," or "bracketed," would cause a zone of fire covering an area two and a half miles in length and a half mile in width.[46]

44 *A Report on the Characteristics, Scope of Utility, Etc. of Railway Artillery, Volume 1* (Washington DC: Government Printing Office, 1921), 723. See also, Henry Miller, *The Paris Gun: The Bombardment of Paris by the German Long Range Guns and the Great German Offensives of 1918* (New York: Jonathan Cape & Harrison Smith, 1930).

45 Royse, *Aerial Bombardment*, 170. Incidentally, this is a book that Spaight favorably reviewed. See J.M. Spaight, "Reviewed Work: *Aerial Bombardment and the International Regulation of Warfare*. by M.W. Royse," *Journal of the Royal Institute of International Affairs* 7:6 (1928); 432–33.

46 Royse, *Aerial Bombardment*, 170–71.

According to the laws of war as they stood, intentionally indiscriminate bombing was never questioned, neither in sieges nor in conventional artillery barrages. A two-and-a-half-mile zone of target calibration was acceptable. What was questionable about long-range bombardment to some was the degree to which it was becoming *unintentionally* indiscriminate. Given the technological limitations of these long-range attacks, i.e. the degree to which bracketing targets was necessary, this meant that zones of operation were exponentially increasing, in many cases at the expense of precision. The theater of conflict was drastically expanding, and the risk of unintentional, promiscuous destruction was increasing.

Bombing from planes did nothing to mitigate these concerns. As many of the early commentators on aerial bombing note, planes could reach deep within the territory of an enemy nation, much farther than even the longest-range guns.[47] As one interwar commentator notes,

> On that continent the increase in the range, carrying capacity, speed, and general efficiency of aircraft, together with the constant growth in their numbers and in the potentialities of production, implies that on the outbreak of another war whole fleets of aircraft will be available for offensive purposes. Each side will at once strike at the heart and nerve centres of its opponent; at his dockyards, arsenals, munition factories, mobilisation centres, and above all at those nerve ganglia of national morale, the great cities. The air raids of the past are no guide as to the nature of future aerial attack, or even of those which could be delivered to-day.[48]

It was this development in long-range strike capability, and the ability to isolate single cities that would typically fall outside of normal military operations, that most needed to be grappled with and incorporated into the norms of warfare. Not only did killing from a distance put a wider swath of civilians in the line of fire, but as it stood, the range that aircraft opened up now threatened to obviate many of the formalities that had governed the well-choreographed norms of nineteenth-century siege war: the logistical demands of the approach and investment prior to laying siege, the duty to give advanced notice prior

---

47 Just to illustrate the extreme leap in range, Francis Arnold Collins reports that German Zeppelins could "sail at a height of one mile, develop a speed of eighty miles an hour with an effective radius of one thousand miles and drop two tons of explosives, meanwhile repelling attacks." Francis Arnold Collins, *The Air Man: His Conquests in Peace and War* (New York: The Century Co., 1917), 134.

48 R.C. Groves, "The Influence of Aviation on International Relations," *Journal of the Royal Institute of International Affairs* 6:3 (1927); 133–152, 144.

to bombardment (unless taking the fort by surprise), the requirement that women and children be allowed (in certain cases) to leave a besieged place, the correspondence between enemy commanders and the readiness to accept a besieged city's surrender, and so on.[49] Aerial approaches vitiated these traditional norms. Targets deep in the interior of a nation could now be hit at any moment, without warning. And even if warning could be given, it was not clear what this would mean or how an interior city could surrender to an airship, even if it wanted to.[50] Unlike those much simpler formulations in future-war fiction, airships could not simply land and occupy a city it threatened to bomb. Nevertheless, despite these important differences, many of the early theorists of aerial bombing were quite explicitly drawing on the permissibility of bombing residential districts of besieged cities to expedite surrender.

Early strategists and theorists were aware of the ways in which air power transformed the spatial dimensions of the battlefield, but it was less clear how the laws of war could accommodate these changes.[51] Throughout the nineteenth century, much as it was the case since the sixteenth century, fortified cities were hubs of economic and political power. The objectives of land war concerned physically occupying or destroying those key pressure points that enabled the enemy to prosecute war. In justifying the bombardment of Charleston during the US Civil War, General Foster explains: "To destroy these means of continuing the war is therefore our object and duty."[52] If one looks at

49 In fact, as will be seen, some of the many accusations against Germany after World War I were that they were bombing without giving advanced warning and as such they were guilty of the "aerial bombardment of open towns without there being any regular siege." See "Commission on the Responsibility of the Authors of the War and on Enforcement of Penalties" *The American Journal of International Law* 14:1/2 (1920); 95–154. 113.

50 These concerns have become increasingly common in late twentieth- and twenty-first century warfare with the rise of drones. In the First Gulf War, Iraqi soldiers attempted to surrender to a drone. See Ted Shelsby, "Iraqi soldiers surrender to AAI's drones," *The Baltimore Sun* 2 March 1991, https://www.baltimoresun.com/news/bs-xpm-1991-03-02-1991061100-story.html (Accessed 15 September 2020). This became a more pointed issue in the Second Gulf War when, as reported through WikiLeaks, Iraqi insurgents were killed after trying to surrender to a US Apache helicopter. Russ Wellen, "WikiLeaks: 'You cannot surrender to an aircraft' (at least not Crazyhorse 18)," *Foreign Policy in Focus* 25 October 2010. https://fpif.org/wikileaks_you_cannot_surrender_to_an_aircraft_at_least_not_crazyhorse_18/ (Accessed 15 September 2020).

51 As William Wherry noted, "Today we find that the newest invention put into the hands of man, namely the airplane, threatens to check the entire development of the law and compel the reversion of civilization to the primitive conditions existing before the fundamental, humane principle was formulated." William Wherry, *Aerial Warfare in International Law* (New York: New York University School of Law, 1939), 1–2.

52 "General Foster's Reply," in *The Rebellion Record: A Diary of American Events, Volume 11*, 592.

the Franco-Prussian War, for instance, the strategic aims of the Prussians were quite clear: they systematically assaulted every major fortified city on their way to Paris, resulting in nearly two dozen sieges. By occupying these places of power, France was simply no longer able to prosecute the war. In this sense, there was a natural correlation between a city's influence on the war effort and the degree to which it was defended. By the twentieth century, the association between a city's economic influence and its defense had weakened.

The complexities of the modern war economy in terms of production and distribution challenged the ascendant norms inasmuch as the strategic logic of targeting was transforming. These early twentieth-century developments in war production exacerbated disagreements about what counted as a defended city, which had always been a consequential, if ambiguous, classification in international law. As seen above, leading up to World War I, there was still a great deal of conceptual and moral ambiguity concerning the acceptable approach to bombing towns from airships versus bombarding towns with siege cannons. Rather than bombarding with the ultimate aim of occupation, aerial bombing was directed at either rendering faraway places – industrial and communication centers, transportation and munitions hubs – inoperable or disrupting production enough to create a tactical advantage – a tradition that can be traced back to the bombing of French coastal cities by the English during the Nine Years' War. Whether a town was technically defended or not seemed to matter less than the degree to which it provided services that enabled the country to continue to prosecute the war. As C.M. Piccoto notes, "It is submitted, then, that the rule of law permits of the dropping of bombs from aircraft upon defended and on fortified places, and (according to the British War Manual) upon the undefended parts of fortified places, and upon any point of military, strategic, or economic importance. The two conditions of the defendedness of the place and its military importance must coexist."[53] In contrast to long-standing norms, bombing was now extended to places that were not, strictly speaking, defended.

Of course, not everyone approved of such permissibility in the bombing of towns. As James Garner points out, there was something unsettling about claiming that *any* defense made a town defended. Did it, for instance, matter if a town could not reasonably defend against an aerial attack? He writes:

53 C.M. Picciotto, "Some Notes on Air-Warfare," *British Institute of International and Comparative Law* 15:2 (1915); 150–155. 153.

> It may very well be doubted, however, whether the distinction between a 'defended' place and an 'undefended' place, as those terms are used in land warfare, is a reasonable and logical test for determining liability to bombardment by an air ship. The test applied in land and naval warfare is reasonable enough and is based on practical considerations. But no place can really be said to be 'defended' as against bomb-dropping from airships, by the mere presence of troops or ordinary artillery or the existence of fortifications. To be defended in any real sense against aerial attacks a town must possess artillery especially constructed for bringing down aircraft.[54]

Very few commentators, however, were willing to affirm the subtleties of Garner's distinction. Most authors assumed that any defense, no matter how marginal or meager, even, for example, small-arms fire, was sufficient to justify aerial bombing. And, of course, many commentators simply elided the question of a place's defense and instead emphasized its strategic value. For instance, J.M. Spaight writes

> I can see nothing in international law to prevent an hostile air craft force from dropping bombs on Chelsea, Wellington, Albany, or Knightsbridge barracks, or on the clothing factory or depot at Pimlico, or on Euston, Kings Cross, Waterloo, and other great railway termini. Many commercial undertakings which hold orders for the war department or admiralty would be liable to bombardment also. So probably would the war office and the admiralty and the headquarters of the eastern command and the London district. The various territorial force headquarters all over London also appear to be possible legitimate objects of attack.[55]

Much like General Foster's justification for bombing Charleston, *any* industry that fed into the war effort was a legitimate pretense for targeting. Under these conditions, just about *every* city could reasonably be construed as a legitimate target.

With the rise of aerial bombing, countries themselves were starting to be conceptually transformed into fortified cities writ large, and as such the tidy spatial demarcations of eighteenth- and nineteenth-century battlefields were no longer sufficient to restrain the totalizing effects of war. The industrialized

---

54 James Garner, *International Law and the World War* (New York: Longmans, Green and Co, 1920), 469.

55 J.M. Spaight, *Air Craft in War* (London: Macmillan, 1914), 20.

European states at war in the early twentieth century more closely resembled besieged cities than countries with designated militarized spaces within them, i.e. the fortified pressure points of military and economic power. Claude Grahame-White and Harry Harper make this transformation explicit:

> not only navies and armies, but entire nations, fight nowadays. The whole of a country has to be organized to supply war materials, seeing that such vast quantities are required. And if hostile aircraft fly over such a country, and drop their bombs, it is practically impossible to frame any rule as to where bombs are permissible and where they are not. The nation is in arms for the prosecution of the war; all the national organization is directed towards that end.[56]

They go on to argue:

> in warfare on its modern scale, almost every man in a country, and a large proportion of women, are doing war work, and are combatants in the sense that if it were possible for the enemy to kill them with a bomb from the air, their death would have an effect more or less prejudicial on the output of some war munition, or the produce of food-stuffs, or in the business of transport and supply.[57]

The ready-made analogy of the nation-state to the fortified city contributed to the widespread assumption that the targeting of residential districts might expedite surrender. And, of course, great feats of military engineering, like the Maginot Line (1928–1938), 480 kilometers (300 miles) of underground defenses built to defend France from German aggression, did nothing to mitigate these concerns. As a *Popular Science* article from just prior to World War II explains: "The frontier presents something new in the science of fortifications. It is a practically continuous defense system, as nearly as possible impregnable, extending along the eastern frontiers of France and Belgium."[58] France itself had become a fortress.[59]

---

56 Claude Grahame-White and Harry Harper, *Air Power: Naval, Military, Commercial* (New York: Frederick A. Stokes Company, 1917), 201.

57 Ibid.

58 Thomas Johnson, "Underground Fortresses Guard France from Invasion," *Popular Science* 129:4 (1936); 14.

59 These trends can be traced back to the French Revolution. As Thomas Carlyle reported, during the revolution "The entire Republic is one vast besieged city." Thomas Carlyle, *The French Revolution: A History, Part 2* (New York: Wiley and Putnam, 1847), 300.

Given these trends, it is not surprising that many commentators started to take the question of regulation much more seriously. In fact, the urgent need to limit the excesses of aerial bombing before the next war runs throughout much of the interwar commentary. Andrew Carnegie, and others involved in the peace movement, led the valiant effort to entirely ban aerial weapons. He writes, "I have hitherto been concerned in regard to the airship becoming an instrument of destruction, like submarines, and have expected a movement for debarring them from being used for projectiles or bombs which could be dropped from above to destroy."[60] The US Ambassador to France, William Sharp, also worried about the direction aerial warfare was going. At a 1918 lecture, he explained: "The very horrors of its abuses will compel the civilized nations hereafter to abandon forever the employment of aircraft for the casting of bombs. As it is a barbarous method of murdering innocents and is lacking in military value, it should be outlawed by the coming league of nations."[61] But there were many challenges to this approach. To begin with, it was not clear that states would willingly refuse a weapon that might be their only chance to deter enemy aggression.

Preliminary post-war efforts to ban aircraft were largely directed at the recently defeated Germany. During one of the "Council of Ten" deliberations at the Paris Peace Conference (1919), Marshal Foch forcefully argued: "no delay should occur in depriving the Germans of their military aircraft which had wrought such evil work throughout the war in connection with the unjustifiable bombing of open towns. The Conference should, therefore, adopt the principle of the Aerial disarmament of Germany."[62] And, indeed, these sentiments were widely shared. Part XI, Articles 313 to 320 of the Treaty of Paris (1919), a section dedicated to aerial navigation, set out multiple provisions that would limit Germany's ability to persist with its air power.[63]

The international community, however, did not stop there.[64] Perhaps the most ambitious attempt to regulate aerial bombing up to that point came in

60 Quoted in Henry Woodhouse, *Textbook of Aerial Laws* (New York: Frederick A. Stokes Company 1920), 149.

61 Ibid.

62 "Minutes of a Meeting of the Supreme War Council Held in M. Pichon's Room at the Quai d'Orsay, Paris, March 12th, 1919 at 3 pm." in *Papers Relating to the Foreign Relations of the United States: The Paris Peace Conference 1919, Volume 4* (Washington D.C.: Government Printing Office, 1943), 337.

63 "Aerial Navigation" Part XI, Articles 313 to 320, in *Papers Relating to the Foreign Relations of the United States: The Paris Peace Conference, 1919, Volume 13* (Washington D.C.: Government Printing Office, 1947), 642–646.

64 For instance, see "Convention Relating to the Regulation of Aerial Navigation, 1919," *Journal of Air Law and Commerce* 1:1 (1930).

1921 at the Conference on the Limitation of Armaments, which took place in Washington D.C. There the attendees made note of several contradictions in international law regarding aerial bombing that required clarification and, if possible, direct action. In fact, Elihu Root, the former US Secretary of State who had taken an active role in the Second Hague Convention, offered a useful preliminary account of why these tactics need greater international scrutiny. He writes:

> when the rule limited bombardment to defended towns, when it prohibited the bombardment of undefended towns, it had reference to military or naval operations against towns that afforded military obstacles to those operations, and as to those towns the provision was that the Commander should notify the defended place, so that the civilians might have an opportunity to withdraw. As to the undefended towns, no one must bombard them at all.[65]

As seen above, many commentators had pointed out that aerial bombings did not fit into this traditional operational framework. Root goes on to explain that even though most cities could not defend against aerial bombing they might have some perimeter defense. By this logic, just about every town in Europe could be bombed. This, of course, led to the exact opposite conclusion of the persistent though problematic efforts to spatially delimit the nineteenth-century battlefield. Additionally, Root continues that the spirit of international law had already provided good reasons to ban certain kinds of aerial bombardment. He points out,

> When one considers these two rules, that a defended town must not be bombarded without notice sufficient to enable the innocent – the women and the children and noncombatants – to withdraw, and that an undefended town must not be bombarded at all, when one considers those two rules, the spirit of them could prevent aircraft from bombarding any town whatever. Bombard a railroad junction, a station crossing? Yes. Bombard a munitions factory? Yes. But the center of an innocent population? No, not under any circumstances at all.[66]

65 *Conference on the Limitations of Armament, Washington November 12, 1921–February 6, 1922* (Washington D.C. Government Printing Office, 1922), 804.

66 Ibid., 804.

Given these tensions, many of the other delegates agreed with Root that the international community needed to formally revisit the laws of war as they pertain to aerial bombing. To get this process started, the committee appointed representatives from the US, Great Britain, France, Italy, and Japan to draft an update to the laws of air war.

Presided over by John Bassett Moore, one of the leading American authorities on international law, the subcommittee responsible for reassessing the rules of air war drafted the (nonbinding) Hague Rule of Air Warfare (1923).[67] These Draft Rules were driven by two principal questions:

(a) Do existing rules of international law adequately cover new methods of attack or defense resulting from the introduction or development, since The Hague Conference of 1907, of new agencies of warfare?

(b) If not so, what changes in the existing rules ought to be adopted in consequence thereof as part of the law of nations?[68]

In terms of their initial findings, the commission noted, much as Root had previously indicated, that the agreement signed at the Hague in 1907 *did* sufficiently prohibit dropping bombs from balloons, but they lamented that the number of signatories to the agreement was too small to be effective, and there was no assurance that countries would honor the agreement if war were to commence. To this, they concluded that "A general agreement, therefore, on the subject of bombardment from the air is much to be desired."[69]

Along these lines, the commission put forward several proposals designed to curtail some of the more devastating implications of air war predicted above. For instance, Chapter IV, Article 22 indicates: "Aerial bombardment for the purpose of terrorizing the civilian population or damaging private property not of a military character, or of injuring non-combatants is prohibited."[70] The policy of intentionally bombing residential districts to terrorize them into

67 This is not to be confused with the International Air Conference of 1923 (London). This conference was dedicated to more technical aspects of flight, and therefore did not touch on the more contentious aspects of the laws of war. Col. W. Lockwood Marsh (ed.), *International Air Congress, London, 1923: Report* (London: Offices of the Royal Aeronautical Society, 1923).

68 "Radio and Aerial Commission of Jurists to Consider and Report upon the Revision of the Rules of Warfare – General Report," *International Law Documents* (Washington D.C.: Government Printing office, 1926), 96.

69 Ibid., 119.

70 The Draft Rules of Aerial Warfare can be found in Part II of the Rules Concerning the Control of Wireless Telegraphy in Time of War and Air Warfare, Drafted by a Commission of Jurists at the Hague, December 1922–February 1923. https://ihl-databases.icrc.org/applic/ihl/ihl.nsf/Treaty.xsp?documentId=B9CA3866276E91CFC12563CD002D691C&action=OpenDocument (Accessed 11 May 2021).

surrender, while traditionally accepted under the conventions of siege warfare, was beginning to be challenged. Article 24 stipulates:

1) Aerial bombardment is legitimate only when directed at a military objective, that is to say, an object of which the destruction or injury would constitute a distinct military advantage to the belligerent.
2) Such bombardment is legitimate only when directed exclusively at the following objectives: military forces; military works; military establishments or depots; factories constituting important and well-known centres engaged in the manufacture of arms, ammunition, or distinctively military supplies; lines of communication or transportation used for military purposes.
3) The bombardment of cities, towns, villages, dwellings, or buildings not in the immediate neighborhood of the operations of land forces is prohibited. In cases where the objectives specified in paragraph 2 are so situated, that they cannot be bombarded without the indiscriminate bombardment of the civilian population, the aircraft must abstain from bombardment.
4) In the immediate neighborhood of the operations of land forces, the bombardment of cities, towns, villages, dwellings, or buildings is legitimate provided that there exists a reasonable presumption that the military concentration is sufficiently important to justify such bombardment, having regard to the danger thus caused to the civilian population.[71]

In certain respects, these regulations do not drastically depart from the original principles that had been set out at the Second Hague Convention. While they certainly seek to foreclose some of the excesses that had been permitted with naval and land artillery, in doing so new ambiguities arose (more on these below).

This document can also be read as a remarkable adaptation of the existing law governing bombardments. Militaries could not bomb for the sole purpose of terrorizing civilians. There needed to be some clearly recognizable military objective that would produce a distinct advantage. And even the notion of "military advantage" was now limited to the area of "the immediate neighborhood of operations." Notice also that militaries are only permitted to bomb towns with sufficient concentrations of militarily relevant infrastructure. Only when such infrastructure exists would the terror that ensues from bombing be permissible. In this respect, terror, as an ancillary benefit, is not prohibited as such. Of course, this distinction would go on to produce its own difficulties. As the scholarship on aerial bombing during World War II attests, one would

71 Ibid.

be hard-pressed to clearly differentiate between precision bombing, which sought to only hit military targets, and terror bombing, which was the deliberate attempt to deplete civilian morale. The Draft Rules failed to anticipate the degree to which limitations of technology and the diffuseness of military infrastructure would make it virtually impossible to make this highly idealized distinction a viable one in practice.

Furthermore, as seen in the early commentaries of bombing, there was always some ambiguity around the meaning of "indiscriminate." Historically, there had been a meaningful distinction (even if inscrutable and obscure to modern sensibilities) between being deliberately indiscriminate, as had been the case with targeting residential districts in siege warfare, and unintentionally indiscriminate, as was the case in the early days of dropping bombs from balloon platforms. As will be seen, documents like this began to contribute to a subtle inversion of this older formulation. Under these proposed rules, targeting now needed to strive for discrimination, which is to say that bombs needed to be directed exclusively at sites of legitimate military interest. In siege warfare, this problem had been simplified; the entire fortified city (including the detached suburbs) is the target. With bombing and the fortification of entire countries, this began to undermine the plausibility of nineteenth-century norms which were designed to limit destruction.

Furthermore, the technological limitations of bombing accuracy made the language of discrimination rather perplexing. Since in most cases there was no way to practically differentiate between those attacks which were striving to discriminate and those that were not, the morally significant variable started to become the *intention* of those engaging in the bombing – i.e. the planners and bombers simply needed to believe that they were dropping bombs on a legitimate targets. Considering all the new epistemic difficulties that bombing posed, it became morally relevant that militaries were making a reasonable effort to hit only military targets or those of strategic value, even if this were technologically implausible and could not be distinguished in practice from terror bombing.

It is also noteworthy that the Draft Rules attempted to do away with the persistent ambiguity that had existed between fortified and non-fortified, and then defended and open towns. These rules recognize the degree to which legitimate military objectives could be diffusely spread throughout an enemy nation. Tacitly, at least, it no longer mattered if the town was defended or not; what mattered was the existence of an object of military interest and the degree of advantage afforded to the belligerent by destroying it.

One of the truly novel features of this formulation is that it attempted to limit bombing to the "immediate neighborhood of operations of land forces."

The premise of this provision seems to be that bombing should only be used for tactical support during conventional assaults and not random bombing raids on cities located in the interior. This may seem to be an excessively restrictive prohibition. Indeed, many believed it was this overly restrictive element that ultimately led the international community to reject these rules. But such concerns fail to appreciate how the increasing range of artillery, not to mention the vast distances land operations now covered, thoroughly undermined any coherent or intuitive meaning of "immediate neighborhood of operations." As William Mitchell would point out in his defense of air war, "The advent of air power has made every country and the world smaller."[72] In practice, given the realities of total war, this supposed limit turned out to be no less vague than the prior distinctions of defended versus open towns.

Along these lines, one of the early commentators on these Draft Rules, Paul Williams, made two important observations about the weakness of the document that highlight some of the murky aspects of these developments that will be explored below. First, he pointed out that despite the specificity of legitimate military targets laid out in Article 24, Section 2, the attempts at distinguishing between civilian and military infrastructure was "a halfway measure, worse than useless, for it permits a belligerent to do under the guise of legality what he might otherwise scruple to do. It allows an interpretation which emasculates the whole series of prohibitions contained in the first part of the article."[73] Since there is no way to differentiate terror bombing from the bombing of legitimate military objects, Williams argues that these rules open the door to *de facto* terror bombing. Militaries just need to *say* that they were intending to hit military targets. These rules incentivized militaries to frame bombing campaigns in the language of positive intentions, giving them exculpatory cover when the practical effect of their bombing was no different than terrorism.

Secondly, Williams points out these rules failed to appreciate the technological capabilities of planes and the limitations of pilots when these formulations were advanced. He continues:

> A pilot may not always be aware of the immediate proximity of a large civilian population; he may overestimate the accuracy of his aim or he may not intend to be accurate, and who shall judge of his intent? In view

72 William Mitchell, *Winged Defense: The Development and Possibilities of Modern Air Power – Economic and Military* (New York: G. Putnam's Sons, 1925), 20.

73 Paul Williams, "Legitimate Targets in Aerial Bombardment," *The Journal of International Law*, 23:3 (1929). 570–581. 577.

> of the fact that intention is generally better than the results, as evidenced by mutual recriminations in the recent war, you cannot in any given case judge with certainty of the intent by the results. The tendency will be for each belligerent to become as convinced of the baseness of his enemies' intentions as of the purity of his own, and the bombed, being in the best position to judge of the damage done to innocent lives and property, may conceivably be goaded to indiscriminate bombardment by way of retaliation.[74]

In other words, these Draft Rules presume a degree of epistemic certainty that did not reflect the reality of bombing as it was practiced. Given the experiences of World War I, it was becoming increasingly clear that a pilot might unintentionally kill civilians in ways that were unlikely in previous eras. Williams goes on to write, "Can a bomber always say beforehand whether or not his operations will result in the indiscriminate bombardment of the civilian population, especially if he must fire from a height of from ten thousand to fifteen thousand feet, which was the ordinary height in the last war?"[75] In short, the conceptual weight placed on the intention to discriminate was vitiated by the practice of bombing. Rather than being deliberately indiscriminate over a limited space, armies could now be unintentionally indiscriminate over a whole country, that is, if they had good intentions.

The moral shift embodied in the Draft Rules that placed greater emphasis on the intention to discriminate – to only target and hit military-related targets – was undermined by the limitations of the technologies being employed. Using spotlights or flares, flying with rudimentary navigation systems, employing primitive bombsights, and attempting to hit targets at night or from altitude all made unintended civilian casualties much more likely, if not inevitable.[76] To Williams it was both a technical problem and a failure of the international community to clearly demarcate civilian from military targets (which, as seen above, was becoming increasingly difficult to do). He writes: "The problem, therefore, is to delineate a way in which aircraft may operate most effectively, and still distinguish combatants and private property which assists in the conduct of the war from non-combatants and private property in general."[77] Indeed, another critical commentator notes, "The Commission's rules would

---

74 Ibid.

75 Ibid., 578.

76 For more on the development of these technologies, see Timothy Moy, *War Machines: Transforming Technologies in the U.S. Military, 1920–1940* (College Station: Texas A&M University Press, 2001).

77 Ibid., 579.

have been put to less strain had its members taken 'judicial notice' of the fact that an aviator at ten thousand feet cannot distinguish an art museum from an armory, or an arsenal from an academy."[78]

At the conclusion of the Conference on the Limitation of Armaments, for reasons that are not entirely clear, "Not one State, not even those represented on the Commission itself, signed any such agreement; not even a conference to discuss the Hague Rules of Air Warfare was arranged."[79] These provisions seemingly went nowhere.[80] A commentator reflecting on the history of these proposals, just as World War II was unfolding, noted that by the time the commissions passed its Draft Rules "the very potency of the airship as an instrument of destruction in war was such that it was obvious States would never totally renounce its employment for bombing purposes."[81] In this sense, the rules can be seen as being underscored with a kind of fatalism. Despite the elasticity and ultimate imprecision of the rules, nations were unwilling to limit the potential use of this technology.

As the interwar period progressed, these issues were becoming more pressing. Beyond the civilian bombing casualties in World War I, aerial bombing had become a staple of police action and military campaigning against colonial populations, much as future-war fiction had predicted. The British were extensively resorting to aerial bombing to police their colonies: to suppress riots in Gujranwala in the Punjab region of India (1919),[82] to overpower local resistance in Kabul and Jalalabad, Afghanistan (1919),[83] to scare off Bolshevik

---

78 K.V.R.T. "Aerial Warfare and International Law," *Virginia Law Review* 28:4 (1942); 516–527. 524.

79 Heinz Marcus Hanke, "The 1923 Hague Rules of Air Warfare," *International Review of the Red Cross* 33:292 (1993); 12–44. 19.

80 Indeed, even as these talks were progressing, the US continued developing its bombing capability. See *Military Airplane Data Prepared for the Select Committee of Inquiry into Operations of the United States Air Services, Volume 5* (Washington D.C.: Government Printing Office, 1925).

81 K.V.R.T. "Aerial Warfare and International Law," 524.

82 In a heated Parliamentary inquiry, Colonel Josiah Wedgewood pressed Edwin Montagu, the Secretary of State for India, if he would "issue directions that such, weapons are not to be used in future against civilian populations?" When Montagu responded that he believed bombing was an effective tool for carrying out martial law, Wedgewood replied: "In view of the fact that these aeroplanes cannot drop bombs accurately, and that therefore in large towns bombs are about certain to hit the wrong people, would it not be more humane to employ the other efficient means of defence in our power?" He would go on to press for details about how many civilians had been killed, to which Montagu could not respond. "Bombing by Aeroplanes," House of Commons Debate 28 May 1919 v.116 cc1183–4.

83 "Afghanistan," House of Lords Debate 28 October 1919 vol. 37 cc43–52.

agitators in Anzali, Iran (1920),[84] to suppress the Dervish fighters of Mohammed Abdullah Hassan in Somaliland (1920),[85] to disperse protesters in Baghdad and Nasiriyah, Iraq (1920, 1922),[86] to punish Hottentot villagers in South Africa (1922),[87] and to discipline resistant tribes in Waziristan, Pakistan (1923).[88] They made persistent aerial raids throughout the Aden Protectorate in present-day Yemen (from 1928 up to World War II), and they regularly used bombing to suppress civil unrest in Palestine (from 1936 up to World War II).[89]

Of course, Britain was not alone in the use of aerial bombing to police colonial holdings. The Italians bombed Harar, Jijiga, and other sites throughout Ethiopia (1936), which included some of the first instances of chemical weapons being dropped from planes.[90] The Germans and Italians engaged in a series of bombing raids during the Spanish Civil War, hitting Guernica, Durango, and Barcelona (1937, 1938).[91] The Japanese had invaded Manchuria

---

84 Houshang Sabahi, *British Policy in Persia, 1918–1925* (London: Frank Class and Co., 1990), 108.

85 Supplement 32107, *The London Gazette*, 29 October 1920, 10589.

86 "Middle Eastern Services," House of Commons Debate 09 March 1922 vol. 151 cc1535–604.

87 "South-West Africa," House of Commons Debate 25 July 1922 vol. 157 cc185–6.

88 "Royal Air Force (Bombing Attacks)," House of Commons, Debate 28 May 1919 vol. 116 cc1183–4. See David Omissi, *Air Power and Colonial Control: The Royal Air Force, 1919–1939* (Manchester: Manchester University Press, 1990).

89 Sebastian Ritchie, *The RAF, Small Wars and Insurgencies in the Middle East, 1919–1939* (Air Media Centre, HQ Air Command, 2011). Shockingly, none of these contentious issues were addressed at the Imperial Conference of 1923. See *Imperial Conference, 1923: Summary of Proceedings* in *Appendix to the Journals of the House of Representatives of New Zealand, 1923, Volume I* (Wellington: 1924).

90 In an official correspondence between the US diplomat to Ethiopia, Cornelius Engert, and the US Secretary of State, Cordell Hull, Engert notes, "reference should perhaps be made here to constant accusations of violations of rules and customs of war by both belligerents. Ethiopian Government and people are thoroughly aroused by what seems deliberate bombing or Red Cross units, of defenseless villages and towns as if the extermination of the civilian population were intended, bombing and burning of churches, use of poison gas, ill-treatment of women." "The Minister in Resident in Ethiopia (Engert) to the Secretary of State," in *Foreign Relations of the United States, Diplomatic Paper 1936, Volume III* (Washington D.C.: Government Printing Office, 1953), 45. See George Baer, *Test Case: Italy, Ethiopia, and the League of Nations* (Stanford: Hoover Institution Press, 1976); and Alberto Sbacchi, "Poison Gas and Atrocities in the Italo-Ethiopian War (1935–1936)," in *Italian Colonialism* eds. Ruthe Ben-Ghiat and Mia Fuller (New York: Springer, 2005); 47–56.

91 Michael Alpert writes, "On the afternoon and early evening of Monday, 26 April 1937, the town of Guernica, with a population of seven thousand swollen to perhaps eleven thousand by civilian refugees and fleeing Basque militia, was destroyed with high explosive and incendiary bombs." Michael Alpert, *Franco and the Condor Legion: The Spanish Civil War in the Air* (New York: Bloomsbury, 2019), 125.

(1931) and Shanghai (1932), which set the stage for further conflict. In 1937, Japanese bombers struck in the regions of Shanghai, Nanjing, and parts of central China, culminating in the terror bombing of Chongqing (1938).[92]

Instances like these reinforced to the international community the need to reconsider the ethical and strategic significance of aerial bombing, particularly as it pertained to civilian casualties.[93] Despite the failure of the Hague Draft Rules to be accepted by the international community, many of these interwar debates about bombing began to adopt its unratified language. For instance, in a British Parliamentary debate over the bombing of Guernica, Anthony Eden, the Foreign Affairs Minister, decried that attack, saying the "direct, deliberate and intentional bombing of non-combatants, as such, is illegal."[94] Given Britain's extensive use of bombing in its colonial holdings, these criticisms no doubt rang decidedly hollow. Nonetheless, because of the disproportionate impact on civilian populations, many commentators, especially those who worked in conjunction with the League of Nations, proposed various schemes to limit the destructive potential of aerial bombing.[95]

In a certain sense, the Hague Draft Rules to limit air war in 1923 could be seen as part of a decade-long process of conventions and conferences designed to culminate in the World Disarmament Conference of 1932.[96] This longer view of treaty negotiations could also be used to explain the lack of immediate concern that concrete progress was not being made and that proposed limits or bans on aerial bombing were not being formally accepted. The idea seems to

92 Diana Lary explains that "Chongqing was the first world capital to by systematically bombed. Between February 1938 and August 1943, the city was bombed over 200 times, on virtually every day that the skies were clear." Diana Lary, *The Chinese People at War: Human Suffering and Social Transformation, 1937–1945* (New York: Cambridge University Press, 2010), 87.

93 For a more extensive history of the developments of international law around aerial bombing in this period, see Paul Goda, "The Protection of Civilians from Bombardment by Aircraft: The Ineffectiveness of the International Laws of War," *Military Law Review* 93 (1966); 93–113; Waldemar Solf, "Protection of Civilians Against the Effects of Hostilities Under Customary International Law and Under Protocol I," *American University International Law Review* 1:1 (1986); 117–135; and Matthew Lippman, "Aerial Attacks on Civilians and the Humanitarian Law of War: Technology and Terror from World War I to Afghanistan," *California Western International Law Journal* 33:1 (2002); 1–67.

94 Quoted in Paul Goda, "The Protection of Civilians from Bombardment by Aircraft: The Ineffectiveness of the International Laws of War," *Military Law Review* 33:97 (1966); 93–113. 97.

95 See Herbert Manisty, "Aerial Warfare and the Laws of War," *Transactions of the Grotius Society* 7 (1921); 33–41.

96 See *Disarmament in 1932: The Year-Book of the Disarmament Information Committee* (Geneva: Published at the Office of Disarmament, 1932).

have been that states would become increasingly acclimated and socialized to the idea that disarmament was in their collective interests. Conferences socialized diplomats to the possibilities of these new norms. In the lead up to the headlining disarmament conference in 1932, a flurry of proposals had been set forth and various negotiations had been taking place, some quite radical in nature, all of which sought to actively limit the scope of aerial bombing.

For instance, at the General Disarmament Conference (Geneva, 1932), the British delegates were willing to go quite far in their commitment to eliminating the threat bombing posed to civilians. In Chapter III of their proposal, in a section on "Air Armaments," Article 34 proposes that the High Contracting Parties (the US, France, Italy, Japan, and the Soviet Union), "accept the complete abolition of bombing from the air (except for police purposes in certain outlying regions)."[97] Article 35(a) proposes the complete abolition of military and naval aircraft. Article 37 seeks to limit the size and weight of non-military airplanes.[98] These ideas were reflected in the Beneš Resolution, a proposal put forward by the Czech delegate, Edvard Beneš, which was itself modeled on the Hoover Proposal.[99] Like the British proposal, the resolution sought to completely abolish airships that could be used for bombing, and, of course, bombing itself.[100] Section C, Article 1 of the proposal states: "Air Forces; a) absolute prohibition of all air attack against civilian populations; b) abolition of all other forms of aerial bombing, under certain conditions, the chief of which is that civil aviation should be subject to international regulation."[101]

---

97 Of course, this is a significant caveat given the extent to which the British were resorting to bombing in their colonial holdings. By permitting colonial bombings, this would have put the British in a position to easily transfer these techniques to a war in Europe, if one should occur.

98 "Draft Disarmament Convention: Proposal Text Prepared in Light of the Modifications Adopted in the First Reading on June 8, 1933. Geneva, September 22, 1933," in *Disarmament and Security: A Collection of Documents, 1919–55* (Washington D.C.: Government Printing Office, 1956). 132–134.

99 The Hoover Proposal built on certain elements of the Briand-Kellogg Pact. It set forth a plan for the "abolition of all bombing airplanes coupled with total prohibition of all bombardment from the air." Malcolm Davis, "Public Opinion and Disarmament," in *Disarmament in 1932: The Year-Book of the Disarmament Information Committee* (Geneva: Published at the Office of Disarmament, 1932), 46.

100 Dick Richardson and Carolyn Kitching, "Britain and the World Disarmament Conference," in *Britain and the Threat to Stability in Europe, 1918–1945* (New York: Bloomsbury, 2016), 48. See also, N.C. Fleming, "Cabinet Government, British Imperial Security, and the World Disarmament Conference, 1932–1934," *War in History* 18:1 (2011); 62–84.

101 "The Benes Resolution," *Disarmament in 1932: The Year-Book of the Disarmament Information Committee* (Geneva: Published at the Office of Disarmament, 1932). 12.

None of these proposals would be ratified.[102] The accumulation of failures in prior talks, the Great Depression which had transformed into a global depression, the rise of Hitler in Germany, and the fear of future war all played some role in the ultimate breakdown of disarmament talks.[103] Furthermore, the clear rift between the true believers in disarmament and bombing realists, those who believed peace and security could only be maintained through defense, could not be greater. Governments were simply unwilling to risk disarmament.

Prior to the outbreak of World War II, there was one final attempt to reign in the destructive potential of aerial bombing. In response to the bombings taking place in Spain (1936–1937), the Cuban delegation to the League of Nations protested the harm being inflicted on noncombatant populations, and in the following session, their delegates put the protection of noncombatant populations on the agenda for debate.[104] Despite its complicated stance on bombing, the British delegates also strongly supported a formal resolution by the League of Nations. Indeed, directly in response to the bombings in Spain and China, the British Prime Minister, Lord Chamberlain, lamented that "there is at present no international code of law with respect to aerial warfare which is the subject of general agreement." He goes on to clarify:

> I think we may say that there are, at any rate, three rules of international law or three principles of international law which are as applicable to warfare from the air as they are to war at sea or on land. In the first place, it is against international law to bomb civilians as such and to make deliberate attacks upon civilian populations. That is undoubtedly a violation of international law. In the second place, targets which are aimed at from the air must be legitimate military objectives and must be capable of

102 Two years later, at the behest of the 7th International Congress of Military Medicine and Pharmacy, which met in Madrid in 1933, a commission of doctors and jurists drafted and approved the Draft Rules on Sanitary Cities and Localities, in Monaco on 27 July 1934, which included similar provisions. Section IV deals with the "Protection of the Civil Population," see Articles 1–10. https://ihl-databases.icrc.org/applic/ihl/ihl.nsf/Treaty.xsp?documentId=206E646653AB8F30C12563CD002D69E3&action=OpenDocument (Accessed 11 May 2021).

103 See Frederick Schuman, *Europe on the Eve: The Crisis of Diplomacy, 1933–1939* (New York: A.A. Knopf, 1939); René Albrecht-Carrié, *France, Europe and the Two World Wars* (Geneve: Librairie E. Droz, 1960); and Philip Noel-Baker, *The First World Disarmament Conference, 1932–1933 and Why it Failed* (New York: Pergamon Press, 1979).

104 Matthew Lippman, "Aerial Attacks on Civilians and the Humanitarian Law of War: Technology and Terror from World War I to Afghanistan," *California Western International Law Journal* 33:1 (2002); 1–68. 13.

> identification. In the third place, reasonable care must be taken in attacking those military objectives so that by carelessness a civilian population in the neighbourhood is not bombed.[105]

Though these rules would, much like the Hague Draft Rules, produce certain difficulties, they would at least be a start.

With the backing of Britain, the League of Nations authorized another committee to consider these issues, which drafted yet another proposal designed to eliminate the "horror of the bombing of civilian populations" along the exact lines set out by Chamberlain in his Parliamentary speech mentioned above.[106] Article I, Section 1 states: "The intentional bombing of civilian populations is illegal;" Section 2 stipulates: "Objectives aimed at from the air must be legitimate military objectives and must be identifiable;" and Section 3 concludes: "Any attack on legitimate military objectives must be carried out in such a way that civilian populations in the neighbourhood are not bombed through negligence."[107] Despite the document's being unanimously approved (30 September 1938), its contents were entirely negated. Exactly a year later, September 1939, aerial warfare was again unleashed on the European continent with the German bombing of Poland. As *Life Magazine* reported, "German bombers rain death and destruction"; in the process "More civilians were killed than soldiers" leaving the streets "cluttered with corpses."[108]

In one sense, the liberal-legal attempts to regulate bombing were an obvious failure. Although none of the international conventions dedicated to limiting or banning aerial bombing generated enough support to restrain state behavior, nevertheless, these efforts were extremely effective in transforming the ethical discourse on bombing. By 1938, the League of Nations had unanimously agreed that the *intentional* bombing of civilians was wrong. At no point since the close of the eighteenth century had the international community been willing to affirm such formulations. This embryonic transformation of norms, limited as it was, ultimately proved to be pyrrhic. By focusing exclusively on

---

105 "Foreign Office," House of Commons Debate 21 June 1938 vol. 337 cc919–1045.

106 Protection of Civilian Populations Against Bombing from the Air in Case of War, Unanimous resolution of the League of Nations Assembly, 30 September 1938. This appears to be modeled on the International Law Association's Draft Convention for the Protection of Civilian Populations Against New Engines of War. Amsterdam, 2 September 1938. https://ihl-databases.icrc.org/applic/ihl/ihl.nsf/Treaty.xsp?action=openDocument&documentId=29DF95181E9CC0FDC12563CD002D6A9B (Accessed 11 May 2021).

107 Ibid.

108 "Polish Conquest: German Bombers Rain Death and Destruction," *Life Magazine* 7:13, 25 September 1939, 21.

the intention to discriminate, militaries employing the tactic, as critics like Paul Williams feared, could justify their behavior by way of good intentions. Since the technological limitations of aerial bombing made it impossible to differentiate between the intention to discriminate and outright terror bombing, the operative moral distinction was ultimately obliterated by the terrible consequences of the tactic.

4 Bombing Realists

The discursive rift between international liberals and military planners could not have been greater. While the former was desperate to circumscribe and limit, if not entirely ban this new technology, bombing realists were not too troubled by the potentialities of air war. In some respect, the worse bombing portended to be the better, that is, if the terror of bombing were to be a reliable restraint. To begin with, they could afford to be optimistic since a proper air war had yet to materialize. Despite the proliferation of minor air conflicts described above, which could be written off as products of asymmetrical air power, there had yet to be an air war between great powers. In fact, given the brutal reality of total mechanized land war, these bombing realists quite reasonably assumed that strategically waged air warfare might shorten the duration of war and expeditiously lead to peace. Even if air warfare were unrestrained, its cruelty meant that it would not be prolonged.

It must be remembered that at the end of World War I strategists and theorists of war were not so far removed from nineteenth-century norms. Many in the French and German military command during World War I had either participated in or had living memory of the Franco-Prussian War.[109] So it is not surprising that nineteenth-century norms would still hold a privileged place in the ethical discourse of this period. It is also important to recognize that the strategic and moral environment of the interwar period among military planners was not an anomaly, but the concrete attempt to apply nineteenth century military ethical norms to airpower. In the post-World War II climate, it has been common to disparage interwar bombing realists due to the fact that they privileged what now may seem like morally dubious assumptions, i.e. the ready-made analogy of aerial bombing to the bombardment of fortified cities, the wide-spread belief that some strategically minimum number of civilians

109 As Douglas Fermer notes, the last veteran of the Franco-Prussian War died in 1955. Douglas Fermer, *Three German Invasions of France: The Summer Campaigns of 1870, 1914, 1940* (South Yorkshire: Pen & Sword Military, 2013), xi.

could be targeted in siege warfare to expedite surrender, and that a quick and decisive air war was preferable to the horrors of conventional war. While it is no doubt true that bombing realists' claims are shocking today, it is clear that they were attempting to give final theoretical expression to nineteenth-century bombing norms.

Among bombing realists, six prominent theorists at the outset of World War I and throughout the interwar period embodied this early posture to rationalize the deliberate aerial bombing of civilians by utilizing the nineteenth-century notion of military necessity: George Aston, J.M. Spaight, Giulio Douhet, William 'Billy' Mitchell, J.F.C. Fuller, and Henry Liddell Hart.[110] While there are many differences in the particular tactics and approaches to air warfare these commentators advocate, they all agree that deliberately targeting some minimum number of civilians through aerial bombing would lead to much less overall suffering in war, and it should therefore be strategically considered. That so many prominent theorists of this era were advocating the bombing of civilians is indicative of the moral and conceptual weight of nineteenth-century norms and what military necessity demanded in the face of what land warfare had become. After all, it was not that these theorists had newly discovered the strategic value of targeting civilians in war.[111] Quite the contrary, as the costs of total war grew more severe, military strategists and ethicists of this time began to recalibrate their moral judgments in favor of minimizing the catastrophic reality of modern war. If the choice were between millions of dead soldiers or merely thousands of dead civilians, the moral calculus clearly favored the killing of civilians. Being willing to accept some strategic minimum number of civilian casualties began to look not only reasonable but morally necessary.

4.1 *Sir George Aston*

In George Aston's *Sea, Land and Air Strategy* (1914),[112] he surmises that "a proud nation, profoundly stirred by the justice of its cause, does not give in to an

110 This list is not exhaustive, but these commentators are some of the most influential in this period. Others could be added. See William Sherman, *Air Warfare* [1926] (Maxwell Airforce Base, Alabama: Air Force Press, 2002); Sir John Cotesworth Slessor, *Air Power and Armies* (London: Oxford University Press, 1936).

111 *Pace* Buckley, targeting civilians themselves did not necessarily represent "crossing the Rubicon"; this was an extension of tactics employed throughout the nineteenth century. It was much more likely the quantity of civilians being killed that began to challenge moral discourse. As he himself indicates, "history is replete with examples of civilians being deliberately targeted in war, especially in sieges and times of blockade." Buckley, *Air Power in the Age of Total War*, 5.

112 Aston was the Professor of Fortification at the Royal Naval College (1896–1899). He served as Secretariat of the English War Cabinet from 1918 to 1919 and was a member of

opponent unless the population can be made to feel the threat of starvation or some equally effective pressure."[113] With this in mind, Aston concludes that it has only been the "recent improvement in aircraft that enable [militaries] to cause considerable discomfort to the crowded inhabitants of towns in a hostile country by dropping bombs promiscuously amongst houses."[114] Aston appeals to the ambiguity in the Hague Convention that allows the bombing of fortified towns to expedite surrender: "If the town is fortified, on the other hand, bombardment is a recognized method, like investing and starving the civil population, which is employed deliberately in order to induce the garrison to surrender. Starvation, combined with bombardment, has frequently brought about this result, but bombardment alone has usually failed."[115] The implication of this line of thinking suggests that even if bombing alone may not be decisive, it may be strategically effective to target civilian infrastructure holistically.

Under current technological conditions, however, Aston concludes that:

> The moral effect which would be produced by highly mobile aircraft dropping bombs first over one city, and a few minutes afterwards over another city a hundred miles away, would no doubt tend to produce local panics, but, given the right spirit, the material effect would not suffice to produce concerted action amongst the population to force their Government to stop the war.[116]

This is why pressuring civilians through aerial bombing should be put off until some point in the future, "When flying has been developed to such an extent that flying combatants can be counted in tens of thousands instead of in hundreds, then the issue between nations may be brought to a conclusion in the air."[117] Aston's objection to aerial bombing amounts to a technological limitation – at some point, when tens of thousands of planes can be set upon a target – only then would it be an effective means to end war dramatically. In other words, the use of bombing in World War I would be problematic not

the Foreign Intelligence Committee of the Admiralty. By the end of World War I, he had become a prominent commentator on war. For an enumeration of his accomplishments, see "Sir George Aston, British General, 77, Had Served in Many Campaigns – Author and Journalist Dies," *New York Times*, 3 December 1938. 19.

113 Aston, *Sea, Land, and Air Strategy* (London: John Murray, 1914), 234.

114 Ibid., 235–36.

115 Ibid., 237.

116 Ibid., 238.

117 Ibid.

because it deliberately targeted civilians, but because it could only do so in a sporadic and limited way.

4.2 *J.M. Spaight*

In his 1914 *Aircraft in War*, J.M. Spaight argues that writers like H.G. Wells, those predicting inordinate devastation from aerial bombing, were being far too pessimistic about the prospects of air warfare.[118] He suggests that "It is unlikely that civilised nations will ever wreck on another's purely residential and commercial cities," though he admits that states would be "within their war rights in dropping dynamite even on the non-defended parts – the civilian quarters – of cities which are defended at other points. They will not be bound to confine their attacks to the perimeter of forts."[119] It is, of course, difficult to justify these arguments through the prism of post-World War II ethical norms. These arguments can only be explained by looking back to the ethical traditions of the nineteenth century.

To justify this approach to aerial bombing, Spaight cites the British Official Manual on Land Warfare, which explains that "A town which is defended by detached forts, though they are at a distance from it, is liable to bombardment, for the town and force form an indivisible whole."[120] And while Spaight didn't believe civilized countries would reduce each other to rubble – surmising that "a foe would probably shrink from exercising his right of destruction"[121] – he goes on to argue that aerial vessels would be justified in threatening the bombardment of undefended cities if they refused to surrender.[122] Furthermore, he concludes that

---

118 Spaight was one of the most prolific commentators on aerial war in Britain. He served in the British Air Ministry from 1918–1937. For more his career and background, see Robin Higham, *The Military Intellectuals in Britain, 1918–1939* (New Brunswick, NJ: Rutgers University Press, 1966), 230.

119 J.M. Spaight, *Aircraft in War* (London: Macmillan, 1914), 10.

120 Ibid., 11. This was a widely accepted interpretation of the Hague Convention. In his commentary on the laws of war, Percy Bordwell writes, "Bombardment of the resident portions of towns – It has been questioned whether the town separated somewhat from the fortifications which defend them are liable to bombardment. This is a part of the larger question as to whether the portions of towns not used for military purposes are exempt from bombardment. It must be admitted, however, that this is one of the rigors of war which belligerents have held to be so useful as not to be given up. Of its hardship there can be no question, though it cannot compare with that of a siege." Percy Bordwell, *The Law of War Between Belligerents: A History and Commentary* (Chicago: Callaghan, 1908), 287.

121 Spaight, *Aircraft in War*, 22.

122 Ibid., 11–13.

> A belligerent is entitled to seek out and destroy the armed forces of his enemy wherever found. If they choose to take up position in a crowded city, he cannot be prevented from attacking them there because to do so would cause damage to building and property; and he cannot be forced, because of his enemy's action, to lay his artillery aside and resort to the costly tactics of street-to-street fighting to clear the other troops out of the city. Were it otherwise, an army threatened with destruction could escape simply by retiring into a large city, which would thus have attributed to it a power of sanctuary extending beyond even that attributed to neutral soil. There are historical precedents for the shelling of occupied, though not "defended," towns, and I do not think that they were cases of belligerents' exceeding their war rights.[123]

The same logic extends to the materials needed to prosecute war. Even though technically undefended, a belligerent would be within his right to bomb those places that make it possible for the enemy to prosecute the war. Borrowing language from the Hague Convention on Naval Bombardments (1907), he writes: "The commandant incurs no responsibility for any damage resulting from such a bombardment."[124]

4.3 *Giulio Douhet*

After World War I, Giulio Douhet published the first edition of *The Command of the Air* (1921).[125] His vision of future air war was, like Spaight's, terrifyingly upbeat. The book is an elaborate argument designed to convince the Italian government to expedite the transformation of its military to one that prioritizes air power over conventional land and naval forces. Such an investment was necessary, he argues, because the airplane would (if it had not already) completely change the nature of the battlefield, and the vulnerabilities that unprepared states would soon face would be quite severe.[126] What is more, these new weapons required a new type of logic that could not, at least at first glance, be reconciled with the classical laws of war.

---

123 Spaight, *Aircraft in War*, 16.

124 Ibid., 17.

125 Douhet was an Italian general and highly regarded as an early theorist of air warfare. See Thomas Hippler, *Bombing the People: Giulio Douhet and the Foundations of Air-Power Strategy, 1884–1934* (New York: Cambridge University Press, 2013). For similar arguments in the Italian tradition, see also Nino Salvaneschi, *Let Us Kill the War: Let Us Aim at the Heart of the Enemy!* (Milan: Art Edition of Bianco e Nero, 1917).

126 These arguments were effective. By the 1920s Italy's air power fell only behind France's.

To begin with, Douhet argues that "the battlefield will be limited only by the boundaries of the nations at war, and all their citizens will become combatants, since all of them will be exposed to the aerial offensives of the enemy. There will be no distinction any longer between soldiers and civilians."[127] Douhet admits that "All of this must inevitably effect a profound change in the form of future wars, because the essential characteristic of those wars will be radically different from those of any previous ones."[128] This radical difference lies in the fact that "aerial warfare admits of no defense, only offense. We must therefore resign ourselves to the offensives the enemy inflicts upon us, while striving to put all our resources to work to inflict even heavier ones upon him. This is the basic principle which must govern the development of aerial warfare."[129]

While the logic of aerial war is to inflict as much damage on the enemy as quickly as possible, Douhet remains open to the question of the most effective targets. Certainly the enemy's civil and military industries are legitimate targets – gaining swift air superiority is crucial – but he also suggests that the deliberate targeting of city centers might expedite the end to war.[130] He explains, "By bombing the most vital civilian centers it could spread terror through the nation and quickly break down [a nation's] material and moral resistance."[131] He continues: "A complete breakdown of the social structure cannot but take place in a country subjected to this kind of merciless pounding

127 Giulio Douhet *The Command of the Air* [1921] trans. Dino Ferrari (Washington D.C.: Air Force History and Museums Program, 1998), 9. Spaight would put forward a similar, though much more qualified, argument in *Bombing Vindicated* (1944). He argues that "To speak of the 'bombing of civilians without qualification is really to confuse the issue. One must define one's terms. The old clear distinction between soldiers and civilian has been obscured. That is not to say that the whole population of an enemy country is subject to attack. Indiscriminate bombing is certainly not justifiable." He goes on to explain that "No chivalrous airman wants to slaughter grandmothers or babies. The tragedy is that he may do so in trying to put the others out of action. It is an unintended, horrible, pitiable incident of war, but to say that is not to condemn air bombardment." J.M. Spaight, *Bombing Vindicated* (London: Geoffrey Bles, 1944), 112.

128 Douhet, *The Command of the Air*, 10.

129 Ibid., 55.

130 Almost the identical argument was put forward by the French theorist J. Boureut-Aubertot: "This is why, for reasons of humanity, we are still trying to restrict the number of unarmed persons against whom air attacks are legitimate. But it is clear that logic no longer allows non-combatants to retain the traditional immunity they enjoyed. Air bombardments certainly make the war much more deadly than in the past. In this ruthless struggle, it is believed that the cruel practices of barbaric times are revived." J. Boureut-Aubertot, *Les Bombardements Aériens* (Paris: Les Presses Universitaires de France, 1923), 99. Much more grimly, Will Irwin explains, "baby-killing had become incidental to military necessity." Will Irwin, *The Next War* (New York: Dutton, 1921), 32.

131 Douhet, *The Command of the Air*, 57.

from the air. The time would soon come when, to put an end to horror and suffering, the people themselves, driven by the instinct of self-preservation, would rise up and demand an end to the war."[132] While this may sound cruel, he assures the reader that this logic is more humane than it seems: "These future wars may yet prove to be more humane than wars in the past in spite of all, because they may in the long run shed less blood, but there is no doubt that nations who find themselves unprepared to sustain them will be lost."[133]

Much like the future wars that Wells imagined, Douhet develops a theory of offensive aerial warfare. But, of course, the logic of Douhet's offensive aerial warfare is still functionally utilitarian – kill a sufficient minimum number of people to bring about a rapid end to hostilities. Much in the same way nineteenth century generals preparing to bombard a fortified place could consider all the inhabitants to be combatants by virtue of the demands of self-defense, Douhet derives a similar theory of aerial bombing.

4.4 *William 'Billy' Mitchell*

In *Our Air Force: The Keystone of National Defense* (1921), William Mitchell sets the stakes for future war.[134] He predicts that "a very conservative estimate of its capabilities ten years from now is such that it can be definitely stated that, in case of war in which the United States is engaged, if the enemy obtains mastery of the air, he will be able to dictate his own terms of peace and any place within the United States that he may desire."[135] In effect, he continues, this means that women and children are now so deeply entangled in a nation's economic and industrial base that entire nations have become combatant forces. He explains, "We must expect, therefore, in case of war, to have the enemy attempt to destroy any or all of our combatant or industrial forces – his attacks being entirely controlled by the dictates of strategy, and the means of bringing the war to a quick conclusion."[136]

Much like the previous commentators, Mitchell believed that a robust investment in an air force would bring about peace. First, they would be a powerful defensive tool: "Interior cities are now as subject to attack as those

---

132 Ibid., 58.

133 Ibid., 61.

134 Mitchell was one of the principal contributors to the founding of the US Air Force. He was one of the leading proponents of US air power. For more, see Roger Miller, *Billy Mitchell: "Stormy Petrel of the Air"* (Washington, D.C.: Office of Air Force History, 2004).

135 William Mitchell, *Our Air Force: The Keystone of National Defense* (New York: E. Dutton & Company, 1921), xxi.

136 Ibid., xxii.

along the coast. Nothing can stop the attack of aircraft except other aircraft."[137] But perhaps more importantly, they would prove to be a powerful offensive weapon:

> No longer will the tedious and expensive processes of wearing down the enemy's land forces by continuous attacks be resorted to. The air forces will strike immediately at the enemy's manufacturing and food centers, railways, bridges, canals and harbors. The saving of lives, man power and expenditures will be tremendous to the winning side. The losing side will have to accept without question the dominating conditions of its adversary, as he will stop entirely the manufacture of aircraft by the vanquished.[138]

In fact, indicative of nineteenth century norms in war, Mitchell conceptualizes the future of air war as an "aerial siege."[139] Rather than needing to advance over rough terrain painstakingly and haltingly, through all manner of obstacles, aerial attacks now resemble artillery bombardments in siege warfare. He writes:

> Every air attack on other aircraft is based on the theory of surrounding the enemy in the middle of a sphere with all our own airplanes around the whole periphery shooting at it. If we attack a city or locality, we send airplanes over it at various altitudes from two or three hundred feet up to thirty thousand, all attacking at once so that if any means of defense were devised which could hit airplanes or cause them to be destroyed from the ground, the efforts would be completely nullified, because they could neither see, hear nor feel all of them. No missile-throwing weapons or any other devices have yet been created or thought of which can actually stop an air attack, so that the only defense against aircraft are other aircraft which will contest the supremacy of the air by air battles. Great contests for control of the air will be the rule in the future.[140]

Air supremacy would give attacking militaries the ability to reduce enemy nations in much the same way besieging armies would reduce fortified places throughout the nineteenth century.

---

137 William Mitchell, *Winged Defense: The Development and Possibilities of Modern Air Power – Economic and Military* (New York: G. Putnam's Sons, 1925), xiv.

138 Ibid., xv.

139 Ibid., 5.

140 Ibid., 9.

By opening this "third dimension" of warfare, wars would be more decisive and briefer. Indicative of the commentary leading up the World War I, Mitchell concludes:

> The menace [of aerial bombing] will be so great that either a state will hesitate to go to war, or, having engaged in war, will make the contest much sharper, more decisive, and more quickly finished. This will result in a diminished loss of life and treasure and will thus be a distinct benefit to civilization. Air forces will attack centers of production of all kinds, means of transportation, agricultural areas, ports and shipping; not so much the people themselves.[141]

Even though Mitchell is reluctant to affirm the bombing of people as such, his approach to bombing would only work, he explains, if "all the important cities would be laid waste, the railroads and bridges destroyed, roadways constantly bombed and torn up so as to prevent automobile transportation, and all seaports demolished."[142] Since there is no feasible way to differentiate civilian and military infrastructure, the strategic targeting of city centers would necessarily lead to civilian casualties. While tragic, he concludes that this is better than the days of old when "Those who were vanquished, were entirely destroyed; their cities looted and burned, and the whole country, laid waste. At that time an entire country went to war."[143]

4.5 *J.F.C. Fuller*

In his 1923 book *The Reformation of War*, J.F.C. Fuller dedicates a chapter to "The Future of Air Warfare."[144] With the rise of new aerial technologies Fuller also points out what appears to be a change in the logic of warfare. Particularly, he laments that "nations have not yet realized that great wars are national wars in which the attack on the will of the so-called non-combatants is the

141 Ibid., 16. Sherman Miles, a critic of what he called "war in the third dimension," that is new war directed against an enemy's noncombatant populations, believed that strategies put forward by Mitchell would not come to pass. He attacked Mitchell by noting, "They believe that the humanizing trend of civilization may be reversed and set back a thousand years." Major Sherman Miles, "War in the Third Dimension," *The North American Review* 223:833 (1926–27); 594–605. 605.

142 Ibid., 11.

143 Ibid., 15.

144 Fuller was a British Army officer who was a prominent theorist about war. He authored dozens of books on numerous issues, especially the incorporation of tanks and aircraft in war. For more on Fuller, see Brian Reid, *JFC Fuller: Military Thinker* (London: Palgrave Macmillan, 1990).

objective."[145] Fuller finds this lack of resolve tragic because air power reflects a humane swiftness that was not possible before. He writes: "a future war can be won at the cost of two or three thousand of the enemy's men, women and children killed, in place of over 1,000,000 men and incidentally several thousands of women and children, as was the case in France during the recent war, surely an aerial attack is a more humane method than the existing traditional type."[146]

Anticipating critics' objections that deliberately targeting civilians is immoral, Fuller points to the very same tension in the laws of war regarding siege warfare:

> As it cannot be more immoral to bomb a town than to bombard it, does the immorality of an aeroplane attack lie in the fact that, while in a bombardment the slaughter of women and children is but an unfortunate incident, in an aerial attack on a town the terrorization of its civil inhabitants becomes the main object? I believe that this is the popular conception, simply because civilians have not yet grasped the fact that: when nations go to war the entire population of each country concerned is ranged against the other, and that the solidarity of their fighting forces is founded on the civil will.[147]

This passage is compelling for two reasons. First, it is striking that Fuller employs a version of the doctrine of double effect to distinguish between naval and land bombardments, on the one hand, and aerial bombing, on the other. The killing of innocents in a bombardment is ancillary, an 'unfortunate event' whereas aerial bombing intends to kill civilians. This is one of the earliest usages of the double effect moral formulation regarding civilian casualties in the twentieth century. However, what is unique about the contrast Fuller makes between the "unfortunate incident" of civilian casualties from bombardment and the "main object" of terrorizing civilian inhabitants with aerial bombing is the absence of any meaningful moral distinction between them. This is so, he argues, because there is no relevant difference between the residents of defended towns and the residents of a nation, especially when the "solidarity of their fighting forces is founded on the civil will."

145 J.F.C. Fuller, *The Reformulation of War* (London: Hutchison and Co., 1923), 149.

146 Ibid., 150. For similar arguments, see also J.F.C. Fuller, *Towards Armageddon: The Defence Problem and its Solution* (London: Lovat Dickson, 1937).

147 Fuller, *The Reformulation of War*, 112.

The second thing to consider regarding this passage is that the characterization of military bombardments that Fuller makes is completely anachronistic. In fact, the argument is entirely inverse. At no time in the nineteenth century was the killing of civilians in defended cities described this way by theorists of war.[148] Indeed, the distinction between naval or land bombardment and aerial bombing that Fuller suggests was always rather tenuous. And, as one can see from the multiple passages indicated above in the discussion on siege warfare, it had long been an accepted practice to bomb the residential districts of fortified or defended towns in order to terrorize the civilian population to expedite surrender. In the bombardment of Vera Cruz, for example, General Scott felt no compunction targeting the civilian population since it was by this means that he made it impossible for the fort to continue its defense. Many theorists simply assumed that civilians lost their noncombatant status by virtue of where they were. Bombing realists like Fuller were now seeking to put this practice into stark relief.

4.6 *Henry Liddell Hart*

In his 1925 *Paris, or The Future of War*, Sir Henry Liddell Hart argues that from 1870 to 1918 militaries suffered from the legacy of the "Napoleonic legend," namely the Clausewitz-formulated doctrine that the true objective of war is "the destruction of the enemy's main forces on the battlefield."[149] According to this tradition, the most effective way to achieve victory is to force the enemy to mount greater battlefield casualties than one's own army, but also, if possible, to do so in a singular blow resulting in a decisive victory. The essence of this position is summarized by Major General James Harbord, who explicitly argues against the aerial bombing of noncombatants in just these terms:

> Apart from the dictates of humanity, it is the fallacy of force misused that condemns attacks on non-combatants. War is decided when the

148 Popular discourse was another matter. Amos (1880) explains how the war correspondent had started to change public perception of the cruelties of war: "But the individual life, given to rapine, and more and more habituated to dispense with the most ordinary moral obligations; reckless of all claims but those of so-called military necessity; blunted and hardened against the commonest sensibilities, and merging all the finer responsibilities of humanity in the one overbearing duty of military obedience, – all this was brought into clear relief by the searching eye of the Special Correspondent" Amos, *Political and Legal Remedies for War*, 32.

149 Henry Liddell Hart, *Paris, or The Future of War* [1925] (New York: Garland Publishing, 1972), 8. Liddell was a prominent historian and military theorist of this period. He was known as the "captain who taught generals." Quoted in John Mearsheimer, *Liddell Hart and the Weight of History* (Ithaca, NY: Cornell University Press, 1988), 50.

> armed forces of one side are destroyed or paralyzed by the other. The first and most fundamental of all military maxims is to concentrate on securing the ultimate victory by such destruction or paralysis. No country's resources are limitless. If you expend your money and your effort on planes – the costliest of munitions – for bombing civilians you weaken to that extent the pressure you can bring upon the enemy's armed forces.[150]

To Hart's way of thinking, the legacy of World War I proved that General Harbord's view (and anyone else who held to the same maxim) was not only false, but colossally delusional; under the conditions of modern war fighting by way of the "Napoleonic legend" might very well undermine all of civilization.[151] While it is true that, as Harbord argues, "no country's resources are limitless," under the conditions of modern industrial societies, war economies can produce tremendous destruction. In this sense, the legacy of Clausewitz was one of extreme bloodletting.[152]

It is not surprising that the extreme suffering and death brought about by World War I would lead many theorists of war to seek ways to ease the pain inflicted on mankind, not by focusing on uncertain humanitarian promises (i.e. those that had failed to humanize and even prevent World War I in the first place), but by seeking more plausible avenues for decisive victory. This impulse is what stands behind Hart's agreement with Marshal Foch's radical break with the traditional 'armed forces objective,' which sees the objective in war as primarily focusing on the defeat of an enemy's military. Foch argues that, "The potentialities of aircraft attack on a large scale are almost incalculable, but it is clear that such attack, owing to its crushing moral effect on a nation, may impress public opinion to the point of disarming the Government and thus become decisive."[153]

For Hart, a tactic like air power reflects the new objective of war that nations ought to pursue: "The aim of a nation in war is, therefore, to subdue the enemy's

---

150 Major General James Harbord, "Review: *The Great Delusion. A Study of Aircraft in Peace and War* by 'Neon,'" *Naval Institute Proceedings*, 53:293 (1927), 906.

151 For more on Hart's legacy, see Phillip Meilinger, "Basil H. Liddell Hart: His Applicability to Modern War," *Air Power Review* 20:3 (2017).

152 This critique of Clausewitz runs throughout his books. In *The Remaking of Modern Armies*, Hart also writes, "Clausewitz and his successors were the false prophets mainly responsible for exalting this superficial deduction to the binding authority of a law of warfare, and for fastening its fatal yoke so firmly on the necks of the people – its slaves and victims." Making the armies the "objective of strike" has turned war into "mechanical butchery." Henry Liddell Hart, *The Remaking of Modern Armies* (Boston: Little, Brown, and Company, 1928), 103–104.

153 Quoted in Hart, *Paris, or The Future of War*, 14.

will to resist with the least possible human and economic loss to itself."[154] Once this is grasped, it becomes clear that targeting the military exclusively is only one of the available means of subduing the enemy. In fact, attacking an enemy's army turns out to be the least sensible way of pursuing war. There are much more effective pressure points that can expedite victory: "a highly organized state is only as strong as its weakest link. In a great war the whole nation is involved, though not necessarily, or wisely, under arms. The fists and the sinews of war are mutually dependent, and, if we can demoralize one section of the nation, the collapse of its will to resist compels the surrender of the whole."[155] In this respect the domestic population might be the Achilles' heel of a belligerent nation. Indeed, "The enemy nation's will to resist is subdued by the fact or threat of making life so unpleasant and difficult for the people that they will comply with your terms rather than endure this misery."[156]

For Hart, the rise of air power was perfectly suited to apply pressure to the domestic population, that is, since

> Aircraft enables us to jump over the army which shield the enemy government, industry, and people, and so strike direct and immediately at the seat of the opposing will and policy. A nation's nerve-system, no longer covered by the flesh of its troops, is now laid bare to attack, and like the human nerves, the progress of civilization has rendered it far more sensitive than in earlier and more primitive times.[157]

154 Ibid., 19.

155 Ibid., 21.

156 Ibid., 31.

157 Ibid., 37. In *The Remaking of Modern Armies*, Hart makes an almost identical claim: "Aircraft enables us to jump over the army which shield the enemy government, industry, and people, and so strike direct and immediately at the seat of the opposing will and policy. Provided that the blow be sufficiently swift and powerful, there is no reason why in a few hours, or at most days, from the commencement of hostilities the nerve-system of one of the contending countries should not be paralyzed. A modern civilized state is such a complex and interdependent fabric that it offers a target highly sensitive to a sudden and overwhelming blow from the air." Hart, *The Remaking of Modern Armies*, 106. Both passages from Hart are reminiscent of a line of thinking found in a staff paper delivered by William Mitchell in 1919 wherein he lays out a similar vision for aerial bombardment: "Probably its greatest value is in hitting an enemy's great nerve centers at the very beginning of the war so as to paralyze them to the greatest extent possible. In the theater of operations of an army, it is used against supply points of all sorts, airdromes, railroad stations, roads and communications, and, last against troops and trains on the roads." "Air Service Tactical Application of Military Aeronautics," 5 January 1919, Maxwell AFB, Ala., USAF Historical Research Agency, document 167.4–1, 3. Quoted in Johnny Jones, *William*

He goes on to write: "Imagine for a moment that, of two centralized industrial nations at war, one possesses a superior air force, the other a superior army. Provided that the blow be sufficiently swift and powerful, there is no reason why within a few hours, or, at most days from the commencement of hostilities, the nerve system of the country inferior in air power should not be paralyzed."[158] The advantage of air power is that it offers the opportunity for a "sudden and overwhelming blow from the air."[159]

Hart anticipated two possible objections to the decisive air attack to achieve rapid victory: one is economic and the other is ethical. The economic consideration stems from the potential obliteration of post-war markets. After all, today's enemies will be tomorrow's new markets. To this Hart suspects that though the amount of damage inflicted by an intense aerial attack would be significant, it would still be less than the destruction generated from a protracted war of attrition.[160] Hart's response to the ethical dimension follows a similar pattern. He concluded that since war has made it impossible to exempt civilians from the cruelties of war, and since protracted wars cause greater destruction, it is more humane to target a nation's weakest point in order to expedite surrender. He writes, "warring nations should endeavor to gain their end of the moral subjugation of the enemy with the infliction of the least possible permanent injury to life and industry, for the enemy of today is the customer of the morrow, and the ally of the future."[161]

5 Strategic versus Terror Bombing

There are a few initial observations to make about these interwar commentators on aerial warfare. First is the degree to which they explicitly articulate certain underlying assumptions about killing the innocent in war that bear much more in common with the thinking of the prior century. Indeed, the cost-benefit analyses that they employed perfectly embodied the utilitarian logic of the "sufficient minimum" amount of death that was ascendant throughout the nineteenth century. The degree to which these authors are entirely unselfconscious about intending harm to civilian populations is no doubt quite shocking to readers acclimated to post-World War II military ethical

---

'Billy' Mitchell's Air Power (Airpower Research Institute College of Aerospace Doctrine, Research, and Education Maxwell Air Force Base, Alabama, 1997), 60n57.

158 Hart, *Paris, or The Future of War*, 40–41.

159 Ibid., 41.

160 Ibid., 43.

161 Ibid., 44.

formulations. As one can infer from the structure of their arguments, they seem to be entirely unconcerned, indeed, unaware, that intentional civilian death represents a moral liability. That so many prominent commentators are oblivious to the moral peril of inflicting intentional harm on civilians is perhaps the biggest indication that such a norm did not yet firmly exist.

The arguments put forward by the above theorists certainly had some impact on the strategic and tactical thinking of military planners and commanders through that period. Despite such a deafening chorus of support for direct assault on city centers, in the interwar period, the US military primarily emphasized the role tactical aerial bombardment would play in combat (i.e. a supporting role to land and sea operations).[162] However, there was a growing awareness that "strategic" bombing, i.e. the targeting of national infrastructure, might expedite victory. Part of the effectiveness of depleting the infrastructure of an enemy nation was that the civilian population's waning resolve and morale would lead them to put pressure on the government to end hostilities.

For instance, the 1920 guide to the US Air Service explains that aerial bombing may be useful "To weaken and possibly destroy the morale of the enemy both civil and military."[163]

In 1925, Benjamin Harmon writes:

> The Air Service must act directly to decrease the fighting power of the enemy forces. Destroying supplies, the means of bringing them up (trains, roads, railroad bridges), or their place of manufacture, assists in accomplishing this object. Attacking the enemy forces directly or bombing his cities in addition to reducing the material power of the enemy to fight reduces his morale or his will, which works to the same end.[164]

The 1926 training guide for the US Air Service Tactical School notes that strategic bombing, if used early, might be an effective tool for bringing a quick end to hostilities.[165]

162 The same was partially true for the British. One interwar bombing manual states: "As a rule, however, bombers should be reserved for trench work, and their chief value in an attack on a large scale will be in clearing trenches." M.V. Campbell, *Practical Bombing as Applied by the Canadian and British Armies* (Detroit: Bartlett Publishing, 1917), 82.

163 Captain Earl Naiden, *Air Service* (Leavenworth, KS: The General Service Schools Press, 1920), 70.

164 Benjamin F. Harmon, "The Past and Future of Defense against Aircraft," *The Coastal Artillery Journal* 63:5 (1925); 450.

165 There were conflicting views as to what role air power was and what kind of air force the US was going to build. The manual mentions but leaves underdeveloped strategic bombing, which is probably because the chief of the US Air Service, Major General

It is often forgotten that the origins of strategic bombing were much more morally ambiguous than the terror/precision bombing distinction that is often made in World War II bombing discourse implies. As Thomas Greer notes, the bombing strategy emphasized in the interwar period differed greatly from that which emerged during World War II. He writes, "Most of the operations of so-called 'strategic' aviation were not truly strategic operations as later conceived and practiced."[166] Limitations in technology, in bomb load capacity, in accuracy, etc., rendered the very idea of "strategic" bombing – where strategic is defined as a calculated strike on a dedicated military objective – well, dubious.[167] In fact, as William Fischer explains, "the overarching purpose of strategic bombardment during the First World War was to damage a belligerent's capacity and willingness to wage war by targeting the military, economic, and civilian segments of society."[168] A strategic strike was *anything* that would weaken the enemy in *any* respect – military infrastructure or civilian morale.[169]

The expansive notion of strategic strike employed in this period touches on some of the tension between terror and morale bombing indicated above. As noted earlier, what we think of as terror bombing today was more closely associated with unintentional, promiscuous bombing, that is to say, the dropping of bombs with no specific agenda or target in mind, i.e. the hope of hitting

---

Charles T. Menoher, defended the "Pershing orthodoxy" which assumed that air power would provide only a supportive role to conventional operations, "to see that the infantryman with rifle and bayonet won the fight." James Cooke, *Billy Mitchell* (London: Lynne Rienner Publishers, 2002), 109.

166 Thomas Greer, *The Development of Air Doctrine in the Army Air Arm, 1917–1941* (Washington, D.C.: Office of Air Force History, 1955), 9.

167 Take, for instance, an *Air Corps Information Circular* from 1920, which states: "Aerial bombardment constitutes one of the most important phases of aerial warfare. Bombing operations, effectively and consistently carried out, accomplish a fourfold purpose. Weakening of the enemy's morale, destruction of enemy materiel and personnel, reconnaissance of the enemy movements in the air and on the ground, and keeps enemy air service back of the lines for home defense." Training and Operations Group, Operations Division, Air Service, *Pursuit and Combat Training Regulations* (Washington DC: Government Printing Office, 1920), Section 569, 83.

168 William Edward Fischer Jr., *The Development of Military Night Aviation to 1919* (Maxwell Air Force Base, Alabama: Air University Press, 1998), 63.

169 This tension can be seen in Sir Frederick Sykes's analysis of British bombing during World War I. He questions what the "true role" of the air force should have been. Rather than being utilized "for attacking purely military objectives in the Army zone" they should have pursued the strategy of "demoralization which pervaded the civil population of the towns attacked." Sir Frederick Hugh Sykes, *Aviation in Peace and War* (London: Edward Arnold, 1922), 91–92. For more on Sykes's contribution to the targeting debate, see Eric Ash, *Sir Frederick Sykes and the Air Revolution, 1912–1918* (London: Frank Cass, 1999).

something of value and, if not, the ancillary hope that civil unrest will ensue. In contrast, morale bombing (i.e. bombing designed to diminish civilian morale) was intentionally promiscuous. Morale bombing was seen as one of many possible benefits to the bombing missions targeting military infrastructure. In other words, it was a foreseen and desirable benefit of bombing campaigns that civilian morale would be diminished. Even though these two approaches to bombing may look identical – in both cases residential districts are bombed and noncombatants are killed – they stem from different assumptions about foresight and the kind of control a military should exert over its use of lethal force, that is, even if such rationales were confused or not always differentiated.

The specific tactical aims of aerial bombing strategy reflected some of this confusion. As the 1926 US bombardment manual indicates:

> The strategical employment of bombardment in stabilized warfare is popularly conceived to be the true role of that class of aviation, chiefly for the reason that such employment offers the most striking opportunities for taking advantage of the great range of bombardment to bring under fire objectives heretofore considered as being absolutely safe. Such was the principal use made of bombardment in the World War, especially by the Independent Air Force, and, without question, such use will be made of it in future wars.[170]

The manual goes on to explain that militaries could employ strategic bombing to bring "fire to bear upon any military objective of sufficient importance to warrant the effort" which included "industrial centers" and "important lines of communications."[171] This is because, especially at the beginning of a war, a bombing campaign is an efficient way "to weaken the morale of the enemy people by attacks on centers of populations."[172] What is more, a nation might also resort to strategic bombing "in retaliation for hostile attacks on similar targets, important political centers."[173] Much like several of the above theorists indicate, it was crucial that a nation's air force commit to the attack: "Sporadic

170 US Air Service Tactical School, Langley Field, VA, *Bombardment* (Washington DC: Government Printing Office, 1926), 72.

171 Ibid., 73. Another manual of this period similarly notes: "One of the most important functions of the air service in warfare is the conduct of bombing raids against the enemy. These are generally directed against important objectives, such as munitions plants, arsenals, lines of communication, trains, convoys, field works, and trenches." Hollis Muller, *Manual of Military Aviation* (Menasha, WI: George Banta Publishing, 1918), 393.

172 *Bombardment*, 62.

173 Ibid.

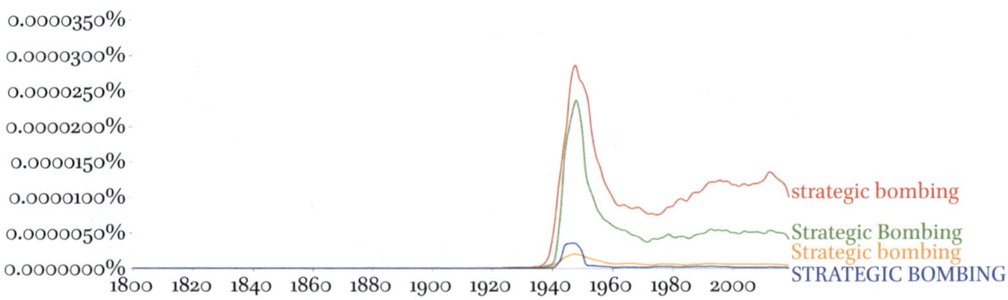

FIGURE 7.1 Ngram search of 'strategic bombing'

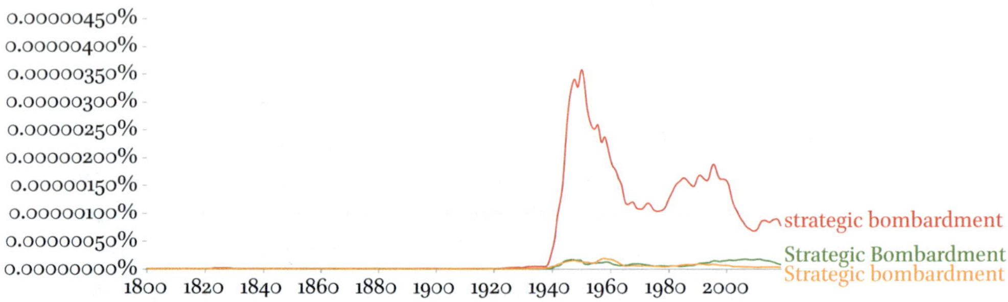

FIGURE 7.2 Ngram search of 'strategic bombardment'

raids may be used to affect morale or to divert the attention of the enemy or to harass some objective, but in the case of a target of vital importance they usually are not sufficient. *A constant pounding is necessary*."[174]

As Figure 7.1 shows, the language of "strategic bombing" started to appear, if only faintly, in the literature of the interwar period. This seems to capture some of the early attempts to incorporate aerial bombing into the ascendant tradition of military necessity, which historically made allowances for a sufficient minimum number of civilian casualties to expedite an end to hostilities. As can be seen in the strategic literature from the outbreak of the war throughout the interwar period, the turn to "strategic bombing" was largely due to the theoretical community's attempt to grapple with the horrors of World War I, the untapped strategic potential of aerial weapons, and the hope that the cruelties of future war might be mitigated through strategically minimal civilian death. Incidentally, it is striking that there was no notion of "strategic bombardment" in the nineteenth century (Figure 7.2). Since fortified cities were *prima facie* strategic, every bombardment was "strategic." This shows that even the language of strategic/non-strategic was an implicit internalization of the

174 Ibid. My emphasis.

developing norms around discrimination that were not entailed in the logic of nineteenth-century war.

In retrospect, it is easy to read into the bombing realists' documents of this period either a cavalier attitude to civilian death or a tragic naiveté that aerial bombing would decisively expedite peace. In 1934 M. Rougeron argues that aerial bombing is "the last trace of certain mystics to whom the War of 1914 was a death blow."[175] In reality, the threat of aerial devastation would provide no real deterrent to the outbreak of war. While Rougeron's argument was something of an exaggeration – in many ways World War I led theorists to redouble their commitment to the "mysticism" of the decisive victory stemming from aerial bombing; it certainly captures something of the mystical spirit of interwar bombing theory. One must remember that looking backward from those of us living in the wake of World War II distorts some of the optimism and confidence that aerial bombing might be a tool of peace. After all, from the perspective of 1918, what could be worse than the millions upon millions of lives lost in conventional warfare? The language of deterrence and the pseudo-scientific assurance that strategic bombing would save lives were difficult to resist.

It is also important to remember that the evolution of the British and US policies on strategic bombing were complex and uncertain.[176] Some have seen the evolution to terror or saturation bombing in World War II as overly determined or specifically engineered by certain key forward-thinking analysts and policymakers like those bombing realists mentioned above. While it is no doubt true that many of the theorists above helped to influence the strategic outlook of that period, it is less likely that what eventually became the reality

175 Quoted in Major C.H. Wash, "Revue de L'Armee de L'air," in *Military Review* 15:57 (1935); 110–112. 111.

176 Some of the major texts that address the evolution of this phenomenon: I.B. Holley, *Ideas and Weapons: Exploitation of the Aerial Weapon by the United States during World War I: A Study in the Relationship of Technological Advance, Military Doctrine, and the Development of Weapons* (New Haven: Yale University Press, 1953); Robert Finney, *History of the Air Corps Tactical School, 1920–1940* (Air Force: Office of Air Force History, 1955); Neville Jones, *The Origins of Strategic Bombing: A Study of the Development of British Air Strategic Thought and Practice u to 1918* (London: Kimber Press, 1973); Maurer Maurer and Albert Simpson *U.S. Air Service in World War I* (Air Force: Office of Air Force History, 1978); Robert Futrell, *Ideas, Concepts, and Doctrines: Basic Thinking in the United States Air Force, 1907–1964* (Maxwell Air Force Base, Alabama: Air University Press, 1989); Tami Davis-Biddle, *Rhetoric and Reality in Air Warfare: The Evolution of British and American Ideas about Strategic Bombing, 1914–1945* (Princeton: Princeton University Press, 2002); John F. Kreis, *Air Warfare and Air Base Air Defense* (Honolulu: HI: University Press of the Pacific, 2002); and Stephen McFarland, *America's Pursuit of Precision Bombing 1904–1945* (Tuscaloosa: University Alabama Press, 2008).

of strategic bombing in World War II is what they had in mind. After all, as Mitchell explains, "The menace will be so great that either a state will hesitate to go to war, or, having engaged in war, will make the contest much sharper, more decisive, and more quickly finished. This will result in a diminished loss of life and treasure and will thus be a distinct benefit to civilization."[177] Mitchell's optimism, which runs throughout virtually all of the literature above, belies a high degree of confidence that aerial bombing, if it were used at all, would be cruel, but quick. In retrospect it is by no means clear that the above theorists were anticipating the multi-year horror of aerial bombing as it unfolded in World War II, Korea, or Vietnam.

It should also be remembered that many features of what became known as strategic bombing were already deeply engrained in the thinking concerning the bombardment of defended and fortified cities, which goes all the way back to the early days of siege warfare. Bombing residential districts was such a well-established norm in the laws of war because it was widely believed to work.[178] The tactic, while a brutal expedient, typically brought sieges to a speedy conclusion without the need for a bloody assault on the battlements. Given the epistemic uncertainty of aerial bombing, the interwar advocates of bombing civilians had every reason to assume that they too were tapping into the utilitarian ethic that had governed so much of military necessity in the prior century.

Despite technological advances, some assumed that the moral and tactical framework was essentially unchanged. For instance, in his 1918 update to the classic *The Principles of War*, General Foch argues that "The present war has, in spite of all its novel features, shown once more that the fundamental principles and tactics remain unchanged."[179] A cursory acquaintance with the military history of killing from a distance reveals the legacy upon which aerial bombing was theorizing. Thus, though the interwar theorists did not necessarily intend for the emergence of what became saturation and terror bombing in World War II, the moral and tactical framework of the nineteenth century

177 Mitchell, *Winged Defense*, 16.

178 Though, some theorists questioned the degree to which bombarding residential districts in a siege really did work. Baker writes, "The exceptional practice of which mention has been made consists in the bombardment … of the houses of the town itself in order to put an indirect pressure on the commandant inducing him to surrender on account of the misery suffered by the inhabitants. The measure is one of peculiar cruelty, and it is not only unnecessary, but more often than not is unsuccessful." Baker, *The Laws of Land Warfare*, 199.

179 Ferdinand Foch, *The Principles of War* (New York: the H.K. Fly Company, 1918), 3.

was clearly being stretched to its limit as militaries started to integrate aerial bombing into their planning.

To be sure, rather than articulating a new ethic, these theorists were doing their best to salvage the normative tradition they had been steeped and educated in. For instance, in Arthur 'Bomber' Harris' *Bomber Offensive* (1947),[180] he did not need the prior two decades of interwar theory on the usefulness of strategic bombing to understand the moral legacy the Allied bombing campaign in World War II was tapping into:

> Whenever the fact that our aircraft occasionally killed women and children is cast in my teeth, I always produce this example of the blockade [which killed an estimated 800,000 Germans], although there are endless others to be got from the wars of the past. I never forget, as so many do, that in all normal warfare of the past, and of the not distant past, it was the common practice to besiege cities and, if they refused to surrender when called upon with due formality to do so, every living thing in them was in the end put to the sword. Even in more civilized times of today the siege of cities, accompanied by the bombardment of the city as a whole, is still a normal practice; in no circumstances were women and children allowed to pass out of the city, because their presence in it and their consumption of food would inevitably hasten the end of the siege. And as to bombardment, what city in what war has ever failed to receive the maximum bombardment from all enemy artillery within range so long as it had continued resistance?[181]

That militaries would bomb at whatever range they could in order to apply pressure to a besieged place was the formative nineteenth-century logic that underwrote strategic bombing campaigns.[182] In any case, at least in the early phase of thinking on strategic bombing, it entailed some of the humanitarian

180 Sir Author Harris was the Command-in-Chief of RAF Bomber Command during World War II. Incidentally, Harris had been deeply involved in Britain's interwar colonial bombing program.

181 Arthur Harris, *Bomber Offensive* [1947] (Barnsley: Pen & Sword Books, 2005), 177.

182 As one interwar commentator notes: "Assuming that aerial bombardment is regulated by the same supposed rule relating to land bombardment, there would be no obligation to give notice – surprise is an essential element in successful aerial attack – not to space civilians in a fortified town. In land bombardment they may be directly fired on and driven back into the besieged place." Rowen Allen Greer, *International Aerial Regulations* (Washington D.C.: Government Printing Office, 1926), 27.

(even if misguided) overtones found in the early formulations of military necessity.[183]

Even among bombing optimists, this sentiment was not necessarily universal. For instance, though a proponent of aerial bombing, at least theoretically, Sir George Aston was somewhat ambivalent about the effectiveness of bombing population centers:

> There is some prospect of producing a serious moral effect upon the population of a country that has not seen hostile military forces within its borders, and has no knowledge of their procedure … Military forces employed on such a venture would, of course, play upon this moral effect to the utmost, and with this end in view might select great industrial centers, or even London, as their objective.[184]

Aston, however, provides a note of caution: "though, there would always be a feeling of doubt at the back of the military commander's mind as to the measures he would take if his threats of vengeance were not responded to."[185] To this Aston develops a middle road strategy. Rather than bombing specifically for the morale effect, it is better to "select an objective of which the destruction or capture of which would produce a definite effect upon the naval or military forces, and trust that at the same time a sufficient moral effect would thereby be produced upon the population to bring about panics, and 'stop-the-war' mobs, to hamper the Government."

Furthermore, Aston provides one additional caution: "There is always the possibility of producing the opposite effect, and exasperating the population with the raiders, rather than with their own Government."[186]

It deserves to be reiterated that even though the arguments of the bombing realists enumerated above sound forebodingly tragic to twenty-first century ears, they reflect an early effort to translate and apply certain aspects of the logic of eighteenth- and nineteenth-century land and naval bombardments to aerial bombing. Despite Douhet's and Fuller's belief that they were articulating a new ethic of war, there is little difference in the formal ethical dimension to what they were arguing for and the claims that were made about fortified cities and defended towns for many centuries prior. And since throughout the early

183 Mark Clodfelter, *Beneficial Bombing: The Progressive Foundations of American Air Power, 1917–1945* (Lincoln: University of Nebraska Press, 2010).

184 Sir George Aston, *Sea, Land, and Air Strategy* (London: John Murray, 1914), 296.

185 Ibid., 297.

186 Ibid.

twentieth century nations themselves were increasingly being seen as fortified entities – the civil and economic components of which underwrite their war-making potential – the more difficult it was to see civilians as exempt from the brutality of war.[187]

As alluded to above, the trends and tensions regarding the degree to which combatants and noncombatants could be distinguished, what responsibility they bore for their own suffering as residents of an enemy nation, and the degree to which one could reasonably distinguish between defended and non-defended cities were obvious and troubling to nineteenth-century commentators on war. In the 1880s, Sheldon Amos had noted:

> During the last quarter of a century, War has been conducted on a scale of magnitude, and with the help of scientific inventions, wholly unknown in former times. The notion of an "armed nation" is no longer confined to the attitude of a country as the exceptional crisis of suffering an invasion. The notion has been extended, and is being increasingly extended, to the chronic condition of all the leading States in times of what is, apparently, the most profound Peace. Thus, whenever War actually takes place, its effects on life, limb, commerce, wealth, and honour radiate to a far wider circumference, and are intensified at every point to which they reach. The provision for the circumscription of War to the combatants professionally engaged, laboriously as it attempted in form, rather express the ever-increasing difficulty of such circumscription, than indicates any natural tendency towards the separation of the fighting from the non-fighting part of the population.[188]

Given the radiating effects of modern war, it was already becoming increasingly difficult to reasonably separate those fighting from those not fighting. Indeed, one early twentieth-century commentator remarked: "it is next to impossible to protect genuine non-combatants unless they and combatants are kept distinct – the only alternatives would be either to herd all the former in towns or camps severely guarded, or to organize an elaborate and probably unworkable system of passes."[189]

187 This argument was also formalized later in Alexander Seversky, *Victory Through Air Power* (New York: Simon & Schuster, 1942).

188 Amos, *Political and Legal Remedies for War*, 324–325.

189 "The Position of Irregular Troops under the Laws of War," *The Law Times Volume 138* (London: Field & Queen, 1915), 527.

Keeping combatants and noncombatants separate was hardly a reasonable prospect, even in the late nineteenth century. In fact, in exasperation, many theorists started to simply give up on the relevance of making such distinctions. Sterling Edmunds writes

> Article XXV of The Hague Regulations declares: "The attack or bombardment, by whatever means, of towns, villages, dwellings, or buildings which are undefended is prohibited." Here is a prohibition apparently designed to spare unfortified places, and the civilian population therein, from land or aerial attack, but it is wholly illusory in the absence of any definition of a "defended" town or village. A town is defended if there is present a military force, but, how numerous must that force be? Must it consist of a half dozen soldiers or a thousand? Professor Garner quotes Colonel Jackson of the Royal Engineers as saying in April 1914, that London was a defended town, and could complain of bombardment only if it were prepared to surrender to enemy aircraft flying over it and making such a demand.[190]

Elvira Fradkin concludes future war will "endanger millions of civilians *formerly considered as non-combatants*. A far greater proportion of the population is liable to extermination and serious casualties from deadly airplane attacks than were involved in past war."[191] Given the reality of how aerial bombing was beginning to unfold, Frank Quindry makes the grim assessment that,

> Historically there seems to be no other conclusion but that there has never been a distinction made between combatants and non-combatants, except as was justified by military circumstances. Commanders of military and naval forces have never hesitated to destroy private property and devastate the lands of the enemy when the military situation permitted and demanded it.[192]

---

190 Sterling E. Edmunds, "The Laws of War: Their Rise in the Nineteenth Century and Their Collapse in the Twentieth," *Virginia Law Review* 15:4 (1929); 321–349. 344.

191 Elvira Fradkin, *The Air Menace and the Answer* (New York: Macmillan Company, 1934), 33–34. My emphasis.

192 Frank Quindry, "Aerial Bombardment of Civilian and Military Objectives," *Journal of Air Law and Commerce* 2:4 (1931); 474–509. 492.

Quindry goes on to give expression to a point that had been made by Royse a few years earlier – that the very idea of noncombatant protection was a fiction created by technological limitations. Twentieth-century theorists beginning to promote the idea of noncombatant immunity had misunderstood the historical basis for protections extended to those not fighting.

In this respect, Douhet's and Fuller's claim, for instance, that in modern wars the distinction between combatant and noncombatant had been obviated bears a striking similarity to Henry Halleck's argument that inhabitants of fortified garrisons "are converted into soldiers by the necessities of self-defense." Once again, this illuminates the pyrrhic victory of nineteenth-century efforts to prioritize the professionalization of their armies and the elimination of irregular soldiers. By the twentieth century, armies were professionalized, and one could much more coherently speak of combatants and noncombatants. However, this achievement was immediately undermined by the advent of aerial bombing. As the industrialized nation-state became a war machine, it was difficult to sustain the increasingly tenuous distinction between legitimate and illegitimate targets, especially since civilians had become deeply implicated in the war economies of nations, a conceptual linkage that had already been forged in the nineteenth century.[193] In a certain sense, theorists like Spaight, Douhet, and Fuller represent some of the last gasp of full-throated support for nineteenth-century ethical norms.

6 Conclusion

World War I failed to resolve the pre-war debates about aerial bombing. Some continued to regard the technology as full of apocalyptic potential. These liberal internationalist commentators argued that the global community should unite to strictly regulate or even completely ban these new weapons. In this period, one sees an array of international conferences and conventions dedicated to mitigating the terrible outcomes that could arise, undoubtedly leading to so much civilian death.

---

193 Amos (1880) argues "The mere fact that it is impossible for a modern civilized State to go to War, or at least to maintain a War, unless the Government can so far conciliate a popularly-constituted Assembly as to procure the necessary supplies, had the effect of launching the vital topic of how the War arose into the field of public discussion of the most serious and responsible kind, and also of making every War, in reality as well as in form, *to be waged by the whole people*." Amos, *Political and Legal Remedies for War*, 36. My emphasis.

In contrast, bombing realists noted that conventional weapons had already decimated continental Europe. They hoped that the strategic shift to aerial bombing would have a humanizing effect on war. By making it so costly, nations would be that much more reluctant to wage war over trivial matters. And even if war did break out, it would likely be quick. Deploying nineteenth-century ethical formulations rooted in necessity, these commentators argued that targeting some strategically minimum number of civilians would lead to a swift end to hostilities.

CHAPTER 8

The Return to Intention: Post World War I

1 Introduction

As seen in the previous chapter, a parallel set of arguments had begun to develop around aerial bombing from the early twentieth century to the end of the interwar period. There were "bombing realists" who advocated for the explicit bombing of civilian populations to expedite peace. In this tradition, many prominent strategists and theorists argued that war could be limited by cruel but quick bombing raids that killed some sufficient minimum number of civilians. This argument represents what can be described as the final effort to subsume aerial bombing under the utility framework that was operative throughout the nineteenth century.

Against this tradition were those liberal-minded internationalists who believed that bombing could be regulated and contained either through more rigorous development and application of the laws of war or, more ambitiously, total disarmament. Through numerous international conventions and negotiations, commentators of this persuasion started to advance moral formulations that privileged the intentions of belligerents. Contrary to nineteenth-century norms, the moral liability of aerial bombing came to be described as the intentional harm directed at civilians. Within this framework, nations that deliberately killed or terrorized civilians were believed to be guilty of violating the laws of war.

Take note of this consequential shift. As seen in the previous chapters, at no point in the nineteenth century did theorists appeal to the good intentions of the belligerent to justify or excuse harm to civilians. Indeed, right up to World War I and throughout the interwar period international law jurists and military planners unselfconsciously argued that militaries could deliberately kill people who might otherwise be considered noncombatants. Some sufficient minimum amount of civilian death may be necessary to bring about an expeditious conclusion to the war. The moral bright line was not the intention to do harm but whether the destruction was excessive or wanton, a distinction left to the judgment of generals and other professionals who might have first-hand experience with such matters. The discursive shift seen in the international documents that sought to regulate bombing by focusing on intentions is remarkable, especially since such arguments had been long abandoned.

 | DOI:10.1163/9789004515482_010

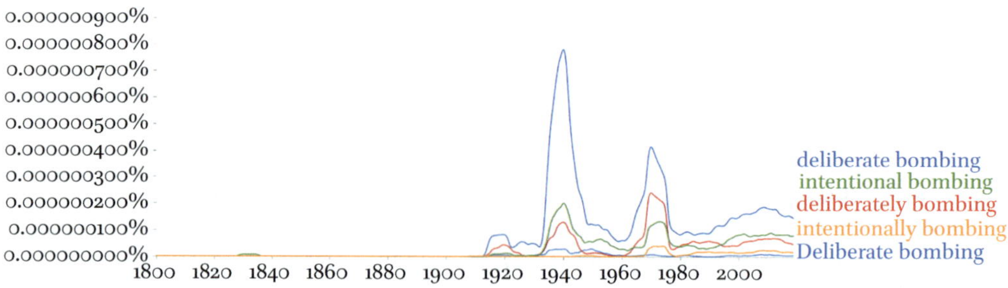

FIGURE 8.1 Ngram search of 'deliberate(ly) bombing,' 'intentional(ly) bombing'

Indeed, to find intention-based justifications for civilian death, one must go back to the remnants of Scholastic just war theory that had faded into obscurity at the end of the eighteenth century.

That this shift was taking place has been clearly demonstrated in the previous chapter. What is less clear are the driving elements of this change. The purpose of this chapter is to more explicitly thematize the factors that contributed to this transformation in moral framing. After all, by the end of World War II no one was theorizing aerial bombing in the way interwar theorists were. Why did utility or necessity-based arguments begin to fail, and why did intention-based arguments begin to flourish?

What will be shown is that there were three primary elements that best explain this shift. The first is the framing of German guilt at the end of World War I. Part of the Allies' rhetorical strategy was to attribute *intentional* harms explicitly to German pilots and implicitly to those who authorized their missions. As seen in Figure 8.1, there was a noticeable surge in references to intentional or deliberate bombing in this period. While this language would become overshadowed in the public discourse around bombing in World War II, as bombing campaigns led to such an unprecedented degree of destruction, it is remarkable that this language finds its origin in the interwar period. Furthermore, while much of this commentary is a sincere attempt to account for German war crimes, it is difficult not to conclude that at least some of it was merely rhetorical. The fact that a) British pilots were in many cases guilty of the same or similar behaviors, and b) the technological limitations of aerial bombing made it difficult to speak coherently about intentions, undermined the disapprobation heaped onto Germany. Despite these contradictions and difficulties, characterizing the behavior of German pilots in the war as intentional created a sufficient conceptual differentiation that began to take hold in popular discourse.

Additionally, while the late nineteenth-century debates on the laws of war generally permitted those actions believed to be necessary to win, haltingly there emerged within this discourse a counter-narrative which emphasized the laws of humanity and public conscience. Going back to Francis Lieber's "General Orders, No. 100" (1863), one finds sporadic appeals to a military authority's "humanity" and sense of justice – the implication being that strict adherence to a rigidly defined sense of necessity should be tempered and ultimately give way to humanitarian impulses. Perhaps a lingering element of what the humanist tradition called "charity," this language was formally embodied in the Hague Conventions (1899, 1907). Specifically, the Martens Clause (named after Friedrich Martens, a Russian delegate) explicitly emphasizes the incompleteness of international law and it presumes that the contracting parties to the Conventions should conceptualize necessity as part of a linear, progressive development. If custom had dictated that certain practices were unnecessary at one period, they should not be thought of as potentially necessary at some later point in the future. The goal was to foreclose on the potential shifts – indeed, like aerial bombing – that would overturn and undermine any agreements about what counted as necessary. The laws of necessity should be thought of progressively; they were making war humane and less violent. Through this redefined conception of necessity, militaries should be seen as striving to honor the progressive customs of war that are grounded in the principle of humanity.

These seeds had been conspicuously planted in the Hague conventions at the turn of the century. By the end of World War I, they started to come into fruition. At that time there was a noticeable shift away from the nineteenth-century language of necessity in favor of the laws of humanity.[1] There is one diplomatic event that dramatized this shift in norms. At the Preliminary Peace Conference (1919), the Commission on the Responsibility of the Authors of the War routinely characterized German war crimes as not only deliberate or intentional, but more specifically as violating the laws of humanity. The report concludes:

> In spite of the explicit regulations, of established customs, and of the clear dictates of humanity, Germany and her allies have piled outrage upon outrage ... Violations of the rights of combatants, of the rights of civilians, and of the rights of both, are multiplied in this list of the most

1 For a broader treatment of this subject matter, see Bruno Cabanes, *The Great War and the Origins of Humanitarianism, 1918–1924* (New York: Cambridge University Press, 2014).

> cruel practices which primitive barbarism, aided by all the resources of modern science, could devise for the execution of a system of terrorism carefully planned and carried out to the end. Not even prisoners, or wounded, or women, or children have been respected by belligerents who deliberately sought to strike terror into every heart for the purpose of repressing all resistance.[2]

Rather than exceeding the *dictates of necessity*, the crimes of Germany were cast as intentionally violating the *dictates of humanity*, that is, the rights of those who should not be subjected to cruelty. This consequential indeed paradigmatic shift did not go unchallenged. In a competing document, the "Memorandum of Reservations," composed by Robert Lansing, the US delegation opposed this shift in the premise of German guilt.[3] Lansing argued that the laws of humanity were far too vague to serve as a basis for liability. These internal debates among the victors are significant because they illuminate important tensions in military ethical discourse at this key transitional moment. The US resistance to the principles of humanity, much like the bombing realists of the previous chapter, represent the last vestiges of nineteenth-century military ethical theory.

The fact that the Commission chose to demonstrate the violation of the laws of humanity through the enumeration of Germany's "cruel practices" is indicative of another shift in conceptualizing war crimes. It is significant that the emphasis on the laws of humanity over necessity coincides with the rise of a novel social scientific war literature that sought to catalogue the cost of war in terms of the hardships meted out to civilians in both qualitative and quantitative terms. While this "cost of war" literature can be traced back to the peace movements and war correspondent literature of the nineteenth century, much of which attempted to quantify the wastefulness of war, in the

2 "Commission on the Responsibility of the Authors of the War and on Enforcement of Penalties," *The American Journal of International Law* 14:1/2 (1920); 95–154, 113.

3 Broadly speaking, the Americans rejected this transition away from strict military necessity. In the Treaty of Peace with Germany Hearings in Congress that took place between 31 July and 12 September 1919, the US representatives set out two reasons that prevented them "from consenting to the tribunal recommended by the Commission. The First of these [was] the uncertainty of the law to be administered in that liability is made to depend not only upon violations of the laws and customs of war, but also upon violations 'of the laws of humanity.' The second of these reasons [was] that Heads of States are included within the civil and military authorities of the enemy countries to be tried and punished for violations of the laws and customs of war and the laws of humanity." *Treaty of Peace with Germany, Hearings before the Committee on Foreign Relations, United States Senate, 66th Congress* (Washington DC: Government Printing Office, 1919), 371.

twentieth century finds its origin in the seminal Carnegie report on the Balkan War (1914). This document, grounded in field work and rigorous methodology, ambitiously sought to catalogue of crimes and atrocities perpetrated during the Second Balkan War. From 1914 through the interwar period there were several similar reports dedicated to recording the plight faced by civilians in war. Much of this literature conspicuously links the plight of civilians to the laws of humanity.

The quantification of human suffering had a poignant impact on the legitimacy of necessity calculations that governed military operations. As civilian body counts started to enter public discourse, it became much more difficult for military planners to write them off cavalierly as merely the necessary costs of doing business, especially when new practices, like aerial bombing, threatened to dramatically increase the number of noncombatant casualties. As the humanity of the innocent victims of war gained new focus, it became much more difficult to disregard anticipated deaths as necessary expedients. Indeed, public discourse could no longer abide the seemingly calloused rhetoric that permitted the intentional killing of innocent women and children to expedite victory. Under this new condition, whereby the humanity of the belligerent was given greater prominence, the only way militaries could honor noncombatants was by not intentionally killing or terrorizing them.

The second key element that explains that shift away from necessity to intention was a return to the Scholastics. With the surge of liberal internationalism in the interwar period, scholars and jurists started to ransack the archives of history to develop a firmer theoretical basis for the emerging international order. This can be demonstrated through the rediscovery and rehabilitation of much older lines of inquiry on the law of nations, including Scholastic formulations which had been largely set aside in the late eighteenth century. Specifically, from just before World War I to the 1930s there was a resurgence of interest in the work of Francisco de Vitoria, one of the intellectual founders of the School of Salamanca, and to a lesser extent, Francisco Suárez.[4]

---

4 The French pacifist Alfred Vanderpol is an early example of this trend. See Alfred Vanderpol, *La Guerre devant le christianisme, suivi d'une traduction du « De Jure Belli » de François de Victoria* (Paris : A. Tralin, 1912) and *La doctrine scolastique du droit de guerre* (Paris : A. Pedone, 1919). See also Herbert F. Wright, *Francisci de Vitoria* De Iure Belli Relectio (Washington: Published by Author, 1916). In 1917, as part of the Carnegie Endowment for International Peace's "Classics of International Law Series," James Brown Scott edited a reprint of Vitoria's *De Indis*. Edwin Dickinson dedicates a chapter to scholastic predecessors to Grotius, see Edwin Dickinson, *The Equality of States in International Law* (Cambridge, MA: Harvard University Press, 1920). For works on Vitoria, see Camilo Barcia Trelles, *Francisco de Victoria, Fundador del Derecho Internacional Moderno* (Cuesta: Talleres Tipográficos, 1928); Charles McKenna, *Francis de Victoria, Founder of International Law* (Washington, D.C.: Catholic

This rehabilitation effort is largely due to the efforts of one man, James Brown Scott, a leading American international legal mind of the early twentieth century. It is no coincidence that with this renewal of interest in Scholastic commentary on international law that intention-based, double effect arguments for justifying the killing of the innocent also started to take on a new relevance.

And finally, as double effect arguments that privileged the intentions of belligerents started to gain traction in the secular international legal discourse, there arose a vociferous Catholic opposition to aerial bombing. Many prominent early twentieth-century Catholic theologians argued that aerial bombing violated the principle of double effect. Since strategic bombing necessarily meant the targeting of population centers, it was impossible to say that the deaths of the citizens who lived there were unintentional. In fact, it appeared as though hitting the military target was the secondary rather than primary effect. Ironically, despite this Catholic opposition, the conceptual linkage of bombing to double effect served to fortify the popular association between bombing and intention. These Catholic writers were forcefully arguing *against* secular strategists and theorists who claimed that bombing was permitted if civilian deaths were not intentional, but this opposition further heightened the centrality of double effect in formulating excuses for noncombatant death in war.

2 German Guilt

Despite the relatively marginal role aerial bombing played in World War I, the cruelty of bombing innocent civilians was specifically singled out as indicative of the fact that Germans were, plain and simple, bad actors. Among the Allies, it was common to decry the bad intentions of German pilots in contrast to the

Association for International Peace, 1930); and James Brown Scott, *The Spanish Origin of International Law: Francisco de Vitoria and His Law of Nations* (Oxford: Clarendon Press, 1934); James Brown Scott, *The Catholic Conception of International Law* (Washington D.C.: Georgetown University Press, 1934); and John Eppstein, *The Catholic Tradition of the Law of Nations* (London: Burns, Oates, & Washbourne, 1935). For works on Suárez, see Joaquín Carreras y Artau, *Doctrinas de Francisco Suárez Acerca del Derecho de Gentes y Sus Relaciones con el Derecho Natural* (Gerona: Carreras, 1921); Adolfo Miaja de la Muela, *Internacionalistas Españoles del siglo XVI, Francisco Suárez (1512–1569)* (Valladolid: Cuesta, 1932); Herbert Wright, *Francisco Suárez: Address in Commemoration of His Contribution to International Law and Politics* (Washington D.C.: Catholic University of America Press, 1933); and James Brown Scott, *Law, the State, and the International Community: A Commentary on the Development of Legal, Political, and International Ideals* (New York: Columbia University Press, 1939).

good intentions of Allied pilots in the interwar period. William Hornaday, for instance, argues:

> The Kaiser and Zeppelin, and the German people, have spent many millions of dollars in *deliberate attempts* to slaughter the unarmed inhabitants of London, and strafe England. All the German talk about attacking 'the fortress of London' is beneath contempt. Rarely indeed has a soldier been injured in London, or any other English city, by a Zeppelin or an airplane bomb. It has been the helpless women, school-children and other non-combatants who have been blown to pieces.[5]

Davison, chairman of the War Council of the Red Cross, reports:

> The outstanding feature of German method at the present time is the effort to terrorize women, children, and old men at home. While the German troops are making their drive on the front, airplanes are bombing nearly every night towns behind the lines, with the *deliberate and declared purpose of terrorizing civilians*. The purpose of the flight behind the lines is to break down the morale of the civilian population to such a point that they will importune their governments for peace. It is the most dastardly, unrighteous, cruel, and devilish plan which could be conceived. It is based upon the theory that the killing of four children out of five will induce the mother to implore her Government to have the war stop that her fifth child may live.[6]

C.C. Turner explains in *The Struggle in the Air, 1914–1919* (1919), "An unexpected phase of the war was afforded by the bombing by German aeroplanes on August 3, 1914, of Lunéville, an 'open' town. This was a foretaste of the indiscriminate bombing of undefended places by the enemy, *who calculated that they would terrorize the civilians* of the Allied nations into a demand for an early peace."[7]

And finally, in his 1924 compendium on *International Law*, Pitt Cobbett also argues that "Apart from genuine mistakes, there is abundant evidence that German aviators in numerous instances deliberately bombed hospitals.

5 William T. Hornaday, *A Searchlight on Germany: Germany's Blunders, Crimes and Punishment* (New York: American Defense Society, 1917), 22. (emphasis mine) See also, John Selden Willmore, *The Great Crime and Its Moral* (London: Hodder and Stoughton, 1917).

6 Quoted in "The Kaiser at Prayer," *The Literary Digest*, 1 June 1918, 42. (emphasis mine)

7 C.C. Turner, *The Struggle in the Air, 1914–1919* (London: Edward Arnold, 1919), 29–30. (emphasis mine)

British, American, French, and Belgian hospitals were so attacked, some of them repeatedly. In these attacks large numbers of hospital personnel and patients were killed or wounded."[8] He goes on to explain: "Apart, however, from any direct evidence of intention, there is every reason to believe that deliberate bombardment of hospitals, as of coastal towns and sinking of hospital ships, was part of the general policy of immobilising the enemy forces."[9] The bombing of hospitals, in particular, was a common accusation throughout this period; however, there is evidence to suggest that the Germans understood the bombing of hospitals differently, namely they believed they were either engaging in a form of in-kind retaliation or merely hit hospitals accidentally.[10]

Despite the frequently and, indeed, formal nature of these claims, it was merely an assumption that the Germans were deliberately using aerial bombing as a form of psychological warfare. David Johnson argues, isolated cases notwithstanding,

> There is very little evidence that any of the belligerents in the 1914–18 war adopted as a sustained policy the making of what in the Second World War became known as 'terror raids.' But that is not to say that the damage done to civilian lives and property even in the first conflict was inconsiderable, or that the potentiality of air raids as a weapon of psychological warfare was not appreciated.[11]

Nevertheless, recriminations against German brutality were persistent throughout the interwar period. While understandable, this language was often confusing and inconsistent. In many cases the Germans could level the same accusations against the British. Moreover, there were widely divergent assessments of what the new aerial bombing technology would allow.

Throughout World War I both sides, as Sterling Edmunds notes,

> advertised each the wickedness of the other in bombing hospitals, in using the Red Cross flag to cover observation posts, in attacks upon hospital ships and otherwise. How much of it was intentional and deliberate will never be known. Yet the Geneva Convention does not guarantee absolute immunity to hospitals and their staffs under all conditions.

8 Pitt Cobbett, *Leading Cases of International Law* (London: Sweet and Maxwell, 1924), 200.

9 Ibid.

10 See Ralph Ellinwood, *Behind German Lines: A Narrative of the Everyday Life an American Prisoner of War* (New York: Knickerbocker Press, 1920), 42; Franklin Henry Martin, *The World War* (Doubleday Doran, 1933); 457.

11 David Johnson, *Rights in Air Space* (Manchester, NH: Manchester University Press, 1965), 30.

> It merely imposes an obligation to respect them as far as the necessities of the military situation allow.[12]

Even though one could not know for sure if the bombings were deliberate, both sides proclaimed their innocence while accusing the other of intentional harm.[13] Ironically, these accusations were being made in a moral context that did not expressly prohibit certain harms when committed because of absolute necessity. As Edmunds notes, militaries could bomb hospitals and civilian structures when such an approach was believed to be absolutely necessary to victory.

To defend his country's honor in World War I, the German Secretary of State for Foreign Affairs, Wilhelm Solf, insisted,

> It is untrue that our troops slaughter wounded, *intentionally* blow up hospitals and schools, rob churches and commit other crimes against enemy prisoners and populations. The work of destruction which in our retreat, as in every operation of retirement in history, was unavoidable, has everywhere been restricted to measures of really indispensable harshness which are intended to prevent a retiring army from leaving in the hands of the enemy bases which to a great extent are of a military character.[14]

Solf goes on to claim that, in fact, "the allies had frequently bombarded French towns on the plea of military necessity and exposed the civil population to great danger and suffering. If such measures by the enemy were justifiable on military grounds, why was the German destruction not equally justifiable?"[15] Solf argues that the only intention operative in German practices were those oriented to legitimate ends. He points out that identical justifications were used by the Allies to defend their actions.

---

12 Sterling Edmunds, *The Lawless Law of Nations* (Washington, D.C.: John Byrne, 1925), 295–96. For similar accusations, see *The Times History of the War, Volume 11* (London: The Times Printing House Square, 1917), 312. For a detailed account of Allied use of propaganda during the war, see Arthor Ponsonby, *Falsehood in War-Time: Containing an Assortment of Lies Circulated throughout the Nations during the Great War* (London: E. Dutton & Company, 1928).

13 These were the same claims made by Germany against Britain. As one German headline suggests: "Nineteen enemy aeroplanes and two captive balloons were brought down yesterday in aerial fighting. A great many French civilians were killed by British aviators who dropped bombs on the hospital at Tourcoing." *Aerial Age*, Volume 7 (1918), 58.

14 Quoted in James Garner, *International Law and the World War* (New York: Longmans, Green and Co, 1920), 327.

15 Ibid.

In another passionate defense of German innocence, Ernst Müller-Meiningen, a member of the Reichstag, argues,

> The English flung bombs not only upon Freiburg, Friedrichshafen and Düsseldorf, all open cities, as well as that portion of Baden known as the Markgraveland with its peaceful villages far removed from the sphere of war operations, not to mention the entire district of South Baden, but they bombarded schools whereby chiefly children were slain. They shelled open towns like Dar-es-Salaam, Victoria in the Cameroons and Swakopmund, and endeavored with the most incredible meanness to carry out their "starving-out policy." Nevertheless in this war of annihilation they adopt a tone of the most revolting cant and moral indignation. This may delude the foolish and the unknowing, but it remains without the slightest effect in Germany.[16]

Müller-Meiningen shows that English accusations of intentional harm ring hollow when both sides were engaged in a "war of annihilation" which included a blockade which starved countless thousands of civilians. In any case, the English could also be characterized as intentionally bombing undefended and open towns.

The volley of recriminations between the Germans and British point to an important shift in ethical discourse that the realities of aerial bombing during World War I helped to initiate. It is significant that the war crimes of the Germans, and their counter claims against the English, were increasingly being characterized as *intentional* rather than, as the norms of the nineteenth century dictated, *unnecessary* or *gratuitous*. This shift in discourse was consequential, even if it was, at best, inconsistent, and, at worst, incoherent. The English aviation pioneers, Claude Grahame-White and Harry Harper, confirm this in their assessment of World War I bombing:

> Aviators are accused of barbarities of which they have no intention. A typical statement, and one quite wrong, is that, say, of an eye-witness who, walking with an ambulance-wagon near a railway-station, declares that a hostile airman, passing overhead at the moment, has dropped a bomb 'deliberately' at the ambulance. What has occurred, really, is that a bomb has fallen near the ambulance, but that the airman above has not

16 Ernst Müller-Meiningen, *Who are the Huns? The Law of Nations and Its Breakers* trans. R.L. Orchelle (Berlin: Georg Reimer Publisher, 1915), 254–55.

> had the faintest wish to place it there. His aim was at the railway-station. From his height he could not even see the ambulance.[17]

Considering the first-hand experiences of pilots, one begins to see how incoherent the language of intentional harm is when applied to aerial bombing. Slowly, however, this language of intentional harm started to percolate through the international legal community, and as seen in the previous chapter, it started to become enshrined in international law.

This legal-political discourse is significant, and it illustrates the extreme disjunction that arose between the public understanding of war-making and that which was acceptable by way of nineteenth-century military ethical norms. After all, reading the interwar bombing realists of the previous chapter, one would be hard-pressed to specify what Germany had been guilty of in their aerial attacks on English cities. These sentiments are evident in a testy exchange between MP Joseph King and Major Baird, Under-Secretary of State at the Air Ministry, during a 1918 Parliamentary discussion of British air warfare. Baird explains,

> I do not know what happens on the German side, but everybody who has been on any kind of front at all knows that within four or five miles on our side of the line there are women and children, and you cannot drop a bomb anywhere on any inhabited place without running the risk of killing women and children. Nothing is more unfortunate. When the hon. Gentleman says that he wants a declaration to the effect that our attacks against the enemy should be carried out "with military objectives," I must point out, of course, they are military objectives. We are at war, and everything that we are doing has a military objective – and in such a manner as to avoid, as far as possible, injury to non-combatants, women and children.[18]

And a bit later, he goes on to say:

> Do not let us bandy words. It is impossible to drop bombs on any towns – German, French, or English – without running the risk of killing women and children. Does the hon. Member mean that we aim not to drop bombs

17 Claude Grahame-White and Harry Harper, *Aircraft in the Great War: A Record and Study* (Chicago: A.C. McClurg, 1915), 179–80.

18 "Estimates Presented by Major Baird," House of Commons Debate February 1918 vol. 103 cc957–1068.

> on German towns? … [Because] it is physically impossible to drop bombs on German towns without running the risk of killing German women and children. I am sorry, and everybody who has anything to do with it is sorry for it. The whole of the towns in the Rhine Valley are towns where munitions are manufactured. It would be difficult to mention one where these are not manufactured. They are all objectives of military interest, and it is perfectly open to us to bomb those places, just as it is open to the Germans to bomb our places. I am sorry I cannot more clearly define the intention and policy of the Government than that, but I think it is sufficiently categorical. We have started bombing German towns. We mean to continue bombing German towns, just as the Germans bomb our towns, and will bomb our towns as long as they can. That is the policy of both countries. It is, unfortunately, one of the new phases of warfare invented by the Germans, and carried out by them.[19]

According to Under-Secretary Baird, all the towns in the Rhine Valley were legitimate military objectives since they produced munitions and other goods used in the war effort, in the same way that many English towns were likewise open to German bombing. Given the expansiveness of what had come to constitute legitimate targets (i.e. "everything that we are doing has a military objective"), and the technical limitations of aerial bombing, it remained unclear how one could ever reasonably condemn a foreign power for harming civilians, much less for striking in-land towns.

William Sherman reiterates these points in his 1926 treatise on *Air Warfare,*

> we do not need the *ex parte* statements of German leaders to prove that the Germans intended to inflict no more damage on the civilian population, than was absolutely necessary to accomplish a direct military purpose. The facts speak for themselves. No military man can regard the sporadic raids on London and Paris as a serious effort to accomplish a decisive military result. They were, after all, quite petty affairs, compared with what the Germans might have accomplished, even with the limited means of that time; there was none of the mass effect, none of the thoroughness, which characterized every major German operation. If sheer "frightfulness" had been the aim, a mass attack with gas bombs on London, for example, would have produced an infinitely more demoralizing effect on the unprotected population. The very aspect of these

19 Ibid.

> air attacks bespeaks their purpose: a diversion, pure and simple, and a remarkably successful one.[20]

Sherman implies that Germany would have been within its rights to inflict much greater damage on Allied cities than they did. In fact, if they had wanted to produce a "mass effect" through extensive terror campaigns, they could have easily done so. Instead, he points out that the German's took a much more limited approach; by engaging in sporadic raids, they obtained a more humane strategic effect.

Lord Trenchard also notes the paradox of attributing bad intentions to aerial bombers, "Whatever is bombed in another war … nothing we can say or do will prevent enemy propaganda from asserting that women and children are bombed intentionally, because, of course, a larger number of women and children will undoubtedly be hit."[21] From the very beginning, the language of intention easily fell into us/them comparisons: we may have killed civilians, but only unintentionally, while attempting to hit legitimate targets; they, on the other hand, deliberately target civilians in an attempt to terrorize the population.[22] While psychologically satisfying, such arguments were merely propaganda.

The emerging consensus among tacticians and theorists was that the bombing of city centers, like London, was justified, plain and simple. As Under-Secretary Baird explains, since manufacturing centers were nestled in and among residential areas, just about every city was transformed into a legitimate target. Ultimately, as Baird suggests, the relevant moral question was not whether to target civilians – given the technical limitations, the death of civilians would inexorably follow from the realities of air warfare – rather the real issue was whether there would be an aerial bombing campaign. If one were willing to allow for aerial bombing, there is very little one could say about civilian casualties. They would necessarily follow.

As can be seen in these passages, the technical limitations of bombing led to some degree of moral confusion around what the specific operational purpose

20 William Sherman, *Air Warfare* [1926] (Maxwell Air Force Base, Alabama: Air Force Press, 2002), 192.

21 Quoted in Spaight, *Bombing Vindicated*, 106.

22 Virginia Held points out this weakness of twenty-first-century moral arguments differentiating acceptable civilian death in war and outright terrorism: "What a lot of discussions of terrorism try of course to do is to come up with a definition such that what they do is terrorism and is unjustified, where what we and our friends do is not terrorism but a justified response to it, or is justified self-defense." Virginia Held, "Terrorism and War," *Journal of Ethics* 8:1 (2004); 59–75, 65.

of bombing campaigns was. Were they, for instance, designed to overwhelm population centers thereby eroding civilian morale through intentionally promiscuous bombing? Were they honest attempts to destroy the enemy's manufacturing capability, and with it the material ability to prosecute the war? Or was it both, a belt-and-suspenders approach to military planning? As seen in the tortured negotiations around bombing regulation in the interwar period, it was never easy to disentangle these seemingly discrete issues, and, in reality, it did not matter. Whatever the intentions of the pilots were, they all led to indistinguishable outcomes.

Take, for instance, the Air Ministry's 1919 report on Royal Air Force bombing in World War I. Trenchard explains that "It was not possible with the forces at my disposal to do sufficient material damage so as to completely destroy the industrial centres in question," which gave rise to the strategy of "attack[ing] as many of the large industrial centres as it was possible to reach with the machines at my disposal."[23] He goes on to explain:

> By attacking as many centres as could be reached, the moral effect was first of all very much greater, as no town felt safe, and it necessitated continued and thorough defensive measures on the part of the enemy to protect the many different localities over which my force was operating. At present the moral effect of bombing stands undoubtedly to the material effect in a proportion of 20 to 1, and therefore it was necessary to create the greatest moral effect possible.[24]

Notice that even within this statement there is a complicated bundle of issues. What made the "moral bombing" effective was that: a) they were striking industrial centers of production, perhaps limiting the ability of the enemy to prosecute the war, b) the unexpectedness and unpredictability of these strikes forced the enemy to take up more extensive defensive positions than they would have otherwise, which diverted crucial resources away from offense (the same point William Sherman makes above), and c) the domestic population was destabilized, perhaps even demoralized, by either the terror resulting from seemingly random air strikes or resignation that they would no longer be able to continue the war effort.

As seen in the early literature on "strategic bombing," it was not at all easy to disentangle the desire to demoralize the enemy from the desire to disable the means of prosecuting war; they were both part-in-parcel to "strategic

23 H. Trenchard, "Air Ministry," *Supplement to the London Gazette* 1 January 1919, 134.

24 Ibid., 135.

bombing." Theories promoting the overwhelming use of force never made tidy distinctions in terms of how the enemy would be demoralized, whether through outright terror or by dismantling and eliminating the means of prosecuting the war; indeed, they often entailed both. A decisive early blow might disable a key component of a country's war-making machine which would also elicit sufficient terror to prevent the enemy from retaliating. If one thinks back to siege bombardments, in many cases artillery fire would be directed at both magazines and residential districts. In much the same way, early bombing theory rarely made such clear-cut distinctions.

These assumptions about morale clearly seem to be operative in the early bombing strategy. If the destruction of industrial centers also led to depleted morale and an expedited surrender, it was not clear that anyone would have the right to complain, especially since the "industrial population" had become implicated in the modern war economy through the radiating effects of war. The sliding scale of targetability, which was operative throughout the nineteenth century, implied that as noncombatants edge closer to those spaces of legitimate attack, they lose their protected status. Plus, it was easy to read into the existing laws of war the permissiveness that privileges incidental harm to noncombatants. Elbridge Colby wryly notes that the laws of war "do not say that the bombs must fall exclusively on military objectives, only that they must be directed exclusively at such. They do not say that the bombardment of the civilian population is prohibited, merely that the indiscriminate bombardment of civilians is prohibited."[25] Of course, it is tempting to read onto this quote a dubious late twentieth-century conclusion about precision. All bombing was indiscriminate, even during World War II.[26]

All of this implies that there was something faintly incoherent about the language of intention being employed at this time. The nature of the technology was such that the intentions of the pilot or military planners were mostly irrelevant. Anything that happened after the bombs were released was largely outside of the pilot's control. Rigorous targeting was wishful thinking; bombs themselves were inherently indiscriminate. Compared to the accuracy of conventional artillery, precision aerial bombardment was far from assured. Pilots routinely thought they had hit their targets when they had not; sometimes they were not even flying over the correct cities that they intended to bomb.

25 Elbridge Colby, "Aërial Law and War Targets," *The American Journal of International Law* 19:4 (1925); 702–715. 715.

26 In 1943 "an average of only 13.6 percent of bombs fell within 1000 feet of the aiming point" (i.e. roughly the length of three football fields). Richard Overy, *The Bombers and the Bombed: Allied Air War Over Europe, 1940–1945* (New York: Viking, 2014). 157.

Frederick Talbot explains how inefficient aerial bombing was. The reason that so many bombs needed to be dropped was that it was imperative in order to increase the chances of actually hitting the target: "The damage inflicted by the bombs was not at all in proportion to the quantity of explosive used."[27] Another commentator just after World War I likewise concurred: "Where perhaps 10 [artillery] shells must be discharged in order to reach the object with mathematical precision, sometimes 100 bombs are required to obtain the same result. The greater the height at which an airship is flying, the harder it is to hit the bull's-eye."[28] And again, Hugh Trenchard reporting about bombing precision in World War I, explained: "I do not think you need to be anxious about our degree of accuracy when bombing stations in the middle of towns … all the pilots drop their eggs into the middle of the town generally."[29]

Given these claims, it is very difficult to speak of the intention to bomb civilians when pilots could not reliably hit their targets.[30] As Raleigh argues,

> Nothing is easier than to drop a detonating bomb, with good intentions, over the side of an aeroplane; the difficulty of hitting the mark lay in determining the flight of the bomb and in devising an efficient dropping gear. To drop a weight from a rapidly moving aeroplane so that it shall hit a particular spot on the surface of the earth is not an easy affair.[31]

Indeed, the good or bad intentions of pilots notwithstanding, the only intention that really seemed to matter was the intention to drop bombs or not. As Under-Secretary Baird noted above in his poignant question about English bombing during the war: civilians are mixed in with military industry – either we bomb towns or not. Much like the use of Congreve rockets in the nineteenth

27 Frederick A. Talbot, *Aeroplanes and Dirigibles of War* (Philadelphia: J.B. Lippincott, 1915), 112; for early analysis of bombing from airplanes, see "Bomb-Dropping from Aeroplanes," *Flight* 18 December 1914, 1213.

28 Houard, "Long-Distance Bombardment from Air-Planes and American Aviation," 329. See also the U.S. Bombing Survey, WWI, 1919 (summary), http://www.au.af.mil/au/awc/awcgate/ww1/bbg-ww1.htm Things were little better during WW2. See also, the post-WWI bomb tests, were still quite inaccurate: "The U.S. Aerial Bombing Experiments on Warships," *Flight* 15 September 1921, 615.

29 Quoted in, Steward Halsey Ross, *Strategic Bombing by the United States in World War II: The Myths and the Facts* (Jefferson, NC: McFarland and Co., 2003), 42.

30 Cyril Hall also notes: "It is a difficult matter to drop bombs with any degree of accuracy from a height of some thousands of feet." Cyril Hall, *Modern Weapons of War, By Land, Sea and Air* (London: Blackie and Son, 1915), 184–85.

31 Walter Raleigh, *The War in the Air: Being the Story of the Part Played in the Great War by the Royal Air Force, Volume 1* (London: Clarendon Press, 1922), 136.

century, even if one did not want to harm civilians the decision to use these weapons was inherently a choice to be functionally indiscriminate.

The very nature of the technology had put aerial bombing into conflict with the conventions designed to limit it. Of course, this did not prevent jurists and legal theorists from trying to subsume aerial bombing under the broader norms around bombing that had persisted up to that time, but this too was often perplexing. In one of the handful of cases referred to in the League of Nation's Greco-German Mixed Arbitration Tribunal after World War I, the ruling of *Kiriadolou v. Germany* (1930) was especially illuminating. In this case, the tribunal was tasked with determining if Germany operated within the bounds of the laws of war in the bombing of Bucharest in 1916.[32] In the end, the court held that Germany had violated international law since they did not give advance notice of their intention to bombard the city, which was the standard requirement regarding bombing fortified cities in land war. The requirement to inform was essential, the Tribunal concluded, "since an air-plane, which very frequently flies by night over the town threatened by bombardment at an altitude of several thousand metres, is not in a position to direct with precision the fall of bombs so as to ensure that they hit only the fortifications and war munitions, without harm being done to the persons and the property of non-combatants."[33] Since aerial bombing was, unlike artillery, unintentionally promiscuous, it needed additional restraint. Since harm was foreseeable, the only way to convey the intention to discriminate was to signal the intention to bombard in advance. This case was, indeed, odd, a vestige of a tradition that was being actively eroded by the technologies of waging war.

While Allied military strategists and theorists were quite sympathetic to the impulse to bomb industrial centers – indeed, many explicitly argued that Germany was within its rights to bomb places like London – popular public discourse was much harsher. Indeed, as seen above, public opinion was ruthlessly anti-German. At the end of World War I, there was a widespread outcry against Germany and, in particular, German pilots who were accused of deliberately bombing civilians *qua* civilians. What made this wrong, for the most part, was the pilots' explicit intention to harm people who should be exempted from the cruelties of war.

In an early reflection on the implications of German bombing practices, H.D. Hazeltine concludes that the spirit of international law up to that point implied that "the civilian population, as non-combatants, are equally immune

32 The claimant's husband was killed in the bombing. Elihu Lauterpacht, *International Law Reports* (Cambridge: Grotius Publications, 1981), 516.

33 Ibid., 517.

from deliberate and intentional attack by bombarding forces."[34] What made German attacks problematic was that they intended to harm noncombatants as noncombatants. H.W. Wilson argues that the crime committed by the German air force was that their "bombs were constantly dropped on small towns and residential districts of large towns. Military works were avoided; in general, the German airships were nowhere near munitions works."[35] An American commentator, on German bombing, Georges Houard, likewise writes:

> At the present time the two belligerent groups are following altogether different tactics. While one group is pouncing upon open towns, seeking only the moral effect, the other, with a military objective in view, seeks to reach its enemy's war-mechanism where it is functioning most actively. The former bombards Paris, Nancy, Calais, and Dunkirk, killing chiefly women and children; the second's targets are behind the enemy's front lines where troops are concentrated, and material and munitions. One seeks effects, the other obtains results, and the true advantage accrues to the last.[36]

According to Houard, the Allies killed civilians incidentally, as an accidental side-effect of a legitimate strike, whereas Germans killed them deliberately as a part of a terror campaign. And in another attempt to distinguish between British and German bombing tactics in this same period, Willis Abbot explains,

> Merely to inflict anguish and death upon a great number of civilians, and those largely women and children, is obviously of no military service. But if such suffering is inflicted in the course of an attack which promises the destruction or even the crippling of works of military character like arsenals, munition plants, or naval stores, it must be accepted as an incident of legitimate warfare.[37]

---

34 H.D. Hazeltine, "Recent Growth on Aerial Law," *Flight International* 10:1 (1919), 295–375, 375.

35 H.W. Wilson, *Convicted Out of Her Own Mouth: The Record of German Crimes* (New York: George H. Doran Company, 1917), 23.

36 Georges Houard, "Long-Distance Bombardment from Air-Planes and American Aviation," *The Living Age* Eighth Series, Volume XII (1918), 329.

37 Willis John Abbot, *Aircraft and Submarines: The Story of the Invention, Development and Present-Day Uses of War's Newest Weapons* (New York: G. Putnam's Sons, 1918), 143.

The German crime was that they inflicted gratuitous suffering on the public, but, of course, to harm civilians as a side effect of legitimate military strikes must be tolerated. And finally, Major General Harbord notes,

> Aerial bombing is from its nature absurdly inaccurate and therefore indiscriminate. Knowing its inaccuracy one cannot acquit the advocates of bombing of intending war on helpless non-combatants, including women and children. The non-combatants are expected to coerce the combatants into surrender, but it will never work that way with the women of any race worth saving.[38]

It is striking that even in these objections to German bombing they all agree that the intention to harm is crucial, even if there is no clear consensus as to what the specific intention was. Each theorist ascribes different problematic intentions to Germany. For Hazeltine, it was merely the intention to kill non-combatants that made German bombing problematic. Wilson sees their crime in the deliberate avoidance of munitions works; instead, they preferred to drop their bombs on open villages and towns, violating the spatial framework of the battlefield. Houard's complaint is that they intended to bomb indiscriminately for the morale effect, with no care for whether the targets are of a military nature or not. Randomness or a cavalier attitude towards discrimination of targets was crucial to their crime. Abbot and Harbord find in German bombing an intention to inflict gratuitous suffering on civilians. In the last three examples, one gets the sense that killing civilians was not only wrong because it was deliberate but because it was ineffectual or excessive. The wrongness of the intention to kill citizens was that it did not advance a clear military objective, which suggests that even while the language of intention was emerging, there was still an impulse to frame bad intentions in relation to military necessity – i.e. the intention to bomb civilians was especially bad because it was also ineffectual.

3 The Laws of Humanity and Intentional Harm

In this same period when public recriminations against German actions during the war reached a high pitch, the language of intention also began to become enshrined in formal legal documents, especially some of the many

38 Major General James Harbord, "Review: *The Great Delusion. A Study of Aircraft in Peace and War* by 'Neon,'" *Naval Institute Proceedings*, 53:293 (1927), 904.

early post-World War I documents assessing German conduct in the war. In the Preliminary Peace Conference of Paris (January 1919), for instance, the fifteen-member Commission on the Responsibility of the Authors of the War believed that there was enough evidence to show that Germany had been guilty of (among other things):

19. Deliberate bombardment of undefended places.
20. Wanton destruction of religious, charitable, educational, and historic buildings and monuments.
21. Destruction of merchant ships and passenger vessels without warning and without provision for the safety of passengers and crew.
22. Destruction of fishing boats and of relief ships.
23. Deliberate bombardment of hospitals.[39]

The commission called for further investigation to study the extent of German guilt regarding these, as it had concluded, credible accusations.

Given the novelty of these accusations – that Germany was guilty of deliberately harming innocent civilians – it is worth considering for a moment the criterion of judgment, which as noted above was intentionality grounded in the nascent principle of humanity. After all, at no point in military ethical deliberations of the nineteenth century had the mental disposition of the belligerent been a point of concern. The central questions had always concerned both a military operation's technicalities – say, whether a place had been fortified or defended or whether advanced notice had been given – and its rationale – whether the act in question was justified by way of military necessity, which, as we have seen, was a rather generous category largely governed by situational factors and a general's professional judgment. So, it is striking that the more conventional language of necessity was set aside in favor of the principles of humanity, which downplayed formal legal prohibitions in favor of an international "social conscience."[40] This shift in focus placed a much greater moral weight on the intention to harm the innocent.

---

39 "Commission on the Responsibility of the Authors of the War and on Enforcement of Penalties," 114. For the German response, see *German White Book Concerning the Responsibility of the Authors of the War* trans. Carnegie Endowment for International Peace (New York: Oxford University Press, 1924).

40 For instance, a few years later Alfred Zimmerman would seek to rehabilitate a Greek notion of the law as an effective analogy for a viable international framework. He argues that rather than relying exclusively on positive formulations, international law should be better understood as a kind of "social consciousness." He writes, "Law for him [the Greek] was not a command imposed from outside by some superior power: it was the formulation of the will of the community, or, better still, an external manifestation of its continuing life." International law is premised on shared values. A bit later, he continues, "for if there is no international society there cannot, except by a very forced and artificial

As indicated above, the principle of humanity had been formalized in the Martens Clause of The Hague Conventions of 1899 and 1907, which was a forward-looking attempt to ground international law in the "respect for the principle of humanity" and the demands of "public conscience."[41] The Preambles of both the 1899 and 1907 Conventions state:

> Until a more complete code of the laws of war has been issued, the High Contracting Parties deem it expedient to declare that, in cases not included in the Regulations adopted by them, the inhabitants and the belligerents remain under the protection and the rule of the principles of the law of nations, as they result from the usages established among civilized peoples, from the *laws of humanity*, and the dictates of the *public conscience*.[42]

Though no conventional interpretation was accepted as to what this clause actually meant – some clearly treated it as rhetorical flourish – it is clear that among certain delegates, this principle sought to establish that international law should not be limited by the letter of the convention but by the spirit underwriting it. The moral force of the document was that it was forward-looking. The trajectory of international law was such that as certain practices came to be seen as inhumane, they too would be proscribed.[43] One could not add new cruelties or more expansive harms. It is significant that the 1919 Commission explicitly refers to this as the basis for their judgment. In fact, the report quotes Baron Marschall's speech at the 1907 conference where he indicates that "Conscience, good sense, and the sense of duty imposed by the principles of humanity will be the surest guides for" military conduct.[44]

Antecedent versions of this "law of humanity" can also be seen in both the Lieber Code (1863) and the St. Petersburg Declaration (1868), both of which suggest that there was an implicit criterion within the laws of war that offered

---

use of the word, be a system of international law." Alfred Zimmerman, "International Law and Social Consciousness," *Transactions of the Grotius Society* 20 (1934); 25–44. 27, 40.

41 Ibid., 118.

42 Preamble, Convention (IV) Respecting the Laws and Customs of War on Land and its Annex: Regulations Concerning the Laws and Customs of War on Land. The Hague, 18 October 1907. https://ihl-databases.icrc.org/ihl/WebART/195-200001.

43 A. Pearce Higgins explains that the "written laws of war must therefore be supplemented by the rules of customary International Law, the evidence of which is to be sought in the works of International lawyers, while the facts on which those rules are based are to be found in historical, judicial, and diplomatic records." A. Pearce Higgins, *Non-combatants and the War* (London: Oxford University Press, 1914), 5.

44 Ibid., 118n65.

some kind of check against the permissiveness of military necessity, which as some commentators feared, gave far too much leeway to generals in the field.[45] In reality, however, these principles can be traced back much farther in military history, indeed, back to the Humanist tradition of the seventeenth century. Grotius decries those committing outrages against "every principle of humanity."[46] Vattel links the duty of sparing those who have surrendered to the "principles of humanity."[47] Even if underdeveloped, there had been a long-standing appeal to humanity in the history of military ethical discourse. As can be seen in Figure 8.2, however, despite the linkage of the laws of war to this clearly expansive notion of humanity, this moral discourse had, in fact, been in steady decline since the beginning of the nineteenth century.

When understood considering these broader historical trends, the Marten's Clause of the late nineteenth century was seeking to rehabilitate and revitalize what had been an increasingly marginal feature of the laws of war discourse. To be sure, the laws of humanity were difficult to square with the utilitarian logic of the period, and, as such, languished. Though they had lost prominence in the nineteenth century, the fact that this vestigial language was still included in some of the documents of this period is significant. It represents a repackaging of these ideas that would subsequently acquire new meaning. Notice, however, that commentators only really started to grapple with the difficult implications of this clause in the context of the cruelties of the early twentieth century. As Figure 8.2 shows, there is a noticeable uptick in the use of these terms around the outbreak of World War I.[48]

The clash between the laws of humanity and military necessity is perhaps the most pointed disagreement that took place among the representatives for

45 These early judgments, in fact, were the basis for what are today called "crimes against humanity." Following World War II, the United Nations War Crimes Commission (UNWCC) grounded its development of the concept "crimes against humanity" in the formulations from the Preliminary Peace Conference of Paris. The UNWCC found in the report two distinct crimes: war crimes or the violations of the norms and laws of war and crimes against humanity, which were seen in the cruelties directed at civilian populations, like the Armenians, during the war. See United Nations War Crimes Commission, *History of the United Nations War Crimes Commission and the Development of the Laws of War* (London: His Majesty's Stationary Office, 1948), 35.

46 Grotius, *Laws of War and Peace*, Bk. II, Chap. XXII, Sec. IV, 268.

47 See Vattel, *The Law of Nations*, Bk. III, Chap. VIII, §148. 551.

48 For more on the developments and the evolution of the principle of humanity, see Robert Kolb, "The Main Epochs of Modern International Humanitarian Law since 1864 and their Related Dominant Legal Constructions," in *Searching for a Principle of Humanity in International Humanitarian Law*, eds. Kjetil Mujezinović Larsen, Camilla Guldahl Cooper, Gro Nystuen (New York: Cambridge University Press, 2013).

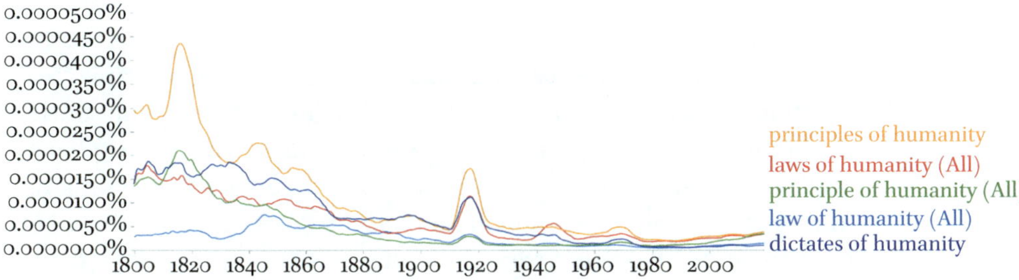

FIGURE 8.2 Ngram search of 'law(s) of humanity,' 'principle(s) of humanity,' and 'dictates of humanity'

the commission dedicated to determining war guilt at the Peace Conference of Paris (1919). The American delegation, led by US Secretary of State, Robert Lansing, argued that the principles of humanity and public conscience were far too vague to be used in something resembling a criminal prosecution. He sought to develop a much more formal set of proposals to judge whether a belligerent had violated the laws and norms governing armed conflict. In effect, he sought to ground the laws of war more explicitly in utilitarian ethical theory. The "Memorandum on the Principles which should Determine Inhumane and Improper Acts of War," authored by Lansing, was designed to serve as just such a guideline in assessing possible German war crimes.[49] As Mark Lewis explains, the Memorandum was explicitly designed to challenge many European legal assumptions: "their reliance on public opinion; their claim that they could modernize international law by criminalizing the failure to prevent violations, their arguments that the 'laws of humanity' represented a valid standard; and their arguments that an international tribunal was necessary to delivering a ringing judgment."[50]

Even though this document sought to temper the much more radical implications entailed by the idea of "crimes against humanity," what is striking about this "standard of justice in measuring the charge of inhuman or atrocious conduct during the prosecution of a war" is that it is one of the earliest attempts to directly grapple with this complex mix of military necessity, intentionality, and disproportionate cruelty. For example, it argues that even though the "slaying and maiming of men" in war is "cruel and contrary to the modern

49 "Commission on the Responsibility of the Authors of the War and on Enforcement of Penalties," 150.

50 Mark Lewis, *The Birth of the New Justice: The Internationalization of Crime and Punishment, 1919–1950* (New York: Oxford University Press, 2014), 47.

conception of humanity," it should be seen as justified inasmuch as those acts of slaying and maiming conform to the "accepted rules of war."[51] Moreover, in the Memorandum, the use of force is affirmed by the underlying principle "[to protect] national safety or to maintain national rights."[52] Much along these same lines, though not in response to these provisions, Norris Hall explains that "it may be remarked that cruelty is not at present *inhuman*, or uncommon, but on the contrary thoroughly human and fashionable so long as it can be practiced at some slight distance from the victim."[53] To engage in killing from a distance is to commit to some level of inhumanity. All these sentiments can be traced back to the nineteenth century.[54]

In one sense, the document is obviously conservative. It places the burden of guilt in a situational context that is largely governed by the principle of military necessity. However, in many ways it too is indicative of the transformations taking place at that time.[55] Summarizing the many variables at play in assessing the guilt of a belligerent, the Memorandum concludes that, "It would appear, therefore, in determining the criminality of an act, that there should be considered the wantonness or malice of the perpetrator, the needlessness of the act from a military point of view, the perpetration of a justifiable act in a needlessly harsh or cruel manner, and the improper motive which inspired it."[56] The memorandum concedes that cruelty "cannot be justly determined by a fixed line of distinction, but one which fluctuates in accordance with the facts in each case," so there is no precise attempt to lay out with exact specification what a violation looks like, but it does set out a complex array of variables that could help assess the degree to which a belligerent had violated the laws of war.[57] The vagaries of war, the need to at times knowingly and deliberately target civilians, required a different kind of judgment. Importantly, according to Lansing, intentions and motives do matter, but only if it can be shown that the commander engages in "wanton" destruction, a criterion that itself could only be determined by way of seasoned and professional military judgment

51 "Commission on the Responsibility of the Authors of the War and on Enforcement of Penalties," 150.

52 Ibid.

53 Norris Hall, "Science in War," in *The Next War* (Cambridge, MA: The Harvard University Press, 1925), 14.

54 No doubt with these provisions in mind that William Schabas claims that this memorandum is "a quite brutal defence of military necessity as a justification for acts of war." William Schabas, *The Trial of the Kaiser* (New York: Oxford University Press, 2018), 146.

55 Lewis, *The Birth of the New Justice*, 47.

56 "Commission on the Responsibility of the Authors of the War and on Enforcement of Penalties," 150.

57 Ibid.

rooted in a framework of military necessity. Bad motives produced bad fruit. In this regard, the framework of Lansing's analysis created some space for intentional harm, but, taking as its starting point nineteenth-century norms, bad intentions were those in which the perpetrator acted out of a spirit of revenge or in such a way that was knowingly gratuitous.

Lansing's Memorandum sets a layer of complexity over the then conventional beliefs about killing in war. To begin with, at no point in the nineteenth century had treatises on the laws of war attempted to weigh or assess the motive or mental state of a combatant, as such. States had to decide for themselves what was necessary for self-defense. Acts of revenge and cruelty had been prohibited, but it was not always clear how revenge differed from justifiable retaliation or retorsion.[58] Military necessity encouraged strategists to pursue some sufficient minimum of cruelty, but this was always difficult to precisely quantify because it largely depended on the experience and judgment of military leaders. In fact, a great deal of latitude had been extended to states in assessing the degree of threat they perceived prior to their resort to violence. It was widely accepted that one state could not adequately judge the level of threat felt by another state, undercutting claims about proper motive, which had been a feature of Scholastic just war theory. What is more, whether civilians of a besieged fortress should be bombed or not typically depended solely on the prudential wisdom of the commander. Under this new set of guidelines, however, the international community was responsible for judging whether the military action was done in a spirit of revenge or with malice, was wanton or excessive, and was justified and militarily necessary. Clearly, even this conservative document carried a surprisingly transformative undercurrent, one which placed normative weight on the interior disposition of military leadership.

Despite the many disagreements between the American delegates and their European counterparts at this preliminary conference – the Memorandum, for instance, was relegated to an appendix due to American reservations against the committee's recommendations – the rejection of the Memorandum can be traced to the declining privilege afforded to military necessity. It is easy to see how the turn to the laws of humanity, with its emphasis on human beings, would serve as a more poignant reminder of who was being trampled upon as the means to the end of military objectives. Though vague, nevertheless the laws of humanity pointed to the dignity of those subjected to the cruelties of war. And in this respect, 'humanity' provided a potentially more robust counterpoint to the perceived excesses of necessity. But this emerging focus on

58 Refer to the section on Reprisal, Retaliation, and Retorsion in Chapter 5.

human dignity and the rights of individual humans, nascent as it was, further expedited the interior movement in the laws of war. As Jens Ohlin and Larry May explain, with the turn to the principle of humanity, "the question that emerges is whether humans can intentionally kill other humans during war and still respect each other's dignity as humans."[59] In short, the rise of the human being as a carrier of rights and bearer of certain protections would eventually demand more humane standards from militaries who killed civilians.

Despite the US delegation's reservations, the final language of the Commission emphasized the cruelties Germany and her allies perpetrated against civilian populations. In fact, Chapter II of the report shows that the Commission was especially attuned to the documentary evidence chronicling atrocities which empirically proved Germany had violated the customs of war and the law of humanity.[60] This empirical feature of the Commission's conclusions was part of a much broader trend in war reporting that sought to provide a more robust account of the atrocities stemming from conflict. Not only were noncombatants receiving much more thematic treatment,[61] but also between 1914 and 1920, there were dozens of methodologically ambitious studies that attempted to quantify more precisely the total number of civilian casualties during conflict, the cumulative effect of which was to robustly account for the costs of war. With the quantification of human suffering, it became much more difficult to explain away harms directed at civilians as simply necessary. At least, it would become much harder to callously instrumentalize human suffering the way that necessity-based arguments tended to do.

While there had been several attempts among nineteenth-century peace activists to chronicle the financial burdens and casualties of war, many of those early reports simply assumed that losses experienced by civilian populations could not be accurately calculated.[62] By the early twentieth century, however, especially with the establishment of well-funded organizations, like the Carnegie Endowment for International Peace (1910), some of these limiting factors in war analytics would be overcome.[63] This is particularly evident in

59 Jens David Ohlin and Larry May, *Necessity in International Law* (New York: Oxford University Press, 2016), 170.

60 "Commission on the Responsibility of the Authors of the War and on Enforcement of Penalties," 113.

61 For an example of this new direct treatment, A. Pierce Higgins republished select chapters of his 1912 book *War and the Private Citizen*, under the new title *Non-combatants and the War* (1914).

62 See note 10 Chapter 4.

63 In a 1913 economic analysis of war, the American economist, Henry Emery, explains that while the loss of life is not a part of his analysis, he writes, "I wish, however, that somebody would make some judicious estimate of the problem of disease in connection with

one of the most robust methodological assessments of the causes and effects of war up to that period, the *Report of the International Commission to Inquire into the Causes and Conduct of the Balkan Wars* (1914).[64] Published just months before the outbreak of World War I, the International Commission's report sought to provide a disinterested and methodologically thorough account of the atrocities that had been widely reported in the media during the Balkan Wars. What is so unique about this report, however, is the degree to which it expressly thematizes the impact of war on noncombatant populations, and quite significantly, it does so with an eye to the principle of humanity.

The ground-breaking Carnegie *Report* gave way to a slew of similar reports at the end of World War I. In one of the first statistical compilations of casualties of that war, Arthur Frothingham (1919) argues that *more* civilians were killed than soldiers: about 9 million civilians to about 7.8 million soldiers.[65] In his report on war casualties, Walter Littlefield (1919) produces very similar numbers: 9,185,523 civilians to 7,781,806 soldiers.[66] In their war casualty totals, Richard Beamish and Francis March (1919) estimate that between 16 and 20 million people were killed in the war – about 7 million military and 10 million civilians: "Of the civilian deaths, over 100,000 were directly caused and nearly 10,000,000 indirectly caused by it."[67] This represents just a small sample of this growing field of inquiry.[68]

---

military operations" because "a very large proportion of the deaths has been due to disease rather than to the shot of the enemy." Henry Emery, *Some Economic Aspects of War* (Washington D.C.: Government Printing Office, 1914), 12.

64 See the *Report of the International Commission to Inquire into the Causes and Conduct of the Balkan Wars* (Washington, D.C.: The Carnegie Endowment for International Peace, 1914).

65 Arthur Frothingham, *Handbook of War Facts and Peace Problems* (New York: National Security League, 1919), 238.

66 Walter Littlefield, "War Casualties of All the Nations," in *The European War*: The New York Times *Current History, Volume XVIII* (New York: New York Times Company, 1919); 239–248, 239.

67 Richard Beamish and Francis March, *America's Part in the World War: A History of the Full Greatness of Our Country's Achievements* (Chicago: John C. Winston Company, 1919), 23.

68 Thomas Jackson, "The Lot of the Innocent Bystander in War," *The Public Health Journal* 7:5 (1916); 276–281; Gaston Bodart and Verson Kellogg, *Losses of Life in Modern Wars* (Oxford: Clarendon Press, 1916); Ernest Bogard, *Direct and Indirect Costs of the Great War: Preliminary Economic Studies of the War, of the Carnegie Endowment for International Peace* (New York: Oxford University Press, 1919); Office of the Adjutant General of the Army, *Summary of Casualties Among Members of the American Expeditionary Forces during the World War, 1919* (Washington DC: Government Printing Office, 1920); Homer Folks, *The Human Costs of the War* (New York: Harper & Brothers, 1920); *Ministère de la Guerre: Données de statistiques relatives à la guerre 1914–1918* (Paris: Imprimerie Nationale, 1922); Great Britain War Office, *Statistics of the Military Effort of the British Empire During the*

What is striking about all this new methodological interest in civilian casualty reporting is that by most estimates it had become exceedingly clear that civilians disproportionately bore the hardships of war. Civilians, not soldiers, had become the most sizable source of casualties. While this fundamental reality was not new – as seen in Chapter 4, historically war almost always had disproportionately affected civilian populations – but those effects were mostly anecdotal. It was now impossible to deny that the radiating effects of war threatened to undermine the tidy logic of necessity that governed nineteenth-century war. It was, after all, hard to overlook the fact that, by way of the principle of necessity, 10 million civilians had been killed. These costs were simply far too high, and as such it became difficult to sustain this nineteenth-century logic of war, which was based on reasonable trade-offs, some minimal amount of noncombatant death for the overall good of the military operation. As the human toll of war was being brought into much clearer focus so too were these much more pressing questions of the humanity and rights of those being killed.

With these trends in mind, it is not surprising that the conclusions of the Preliminary Peace Conference of Paris based German guilt on the degree to which they violated the principle of humanity. In making this conceptual shift to humanity, however, this initiated a turn to the interior world of conscience, intentions, and motives to assess whether Germany had violated not merely the rules of war but the unwritten rules of humanity. Indeed, as seen in the previous chapter, throughout the interwar period the international legal community increasingly sought to prevent the excesses of aerial bombing with more sophisticated arguments that not only emphasized the humanity of the victims, but also the mental state of the belligerent.

In a certain sense, one might also say that the appeal to humanity became much more attractive given the failure of the doctrine of necessity to humanize war. Since the nineteenth century it had been routinely advocated that a properly calibrated use of violence would make war more humane. As World War I had shown, however, as the conditions under which one fought became more desperate, necessity knew virtually no limit. What the war had laid bare is that the very idea of the noncombatant was a useful fiction generated by the

---

Great War, 1914–1920 (London: H.M. Stationery Office, 1922); Samuel Dumas and Knud Otto Vedel-Petersen, *Losses of Life Caused by War* (Oxford: Clarendon Press, 1923); Harry L. Gilchrist, *A Comparative Study of World War Casualties from Gas and Other Weapons* (Washington DC: Government Printing Office, 1928); Edward B. Ashmore, *Air Defence* (New York: Longmans, 1929); Thomas Mitchell, *Casualties and Medical Statistics of the Great War* (London: H.M. Stationary Office, 1931); and William W. James and Benjamin W. Wickling, *Injuries of the Jaws and Face: With Special Reference to War Casualties* (London: John Bale & Staples, 1940).

spatial realities and conditions of the nineteenth-century battlefield. If states believed that killing civilians would expedite victory, and that they could get away with it, they would do it.[69]

This sentiment is also captured in John Spencer Bassett's *The Lost Fruits of Waterloo* (1919). In this impassioned defense for a league of Europe, he laments the naïve optimism that positive agreement of nations could rein in the violent excesses of the state. The "terrible realism" of war had undermined the premises of restraint and cooperation that had been meticulously constructed throughout the nineteenth century. He writes, "We had adopted many specific rules to restrain needless barbarity in war. For example, we would not use dum-dum bullets, not drop bombs on non-combatants, nor shell the homes of innocent dwellers on the seashore … In three and a half horrible years we have drifted a long way from these protestations."[70] In short, necessity had proved to be far too elastic. In this sense, the appeal to humanity can also be understood to be fueled by the nostalgia for the orderly demarcation between military and non-military spaces that had characterized the nineteenth-century battlefield.

Notwithstanding the ease with which international law had been abandoned in the process of war-making, Bassett and many other commentators of this period, believed that the brutality of war had finally primed the international community to take the necessary steps to end war. He believed the crucible of World War I had opened up an arena of possibility, namely, the regulation of the international sphere through the federative action of states: "It has been a hundred years since the world had another opportunity like this that faces us, and if it is not now tried out to the utmost possibility, there is little hope that the next century will be as bloodless as the past has been."[71] In addition to the return to intention and the ascendancy of the laws of humanity as a standard to be applied even during war, it is this transformative spirit of international cooperation that framed thinking on why killing the innocent is wrong.

4 The Return of the Scholastics

Though inchoate, through the interwar period we have seen how the language of intentional harm slowly gained wider purchase. The crimes attributed to the

69 Royse, *Aerial Bombardment*, 241.

70 John Spencer Bassett, *The Lost Fruits of Waterloo: Views on a League of Nations* (New York: Macmillan, 1919), 4.

71 Ibid., 16.

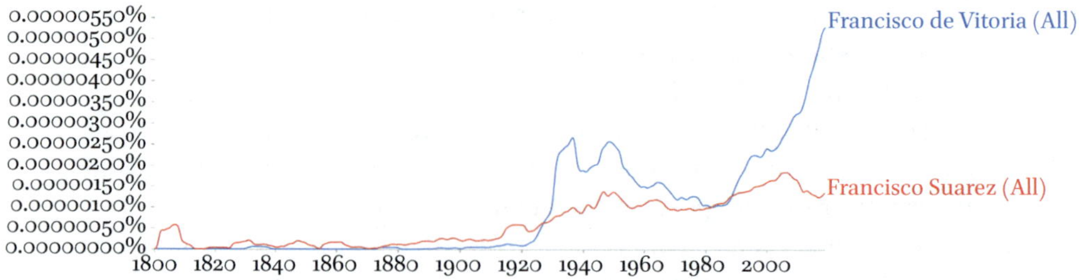

FIGURE 8.3 Ngram search of 'Francisco de Vitoria,' and 'Francisco Suarez'

Germans were increasingly couched in the language of deliberation and intention, even though aerial bombing was well understood to be, by its very nature, imprecise and indiscriminate. Despite the bombing realists' attempts to couch civilian casualties from aerial bombing in the language of necessity, the fact that new technologies put whole nations at risk made the logic of necessity that much more difficult to accept. As human suffering started to be quantified in the numerous casualty assessments of this period, it became much more difficult to sustain necessity-based arguments.

There are two additional elements that help to explain this discursive shift. The first is the rediscovery of Scholastic theory.[72] In the interwar period, at a moment in history when internationalism and the desire to reconceptualize the society of nations began to take root, academics and theorists started to scavenge the archives of history, looking for compelling frameworks of description. At this same time, one sees the resurgence of books and articles dedicated to Francisco de Vitoria and Francisco Suarez. Note in Figure 8.3 that these theologians go from near complete obscurity in the nineteenth century to rapid prominence in the twentieth. Specifically, there is a dramatic spike in references to Vitoria throughout the 1920s and 30s, and to a lesser extent a gradual increase of references to Suarez in this same period. The renewed

72 In the early twentieth century there was a resurgence of Catholic writing on the Scholastics, see Désieré Mercier, *A Manual of Modern Scholastic Philosophy, Volume 2* (K. Paul, Trench, Trubner & Company, 1917); Sylvester J. Hartman, *A Textbook of Logic: A Normative Analysis of Thought* (American Book Company, 1936); Charles Reinhard Baschab, *A Manual of Neo-Scholastic Philosophy* (B. Herder Book Company, 1937); and John Elliot Ross, *Ethics from the Standpoint of Scholastic Philosophy* (Devin-Adair Company 1938). For recent analysis, see John H. Yoder, "A Consistent Alternative View within the Just War Family," *Faith and Philosophy: Journal of the Society of Christian Philosophers* 2:2 (1985); 112–120; John H. Yoder, *When War is Unjust: Being Honest in Just-War Thinking* (Eugene, OR: Wipf and Stock Publishers, 1996), 41; and James Turner Johnson, *The Quest for Peace: Three Moral Traditions in Western Cultural History* (Princeton: Princeton University Press, 1987).

focus on these scholars is largely with regards to their contribution to international law, but it is no coincidence that in this period one also finds a rise in intention-based arguments for killing the innocent.

Secondly, as arguments for aerial bombing from bombing realists and others started to become more prevalent, one finds that Catholic commentators began deploying double effect arguments to show that bombing *violated* just war principles. As indicated above, throughout the nineteenth century, Catholics continued to employ the moral formula of double effect; however, these references were insular and never made much impact outside of those intellectual communities.[73] By the interwar period these dynamics had changed. Of course, this very likely had something to do with the rediscovery of the late Scholastics, but it was also driven by the degree of interplay between Catholic and non-Catholic writers. Since an argument of Catholic provenance was increasingly being deployed in public debates on military ethics, it makes sense that those commentators would publicly challenge secular interpretations. The paradoxical effect of Catholic engagement in these debates, however, was to link double effect arguments more firmly with aerial bombing.

5 The Rediscovery of Vitoria and Suarez

As noted above, in the 1920s and 30s there was a resurgence of interest in the late Scholastics, particularly Francisco de Vitoria and Francisco Suarez.[74] As Ignacio de la Rasilla de Moral has pointed out, the "rediscovery" of Vitoria really began in the mid-1870s.[75] The work of James Lorimer (1818–1890) and Ernest Nys (1851–1920) both anticipate and foreshadow the wider interest that Vitoria would garner in the early part of the twentieth century. While there were several scholars addressing this subject matter in this period, this renaissance can be largely attributed to the effort of one man: James Brown Scott.[76] Scott was

73 See note 122 in Chapter 4.

74 Not all this interest was flattering. For instance, Waldo Frank writes, "But the abstract logic of Vitoria was more useful to the State of Spain than the ethical conclusions of Las Casas. That supreme apologia for villainy and greed – International *Law* – is born and baptized under Christ." Waldo Frank, *Virgin Spain: The Drama of a Great People* (New York: Boni & Liveright, 1926), 176.

75 Ignacio de la Rasilla del Moral, *In the Shadow of Vitoria: A History of International Law in Spain (1770–1953)* (Boston: Brill, 2018), 80.

76 The impact Scott had on the development of international law in this period has become a subject of increased interest in recent years; see Christopher Rossi, *Broken Chain of Being: James Brown Scott and the Origins of Modern International Law* (Boston: Brill, 1998); John Hepp, "James Brown Scott and the Rise of Public International Law," *The Journal of*

a momentous figure in international law. He commanded a huge influence for about half a century.[77] He was a US delegate to the 1907 Hague Convention and later he served as the Secretary for the Carnegie Endowment for International Peace, among many other official posts and accolades. Perhaps most significantly, Scott would serve as the editor of what would be known as *The Classics of International Law* series, a book series sponsored by the Carnegie Foundation. Under Scott's leadership, this book series would reintroduce "forgotten" texts in international law to a wider audience, eventually publishing some twenty-two volumes set out in nearly forty books (1911–1950).

In a 1906 letter to R.S. Woodward, the President of the Carnegie Foundation of Washington, Scott set out his vision for this series. He explains that part of the foundational logic of this series would be to educate students of international law about the pre-Grotian origins of the discipline. Even though, he writes, Grotius was "its first and its greatest expounder," for proper context one needed to understand his predecessors.[78] In fact, Scott noted that the inordinate emphasis placed on Grotian international theory had had a stultifying effect on international legal development, in part because "the international law of to-day is rooted in a more remote past."[79] Getting back to international law's pre-Grotian roots was necessary for future development.

In this letter Scott does not explain why international law had reached an impasse with its appeal to Grotian principles, but clearly, he believes that *something* had been lost in the discipline's development. There was a discontinuity between Grotius's development and the discipline's origins. He writes, "If international law be considered as undergoing a constant growth and development, it necessarily follows that the law of nations must be studied in the light of its origin and history in order to obtain an adequate knowledge of the system at present and to predict with certainty the future development of the science."[80] By appealing to the tradition prior to Grotius one might be able to resolve certain problems that had arisen in the development of international law.

---

the Gilded Age and Progressive Era 7:2 (2008); 151–179; Thomas Cooley, George Finch, and William Butler, *Adventures in Internationalism: A Biography of James Brown Scott* (Clark, NJ: Lawbook Exchange, 2012); and Paola Amorosa, *Rewriting the History of the Law of Nations: How James Brown Scott Made Francisco de Vitoria the Founder of International Law* (New York: Oxford University Press, 2019).

77 For more on what a giant in the field of international law Scott was, see George Finch, "James Brown Scott, 1866–1943," *The American Journal of International Law* 38:2 (1944); 183–217.

78 James Brown Scott, *The Classics of International Law* (Washington D.C.: Printed for the Carnegie Institution of Washington, 1907), 2.

79 Ibid.

80 Ibid.

At some point in the process of editing this series, Scott became convinced that Vitoria, rather than Grotius, was the true intellectual father of international law. The series retranslated and reprinted Vitoria's *De Indis* and *De Jure Belli* in volume 7 (1917). In the Preface of this edition Scott declares that these texts are sufficient "to show that International Law is not a thing of our day and generation or of The Hague Conferences, nor indeed the creation of Grotius, but that the system is almost as old as the New World."[81] The printing of volume 7 appears to be the origin of Scott's decades-long interest in the revitalization of Scholastic thought. What is more, symbolically speaking, Scott's interest in Vitoria was also rooted in the Scholastic interest in the Americas. In fact, many contemporary critics looking at Scott's legacy today see this shift as part of a strategy to promote the centrality and supremacy of America in the domain of international law. That Scott found the origins of International Law in reflections on the "New World" was a sign that the so-called universal moral framework Scott promoted was distinctly American.[82]

From the 1917 reprinting of Vitoria's texts, Scott's opinion of the Scholastics seems to have grown exponentially positive. For instance, an entry in the Bulletin Report of the 1928 Pan American Union Conference (Havana, Cuba) explains that the progressive nature of Latin American laws finds its origin in "the glorious 'Peninsular' tradition, particularly of Spain, which, according to James Brown Scott, played such a great part in the persons of her pioneering figures, Victoria and Suarez, in the establishment of international law."[83] By the late 1920s, Brown was promoting Vitoria's thought at international conferences in his capacity as a US delegate. And in 1933, at the seventh International Conference of American States (Montevideo, Uruguay), Brown led an effort to have a bust of Vitoria placed in the headquarters of the Pan American Union in Washington DC "as a tribute to the professor of Salamanca who, in the sixteenth century, established the foundations of modern international law."[84] These are just some of the small, but powerfully symbolic indications of the degree to which Scott participated in the establishment of Vitoria's status as the founder of international law.

---

81 James Brown Scott, "Preface," in Francisco de Victoria, *De Indis et De Iure Belli Relectiones* (Washington DC: Carnegie Institution of Washington, 1917), 5.

82 See Mark Somos and Joshua Smeltzer, "Vitoria, Suárez, and Grotius: James Brown Scott's Enduring Revival," *Grotiana* 41:1 (2020); 137–162.

83 Union of the American Republics, *Bulletin of the Pan American Union* (Washington DC: US Government Printing Office, 1928), 878.

84 James Brown Scott, "The Seventh International Conference of American States," *The American Journal of International Law* 28:2 (1934); 219–230, 229.

Part of what explains Scott's interest in the natural law theory of the late Scholastics is the critical leverage they provided against the principle of necessity. What he finds attractive in their philosophy is the view that international law is derived not from the expediency of perceived necessity but from objective principles grounded in the natural law. The Scholastics appealed to fixed and immutable standards of right. In other words, "the natural law forbids a thing because it is evil or commands it because it is good. Not so, however, with human law, including the *jus gentium*."[85] While the natural law "is necessary in the nature of things; the *jus gentium* is advantageous, expedient, and even necessary, but only from the human point of view."[86] If international legal norms are grounded on the principle of expediency, stemming from "the human point of view," as was largely the case with nineteenth-century international law, it would inevitably be a challenge to clearly set out the parameters of prohibited behavior. The principle of humanity itself required something more fixed and permanent. Necessity and expediency are far too elastic to be a real restraint. The appeal to a fixed law underwrites the essential difference between the Scholastic tradition and the ascendant system of international law: "In the law natural, a thing or action is prohibited because it is inherently wrong in itself, whereas, as we have said, … the negative precepts of the *jus gentium* do not forbid anything on the ground that it is evil in itself"; they are merely wrong *because* they are prohibited.[87] This is a fundamental challenge to the existing order.

In his *Law, the State, and the International Community* (1939), Scott makes this point even clearer:

> It is only because of the so-called "enlightenment" of the modern world that an effort has been made to separate law from religion and morality and place it, so to speak, in a water-tight compartment, where it may not be "tainted" by the spiritual standards and aspirations of man. The result is that law and politics have been too long without a standard; and that law, government, and international relations are founded upon expediency instead of upon the bedrock of principle.[88]

---

85 Scott, *The Catholic Conception of International Law*, 164.
86 Ibid., 164.
87 Ibid., 168.
88 James Brown Scott, *Law, the State, and the International Community* (New York: Columbia University Press, 1939), 11.

The moral foundations of international law were essential to Scott's appeal to Vitoria and Suarez. In fact, in his dedication to Vitoria and Suarez at the 1933 International Conference of the Americas, he explains that Vitoria's international law has "a very definite relation to morality; that the moral standard of the individual would be the standard of the state; and that therefore a law, national or international, which is not moral should not be legally binding."[89] What made their conception of international law so compelling was that they presumed an "international morality" derived from concrete principles.[90] This return to the moral foundations of international law is crucial in order to understand the renewed appeal to intentions and the prohibitions against deliberately killing the innocent.

For Scott, grounding the rules of war in fixed principles of the natural law would not only eliminate the recourse to unjustified violence but also limit the scope of violence itself, particularly that which could permissibly be inflicted upon noncombatants. Perceived slights and ambiguous or minor threats could never count as justified causes of war, much less the hope of economic or territorial gains. One might recall that by the nineteenth century, the principle of necessity and the duties of self-defense provided states with significant latitude in articulating to themselves what self-preservation requires. Of course, this conception of morality grounded in "the human point of view" was problematic for just this reason. If the right to wage war were not rooted in objective facts and concrete principles about right and wrong which could be definitively pointed to in each case, subjective perspectivism would continue to lead to a cascade of violence. An international law premised on necessity and expedience would undermine any attempt to articulate fixed principles. This, in fact, is what made international agreement about the laws of war so difficult – no one could agree on the fundamental premise that, say, innocent people should not be killed since there were too many cases when it would be perceived as necessary and expedient to do so. These principles also explain why there could be no fixed principle like noncombatant immunity. Since organizational principles of necessity governed what was just or wanton in any particular case, one could not say in advance that killing the innocent was wrong since the material conditions would have to determine the moral requirements of each specific case. This moral ambiguity is exactly what Scott was reacting against.

89 Scott, "Seventh International Conference of American States," 229.

90 Ibid., 230.

What attracted Scott to the late Scholastics is that they offer a firm principle of right and wrong, specifically regarding the killing of the innocent. Scott agrees with Vitoria when he asserts, "The deliberate slaughter of the innocent is never lawful."[91] Even in a just war, one dedicated to redressing an objective wrong, the belligerent is never permitted to intentionally kill the innocent as a means to an end. When committed accidentally or unintentionally, the killing of an innocent still bears the mark of a moral violation – it is still an evil, one that must be *excused* as opposed to justified. This especially applies to assaults made on fortified cities. Scott explains:

> If its storming be required for the conduct of a just war, it may indeed to be taken and the innocent must suffer with the guilty the consequences of the assault. But Victoria [sic] the humanitarian comes to the aid of the innocent, even in active combat, applying the principle which he has established 'to see that greater evils to not arise out of the war than the war would avert'; for, he says, summing up the matter, 'it is never right to slay the guiltless, even as an indirect and unintended result, except when there is no other means of carrying on the operations of a just war.'[92]

The point Scott is making here is that the killing of the innocent does not *become* right because it is unavoidable and necessary.[93] An attacking force is, in fact, forced to weigh and decide between two mortal wrongs – permitting a wrong to go unpunished and the inadvertent death of killing the innocent that would occur when punishing the original wrong. This is a direct challenge to the principle of necessity where the very means to obtain a good not only transforms into a good *itself* but may perhaps even become a moral obligation. If one has a *duty* to survive, then anything necessary to that end transforms into a moral necessity, including killing the innocent.

Scott, in fact, points to the example Vitoria provides to illustrate the dangers of necessity, namely when Christian armies were deliberating about whether to kill the children of the Saracens since there was "'good reason to fear' that if they were spared, later when grown up they would attack the Christians."[94] As Vitoria implies, if unchecked, necessity might demand a much more brutal measure than is actually permitted. Scott praises Vitoria for his conclusion:

---

91 Scott, *The Catholic Conception of International Law*, 50.

92 Scott, *The Spanish Origins of International Law*, 231.

93 Notice that Scott employs a similar formulation to this much earlier in his career. See James Brown Scott, "Bombardment by Naval Forces," *The American Journal of International Law* 2:2 (1908); 285–294. 285–86.

94 Scott, *The Catholic Conception of International Law*, 51.

"'I believe that it is no wise right, seeing that evil is not to be done in order to avoid greater evil still, and it is intolerable' – he adds – 'that anyone should be killed for a future fault.'"[95] Again, the point here is that evil cannot be permitted just because it would avoid greater evil in the future. A blanket justification based on the principle of the lesser evil would lay the ground for all manner of cruelty in retaliation for even speculative future threats that had not yet materialized.[96]

If, however, one is forced to treat even the unintended harm to noncombatants as a moral liability, this would give an attacking army greater pause to determine whether the evil that underwrites the just cause of their war is so great that they must necessarily accrue some moral liability in the process of punishing the wrong. In other words, killing the innocent is not automatically excused in a just war. One must assess what the justifying logic of the use of force was, whether the particular action was truly necessary, and if so what the results of the use of force would be. It may very well be that a just cause could employ an *unjust* amount of force while storming a city. In this case, they would vitiate the strength of their moral claim. Affirming Suarez, Scott argues that a just war "'may have some evil aspect' which will nullify, so to speak, what would otherwise be a just war and make it contrary to charity."[97]

By appealing to a single standard of morality, it would no longer be acceptable to appeal to a crude utilitarian calculus that was permitted by the nineteenth-century formulations of the *raison d'état*. Since states are morally liable in the way people are, it should be possible to set out standards of conduct in advance that should be morally binding for states. *Pace* Machiavelli, Scott argues that there is no division in private morality and the morality of states. Thinking otherwise had been a barrier to international moral development.[98]

95 Ibid., 51.

96 In another example that illustrates the dangers of necessity, Scott points to the infamous Dudley case decided by Lord Chief Justice Coleridge (1884) in which survivors of a shipwreck killed and ate one of their number, a young boy. Coleridge rules that 'It is further admitted that there was in this case no such excuse, unless the killing was justified by what has been called 'necessity.' But the temptation to the act which existed here was not what the law has ever called necessity.' To this Scott concludes, "I submit ladies and gentlemen, that this single moral standard applicable to the individual is applicable to the group of individuals united in society in a body politics, whether we call it commonwealth, state, or nation." James Brown Scott, "A Single Standard of Morality for the Individual and the State," *Proceedings of the American Society of International Law at its Annual Meeting* 26 (1932); 10–29. 28–29.

97 Scott, *Catholic Conception of International Law*, 474.

98 Scott is clearly positioning his theory of international politics as an attack against Machiavelli. In fact, Scott writes of Machiavelli: "For however much in accord with certain

6 Catholics against Bombing

By World War II, the good intentions of a belligerent increasingly came to override necessity in terms of justifying the terrible side-effects of military action.[99] The unselfconscious appeal to a sufficient minimum of civilian deaths that were prevalent in the interwar period among the bombing realists simply disappeared. As seen in Figure 8.4, the references to intentionally or deliberately killing civilians or noncombatants only became a feature of military ethical discourse in the post-World War II era.[100] The same goes for "bombing civilians," as Figure 8.5 shows. To be sure, among the commentators of this period it was widely argued that while US bombers may have foreseen the harm bomb dropping posed to civilians, they did not intend those noncombatant deaths. Their bombing raids were part of a strategic campaign to disable the means for Germany to continue waging war.

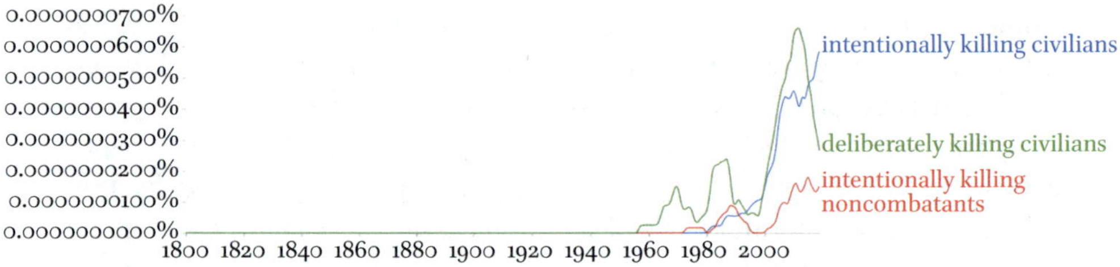

FIGURE 8.4 Ngram search of 'intentionally killing civilians,' 'intentionally killing noncombatants,' and 'deliberately killing civilians'

---

usages of his day *The Prince* may have been, and however ably written, it is not in harmony with the normal development of law and political institutions." He goes on to explain that he only includes a section on Machiavelli "since his book has unhappily exerted considerable influence on political theory and practice." He addresses Machiavelli but refuses to include any excerpts of his work. Scott, *Law, the State, and the International Community*, 278.

99 In fact, these formulations were finding their way into war manuals. Take, for instance, the 1940 US *Rules of Land Warfare*. The manual states, that military necessity permits "All direct destruction of life or limb of *armed* enemies, and of other persons whose destruction is *incidentally* unavoidable in the armed contests of war." *Basic Field Manual, Rules of Land Warfare* (Washington, D.C.: Government Printing Office, 1940), 7. Though, these provisions are not included in the 1956 edition, which repeats the traditional formulation of bombardments, assaults, and sieges. See United States, Department of the Army, *The Law of Land Warfare* (Washington DC: 1956), Section IV. 39–55.

100 In fact, most of these references stem from the turbulent public discourse around the Vietnam War.

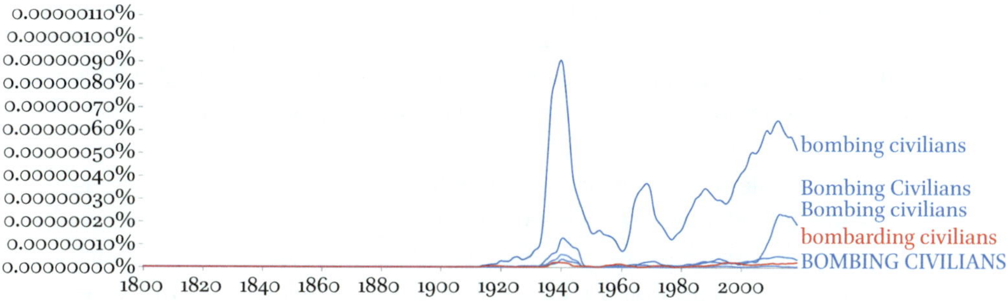

FIGURE 8.5 Ngram search of 'bombing civilians' and 'bombarding civilians'

For instance, in describing the Allied bombing campaign during World War II, Arthur Harris, the head of RAF bomber command, writes,

> At least by the early summer of 1941 everyone who had anything to do with staff policy knew that the great majority of our attacks on Germany were, in fact, *though not in intention*, area attacks; the area concerned was, more often than not, a very large one in open country ... The principle of attacking morale was at any rate half-admitted, though proposals to concentrate exclusively on this were turned down; the idea was to keep on at small targets for their strategic importance but, to put it crudely, not to mind when we missed them, or at any rate to regard a miss as useful provided that it disturbed morale.[101]

And in an assessment of the US Army Air Forces' bombing raids during World War II, one report indicates "While the Allies would undoubtedly have been pleased if German civilian morale had broken under the bombings, *this morale had not been the primary target* ... the hideous ruin visited on German cities by Allied bombers was in both purpose and result not a matter of breaking morale but of depriving the enemy of the means to produce and transport the materials of war."[102] Commentators throughout the World War II era make repeated appeals to the same tenuous, exculpatory line of good intentions – like that set out in the Hague Draft Rules – they could foresee and even hope for a powerful morale effect, but they were not intentionally bombing with that aim.

101 Harris, *Bomber Offensive*, 77. My emphasis.

102 John Fagg, "Mission Accomplished," in *The Army Air Forces in World War II: Europe, Argument to V-E Day, January 1944 to May 1945* eds. Wesley Craven and James Cate (Chicago: University of Chicago Press, 1951), 802. My Emphasis.

In fact, at the end of the war, many commentators recognized the challenge that the ever-expanding conception of "military objective" posed in conjunction with good intentions. In Robert Tucker's *The Law of War and Neutrality at Sea* (1955), he notes that given the elasticity of legitimate targets,

> the problem of determining the limits of the 'incidental' injury that may be inflicted upon the civilian population in the course of attacking military objective becomes crucial. The failure to provide a concrete solution to this problem may well mean that from a practical point of view the general prohibition against making non-combatants the direct object of attack will prove no more than nominal.[103]

What this concern about incidental injury inadvertently shows is the extent to which intention-based justifications for aerial bombing had started to become internalized by military planners and commentators in this period. Moreover, these assessments of British and American approaches to bombing in World War II reflect Paul Williams's concern about the problematic turn to intentions that the 1923 Hague Draft Rules represented. Since it would be a moral violation for a nation to deliberately terror bomb the enemy, they could still elicit the same desired effect by simply intending to hit strategic targets and not worrying when they missed. Thereby nations could proclaim their innocence – their good intentions – by maintaining that they were only aiming at sites with military value. If bombs just so happened to go off course, which given aerial bombing's inaccuracy even in World War II was inevitable, they might still produce a desired morale effect.

It is no coincidence that just as intention-based justifications for killing the innocent were reentering the broader secular public discourse, that a Catholic opposition to such arguments began to emerge. As secular theorists started to deploy intention-based justifications for bombing the innocent – i.e. that scores of civilians may be killed as long as those deaths are not intended – the more one sees Catholic theorists explicitly arguing that aerial bombing *violated* the principle of double effect as it had been articulated by the late Scholastics. Paradoxically, the counterintuitive result of this debate was to reinforce the linkage between intentions and aerial bombing. In any case, as seen in Figure 8.6, by the end of World War II double effect arguments begin to take on the prominence that they would enjoy in the latter half of the twentieth century.

103 Robert Tucker, *The Law of War and Neutrality at Sea* in *International Law Studies, 1955*, Volume 50 (Washington DC: Government Printing Office, 1957), 149.

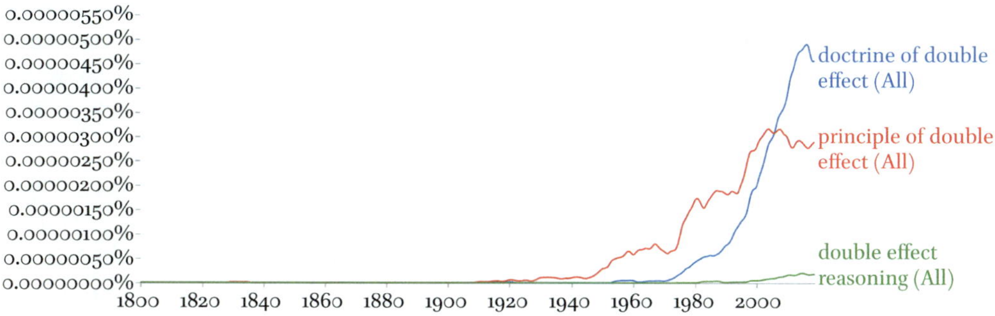

FIGURE 8.6 Ngram search of 'doctrine of double effect,' 'principle of double effect,' and 'double effect reasoning'

It is in the wake of these war-time debates about bombing that the principle of double effect became further enshrined in military ethical discourse.

One of the earliest documented cases of a Catholic writer rejecting the aerial bombing of populous towns by virtue of double effect thinking can be found in an editorial on reprisal, published in a 1915 edition of *The Month*.[104] There had been extensive public debate as to whether England would be justified in intentionally bombing German towns as retaliation for the Zeppelin attacks on the English. The author of the editorial comment, identified only by the initials J.K., vociferously rejects the view that England was within its rights to retaliate indiscriminately. The Catholic doctrine suggests, he writes, that even if "by the direct and intentional killing by the Allies of a single innocent German, the Zeppelin raids could be prevented, nay, the whole war brought to a triumphant close, such an act would still be unlawful, an act of murder, always and everywhere, because it is intrinsically, wrong."[105] He continues,

> I am taking for granted that the teaching of Catholic theology on the subject of the Fifth Commandment is known and accepted. The Catholic doctrine is that the direct killing of innocents is intrinsically unlawful, i.e. can never be justified on any plea ... But the innocent may often lose their lives in warfare, indirectly or accidentally. By a principle well-established in morality, if I have a right to do a certain act which has evil effects as well as good, I may exercise that right, provided the good effects are at

104 *The Month* was a Catholic magazine founded by the Society of Jesus in 1864. Also in 1915, though not dedicated to aerial bombing, Thomas Slater produced a rousing defense of the principle of double effect against Walter McDonald's (1903) criticisms. See Thomas Slater, *Questions of Moral Theology* (New York: Benziger Brothers, 1915), 306.

105 J.K., "Miscellanea: Critical and Historical Notes, Reprisals" in *The Month*, Vol. CXXVI, No. 617 (London: Longmans, Green, and Co., 1915), 531.

> least as immediate as the evil, and are of sufficient importance to warrant the evil being permitted. Thus a commander may lawfully bombard a fortified enemy position, in spite of the fact that his shells will destroy life and property which do not belong to the combatants.[106]

J.K. repeats the well-known formula of the late Scholastics that permits the bombing of fortified cities, that a certain amount of civilian death is permitted "indirectly or accidentally," but reprisal bombings would run afoul of this doctrine because it would require the intentional attack on innocent people.

While arguments like those advanced by J.K. can be found sporadically in the Catholic editorials of that period, the first systematic treatment of the subject came in 1933, when John Kenneth Ryan published the first edition of *Modern War and Basic Ethics*. This was perhaps the earliest holistic treatment of aerial bombing through the lens of double effect. Ryan updated and reprinted this book in 1940, just as the horrors of modern war had again been unleashed on the Continent.[107] In this book, Ryan affirms the classical Scholastic view that "a direct attack upon the civil population of a nation at war is something intrinsically evil" and, as such, "it cannot be justified on the plea of military necessity."[108] Much like Scott's arguments above, Ryan argues that it was a mistake to premise war on the demands of necessity due to the fact that "when it breaks forth with all its natural force, when it appears in its absolute perfection, it sweeps aside all the laboriously constructed restrictions of positive international law."[109] Lofty sentiments, like noncombatant immunity are quickly trampled when it is expedient to do so.

One of the fundamental premises of Ryan's attack on modern war is that it is no longer possible to assess whether a war is just or not. This is because the self-deception at work in most arguments from necessity, rooted in subjective perspectives and dubious notions of 'national interest,' had given states a great deal of latitude to articulate what was right and good to them. As Robert Lansing (as seen above, the last great public defender of nineteenth-century ethical theory) had argued, "Only world public opinion and history yet to be written can determine which party was in the wrong, and that an appeal to force was in violation of legal right and moral obligation," therefore, the concept of "Outlaw War" should be abolished.[110] Under modern conditions, it

106 Ibid., 533.

107 Ryan reprises some of these arguments in John K. Ryan, "Unlimited War and Traditional Morality," *The Modern Schoolman* 22:1 (1944); 24–32.

108 Ryan, *Modern War and Basic Ethics*, 114–15.

109 Ibid., 59.

110 Quoted in Ibid., 83.

had become impossible to distinguish between defensive and aggressive war because no state would admit to waging an aggressive war, even if they were doing so. Even offensive wars were framed in the language of self-defense. As Ramsay MacDonald had argued at the League of Nations "Reduction of Armaments" conference (1924), "the ability to assign responsibility for aggression is always about the last thing that comes," and, he continues, it "belongs to the historian who studies and writes fifty years after the war."[111] Within the framework of necessity, the perception of state security mattered the most. Objective wrongs were not only epistemically fraught, but largely irrelevant to the practice of waging war.

The inability to coherently assess whether a nation possesses a just cause to wage war had a profound effect on whether the killing of the innocent could be excused in modern war. But even if one could distinguish between just and unjust wars, Ryan goes on to ask, "What is the evil to be avoided or the good to be gained that is proportional to the means that is modern war?"[112] The challenge had become that now the costs of war were so immense and the degree of devastation was so vast, that they seemed to negate and outweigh any reasonable ends that modern war purports to pursue. War itself had become fundamentally irrational.[113] Ryan writes, "Aggressive modern warfare is no longer a morally lawful means to an end, because there is no end that is proportionate to it, no good to be gained, no evil to be avoided, which can justify the evil inseparable from modern war."[114] Indeed, modern war more closely resembles suicide than the retributive and restorative violence it was originally meant to represent.

Ryan argues that the conditions of modern war completely nullify the analogies and parallels made by comparing aerial bombing to siege warfare previously used by the late Scholastics to rationalize noncombatant death. He explains, "when a fortified town is attacked, for instance, it is neither expected nor possible so to direct shells that they will entirely avoid the quarters of noncombatant populations"; however, this was not a moral problem in the

111 Quoted in Ibid., 85.

112 Ibid., 88.

113 Charles E. Raven argues "whatever excuses Thomas Aquinas or Luther might find for local and professional conflicts of their days are wholly irrelevant to the mass-murder, the bestiality and terrorism of war as it is today." This is because "War as we now prepare for it is not a matter of heroism, but of butchery, the smashing into poisoned pulp of the innocent and the defenceless, the stamping out of all semblances of humanity by torture and panic, the destruction of every decent human quality in victors and vanquished alike." Charles Earle Raven, *War and the Christian* (Universal Christian Council for Life and Work, 1938), 143–144.

114 Ryan, *Modern War and Basic Ethics*, 94–95.

traditional sense because "the loss of these concomitant lives was admitted to be an evil, but it was unavoidable and indirectly willed."[115] He seeks to make it clear that the killing of civilians remains an evil, one that must be factored into the decision to punish a recalcitrant fort. But it would be entirely disingenuous to compare a fort to a country writ large. In fact, one of the true evils of modern war is that it no longer affords militaries the same room to make these kinds of reasonable distinctions; "The nation in arms was becoming more and more of an actuality, with the result that a larger and larger part of the population was becoming combatant or so closely connected with the fighting forces as to be considered almost a part of them."[116]

Of course, the conflation of combatants and noncombatants was one of the emerging problems of modern war that made the appeal to right intention so disingenuous.[117] If *everyone* has become a combatant due to the realities of industrial war, then there is no longer a need for these types of morally cautious distinctions. This is what was so odd about the justifications for aerial bombing that were starting to emerge. Ryan writes, "Both sides were careful to maintain the official fiction that they were attacking fortified cities, munition factories, transportation systems, and the like, and that they had no direct intention of destroying noncombatants."[118] What made these claims so dubious is that given the development of the modern war economy, and with it the

115 Ibid., 97.

116 Ibid., 99.

117 The Catholic opposition to modern war was largely premised on the inability to distinguish between combatants and noncombatants. Keating writes, "now more than ever the distinction between combatant and non-combatant has become confused." Joseph Keating, "The Ethics of Bombing," *The Catholic Mind*, XXXVI (1938), 279; and E.J. Mahoney argues "The innocent, that is to say, the harmless (if, apart from infants there are any) are of course immune from direct attack on their lives; but in modern conditions the theologian cannot tell *who* they are, and the attacking airman does not know *where* they are." Lawrence L. McReavy, "Reprisals: A Second Opinion," *Clergy Review*, XX (1941), 138.

118 Ryan, *Modern War and Basic Ethics*, 100. By the 1950s and 60s, especially with the rise of nuclear weapons, the sophistry of intention-based justifications for aerial bombing were attracting more critical attention. The great legal theorist, Julius Stone, for instance, writes "The protection afforded to civilians by a verbal prohibition of deliberate attack on morale, when such 'incidental' attack is clearly licensed, is verbal merely." Julius Stone, *Legal Controls of International Conflict* (London: Stevens, 1954), 630. William Nagle also cautioned against "overly facile applications of the principle of double effect, a 'principle' which is fast becoming a moral slide rule by which almost any act of war can be justified." William Nagle, *Morality and Modern Warfare: The State of the Question* (Baltimore, MD: Helicon Press, 1960), 111. As an example of those willing to employ double effect more extensively, Austin Fagothey argues that nuclear fallout, among other things, can be justified by the principle of double effect. See Austin Fagothey, *Right and Reason: Ethics in Theory and Practice* (St. Louis: The C.V. Mosby Company, 1963), 465.

collapse of the distinction between combatant and noncombatant, the attack on population centers had become *the* defining feature of war. As such, why should intentions matter?

In the end, Ryan argues that the principle of double effect only holds in the following circumstances:

> When the two sets of effects, one good and the other evil, follow from an act, that act may be performed if certain conditions are fulfilled. These are: first, the original act is morally good or indifferent in the abstract; second, there must be an adequate cause and a proper proportion between the good sought and the evil resulting: a trifling good cannot be sought at a cost of great evil; third, both good and evil effects must follow directly and with equal immediacy from the original act; the good effects cannot come by means of the evil; finally, the evil effects are merely permitted, not willed.[119]

Aerial bombing does not satisfy these conditions because – the official line of good intentions, notwithstanding – civilians had become direct targets of attack. In fact, Ryan argues, in modern war it appears as though the *military objects* are destroyed incidentally to the primary intention, which is the destruction of the city center. In this sense, it is the evil that is immediate (the killing of civilians) and the desired effect (military advantage) which may (or may not) come later. For these reasons, Ryan concludes that "a general bombardment of a great city cannot be morally defended under the principle of double effect."[120]

Ryan's position on aerial bombing had been set out before the truly dark turn it would take as World War II progressed. As the war progressed, subsequent Catholic commentators would take up Ryan's opposition to aerial bombing.[121] In fact, these ideas would obtain even greater prominence in the thought of G.E.M. Anscombe. In 1939, while still an undergraduate student at Oxford, Anscombe co-authored a pamphlet assessing the justice of the UK's

119 Ryan, *Modern War and Basic Ethics*, 104–05.

120 Ibid., 107.

121 These double effect considerations were also explored in The Church of England, Atomic Power Commission, *The Church and the Atom: A Study of the Moral and Theological Aspects of Peace and War* (Press and Publications Board of the Church Assembly, 1948). For more early Christian opposition to the conditions of modern war, especially the bombing of population centers, see George Barry O'Toole, *War and Conscription at the Bar of Christian Morals* (Catholic Worker Press, 1941), 35–37; and Franziskus Stratmann, *War and Christianity Today* (Westminster, MD: Newman Press, 1956), 35.

involvement in World War II. While she addresses many ethical aspects of the war, it is significant that Anscombe spends time assessing whether the bombing of civilians could be justified by virtue of double effect. Specifically, she feels compelled to address this issue because, as Anscombe explains, "It has been argued that it is justifiable to attack civilians because their death is an example of 'double effect,'" another indication, by the way, that these intention-based justifications were gaining in prominence.[122] To this she writes,

> There is a great distinction between attacking a group of persons directly, and killing them accidentally in the course of attack on others. But yet another distinction must be made. It is a different thing, while making one group of persons a target, to kill others by accident, and to make a group of persons a target, in order – by attacking them all – to attack some members of the group who are persons who may legitimately be attacked. The first case involved no sin; the second involved murder and is not an example of double effect.[123]

On these grounds, she argues that the aerial bombing campaign and the UK blockade of Germany *violate* the principle of double effect since both make a group of persons a target (the German population) in order to attack some members of the group (those who work in German war-related industry).

Anscombe extends these arguments in another self-published pamphlet, "Mr. Truman's Degree" (1957). Like Ryan, she also argues that the prohibition of killing civilians "does not depend on its promulgation as part of positive law, written down, agreed upon, and adhered to by the parties concerned."[124] So this is not a prohibition that can simply be violated at will. She emphatically concludes that "to choose to kill the innocent as a means to the end is always murder, and murder is one of the worst human actions."[125] There is a great deal of nuance to this formulation. She argues that under certain conditions it is possible to pursue actions that will lead to some statistically certain noncombatant death, for instance, "if you attack a lot of military targets, such as munition factories and naval dockyards, as carefully as you can."[126] In these cases, the killing of the innocent would not necessarily violate the principle of

122 G.E.M. Anscombe, "The Justice of the Present War Examined," [1939] in *War in the Twentieth Century: Sources in Theological Ethics* ed. Richard Miller (Louisville, KY: Westminster, 1992), 133.

123 Ibid., 134.

124 G.E.M. Anscombe, *Mr. Truman's Degree* (Oxford: Published by author, 1957), 64.

125 Ibid.

126 Ibid., 66.

double effect. However, if it is only possible to hit the target "by taking as the object of your attack what includes large numbers of innocent people, then you cannot very well say that they die by accident."[127] Much like Ryan, Anscombe also concludes that the military objective had dubiously transformed into a secondary effect.

Years later, in an essay titled "War and Murder" (1961), Anscombe states these arguments much more bluntly. She begins by pointing out that the doctrine of double effect has been unscrupulously used "to prevent any doubts about the obliteration bombing of a city. The devout Catholic bomber secures by a 'direction of intention' that any shedding of innocent blood that occurs is 'accidental.'"[128] In an almost off-handed way, she dismisses this argument: "It is nonsense to pretend you do not intend to do what is the means you take to your chosen end."[129] What is more, she simply rejects as "pure nonsense" the argument that "the distinction between combatants and non-combatants is meaningless, so an attack on anyone on the enemy side is justified."[130]

Vera Brittain,[131] a remarkable woman who served as a nurse during World War I and who provided a detailed account of her experiences under Nazi bombardment in London during World War II,[132] argues that contrary to the appeals to good intentions, the facts show that the Allies were deliberately bombing residential districts to either expedite the war or exact revenge. Appeals to double effect or good intentions were undermined by the practices the Allies pursued. What is more, she explains,

> The cumulative growth of civilian bombing to its present nightmare state seems, on present information, to be an outstanding instance of the tragic fashion in which wartime cruelty grows like a snowball by its own momentum once the power Juggernaut has taken control. Some

---

127 Ibid.

128 G.E.M. Anscombe, "War and Murder," in *Nuclear Weapons: A Catholic Response* [1961] ed. Walter Stein (New York: Steed and Ward, 1962), 59.

129 Ibid.

130 Ibid., 60. The issue of double effect was further popularized in an exchange between Anscombe and Philippa Foot. See Philippa Foot, "The Problem of Abortion and the Doctrine of the Double Effect," *Oxford Review* 5 (1967); 5–15; and G.E.M. Anscombe, "Who is Wronged? – Philippa Foot on Double Effect: One Point," *Oxford Review* 5 (1967); 16–17.

131 Though Anglican and not Catholic, Brittain married a Catholic man, George Catlin and raised their children Catholic. Brittain's daughter, MP Shirley Williams, explained in an interview that her parents taught her the Catholic just war theory. See the interview by Tony Padman, "Shirley Williams: My Family Values," *The Guardian* 15 May 2015 https://www.theguardian.com/lifeandstyle/2015/may/15/shirley-williams-my-family-values.

132 See Vera Brittain, *England's Hour* [1941] (New York: Continuum, 2005).

> accidental violation of international law, assumed to be deliberate, is repaid by a reprisal 'in kind.' The enemy 'hits back'; we retaliate harder still; in each case the accidental consequences (such as the bombing of a church in mistake for a factory) are advertised by the victim as intentional, for propaganda purposes. So the competition goes on until the mass murder of civilians becomes part of our policy, a descent into barbarism that we should have contemplated with horror in 1939.[133]

The veneer of good intentions ultimately provided the Allies with sufficient moral cover so that they could pursue a policy that amounted to mass murder.

In a similar manner, also in 1944, John Ford, one of the most highly regarded Catholic moralists of the twentieth century, also repudiated aerial bombing as a legitimate tactic in a just war. He notes, "The leaders have clearly declared their purpose to bomb very large sections of ninety German cities, with the direct intent of wiping out, if possible, not only the industrial but also the residential built-up districts of these cities."[134] This form of "obliteration bombing," Ford argues, poses a serious moral liability to those belligerents who employ the tactic. In fact, it is wrong because it necessarily requires the targeting of innocent civilians.

For those who were quick to deploy intention-based justifications, he cautioned against treating double effect as a "mathematical formula."[135] Instead, it should be viewed as a summation of the practical wisdom about killing in war, which relies on a good-faith effort to reflect on how intentions work. In this respect, he recommends supporters of this tactic to honestly consider the psychological plausibility of employing obliteration bombing without directly intending to kill innocent civilians. In other words, just because one *can* frame the killing of ordinary citizens in the language of double effect, does not mean it is the proper use of the principle. On this point, Ford believes that most of the public discourse is disingenuous. He writes, "[many] will consider it merely playing with words to say that in dropping a bomb on a man's house, knowing he is there with his family, the intent is merely to destroy the house and interfere with enemy production (through absenteeism), while permitting the

---

133 Vera Brittain, *Massacre by Bombing: The Facts Behind the British-American Attack on Germany* (New York: Fellowship of Reconciliation, 1944), 50. This book was originally published in the UK as Vera Brittain, *Seeds of Chaos: What Mass Bombing Really Means* (London: Bombing Restriction Committee, 1944).

134 John C. Ford, "The Morality of Obliteration Bombing," *Theological Studies* 5:3 (1944); 261–309. 264.

135 Ibid., 289.

injury and death of the family."[136] Furthermore, the degree to which the Allies assented to a policy of "terror and devastation" of civilian morale also vitiated their appeals to good intentions. Ford writes, "it is impossible to make civilian terrorization or the undermining of civilian morale, an object of bombing without having a direct intent to injure and kill civilians."[137] As Anscombe and Brittain had shown, the appeal to good intentions could not withstand the practical effects of targeting entire populations.

7 Conclusion

This chapter shows that the shift to intention-based language can be explained by several important factors. First, German War crimes were popularly decried as being intentionally inflicted on civilian populations. It was widely argued that German pilots deliberately bombed hospitals and civilian structures, and these kinds of accusations made their way into official documents like those set out in the Paris Peace Conference (1920). The debates that took place at this conference were significant, and it shows the degree to which the laws of humanity were beginning to outstrip and overrule the language of necessity.

Additionally, the spirit of internationalism that overtook much of the globe at the end of World War I corresponded with the resurgence of the late Scholastic work of Vitoria and Suarez. Largely popularized by the efforts of James Brown Scott, the rediscovery of their thought renewed the appeal for principled opposition to the intentional killing of the innocent. Along these same lines, as secular commentators and theorists began to justify the unintentional killing of noncombatants in aerial bombing campaigns, Catholic commentators increasingly argued that bombing violated the principle of double effect. This commentary, however, had the paradoxical effect of further centralizing intention-based rhetoric in the defense of aerial bombing.

136 Ibid., 291.

137 Ibid., 294. For an incisive argument against both saturation and atomic bombing based on the principle of double effect, see Joseph Scully, "The Ethical Implications of Atomic Energy," in *The Implications of Atomic Energy* (Brooklyn: St. John's University, 1950). Arguments for and against double effect justifications for killing the innocent would spring up periodically in the major Catholic journals from 1950–1960: *The Catholic World, The Catholic Worker,* and *Commonweal.* Over time they even start to appear in *Atomic Scientists Journal.* These arguments became increasingly popular in the 60s and 70s. See Jonathan Bennett, "Whatever the Consequences," *Analysis* 26 (1966); 83–102, and Leonard Geddes, "On the Intrinsic Wrongness of Killing Innocent People," *Analysis* 33:3 (1973); 93–97.

CHAPTER 9

Postscript: Intention in the Twenty-First Century

1 Introduction

The role good intentions have played in explaining and rationalizing civilian death goes back to at least the late Scholastic period. In the sixteenth century, Vitoria and his followers adapted and repurposed Aquinas's double effect formulation to frame the bombing and burning of those who were known to be innocent but who were living in cities amongst the guilty. They argued that those deaths were permissible if they were an unintended or incidental consequence of a necessary military action in a just war. While certain cruelties could be permitted, they argued one must consider the risks to the innocent in relation to the overall strategic goal of the attack. Major casualties for minor victories were unacceptable, even if unintended.

While this ethical formula would enjoy wide currency for several centuries, it precipitously went out of fashion at the end of the eighteenth century, at which time one sees the sharp moral ascendance of legal positivism and utilitarianism coupled with a post-Westphalian framework that privileged military necessity and self-preservation as components of a newly calcified sovereignty structure. While appeals to necessity had also been very much part of the antecedent humanist tradition, those arguments had been tempered by a broader set of duties to the natural law, like charity and moderation, not sheer utility. By the nineteenth century, necessity-based justifications for killing the innocent grounded in some cost-benefit metric had gained clear predominance. The question of killing noncombatants, as they were then newly classified, was framed in terms of military necessity. Under this framework, certain harms could be both foreseeable and intended if they were deemed strategically necessary. Within this tradition there arose an illusion of immunity. In most cases, wasting money on killing noncombatants was not believed to be an effective way to wage war; indeed, it was thought to be inefficient and gratuitous. In fact, some cruelties might have made it harder to prosecute a war, to collect contributions or provisions. In certain circumstances, however, immunity could be withheld or lost. There was no question, for instance, that ravaging border lands or bombing the residential district of a fortified city were effective ways to expedite surrender. In fact, residential parts of the city were usually bombed *because* non-fighting residents lived there. Their suffering might induce the non-fighters to pressure the city's officials to press for an end to hostilities.

 | DOI:10.1163/9789004515482_011

These nineteenth-century arguments from necessity held right up to World War I, making their final appearance in the interwar debates about aerial bombing. Bombing realists after the war put forward the last strict necessity-based arguments for killing the innocent. Some sufficient minimum number of civilians, if struck quickly and deliberately, could bring war to a swift end. It was only with the emergence of twentieth-century warfare, which relied predominately on airpower and other weapons of mass destruction, that necessity-based arguments for killing the innocent came to be seen as so morally unpalatable that they were no longer sustainable. Few people were willing to admit that potentially *millions* of civilian casualties could be "worth it," even if the war were believed to be just. It was in the face of mass civilian death that good intentions reemerged as the moral bright line that allowed militaries to differentiate between tragic but acceptable civilian death and indiscriminate violence, which after World War II eventually came to be seen as a war crime, if not outright terrorism.[1] Only war criminals and terrorists deliberately targeted civilians.

Like siege warfare in the sixteenth and seventeenth centuries, aerial bombing led to a profound shift in military ethical discourse. In the same way that the late Scholastics needed to provide a moral account of the devastation unleashed by artillery and incendiary bombs launched into cities, early twentieth-century commentators grappled with the moral implications of aerial bombing. In fact, aerial bombing precipitated a return to Scholastic formulations.[2] Though not developed here, this moral discourse anticipated and, to a large extent, continued with the development of nuclear weapons. In fact, many early twentieth-century commentators who advocated robust air warfare preparation employed the language of deterrence in much the way it

1 The intention to kill civilians has emerged as one of the defining features of terrorism in the twentieth century. Michael Walzer argues that "terrorism, in the strict sense, the random murder of innocent people, emerged as a strategy of revolutionary struggle only in the period after World War II, this is, only after it had become a feature of conventional war." Walzer, *Just and Unjust Wars*, 198.

2 Several contemporary scholars have pointed to the transformational quality of this period. Brian Orendt argues that "By 1941, the world no longer dismissed just war theory as sterile and of use only to European scholars of the history of Catholicism. Systematic works appeared, including English-language works by John Ryan and John Ford. Extensive effort was placed into the justification of the war on the part of the Allies. But this rebirth was not merely one-sided, superficial Allied propaganda. It was the just war community, after all, who first raised the questions and criticisms which still haunt historians today." Brian Orendt, *The Morality of War* (Peterborough, ON: Broadview Press, 2013), 23. Ezio di Nucci also sees the double effect debates emerging in the interwar period. Ezio di Nucci, *Ethics Without Intention* (London: Bloomsbury, 2014), 158.

would be used by Cold War planners about nuclear weapons. At the turn of the century, air enthusiasts argued that thousands of airships launched against an enemy would prove catastrophic, and, as such, this would prevent war. Those arguments, hyperbolic and fantastical as they were at the time, really *did* apply to nuclear missiles. In the end, the rise of nuclear weapons to an even greater extent precipitated this shift to intention-based justifications for killing the innocent.

As Matthew Evangelista explains, in the early stages of planning a war with the Soviet Union during the Cold War, the US was prepared to attack Moscow with *seventy* atomic bombs. Officials insisted the targets were legitimate. They were aiming at military, economic, and political sites dedicated to the Soviet ability to wage war even though if hit millions of civilians would die. Supposedly those anticipated civilian deaths would have been "unintended," but the fear induced by those massive civilian casualties, not to mention the fact that large swaths of the world (if not the world itself) would become uninhabitable, was calculated to deter Soviet aggression.[3] Evangelista concludes that "focusing on intentions rather than behavior serves as a way of reconciling putative American values with actions that clearly violate them. If one focuses on *consequences* rather than intentions, the distinction begins to blur between terrorism committed by non-state actors and violence perpetrated against civilians by states at war."[4]

The devastation of aerial bombing in World War II[5] and the anticipated annihilation that would result from nuclear war,[6] generated a moral environment that incentivized the appeal to intentions. If states were to continue using such

---

3 Matthew Evangelista, "Introduction: The American Way of Bombing," in *The American Way of Bombing: Changing Ethical and Legal Norms, From Flying Fortresses to Drones* eds. M. Evangelista and H. Shue (New York, 2014). 11–12.

4 Matthew Evangelista, *Law Ethics, and the War on Terror* (Malden, MA: Polity, 2008). 36. Michael Sherry also argues that since claiming so many civilian casualties were justified due to military necessity was morally repugnant, "some Americans resorted to another, slipperier defense of bombing: that civilian deaths were acceptable as the accidental by-product of the destruction of legitimate targets. That is, lack of the intention to harm absolved the destroyer of moral responsibility." Michael Sherry, *The Rise of American Air Power: The Creation of Armageddon* (New Haven: Yale University Press, 1987), 141.

5 John Keegan estimates that 600,000 civilians died in Germany from aerial bombing (see John Keegan, *The Second World War* (New York: Random House, 1989); Richard Frank estimates that an additional 425,000 civilians were killed in Japan. Richard *Frank, Downfall: The End of the Imperial Japanese Empire* (New York: Penguin, 1999).

6 In 1959, Ralph Lapp estimated that in a hypothetical nuclear attack "involving 263 nuclear weapons with a total yield of 1,446 megatons," hitting multiple targets throughout the United States, would kill 50 million Americas. Ralph La "What is the Price of Nuclear War?" *Bulletin of the Atomic Scientists* 15:8 (1959), 340–343. 340.

weapons, they needed palatable ways of justifying their use. Hannah Arendt also notes how the emerging framework of war crimes had fundamentally changed with the rise of these war-related technologies:

> For the truth of the matter was that by the end of the Second World War everybody knew that technical developments in the instruments of violence had made the adoption of "criminal warfare" inevitable. It was precisely the distinction between soldier and civilian, between army and home population, between military targets and open cities, upon which the Hague Convention's definitions of war crimes rested, that had become obsolete. Hence it was felt that under these conditions war crimes were only those outside of all military necessities, *where a deliberate inhuman purpose could be demonstrated.* The factor of gratuitous brutality was a valid criterion for determining what, under the circumstances, constituted a war crime.[7]

Since military technologies obliterated the old distinctions and categories, the international community was pressed to adopt some new criterion, like "deliberate inhumane purpose," to differentiate those who merely use terrible weapons and those who use them with some nefarious design. Indeed, as the number of civilians who would be killed in military attacks became unfathomably high, it became harder to sustain the fiction that those deaths were somehow necessary for just military ends. In this respect, the return to intentions provided a great deal of moral comfort; it allowed Western states the ability to decouple tactical decisions from the outcomes and consequences that would naturally ensue. Sahr Conway-Lanz argues that during World War II, Americans "tenaciously clung to the optimistic assumption that violence in war could still be used in a discriminating manner despite the increased destructiveness of weapons."[8] To be sure, under the conditions of modern war, one could no longer determine the legitimacy of a military strike based on consequences alone, that is whether a strike was (in the language of the nineteenth century) wanton or excessive. The appeal to intentions emerged as a psychologically plausible way to differentiate between good and bad behavior.

Much as was the case in the interwar period, where intermittently one begins to see intention-based prohibitions of military violence, by the end of World War II, in the flurry of protocols and conventions that sought to capitalize on

7 Hannah Arendt, *Eichmann in Jerusalem: A Report on the Banality of Evil* (New York: Penguin Books, 1977), 256. My emphasis.

8 Sahr Conway-Lanz, *Collateral Damage* (New York: Routledge, 2006), 19.

the postwar internationalist spirit, the language of "intention" started to officially enter the many transnational agreements that would come to make up International Humanitarian Law (IHL). The Nuremberg Principles of 1945, prohibits the deliberate murder of civilian populations.[9] The Geneva Conventions of 1949 stipulates that "persons taking no active part in the hostilities" should not be directly targeted.[10] The 1968 Resolution on Human Rights forbids the direct "attacks against the civilian population."[11] The 1977 Additional Protocol I states, "Civilian objects shall not be the object of attack."[12] The 1980 Protocol III to the Convention on Certain Conventional Weapons declares, "It is prohibited in all circumstances to make the civilian population as such, individual civilians or civilian objects the object of attack by incendiary weapons."[13] The 1996 Amended Protocol II to the Convention on Certain Conventional Weapons asserts, "It is prohibited in all circumstances to direct weapons to which this article applies, either in offence, defence or by way of reprisals, against individual civilians or civilian objects."[14] The 1998 ICC Rome Statute states, "Intentionally directing attacks against civilian objects, that is, objects which are not military objectives" is a violation of the laws and customs of war.[15] One of the "Precautionary Principles" of the celebrated Responsibility to Protect Summit (2005) is that when militaries proceed with intervention they must do so with "right intention."[16] The US Military Commissions Act (2006)

9 Nuremberg Principles, Charter of the International Military Tribunal, 8 August 1945, Article VI, Section b.

10 Geneva Convention Relative to the Protection of Civilian Persons in Time of War of 12 August 1949, Section IV, Article 3. https://www.icrc.org/en/doc/assets/files/publications/icrc-002-0173.pdf.

11 Resolution on Human Rights in Armed Conflicts, United Nations General Assembly Resolution 2444, 19 December 1968. XXIII. 244. 1(b).

12 Protocol Additional to the Geneva Conventions of 12 August 1949 and relating to the Protection of Victims of International Armed Conflicts (Protocol I), 8 June 1977. Article 52(1) https://ihl-databases.icrc.org/ihl/WebART/470-750067.

13 Protocol on Prohibitions or Restriction on the Use of Incendiary Weapons (Protocol III). Geneva, 10 October 1980. Article 2(1). https://ihl-databases.icrc.org/applic/ihl/ihl.nsf/ART/515-830002?OpenDocument.

14 Protocol on the Prohibitions or Restrictions on the use of Mines, Booby-Traps, and Other Devices to the Convention on Prohibitions or Restrictions on the Use of Certain Conventional Weapons Which May be Deemed to Be Excessively Injurious or to Have Indiscriminate Effects, Geneva, 10 October, 1980, Article 3(7) of https://treaties.un.org/doc/source/docs/CCW_CONF.I_16_Part%20I-E.pdf.

15 Rome Statues of the International Criminal Court, 17 July 1998, Article 8(2)(b)(ii). https://www.icc-cpi.int/resource-library/documents/rs-eng.pdf.

16 Report of the International Commission on Intervention and State Sovereignty, *The Responsibility to Protect* (Ottawa, ON: International Development Research Centre, 2001), xii.

states, "Any person subject to this chapter who intentionally engages in an attack upon a civilian object that is not a military objective shall be punished as a military commission under this chapter may direct."[17] This exact language is repeated in a section titled, "Crimes and Elements," of the US Manual for Military Commissions (2007) and the US Military Commissions Act (2009). This is just a small sample of the proliferation of intention-based prohibitions against killing the innocent that emerged after World War II.

As intentional harm to civilians came to be seen as a war crime, this provided a conceptual foothold to differentiate between types of war crimes. In fact, as the nascent principle of noncombatant immunity and the prohibition against intentional harm was becoming standardized, international agreements started to deploy this same language to define terrorism. Terrorists were those who *deliberately targeted civilians*. In 1994, the UN General Assembly resolution defined terrorism as "criminal acts *intended* or *calculated* to provoke a state of terror in the general public."[18] According to the 1997 UN Convention for the Suppression of Terrorist Bombings, a terrorist attack occurs when a

> person unlawfully and *intentionally* delivers, places, discharges or detonates an explosive or other lethal device in, into or against a place of public use, a State or government facility, a public transportation system or an infrastructure facility, *with the intent* to cause death or serious bodily injury, or extensive destruction likely to result or actually resulting in major economic loss.[19]

The 2004 UN Security Council Resolution 1566 also defines terrorism as follows:

> criminal acts, including against civilians, committed with the *intent to cause* death or serious bodily injury, or taking of hostages, with the purpose to provoke a state of terror in the general public or in a group of persons or particular persons, intimidate a population or compel a

17 Military Commissions Act of 2006, 120 State. 2600, Public Law 109-366, 17 October 2006. §950. A(4)(b)(1) https://www.loc.gov/rr/frd/Military_Law/pdf/PL-109-366.pdf.

18 1994 United Nations Declaration on Measures to Eliminate International Terrorism annex to UN General Assembly resolution 49/60, "Measures to Eliminate International Terrorism," December 9, 1994, UN Doc. A/Res/60/49.

19 United Nations, "International Convention for the Suppression of Terrorist Bombings," New York, 15 Dec. 1997. https://treaties.un.org/Pages/ViewDetails.aspx?src=IND&mtdsg_no=XVIII-9&chapter=18&lang=en.

> government or an international organization to do or to abstain from doing any act.[20]

These are just a few of the many examples that illustrate the prominence intention has come to play in contemporary moral discourse in war and other types of militarized violence, and it reinforces Maja Zehfuss's claim that the intentional killing of the innocent has become the central question in scholarship on the ethics of war.[21]

There are, however, a few significant observations to be made about the historical development set out above. First, it was only in the moral environment that arose in the wake of certain technological innovations that intention-based formulations gained renewed attention. As a kind of ethical technology itself, the moral vocabulary of intention was adapted and redeployed in response to concrete material conditions to excuse the killing that would inevitably follow the widespread use of new weapons. Even though future-war fiction anticipated the trauma of aerial bombing, it is striking that intention-based arguments were only adapted after the technology materialized. In this sense, the appeal to intentions was not a rational-deliberative moral formulation *preceding* aerial bombing; rather, it was adapted to the historical development of aerial bombing post hoc. Secondly, in their current form, intention-based arguments against killing the innocent are relatively young. While they borrow some of the structure of Scholastic formulations, the contemporary formulations are stripped from the natural law framework in which they were originally deployed. This is useful to remember since, in reading the contemporary literature on the laws of war, one gets the impression that there has been a long-standing and indeed unbroken ethical tradition that privileges both the immunity of noncombatants and the prohibition against intentional harm going all the way back to the thirteenth century. As demonstrated in the preceding chapters, for more than a century these arguments almost entirely disappeared. From the Napoleonic Wars to World War I, virtually no major commentator on international law or the ethics of war made an appeal to the good intentions of the attacking army in the process of killing the innocent. The return to intentions in the interwar period was driven by prominent theorists, like James Brown Scott, who lamented the loss of a fixed conception of

---

20 UN Security Council Resolution 1566 (2004) https://www.un.org/ruleoflaw/files/n0454282.pdf.

21 Maja Zehfuss, "Killing Civilians: Thinking the Practice of War," *The British Journal of Politics and International Relations* 14 (2012); 423–440. 423.

right and who argued that without such agreement the international community would never secure a norm like noncombatant immunity.

These moral legacies are important. They reveal that, rather than reflecting indubitable truths, the laws of war are adaptive and reactive inasmuch as they respond to the material conditions of international conflict. If this is true, it is worth considering the suitability and, indeed, sustainability of an intentions-centric contemporary military ethical discourse. In the late-twentieth- and early twenty-first-centuries, transformations in military technologies have started to undermine the appeal of intention-based formulations for killing the innocent. To begin with, late-twentieth- and early twenty-first-century moral psychologists have profoundly problematized intention as a discrete mental state. "Intention" is better described as a folk-psychological tool that humans use to anticipate and predict behavior rather than a mental process. This characterization not only undermines the language of intention as a feature of interpersonal morality, but it also renders the idea of state or military (collective) intentions increasingly incoherent. If intention is a cognitively questionable process, perhaps it would be best to conceive of state behavior in different terms, too. Secondly, even if one wishes to persistently hold onto the notion of intentions, contemporary war technologies are precipitously rendering this language outmoded. The rise of precision strike capability, autonomous weapons, and various casualty-mitigation techniques have all problematized the way one might think about intentions in war. While one can still describe programs and codes in the language of intention – that is, even learning algorithms can be described as operating within the parameters and "intentions" of their coders – it is by no means clear that this language is the most effective way to talk about the behavior of machines.

For instance, when a self-driving car killed a pedestrian in Arizona in March 2018, it was difficult to see how intentions mattered at all.[22] Everyone involved in the crash was "innocent" in the sense that no one wanted the pedestrian to be killed. The team of programmers who developed this system certainly did not intend for this to happen; the car's lidar sensors were designed exactly for the purpose of detecting pedestrians, but in this instance failed. As this case suggests, one gets the impression that as automated systems begin "making decisions" based on the sensors and algorithms that help them interface with the world, the role intentions will play in our assessments of these harms will certainly be attenuated if not completely diminished. In fact, the

22 Troy Griggs and Daisuke Wakabayashi, "How a Self-Driving Uber Killed a Pedestrian in Arizona," *New York Times*, 21 March 2018. https://www.nytimes.com/interactive/2018/03/20/us/self-driving-uber-pedestrian-killed.html (Accessed 16 April 2021).

only intention that really seemed to matter in this case was not the one that caused the harm, but what the company did in response to the harm, how they attempted to compensate the victim and correct the problem so that it does not happen again. The same can be said for military technologies. Military programmers and the human attendants to automated systems may sincerely intend that their weapons only strike legitimate targets.[23] To be sure, even though lethal autonomous systems likely will be governed by the good intentions of their programmers, inevitably they will still lead to unwanted civilian death. This is already the case with the elaborate casualty mitigation protocol used by the US military. Noncombatants can still be killed even though everyone involved are following elaborate protocols designed to prevent unwanted death.[24] Due to these potentialities, there are signs in contemporary ethical discourse that a revised moral framework for understanding killing the innocent is necessary.

2 Intention and Folk Psychology

A great deal of military ethical theory is predicated on what Hedley Bull describes as the domestic analogy: "the argument from the experiences of individual men in domestic society to the experience of states, according to which states are like individuals."[25] Accordingly, whether a state is justified or excused in its motivation to use violence stems from domestic examples, such as self-defense. As demonstrated above, the very first justifications for killing the innocent were based on Aquinas's excuse for killing an attacker in self-defense. These arguments gradually were adapted and repurposed to explain collective state behavior. The point here is to recognize that collective intention is analogical, and the strength of the analogy is weakened if the grounding idea of individual intention is undercut.

The first and most obvious problem with this domestic analogy is that neuroscience and moral psychology are painting a picture of the mind that increasingly vitiates classical notions of intention – a problem that is already well

23 The DOD Directive on Autonomous Systems says that (semi) autonomous systems will be equipped with the ability to discriminate and they will be overseen with a sufficient among of human oversight, so that "unintended engagements" do not occur. Department of Defense Directive, "Autonomy in Weapons Systems," Number 3000.09. 21 November 2012.

24 See Brian Smith, "Civilian Casualty Mitigation and the Rationalization of Killing," *Journal of Military Ethics* 20:1 (2021); 47–66.

25 Hedley Bull, *The Anarchical Society: A Study of Order in World Politics* (New York, 1977). 44.

understood in legal theory and the discipline of applied ethics more broadly.[26] In *The Illusion of Conscious Will*, Daniel Wegner shows that in many cases what is perceived as intention is a revisionist, post hoc construction.[27] He would go on to elucidate the illusion of intentionality, understood as some form of mental causation, is at best a "mental trick."[28] While this doesn't eliminate the possibility of intention being relevant to our moral considerations as a kind of guideline or heuristic, it does undermine its centrality.

Today, moral psychologists typically characterize the ascription of intention as a form of "folk psychology." As Daniel Hutto explains, "Our primary modes of interpersonal engagement are not driven by mentalistic predictions or explanations; rather they are characterized by the possession of embodied expectations. Such expectations are not intellectual products; arguably they are not the outcomes of manipulation of representations by inferential operations at all."[29] In other words, as a form of "folk psychology" the attribution of intention is a social evolutionary adaptation that has emerged to aid us in predicting and regulating social behavior. Daniel Dennett calls this the "intentional stance" – i.e. the attribution of beliefs, desires, and interests to a system in order to predict its behavior.[30] It is an evolutionary or adaptive tool that humans have developed to manage expectations and control outcomes. Moral psychologists see intention, to some extent, as a sociocultural construct and therefore a malleable product of human socialization rather than an empirical denotation of a culpable or suspect mental state. As Malle & Knobe conclude, "Whether or not intentionality is truly an objective attribute of the mind, people certainly ascribe intentions to each other."[31]

Moral psychologists seek to descriptively understand the rules that govern "folk intention," which, according to Malle & Nelson, can be reliable and

---

26 See, for instance, *Law and Neuroscience: Current Legal Issues, Vo. 13* ed. Michael Freeman (New York: Oxford University Press, 2011); Michael Pardo & Denniss Patterson, *Minds, Brains, and Law: The Conceptual Foundations of Law and Neuroscience* (New York, Oxford University Press, 2013); and *Neuroscience and Legal Responsibility* ed. Nicole Vincent (New York: Oxford University Press, 2013).

27 Daniel Wegner, *The Illusion of Conscious Will* (Cambridge, MA, 2003). See also, Ruud Custers & Henk Aarts, "The Unconscious Will: How the Pursuit of Goals Operates Outside of Conscious Awareness," *Science* 329 (2010). 47–50.

28 Daniel Wegner, "The Mind's Best Trick: How We Experience Conscious Will," *Trends in Cognitive Sciences* 7:2 (2003); 65–69.

29 Daniel Hutto, *Folk Psychological Narratives: The Sociocultural Basis of Understanding Reasons* (Cambridge, MA: MIT Press, 2012). 3.

30 See Daniel Dennett, *The Intentional Stance* (Cambridge, MA: MIT Press, 1989), 17.

31 Bertram F. Malle and Joshua Knobe, "The Folk Conception of Intentionality," *Journal of Experimental Social Psychology* 33 (1997). 101–121. 102.

systematic.[32] In other words, if one can isolate the social rules, norms, and values that help one to attribute intentionality to a certain behavior, one might be able to clarify how intentions operate in a particular moral context. The growing body of experimental literature on this subject paints a compelling picture of how various – indeed sometimes counterintuitive – conclusions about intention are derived within a particular moral context. For instance, Joshua Knobe, offer some evidence to suggest that prior moral judgments (or social intuitions) about what counts as an acceptable outcome influence the degree to which we consider an action to be intentional or not. In one study, participants were asked whether a corporate executive's investment behaviors were intentional or not. In one case his investments helped and in the other they harmed the environment. In both cases the investors said he was interested not in the means but rather only in the monetary return on his investment. Participants in the experiment observing the executive's behavior routinely attributed intention to the second scenario, that is, the executive was judged as having intentionally harmed the environment even though the observers had been told that the executive's only focus was on profit, not the means to obtain it.[33] This experiment demonstrates that within a particular value context, an agent's indifference to a foreseen harmful outcome can register as intentional, i.e. observers will attribute intention to the perceived wrongdoer.

As Cushman & Knobe demonstrate, the greater the moral disapprobation there is for a foreseeable consequence of an action, the more likely the "unintended" consequence will be interpreted as intentional; i.e. "morally bad behavior is more likely to be construed as actively 'doing' than as passively 'allowing.'"[34] And finally, Cushman argues that the degree to which behavior neatly maps onto double effect reasoning is more likely and "accidental byproduct of cognitive architecture," and, for this reason, one should "question its normative significance."[35] These studies are, of course, experimental and descriptive in nature. What they reveal, however, is that it is difficult to precisely identify what it means for an action to be intentional independent

32 Bertram Malle & Sarah Nelson, "Judging *Mens Rea*: The Tension between Folk Concepts and Legal Concepts of Intentionality," *Behavioral Sciences and the Law* 21 (2003). 563–580.

33 See Joshua Knobe, "Person as Scientist, Person as Moralist," *Behavioral and Brain Sciences* 33:4 (2010). 317–19.

34 F. Cushman, J. Knobe, W. Sinnot-Armstrong, "Moral appraisals affect doing/allowing judgment," *Cognition* 108 (2008), 281–289. 283. See also, Joshua Knobe, "Intentional Action and Side Effects in Ordinary Language," *Analysis*, 63:3 (2003). 190–194.

35 Fiery Cushman, "The Psychological Origins of the Doctrine of Double Effect," *Crime Law and Philosophy* 10 (2014); 763–776.

of the context of assumptions, moral values, and judgments used to ascribe intention.

These observations not only highlight the dubious quality of individual intentional acts in war but also question the analogical intention-based framework as it applies to states and their militaries. As moral psychology has problematized intention as a discrete mental state, it has become even more difficult to speak coherently about something like intentional collective action, i.e. the intention of states, sub-national entities, and militaries. Many international relations theorists have pointed out the conceptual ambiguity in talking about state intention as though it were directly analogous to the intention of an individual. As Helen Frowe argues, "a state is an abstract entity that has no intention at all."[36] Frowe suspects that "state intention" is really shorthand for the intention of a state's leaders.[37] Christopher Kurtz sees "joint intention" as the "overlap among individual participatory intentions."[38] The findings in moral psychology indicated above would only seem to further problematize the characterization of state intention as the aggregation of individual intentions.

The ways militaries speak about intention in their internal documents is also obscure. For instance, military manuals will speak of the "difficulty in translating the commander's intent into actions."[39] While these descriptions help to illuminate the complex task of translating and achieving operational goals within sophisticated networks of bureaucratic actors and agents, reducing military intention to "commander's intent" itself conceals the sophisticated information-gathering, planning, decision-making, and command-processes that determine how "intent" is developed, promulgated, and interpreted by subordinates. Indeed, it is not clear that all actors in the chain of command need a shared belief about intent for military action to be undertaken.

---

36 Helen Frowe, "Judging Armed Humanitarian Intervention," in *The Ethics of Armed Humanitarian Intervention* ed. Don Scheid (New York, Cambridge University Press, 2014). 99.

37 This is a point that has been made by many theorists: see Manuel Velasquez, "Why Corporations Are Not Morally Responsible for Anything They Do," Collective Responsibility: Five Decades of Debate in Theoretical and Applied Ethics eds. Larry May and Stacey Hoffman (Savage, MD: Rowman & Littlefield, 1991).

38 Christopher Kurtz, *Complicity* (New York: Cambridge University Press, 2001). 192. Many others have grappled with this issue: See Dwight Newman, *Community and Collective Rights: A Theoretical Framework for Rights Held by Groups* (Oxford: Bloomsbury, 2011).

39 David Alberts, John Garstka, and Frederick Stein, *Network Centric Warfare: Developing and Leveraging Information Superiority* (CCRE, 2000), 71. See also, U.S. Army, *Field Manual 5.0: The Operations Process* (Washington DC: US Department of the Army 2010).

Elaborate information and decision-making systems can produce their own logic that attenuate classical notions of intentionality.[40]

The collective and indeed institutional nature of military decision-making all serve to weaken the domestic analogy. As such this poses clear challenges to describing the moral quality of political violence perpetrated by states in the language of intentionality. Brian Orend argues that "we have always employed the domestic analogy in our moral and legal discourse about the ethics of war and peace; it is inherent in the deepest structure of our talk about morality and war; and we all understand what is meant by it. It is an irreplaceable aid to understanding in this regard."[41] However, Orend continues that the loss of a concrete mental-procedural account of intention on an individual level does not have to be a fatal blow to ethical reasoning about state violence. In fact, Orend seeks to preserve intent-based reasoning by arguing that "we know an agent's intent through his conduct."[42] This characterization does not rely on an ambiguous and problematic appeal to a mental state; instead, it looks at repeated outcomes and the persistent effects of action. He continues, "Though difficult, it is possible to tell whether a state is prosecuting a war out of ethnic hatred, for example, as opposed to vindicating its right of self-defense. Dark motivations produce distinctive and noticeable results, such as torture, massacres, mass rapes, and large-scale displacements."[43] If torture, massacres, and mass rapes are reliably produced by states, they can be treated as intended.

In her bracing analysis of collateral damage, Neta Crawford also shows that intention matters less than what a collective's operational processes allow and how its decision-making structures respond to unwanted procedural outcomes. Like Orend's analysis above, Crawford shows that the refusal or failure of a collective's decision-making structures to respond to problematic operational outcomes – casualty mitigation protocols that are too permissive or the tolerance of "normal accidents," i.e. accidents that are normal enough to be foreseen and which could be mitigated by way of modified processes or procedures – is the relevant locus of moral and legal liability. She writes,

> I am not focused on a single incident of inadvertent/accidental civilian killing, but rather on how the harm to noncombatants is foreseen, sometimes calculated, and "adds up" in the context of the war so that

40 For an ideal-typical formulation of what these processes entail, see Gary Klein, *Sources of Power: How People Make Decisions* (Cambridge, MA: MIT Press, 1999).

41 Orend, *The Morality of War*, 48.

42 Ibid., 47.

43 Ibid., 47–48.

> the conditions that will produce such harm are predictable and avoidable. Therefore, incidents that may be perceived as genuine accidents may actually be "normal" or routine accidents. Everyone may genuinely be trying their best, but the goals and the definitions or specific criteria of military necessity and proportionality or a failure to distinguish between combatant and noncombatant may predictably lead to civilian killing. Thus, moral and perhaps legal categories should be expanded to include moral responsibility for a pattern of unintended yet foreseen harm.[44]

Genuine accidents might be excusable, but when military strategy reliably generates the same results, to the point that a certain number of civilian casualties can be systemically anticipated and accepted, moral liability emerges.

It is becoming increasingly common for commentators writing on military ethics to jettison or at least downplay moral formulations that rely primarily on intentions. In fact, the ethical arguments of Orend and Crawford above resemble, in a vague way, the moral formulations of the nineteenth century, which emphasized gratuitous harm and the wantonness of the violent act in question. At that time, militaries were not guilty of intentional harm but by pursuing actions that appeared excessive to fellow military professionals. Though contemporary scholars like Orend and Crawford approach this issue from a slightly different angle than nineteenth-century military ethicists, both the former and the latter argue that moral liability manifests in the repeated and systemic indifference to the harms perpetrated against the innocent. Dark motives (whether such motives exist is irrelevant) can be reasonably inferred and attributed to an aggressor who repeatedly kills the innocent. The systemic indifference to foreseen harm produces a moral liability, even if those harms are not, strictly speaking, intended. One of the strengths of Crawford's argument is that it provides a framework to morally assess persistent, unintended harm. If civilian casualties continue to occur despite a military's use of sophisticated casualty mitigation protocols, which signal their "good intentions" to minimize harm and discriminate, such harm can still be understood as "intentional" inasmuch as they are indifferent to systemic outcomes.

Indeed, the number of commentators who now argue for consequence-based prohibitions for killing the innocent in war is striking. David Rodin argues that the framework of double effect should be expanded to include not just intended harms but "reckless and negligent" behaviors, i.e. excessive outcomes that are not, strictly speaking, intended.[45] Peter Olsthoorn argues

---

44 Crawford, *Accountability for Killing*, 44.

45 David Rodin, "Terrorism without Intention," *Ethics* 114:4 (2004); 752–771, 762.

for the rehabilitation of the utilitarian approach to killing in war: "there is something to be said for a military ethic that attaches less weight to intentions, and more to the consequences."[46] Harry Gould argues that the debates about intended versus foreseen harm are different than the question of responsibility. He writes, "No matter the intention of the agent, she or he brought about the bad state of affairs … and is thus responsible for its occurrence."[47] This is just a small sample of the rise in consequence-based arguments for killing the innocent in war. To be sure, the language of intention is still predominant in ethical discourse. Many more articles could be cited showing that military ethicists still rely on the moral language born out of the trauma of aerial bombing in the early twenty-first century to explain civilian casualties; however, the rise of these alternative formulations is suggestive of subtle transformation in the ethical landscape that may be indicative of an inchoate development in military ethical discourse.

3 Are Intentions Relevant for Twenty-First Century War?

If it is true that military ethical discourse is facing a transitional moment, one should ask which material, institutional, or technological shifts are precipitating this transformation. On this account, it is worth assessing if the way twenty-first century war is fought is forcing ethicists to reevaluate the ascendant moral vocabulary governing conflict. If aerial bombing and nuclear weapons were instrumental in precipitating the shift to intention-based justifications for killing the innocent in the period just after World War II, it is worth considering whether the rise of precision-strike capability, autonomous weapons, casualty mitigation protocols, and artificial intelligence are producing similar pressures today.

Some of these trends have recently been noted by Henry Kissinger, who indicates with concern that we are on the precipice of a "sweeping technological revolution whose consequences we have failed to fully reckon with, and whose culmination may be a world relying on machines powered by data and algorithms and ungoverned by ethical or philosophical norms."[48] He cautions that in this era of data and processing power "Information threatens to

46 Peter Olsthoorn, "Intentions and Consequences," *Military Ethics* 10:2 (2011); 81–93, 81.

47 Harry Gould, "Rethinking Intention and Double Effect," in *The Future of Just War: New Critical Essays* (Atlanta: The University of Georgia Press, 2014), 142.

48 Henry Kissinger, "How the Enlightenment Ends," *The New Atlantic* (2018); https://www.theatlantic.com/ magazine/archive/2018/06/henry-kissinger-ai-could-mean-the-end-of-human-history/559124/ (Accessed 26 September 2020).

overwhelm wisdom."[49] Despite the desire to keep humans "in the loop," it seems increasingly likely that decisions about the use of violence will be gradually outsourced to automated systems.[50] This transformation will place pressure on and perhaps even undermine the extant norms and moral formulations that underwrite killing in war.

Many twenty-first century commentators have been attuned to these significant shifts. Brad Allenby points out that "A military leadership class that has developed in traditional combat environments will not have the same values, nor behave in the same way, as a military leadership class selected for its ability to play video games in high school."[51] He goes on to point out that when many simultaneous transformations are taking place in military technology, institutional structures, and operational culture, "one must ask whether the laws of war, conceptually and as embodied in existing treaty and agreement, are adequate."[52] It seems likely that as militaries more readily come to rely on automated systems, and those systems begin to kill civilians, it is not clear how good intentions will factor into the moral judgment of the international community.[53]

---

49 Ibid.

50 In fact, it has recently been reported that Turkey's Kargu-2 drone quadcopter hunted down and killed humans autonomously. David Hambling, "Drones May Have Attacked Humans Fully Autonomously for the First Time," *New Scientist* 27 May 2021. https://www.newscientist.com/article/2278852-drones-may-have-attacked-humans-fully-autonomously-for-the-first-time/?fbclid=IwAR295rHSzOp31xkv46aFJyx7bNZzihGZlT8ZjpCgRlaV 6v8XmPkv-jYWlDg (Accessed 1 June 2021).

51 Brad Allenby, "The Implication of Emerging Technologies for Just War Theory," in *The Applied Ethics of Emerging Military and Security Technologies* ed. Braden Allenby (New York: Routledge, 2015), 3.

52 Ibid., 15. For similar arguments, see *Military Ethics and Emerging Technologies* eds. Timothy Demy, George Lucas Jr., Bradley Strawser (New York: Routledge, 2014); Peter Singer, "The Ethics of Kill Applications: Why Is it so Hard to Talk about Morality when it comes to New Military Technology," in *The Applied Ethics of Emerging Military and Security Technologies* ed. Braden Allenby (New York: Routledge, 2015); Peter Singer, "The Five Deadly Flaws of talking about Emerging Military Technologies and the Need for New Approaches to Law, Ethics, and War" in *Drone Wars*, eds. Peter Bergen and Daniel Rothenberg (New York: Cambridge University Press, 2015); Rain Liivoja, "Technological Change and the Evolution of the Law of War," *International Review of the Red Cross* (2015); 1157–1177; and Armin Krishnan, *Military Neuroscience and the Coming Age of Neurowarfare* (New York: Routledge, 2017).

53 The discourse surrounding this transformation bears remarkable similarities to ethical transformation that occurred with the rise of aerial bombing. There are many techno-optimists who claim that these (semi-)automated technologies will make war more moral discriminating and less cruel. Ronald Arkin argues that automated systems will enable militaries to "conform as well or better than our soldiers with respect to adherence

In many respects, at the dawn of the twenty-first century, we are like those early commentators theorizing the significance of air power at the beginning of the twentieth century. The technologies we now possess will likely change exponentially – indeed, in ways that we can only faintly imagine. This is significant because, as the history of technological development has shown, new systems of violence transform the military's institutional structures, operating systems, and, ultimately, the threat environment itself. In the end, these developments will change not only how one kills, but also how one explains unwanted or undesirable killing. Despite the creative and rather prescient predictions by future-war fiction writers, the particular facts of history produce their own causal sequence. Ultimately, the exact impact of these killing systems will only be understood in some future moment, that is, even if we are already anticipating and imagining that trauma in our late twentieth–early twenty-first-century science fiction literature. The tragedy of autonomous systems is psychically overdetermined in the public imaginary. While we cannot anticipate the exact historical expression of these trends, we should expect the moral language governing killing in war to mutate and evolve to meet these new challenges.

---

to the existing Laws of War." Ronald Arkin, *Governing Lethal Behavior in Autonomous Robots* (New York: Taylor & Francis, 2009), 29; Jean-François Caron argues that we have a moral obligation to pursue these new technologies for the numerous advantages they produce, particularly, that these technologies will allow states to live up to International Humanitarian Law more effectively. Jean-François Caron, *Contemporary Technologies and the Morality of Warfare: The War of the Machines* (New York: Routledge, 2020).

Bibliography of Primary Sources

Prior to the Sixteenth Century

Alberti, Leon Battista, *On the Art of Building in Ten Books* [1452], trans. Joseph Rykwert, Neil Leach, Robert Tavernor (Cambridge, MA: MIT Press, 1988).

Alighieri, Dante, *De Monarchia* [1312], trans. Aurelia Henry (New York: Houghton Mifflin, 1904).

Aquinas, Thomas, *Summa Theologica* Volume 3, trans. Fathers of the English Dominican Province (New York, 2007).

Aristotle, *A New Translation of Aristotle's Rhetoric*, trans. John Gillies (London: Printed for T.Cadell ..., 1823).

Augustine, *On Free Choice of the Will* trans. Thomas Williams (Indianapolis: Hackett, 1993).

da Legnano, Giovanni, *De Bello, De Pepresaliss et De Duello* [1360], trans. Thomas Holland (London: Oxford University Press, 1917).

Gratian, *Concordia discordantium canonum* [c. 1150] (Lipsiae: Ex Officina Bernhardi Tauchnitz, 1879).

Ockham, William, *A Short Discourse on the Tyrannical Government*, trans. John Kilcullen, ed. Aurthur Stephen McGrade (New York: Cambridge University Press, 1992).

Plato, *Gorgias* trans. Donald Zeyl (Indianapolis: Hackett, 1987).

The Sixteenth Century

Ayala, Balthasar, *On the Law and the Duties Connected with War and on Military Discipline* [1582] trans. John Pawley Bate (Washington DC: Carnegie Institute, 1912).

Bañez, Domingo, *Commentaria in secundam secundae D. Thomae* (Venetiis: Apud Bernardum Iuntam, 1587).

Barret, Robert, *The Theorike and Practike of Moderne Warres Discoursed in Dialogue Wise* (London: for William Pasonby, 1598).

Barwick, Humfrey, *A Breefe Discourse, Concerning the Force and Effect of all Manuall Weapons of Fire* (London: for Richard Oliffe ..., 1592).

Belli, Pierino, *A Treatise on Military Matters and Warfare, Volume 2* [1563] (Oxford: Clarendon, 1936).

Benzoni, Rutilio, *Speculum episcoporum et curatorum* (Venetiis: Apud Minimam Societatem, 1595).

Bourne, William, *Inventions or Devises Very Necessary for All Generalls and Captaines, or Leaders of Men* (London: Printed for Thomas Woodcock, 1590).

Cajetan, Thomas, *Commentarium super Thomae Aquinatis Summa Theolgoiae, Volume 4* [1540] (Rome: Polyglot Press, 1882).

Clayton, Gyles, *The Approved Order of Martiall Discipline* (London: By I.C. for Abraham Kitsonne ..., 1591).

Corradus, Joannes Baptista, *Responsa ad cujuscunque pene generis casuum Conscientiae, Volume 1* (Venetiis: Apud Societatem Minimam, 1598).

Carbone de Costacciaro, Lodovico, *Tractatus de omnium rerum restitutione amplissimus* (Venetiis: Apud Haeredes Ioan Baptistae Somaschi, 1592).

Digges, Thomas, *An arithmetical warlike treatise named Stratioticos* (London: R. Field, 1590).

Elizabeth I, "Vol. CCLXIII, 1597" *Calendar of State Papers, Domestic Series, Elizabeth, 1695–1597*, ed. Mary Anne Everett Green (London: Longmans, Green, Reader and Dyer, 1869).

de Alava y Viamont, Diego, *El Perfecto Capitán* (Madrid: Por Pedro Madrigal, 1590).

de la Noue, François, *The Politicke and Militarie Discoverses of the Lord de la Nouue* trans. E.A. (London: Printed for T.C. and E.A ..., 1587).

de las Casas, Bartolomé, *Brevissima relación de la destruycion de las Indias* ([s.n.: n.p], 1552).

de Loque, Bertrand, *Discourses of Warre and Single Combat*, trans. John Eliot (London: Printed by John Wolfe ..., 1591).

de Mendoza, Don Bernardiono, *Theorique and Practise of Warre* trans. Sir Edwarde Hobt (Middelburg: Printed by Richard Schilders, 1597).

de Molina, Luis, *De justitia et jure opera omnia* [1593–1609] (Coloniae: Sumptibus Marci Michaelis Bousquet, 1733).

de Soto, Dominicus, *De justitia et jure libri X* (Venetiis: Apud Minimam Societatem, 1594).

de Sylvestris Ferrariensis, Franciscus, *Commentaria in Summa contra Gentiles* (Romae: Sumptibus et typis Orphanotrophii a S. Hieronymo Aemiliani, ca. 1516).

de Valentia, Grégoire, *Commentariorum theologicorum tomus tertius* (Lugduni: Sumptibus Horatii Cardon, 1609).

de Valdes, Francisco, *Espeio, y disciplina military* (Brussellas: En casa de Roger Velpius ..., 1596).

de Vitoria, Francisco, "On the Law of War," in *Political Writings*, eds. Anthony Pagden and Jeremy Laweance [1532] (New York: Cambridge University Press, 1991).

di Porcia, Jacopo, *The Preceptes of Warre* trans. Peter Betham (London: In the Olde Lewery ..., 1544).

Erasmus, Desiderius, *Enchiridion Militis Christiani* [1503] (London: Methuen, 1905).

Erasmus, Desiderius, *Against War* [1515] (Boston: Merrymount Press, 1907).

Erasmus, Desiderius, "Dulce bellum inexpertis: War is a treat for those who have not tried it" [1515], in *Collected Works of Erasmus, Volume 35, Adages III iv 1 to IV ii 100* trans. Denis Drysdall, ed. John Grant (Toronto: University of Toronto Press, 2005).

Erasmus, Desiderius, *The Education of a Christian Prince* in *The Education of a Christian Prince with the Panegyric for Archduke Philip of Austria* [1516] (New York: Cambridge University Press, 1997).

Fourquevaux, Raimond, *Instructions for the warres. Amply, learnedly, and politiquely, discoursing the method of militarie discipline*, trans. Paul Iue (London: Printed for Thomas Man and Tobie Cooke, 1589).

Frontinus, Sextus Juliys, *The Strategemes, Sleyghtes, and Policies of Warre*, trans. Rycharde Morysine (London: Thom. Bertheleti typis impress, 1539).

Garrard, William, *The Arte of Warre* (London: Printed for Roger Warde ..., 1591).

Gates, Geffrey, *The Defence of Militarie Profession* (London: By Henry Middleton, for Iohn Harison, 1579).

Gentili, Alberico, *De Jure Belli Libri Tres, Volume 2* [1598], trans. John Rolfe (London: Humphry Milford, 1933).

Gentillet, Innocent, *A Discourse upon the Means of Wel Governing and Maintaining in Good Peace, a Kingdome* [1576] (London: Printed by Adam Flip, 1608).

Ive, Paul, *The Practise of Fortification* (Imprinted at London by Thomas Orwin for Thomas Man and Toby Cooke, 1589).

Lipsius, Justus, *Politica: Six Books of Politics or Political Instruction* [1589] ed. Jan Waszink (Assen, The Netherlands: Royal Van Gorcum, 2004).

Londoño, Sancho, *El Discurso sobre la forma de reducir la disciplina militar y antiguo estado* (Madrid: Por Luys Sanchez, 1593).

Luther, Martin, *Secular Authority: To What Extent it should be Obeyed* [1523], in *Martin Luther: Selections from His Writings* ed. John Dillenberger (New York: Anchor, 1961).

Machiavelli, Niccolò, *The Prince* [1532] ed. Quentin Skinner (New York: Cambridge University Press, 1988).

Machiavelli, Niccolò, *Discourses on Livy* [1531] trans. Harvey Mansfield and Nathan Tarcov (Chicago: University of Chicago Press, 1996).

Machiavelli, Niccolò, *The Art of War* [1521] in *The Works of the Famous Nicholas Machiavel* trans. Henry Neville (London: Printed for John Starkey ..., 1680).

Mair, John, *In Quarum Sententiarum quaestiones utilissimae suprema ipsius lucubratione enucleatae* ([Paris?]: Venundantur a sui impressore Iodoco Badio, 1519).

More, Thomas, *Utopia* [1516], trans. G. Logan and R. Adams (New York: Cambridge University Press, 2003).

Mosquera de Figueroa, Cristóbal, *Comentario en breve compendio de disciplina military* (Madrid: Por Luis Sanchez, 1596).

Norden, John, *The Mirror of Honor* (London: By the Widowe Orwin ..., 1597).

Platonico, Onosandro, *Of the Generall Captaine* trans. Fabio Cotta (London: By William Seres ..., 1563).

Renatus, Flavius Vegetius, *The Foure Bookes of Flavius Vegetius Renatus* (London: Fletestreate ..., 1572).

Rich, Barnabe, *A path-way to military practice* (London: Printed by Iohn Charlewood ..., 1587).

Smythe, Sir John, *Certain discourses, written by Sir John Smythe, Knight: concerning the formes and effects of divers sorts of weapons, and other verie important matters militarie* (London: Printed by Richard Iohnes ..., 1590).

Statues and Ordynances for the Warre (London: Fleetstreet by Thomas Barthelet ..., 1544).

Styward, Thomas, *The Pathwaie to Martiall Discipline* (London: Printed by T.E. for Myles Jenyn ..., 1581).

Suarez, Francisco, *A Work on the Three Theological Virtues: Faith, Hope, and Charity* in *Selections from Three Works* trans. Gwladys Williams, Ammi Brown, and John Waldron, ed. Thomas Pink (Indianapolis: Liberty Fund, 2015).

Suarez, Fransisco, *A Treatise on the Laws of God the Lawgiver* in *Selections from Three Works* trans. Gwladys Williams, Ammi Brown, and John Waldron, ed. Thomas Pink (Indianapolis: Liberty Fund, 2015).

Sutcliffe, Matthew, *The practice, proceedings, and laws of armes* (London: by the deputies Christopher Barker ..., 1593).

Vazquez, Gabriel, *Commentariorum, ac disputationum in tertiam partem sancti thomae, tomus tertius* [1598–1615] (Antverpiae, Apud Petrum & Ionnem bellerso, 1621).

Whithorne, Peter, *Certain Wayes for the Ordering of Soldiours in Battleray* (London: by W. Williamson ..., 1573).

Williams, Sir Roger, *A briefe discourse of warre* (London: Printed by Thomas Orwin ..., 1590).

The Seventeenth Century

An Abridgment of the English Military Discipline (London: Printed for Robert Harford, 1680).

An Account of the Bombarding of Granville (London: Printed by Edw. Jones in the Savoy, 1695).

An Account of the Bombarding of St. Malo, by the Fleet under the Command of the Lord Berkeley, Admiral of the Blue (London: Printed by Edw. Jones in the Savoy, 1695).

An Account of the Burning of Havre de Grace (London: Printed for Richard Baldwin, in Warwick-Lane, 1694).

Achesone, James, *The Military Garden, Or Instructions for All Young Souldiers and Such who are Disposed to* Learne (Edinburgh: Printed by Iohn Wreittoun, 1629).

Ames, William, *De conscientia, et eius iure, vel casibus, libri quinque* (Amstelodami: Apus Ioan Ianßonium, 1643).

Anderson, Robert, *Of Granadoes, Carcasses, and Fireballs* (London: Printed for Robert Morden ..., 1690).

Andrewes, Lancelot, *The Patterns of the Catechistical Doctrine at Large* (London: Imprinted by Roger Norton ..., 1650).

Baxter, Richard, *The Catechizing of Families* (London: Printed for T. Parkburst ..., 1683).

Becanus, Martin, *Summa Theologiea Scholasticae* (Lugduni: Sumptibus Antonii Pillehotte ..., 1621).

Berard, Serapi, *Manifestacion en que se publican muchos y relevantes* (Barcelona: En Casa de Cormellas ..., 1697).

Bernard, Richard, *The Bible-battells, Or The sacred art military For the rightly waging of warre according to Holy Writ* (London: Printed for Edward Blackmore ..., 1629).

Bland, Francis, *The Souldiers March to Salvation* (York: [n.p.], 1647).

Bonacina, Martino, *Sacrae theologiae* [1623] (Lugduni: Sumptibus Anisson & Joannis Posuel, 1700).

Bingham, John, *The Tactics of Ælian: Or, Art of Embattailing an Army After ye Grecian Manner* (London: for Laurence Lisle ..., 1616).

Busenbaum, Hermann, *Medulla theologiae moralis, facili ac perspicua methodo resolvens casus conscientiae* [1645] (Venetiis: Apud Hieronymus Albriccium ..., 1689).

Capizucchi, Raimondo, *Ordinis Praedicatorum Sacrae Theologiae Magistri. Controversiœ Theologicœ Selectae Scholasticœ Morales, Dogmaticœ Scripturales, ad Mentem D. Thomae Aquinatis* (Rome: Typis Angeli Bernabo, HH. Manelphi Manelphij, 1670).

Caspari, Johannes, *Decas Qvaestionum Illustrium Juridico-Politicarum* (Jenae: Typis Tobiae Steinmanni, 1623).

Cooke, Edward, *The character or vvarre, or The image of martiall discipline* (London: Printed by Tho. Purfoot, 1626).

Crucé, Émeric, *The New Cyneas* [1623] trans. Thomas Willing Balch (Philadelphia: Allen, Lane, and Scott, 1909).

Cruso, John, *Military instructions for the cavallrie* (Cambridge: Printed by the Printers to the University of Cambridge, 1632).

D'Auvergne, Edward, *The History of the Campagne in Flanders, for the Year 1695* (London: Printed for Mat. Wotton and John Newton, 1696).

Davies, Edward, *The art of vvare, and Englands traynings* (London: Printed by Edward Griffen ..., 1619).

Dawson, George, *Origo Legum: Or A Treatise of the Origin of Law, and Their Obliging Power* (London: Printed for Richard Churchill ..., 1694).

de Aragón, Pedro, *Ordinis eremitarums augustini, atrium, & Sacrae Theologiae Magistri, & in Clarissima Salmaticensi Academia publici professoris* (Venetiis: Apud Minimam Societatem, 1608).

de Gaya, Lewis, *A Treatise of the Arms and Engines of War* (London: Printed for Robert Harford …, 1678).

de la Cruz, Johannes, *Directorium conscientiae in duas partes divisum* (Parisiis: Apud Edmundum Couteror, 1673).

de Ledesma, Pedro, *Segunda Parte de la Summa* (Lisboa: Impresso por Pedro Crasbeeck, 1617).

della Croce, Flaminio, *Theatro Militare* (In Anversa: Appresso Henrico Aertssio, 1617).

de Tapia, Pedro, *Catenae moralis doctrinae* (Hispali: in aedibus archiepiscopalibus excussit Ioannes Gomez a Blas, 1657).

de Torres, Luis, *Disputationum in secundam secundae D. Thomae* (Lugduni: Sumptibus Horathii Cardon, 1617).

de Vauban, Sébastien le Prestre, *The New Method of Fortification* (London: Abel Swall, 1691).

Devereux, Robert, *Camp Discipline, or The Soldiers Duty* (London: Printed for Jo. Rothwell …, 1642).

de Villalobos, Enrique, *Manual de Confessores* (Salamanca: En casa de Diego Cyssio, 1625).

Downame, John, *The Christian Warfare Against the Devill, World, and Flesh* (London: Printed by William Stansby, 1634).

duc de Rohan, Henri, *The complete captain* (Cambridge: Printed by Roger Daniel …, 1640).

Du Pin, Louis Ellies, *A New Ecclesiastical History of the Sixteenth Century* (London: Printed by H. Clark, for Tim Child …, 1699).

Du Praissac, Sieur, *The art of warre, or Militarie discourses of leavying, marching, encamping; and embattailing an armie* (Cambridge: Printed by Roger Daniel …, 1639).

Elton, Richard, *The Compleat Body of the Art Military* (London: Printed for Robert Leybourn, 1650).

An Exact and Complete Journal of all the Cosiderable Actions of the Confederate Fleets and the Armys from the Opening of the Campaigne, for this Present year 1695, until the Conclusion Thereof (London: Printed for John Whitlock, 1696).

An Exact Diary of the Siege of the City of Ments (London: Printed by R. Janeway, 1689).

Fernandez, Don Francisco Antonio, *Estoque de la Guerra y Arte Militar* (Barcelona: En casa Cormellas, por Thomas Loriente, 1699).

Fernández de Medrano, Sebastián, *El perfecto artificial, bombardero y artillero* (Brusselas: En Casa de Lamberto Marchant, 1699).

Freke, William, *Art of War; Summary of the Martial Precepts Necessary for an Officer* (London: Printed for Tho. Minors ..., 1693).

Fulbecke, William, *The Pandectes of the Law of Nations* (London: Imprinted by Adam Islip for Thomas Wight, 1602).

Gibson, Abraham, *Christiana-Polemica, or A Preparatiue to Warre Shewing the Lawfull Use Thereof* (London: Printed by Edw. Griffin ..., 1619).

Gouge, William, *The Whole Armor of God: or a Christians spiritual furniture to keepe him safe from all the assaults of Satan* (London: Printed by Iohn Beale, 1619).

Grotius, Hugo, *The Rights of War and Peace* [1625] trans. A.C. Campbell (New York: M. Walter Dunne, 1901).

Gurnall, William, *The Christian in Compleat Armour; Or, A Treatise of the Saints War Against the Devil* (London: Printed for Ralph Smith ..., 1655).

Hexham, Henry, *The first part of the principles of the art military practiced in the warres of the United Netherlands* (Delf, Holland: for the lovers of the noble art military, 1642).

Hobbes, Thomas, *The Elements of the Law* [1655] in *The Moral and Political Works of Thomas Hobbes* (London: [n.p.], 1750).

Hurtado, Gaspar, *Tractatus de iustitia et iure* (Matriti: Apud Ioannem Sanches, 1637).

Industrious Souldiers Golden Treasury of Knowledge in the Art of Making War (London: Printed for Nicholas Boddington, 1690).

Jordan, Thomas, *The Christian Soldier, or Preparation for Battaile* (London: Printed for Edward Christopher, 1642).

Kellie, Sir Thomas, *Pallas armata, or Militarie instructions for the learned* (Printed at Edinburgh: By the heires of Andro Hart, 1627).

Lawes and Ordinances of Warre (London: Printed by Robert Barker ..., 1640).

Lawes and Ordinances of Warre: Established for the Better Conduct of the Army (London: Printed for Luke Fawne, 1643).

Leighton, Alexander, *Speculum Belli Sacri: Or, The Lookingglasse of the Holy War* ([Amsterdam]: Printed [by the successors of Giles Thorp], 1624).

Lessius, Leonardus, *De iustitia et iure ceterisque virtutibus cardinalibus* (Lugduni: Sumphibus Ludovici Prost Haeredis Roville, 1622).

Littleton, Edward, *A Preservative for our Money* (London: Printed by B. Motte ..., 1696).

Locke, John, *The Second Treatise of Government* [1690] in *The Works of John Locke, Volume 4* (London: Printed for C. and J. Rivington ..., 1824).

Lupton, Donald, *A Warre-like Treatise of the Pike* (London: Printed by Richard Hodgkinsonne ..., 1642).

Marcelline, George, *Vox Militis* (London: By B.A. for Thomas Archer ..., 1625).

Márquez, Juan, *El governador christiano* (En Amberes: En casa de Jacobo Meursio, 1664).

Mayne, Jasper, *The Amorous Warre* (Oxford: Printed by Henry Hall ... for Ric. Davis, 1659).

Military Orders and Articles Established by His Majesty, For the Better Ordering and Government of His Majesties Army (Oxford: By Leonard Lichfield ..., 1642).

Molloy, Charles, *De Jure Maritimo et Navali; Or, A Treatise of Affairs Maritime and of Commerce* (London: Printed for John Bellinger ..., 1682).

Monro, Robert, *The Scotch Military Discipline Learned from the Valiant Swede* (London: Printed for William Ley at Pauls-Chaine, 1644).

Moore, Jonas, *Modern Fortification; Or, Elements of Military Architecture* (London: Printed for Obadiah Blagrave ..., 1689).

Moretti, Tomaso, *A General Treatise of Artillery* (London: Printed by A.G. and J.P ..., 1683).

Moxon, Joseph, *An Epitome of The Whole Art of Warre* (London: Printed for J. Moxon ..., 1692).

Nicholetts, Charles, *The Great Work of God in this Present Dispensation of Peace* (London: Printed by Hugh Flemman ..., 1698).

Norton, Robert, *Of the Art of Great Artillery* (London: Printed by Edw. Allde ..., 1624).

Norton, Robert, *The Gunner Shewing the Whole Practise of Artillery* (London: Printed by A.M. for Humphry ..., 1628).

Norwood, Richard, *Fortification or Architecture Military* (London: Printed by Tho. Cotes, for Andrew Crooke ..., 1639).

Noydens, Benito Remigio, *Decisiones practicas y morales para curas, confesores y capellanes de los exercitos, y armadas* (Madrid: Por Andres Gareta de la Inglesia, 1665).

Orrery, Roger Boyle, *A Treatise of the Art of War* (Savoy: Printed by T. N ..., 1677).

Papillon, David, *A Practicall Abstract of Fortification and Assailing* (London: Printed by R. Austin ..., 1645).

Pascal, Blaise, "Letter VII" *Provincial Letters* [1656], trans. Rev. Thomas McCrie (New York: Derby & Jackson, 1859).

Perrault, Nicholas, *The Jesuit Morals*, trans. Ezerel Tonge (London: Printed for John Starkey, at Miter in Fleet-street, 1670).

Perron, W., *Galliæ Flagellum: The Reduction of France Made Easie* (London: [n.p.], 1696).

The Present State of Europe: or, The Historical and Political Monthly Mercury, Volume VII (London: Printed for Henry Rhodes ..., 1696).

Price, Daniell, *Prœlium & Prœmium: The Christians War and Rewarde* (Oxford: Printed by Joseph Barnes, 1608).

Pufendorf, Samuel, *Of the Law of Nature and Nations* [1672] trans. Basil Kennett (London: Printed for I. Walthoe ..., 1729).

Ram, Robert, *The Souldiers Catechisms: Composed for the Parliaments Army* (London: Printed for J. Wright in the Old-Baily, 1644).

Rich, Barnabe, *A Looking-Glasse for Warre* (London: By Thomas Creede …, 1604).

Rutherford, Samuel, *Lex, Rex: The Law and the Prince. A Dispute for the Just Prerogative of King and People* (London: Printed for Iohn Field …, 1644).

Sánchez, Tomás, *Disputationum de Matrimonii Sacramento* (Antverpiae: Apud Ioannem Hasrey, 1614).

Sancroft, William, *Modern Policies Taken from Machiavel, Borgia, and Other Choice Authors* (London: Printed for Tho. Dring …, 1655).

Sayers, Gregory, *Clavis Regia Sacerdotum* (Monasteri Westfaliæ: Samptibus Michaelis Dalii, & Bernardi Raesfelt, 1628).

Shields, Alexander, *A Hind Let Loose* (Edinburgh?: [n.p.], 1687).

Siemienowicz, Kazimierz, *The Great Art of Artillery* [1650], trans. George Shelvocke (London: Printed for J. Tonson …, 1729).

Smith, Thomas, *The Art of Gunnery* (London: Printed for William Ponsonby, 1600).

Sylvius, Francis, *Commentarius in Totam Secundam Secundae S. Tho. Aquinatis* (Duaci: Apud Gerardum Patte …, 1628).

Taylor, Jeremy, *Ductor Dubitantium, Or, The Rule of Conscience in All Her Generall measures* (London: Printed by James Flesher …, 1660).

Thims, Gregory, *The Protestant Informer, or Information to all Protestants: Shewing the Causes and End, of this Unjust Warre* (London: [n.p.], 1643).

Turner, Sir James, *Pallas armata, Military essayes of the ancient Grecian, Roman, and modern art of war* (London: M.W. for Richard Chiswell …, 1683).

Ward, Richard, *The Anatomy of Warre* (London: John Dalham and Richar Lownds, 1642).

Ward, Robert, *Adimadversions of Warre* (London: Iohn Dawson …, 1639).

Watts, William, *The Swedish Discipline, Religious, Civile, and Military* (London: Iohn Dawson …, 1632).

Weemse, John, *An Explanation of the Ceremonial Lawes of Moses* (London: T. Cotes for Iohn Bellamie …, 1632).

Whitehall, August 19, 1695. This morning came in an express from my Lord Berkeley with an account of the Bombarding of Calais (London: Edw. Jones …, 1695).

Whitehall, July 23, 1696. This evening arrived here on express from my Lord Berkeley with letters dated the 20th of this month, off Ushant (London: Edw. Jones …, 1696).

Wiggers, Joannis, *Commentaria de iure et iustitia* (Lovanij: Sumptibus Georgii Lipsii, 1662).

Wilkins, John, *Mathematical Magick; Or the Wonders That May Be Performed by Mechanichal Geometry* (London: Edw. Gellibrand at the Golden Ball …, 1680).

Wiseman, Sir Robert, *The Law of Laws: or, The Excellency of the Civil Law, Above all other Humane Laws* (London: R. Royston …, 1686).

Zepeda y Adrada, Alonso, *Epitome de la fortificacion moderna* (Brusselas: Por Francisco Poppens, Mercader de Libros, 1669).

The Eighteenth Century

"An Account of the First Aerial Voyage in England," in *The Gentleman's Magazine, Volume 54* (London: John Nichols ..., 1784).

Adye, Stephen Payne, *A Treatise on Courts Martial* (London: J. Murray and S. Highley, 1796).

The Air Balloon: Or, a Treatise on the Aerostatic Globe (London: G. Kearsley ..., 1783).

Alcock, Mary, *The Air Balloon; Or, Flying Mortal. A Poem* (London: E. Mackley ..., 1784).

The Analytical Review, Volume 10 (London: J. Johnson ..., 1791).

Anonimo, *A Dissertation on the Law of Nature, The Law of Nations, and the Civil Law* (London: J. Roberts, 1723).

Baxter, Richard, *The Practical Works of the Late Reverend and Pious Mr. Richard Baxter, Volume 4* (London: Thomas Parkhurst ..., 1707).

Billuart, Charles-René, *Summa S. Thomae hodiernis academiarum moribus accomodata* (Leodii: Apud Everardum Kints, 1751).

Blackstone, William, *Commentaries on the Laws of England, Volume 1* [1765] (New York: E. Duyckingck ..., 1827).

Bland, Humphrey, *A Treatise of Military Discipline* (London: Daniel Midwinter ..., 1743).

Bruce, Alexander, *The Institutions of Military Law, Ancient and Modern* (Edinburgh: Printed by the Heirs and Successors of A. Anderson, 1717).

Burlamaqui, Jean Jacques, *The Principles of Natural and Politic Law, Volumes 1–2* [1751] (Philadelphia: Carey and Lea, 1830).

Bynkershoek, Cornelius van, *A Treatise on the Law of War* [1737], trans. Peter Stephen Du Ponceau (Philadelphia: Farrand & Nicholas, 1810).

Campbell, John, *Lives of the Admirals, and other Eminent British Seamen, Volume 3* (London: John Applebee, 1744).

Coke, Roger, *A Detection of the Court of and State of England, Volume II* (London: J. Brotherton and W. Meadows ..., 1719).

Considerations Upon War, Upon Cruelty in General, and Religious Cruelty (London: T. Osborne, 1758).

Cuniliati, Fulgentius, *Universae theologiae moralis accurata complexio instituendis candidatis accommodata* (Venetiis, Bettinelli, 1772).

Dalrymple, Sir John, *Memoirs of Great Britain and Ireland* (Edinburgh: John Bell and William Creech ..., 1788).

Defoe, Daniel, *An Ecclesiastical History of Scotland* (London: Charles Marsh, at Bible in Angel-Court ..., 1734).

de Guibert, Comte, *A General Essay on Tactics* (London: J. Millan ..., 1781).

de Saxe, Maurice, *Reveries, or Memoirs Concerning the Art of War* (Edinburgh: Sands, Donaldson, Murray, and Cochran ..., 1759).

de Vattel, Emir, *The Law of Nations* [1758], eds. Béla Kapossy and Richard Whatmore (Indianapolis: Liberty Fund, 2008).

Douglas, Neil, *Britain's Guilt, Danger, and Duty* (Dundee: [n.p.], 1795).

Feuquières, Antoine de Pas de marquis, *Memoirs Historical and Military: Containing a Distinct View of All the Considerable States of Europe* (London: T. Woodward …, 1735).

A General History of Sieges and Battles, by Sea and Land, Volume V (London: J. Curtis in Fleet-street …, 1762).

Gittins, John, *A Complete System of Military Discipline* (London: J. Humfreys …, 1735).

Groddeck, Gabriel, *De eo quod justum est circa tormenta bellica* (Lispsiae: In Collegio Pavlino Prope Avditorium Theologicum, 1710).

Gros, John Daniel, *Natural Principles of Rectitude: For the Conduct of Man in All States and Situations of Life* (New York: T. and J. Swords, 1795).

Grose, Francis, *Military Antiquities Respecting a History of the English Army, Volume 2* (London: S. Hooper …, 1786).

J.H., *The Complete Gentleman Soldier; Or, a Treatise of Military Discipline* (London: Thomas Ballard …, 1702).

Harris, Walter, *The History of the Life and Reign of William-Henry* (London: Edward Bate in George's Lane …, 1749).

Hervey, Frederic, *The Naval History of Great Britain, Volume 3* (London: William Adlard …, 1779).

A History of All the Invasion of England (Windsor: C. Knight …, 1794).

Hutcheson, Francis, *A System of Moral Philosophy, Volume 2* (London: A. Miller …, 1755).

An Introduction to the Art of Fortification (London: John Brindley …, 1745).

Kant, Immanuel, *To Perpetual Peace: A Philosophical Sketch* [1795], trans. Ted Humphrey (Indianapolis: Hackett, 2003).

Lacroix, Claude, *Theologia moralis* [1704–1714] (Parisiis: Apud ludovicum vives …, 1866).

Lee, Richard, *A Treatise of Captures in War* (London: W. Sandby …, 1759).

Liguori, Alphonsus, *Theologia moralis* [1748] (Parisiis: Apud Adrianum le Clere et Soc, 1857).

Ludovici, Jakob Friedrich, *De limitibus defensionis in bello defensive* (Halae: Apud Christoph, Andr. Zeitlerum, 1706).

Mackintosh, Sir James, *A Discourse on the Study of Law of Nature and Nations* (London: T. Cadell …, 1799).

Martens, G.F. von, *The Laws of Nations Being the Science of Natural Law, Covenants, Power, Founded Upon the Treaties and Customs of Modern Nations in Europe* [1789] trans. William Cobbett (London: Fleet St, 1829).

A Military Dictionary Explaining and Describing the Terms, Phrases, Works, and Machines Used in the Science of War (London: G. Robinson …, 1778).

Montesquieu, Baron, *Spirit of the Laws* in *The Complete Words of M. de Montesquieu, Volume 1* [1748] (London: T. Evans ..., 1777).

Muller, John, *The Attack and Defence of Fortify'd Places* (London: J. Millan ..., 1757).

Muller, John, *Treatise of Artillery* (London: John Millan ..., 1768).

A New and Complete Dictionary of Arts and Science, Volume 1 (London: W. Owen ..., 1763).

Pilon, Frederick, *Aerostation; Or the Templar's Stratagem, A Farce* (Dublin: J. Exshaw ..., 1785).

Ralph, James, *The History of England, During the Reigns of King William, Queen Anne, and King George the First, Volume 2* (London: Daniel Browne ..., 1746).

Rapin de Thoyras, Paul, *The History of England, Volume 3* (London: John and Paul Knapton ..., 1744).

Raspe, Rudolf, *Surprising Adventures of Baron Munchausen* [1786] (London: Plummer, 1809).

Rutherford, Thomas, *Institutes of Natural Law* [1754–6] (Baltimore, MD: William and Joseph Neal, 1832).

Schiara, Antonius, *Theologia bellica* (Romae: Typis, & Fusoria Cajetani Zenobiii Typographi ..., 1703).

Sharp, Granville, *Remarks on the Opinions of Some of the Most Celebrated Writers on Crown Law Respecting the Due Distinction between Manslaughter and Murder* (London: B White and R. Horsfield, 1773).

Shovel, Sir Cloudsley, *Secret Memoirs of the Life of the Honerable Sir Cloudsley Shovel* (London: B. Bragge ..., 1708).

Simes, Thomas, *A Treatise on the Military Science* (London: John Millan, 1780).

Swięcicki, Konstanty, *De jure naturae et gentium in genere; et de jure belli et pacis in specie* (Venetiis: Apud Antonium Bortoli, 1763).

A System of Camp-Discipline (London: J. Millan ..., 1757).

Taylor, John, *Elements of the Civil Law* (London: Ioseph Bentham ..., 1755).

Thomasius, Christian, *Foundations of the Law of Nature and Nations* [1705] in *Institutes of Diving Jurisprudence*, trans. Thomas Ahnert (Indianapolis, IN: Liberty Fund, 2011).

Urban, Sylvanus, "Historical Account of the Town of St. Malo," in *The Gentleman's Magazine, Volume 75* (London: John Nichols ..., 1794).

Voltaire, "Guerre, War [1764]," in *Voltaire: Political Writings* trans. David Williams (New York: Cambridge University Press, 1994).

Ward, Robert, *An Enquiry into the Foundation and History of the Laws of Nations in Europe, Volume 1* (London: A. Strahan and W. Woodfall ..., 1795).

Washington, George, *The Writings of George Washington, Volume 5* ed. John C. Fitzpatrick (Washington, D.C.: United States Government Printing Office, 1932).

Wolff, Christian, *The Law of Nations Treated According to the Scientific Method* [1749] ed. Thomas Ahnert, trans. Joseph Drake (Indianapolis: Liberty Fund, 2017).

The Nineteenth Century

"About the War and Wounded," *Medical Press & Circular: A Weekly Journal of Medicine and Medical Affairs, From January to June 1871* (London: Strand, 1871).

Adams, W.H. Davenport, *The Franco-Prussian War: Its Causes, Incidents, and Consequences* (London: William Mackenzie, 1871).

Alfred-Henty, George, *The Young Francs-Tireurs, and their Adventures in the Franco-Prussian War* (London: Griffith and Farran, 1873).

Amos, Sheldon, *Political and Legal Remedies for War* (London: Cassell, Peter, Galpin & Co., 1880).

Annual Reports of the War Department: Report of the Secretary of War (Washington DC: Government Printing Office, 1897).

Arnold, Thomas, *Introductory Lectures on Modern History* (New York: D. Appleton & Co., 1862).

Atkinson, George, *International Morality; or the Touchstone of the Law of Nations* (London: G. Woodfall and Son, 1851).

Austin, John, *The Province of Jurisprudence Determined, Volume 1* [1832] (London: John Murray, 1861).

Ballou, Adin, *Christian Non-resistance* (London: Charles Gilpin ..., 1848).

Bancroft, Hubert, *History of Mexico, Volume 4* (San Francisco: A.L. Bancroft & Co, 1885).

Barton, Samuel, *The Battle of the Swash and the Capture of Canada* (New York: Charles T. Dillingham, 1888).

Beawes, Wyndham, *Lex Mercatoria; Or a Complete Code of Commercial Law, Volume 1* (London: Printed for F.C. and J. Rivington ..., 1813).

"The Beagle Affair," *The Law Journal, Volume XIII* (London: E.B. Ince, 1879).

Bello, Andres, *Principios de Derecho Internacional* (Caracas: Almacen de J.M. de Rojas ..., 1847).

Bellows, John, *The Track of the War around Metz and the Fund for the Non-combatant Sufferers* (London: Trübner and Co ..., 1871).

Bentham, Jeremy, "A Plan for an Universal and Perpetual Peace," in *The Works of Jeremy Bentham, Volume VIII* (Edinburgh: William Tait, 1839).

Bentham, Jeremy, *Principles of Penal Law* in *The Works of Jeremy Bentham, Volume 1*, ed. John Bowring (Edinburgh: William Tait, 107 Prince's Street, 1843).

Bentham, Jeremy, *The Constitutional Code* in *The Works of Jeremy Bentham, Volume 9* (Edinburgh: William Tait, 1843).

Bernard, Mountague, *On the Principle of Non-intervention* (Oxford: J.H. and Jas. Parker, 1860).

Bolles, John A., *Essay on a Congress of Nations for the Adjustment of International Disputes* (Boston: Whipple and Damrell, 1839).

Botts, Barnet, *The Problem of Aerial Navigation* (Paso Robles, CA: Botts, 1894).

Boutros, F., *Principles of the Law of Nations* (Serampore: Ostell and Lepage ..., 1845).

Brownbill, John, *Principles of English Canon Law, Part 1* (London: Kegan Paul, Trench & Co., 1883).

Brussels Déclaration 1874. https://web.ics.purdue.edu/~wggray/Teaching/His300/Handouts/Brussels-1874.html.

Buckle, Henry Thomas, *History of Civilization in England, Volume 1* (London: Parker, Son, and Bourn, 1861).

Callwell, Colonel C.E., *Small Wars: Their Principles and Practice* [1896] (London: H.M. Stationery Office, 1906).

Carlyle, Thomas, *The French Revolution: A History, Part 2* (New York: Wiley and Putnam, 1847).

Caulkins, Daniel, *Aerial Navigation: The Best Method* (Toledo, OH: Blade Printing & Paper, 1895).

Chanute, Octave, *Progress in Flying Machines* (New York: The American Engineer and Railroad Journal, 1894).

Chanute, Octave, *Aerial Navigation* (New York: The Railroad and Engineering Journal, 1891).

"A Chapter on Balloons," *Fraser's Magazine for Town and Country, Vol. XL July to December 1849* (London: John W. Parker, 1850).

Chitty, Joseph, *A Practical Treatise on the Law of Nations* (Boston: Bradford and Read, 1812).

Clarke, Edward, *A Treatise upon the Law of Extradition* (London: Stevens and Haynes, 1874).

Clausewitz, Carl von, *On War* [1832] trans. Michael Howard and Peter Paret (New York: Oxford University Press, 2007).

Coffin, Charles Carleton, *Four Years of Fighting* (Boston: Arno, 1866).

Coggeshall, George, *History of the American Privateers and Letters-of-marque* (New York: Geo. P. Putnam, 1861).

Colomb, P., J.F. Maurice, F.N. Maude, Archibald Forbes, Charles Lowe, D. Christie Murray, and F. Scudamore, *The War of 189-: A Forecast* (London: William Heinemann, 1895).

Cosgreve, Sir William, *A Concise Account of the Origin and Progress of the Rocket System* (Dublin: Printed by A. O'Neil ..., 1817).

Coxwell, Henry, *My Life and Balloon Experiences* (London: W.H. Allen & Company, 1889).

Coxwell, Henry, *A Knight of the Air; Or, The Aerial Rivals* (London: Digby, Long, and Company, 1895).

Creasy, Sir Edward, *First Platform of International Law* (London: John van Voorst, Paternoster Row, 1876).

Cutbush, James, *A System of Pyrotechny: Comprehending the Theory and Practice, with the Application of Chemistry; Designed for Exhibition and for War* (Philadelphia: Clara F. Cutbush, 1825).

Dahlgren, John, *Maritime International Law* (Boston: B.B. Russell, 1877).

Davis, George B., *The Elements of International Law* [1887] (New York: Harper and Brothers, 1900).

Davis, George B., "Les Lois de la Guerre Continentale," in *The American Journal of International Law* 7:4 (1913); 920–923.

Deane, Sir James P., *The Law of Blockade: As Contained in the Report of Eight Cases Argued and Determined in the High Court of Admiralty* (London: Butterworths ..., 1855).

de Bausset, Arthur, *Aerial Navigation* (Chicago: Fergus Printing Press, 1887).

de Burgh, William, *The Elements of Maritime International Law* (London: Longmans, Green, and Co., 1868).

Declaration Renouncing the Use, in Time of War, of Explosive Projectiles Under 400 Grammes Weight. Saint Petersburg, 29 November/11 December 1868. https://ihl-databases.icrc.org /ihl/full/declaration1868.

d'Ivernois, François, *Effects of the Continental Blockade Upon the Commerce, Finances, Credit, and Prosperity* (London: Cox, Son, and Baylis ..., 1810).

Dodge, David Low, *War Inconsistent with the Religion of Jesus Christ, as it is Inhuman, Unwise, and Criminal* (New York: Dodge and Sayer, 1815).

Douglas, Sir Howard, *A Treatise on Naval Gunnery* (London: John Murray ..., 1855).

Duane, William, *The Law of Nations, Investigation in a Popular Manner* (Philadelphia: William Duane, 1809).

Duane, William, *The American Library; or Compendium of the Modern Tactics, Volume 2* (Philadelphia: Market Street ..., 1809).

Duncan, Captain Francis, *History of Royal Regiment of Artillery* (London: John Murray, 1872).

Duparcq, Edouard le Barre, *Elements of Military Art and History* trans. George Cullum (New York: D. Van Nostrand, 1863).

Durant, Henry, *Un Souvenir de Solferino* (Genève: Imprimerie Jules-Guillaume Fick, 1862).

Dymond, Jonathan, *An Inquiry into the Accordancy of War with the Principles of Christianity* (Philadelphia: I. Ashmead & Co., 1834).

Early, Jubal A., *A Memoir of the Last Year of the War for Independence, in the Confederate States of America* (New Orleans: Blelock & Co., 1867).

Easterley, Robert and John Wilbraham, *The Germ Growers: An Australian Story of Adventure and Mystery* (Melbourne: Melville, Mullen, & Slade, 1892).

Ellis, T. Mullett, *Zalma* (London: Tower Pub. Co., 1895).

Elmendorf, John, *Elements of Moral Theology: Based on the Summa Theologiae of St. Thomas Aquinas* (New York: James Pott & Company, 1892).

Facts and Illustrations in Reference to War, Peace, and International Arbitration (London: London Peace Society, 1872).

"The Fall of Vera Cruz," *The Ladies' National Magazine, Volume XIII* ed. Ann Stephens (Philadelphia, PA: Charles Peterson, 1848).

Farrer, James, *Military Manners and Customs* (London: Chatto & Windus, 1885).

Fawcett, E. Douglas, *Hartmann the Anarchist: Or, The Doom of the Great City* (London: Edward Arnold ..., 1893).

Ferguson, Jan Helenus, *Manual of International Law, Volume 2* (London: W.B. Whittingham & Co., 1884).

Field, David Dudley, *Outlines of an International Code* (New York: Baker, Voorhis, and Company, 1876).

Flagg, Edmund, *Venice: The City of the Sea, Volume 2* (New York: Charles Scribner, 1853).

Forbes, Archibald, *My Experiences of the War between France and Germany, Volume 1* (London: Hurst and Blackett, 1871).

Forney, M.N. (ed.), *Aeronautics* (New York: The American Engineer and Railroad Journal, 1893).

Furber, George, *The Twelve Months Volunteer* (Cincinnati: J.A. & U.P. James, 1848).

Gallaudet, Edward, *A Manual of International Law* (New York: A.S. Barnes & Co, 1879).

Génicot, Edward, *Theologiæ Moralis Institutiones* (Lovanii: Typis et sumptibus polleunis et ceuterick, 1898).

Gentz, Friedrich von, *Fragments Upon the Balance of Power* (London: C&R Baldwin Printers, 1806).

Glenn, Edwin, *Hand-book of International Law* (St. Paul: West Publishing, 1895).

Gracey, Samuel Levis, *Annals of the Sixth Pennsylvania Cavalry* (Philadelphia: E.H. Butler and Co., 1868).

Great Britain Admiralty, *The Queen's Regulations for the Royal Nave: Revised* (London: Harrison and Sons, 1862).

Green, Nathan, *The War with Spain* (Baltimore: International News and Book Co., 1898).

Greer, Tom, *A Modern Dædalus* (London: Griffith, Farran, Okeden & Welsh, 1885).

Griffith, George, *The Angel of the Revolution: A Tale of the Coming Terror* (London: Tower Publishing, 1894).

Griffith, George, *Olga Romanoff: Or, the Syren of the Skies* (London: Tower Publishing, 1894).

Griffith, George, *The Outlaws of the Air* (London: Tower Publishing, 1895).

Griffiths, Major F.A., *Artillerist's Manual and British Soldier's Compendium* (London: W. Clowes & Sons ..., 1868).

Grover, Lieutenant G.E., "On the Uses of Balloons in Military Operations," in *Papers on Subjects Connected with the Duties of the Corps of Royal Engineers, Volume XII* (Woolwich: W.P. Jackson, 1863).

Gury, Jean Pierre, *Compendium theologiae moralis* (Ratisbonae: Typis et sumptibus Georgii Josephi Manz, 1868).

Hall, William, *The Rights and Duties of Neutrals* (London: Longmans, Green and Co., 1874).

Hall, William, *International Law* (Oxford: Clarendon, 1880).

Halleck, Henry W., *International Law or Rules regulating the intercourse of states in peace and war* [1861] (London: Kegan, Paul, Trench, Trubner, 1893).

Harrison, Frederic, "Bismarkism," *The Fortnightly Review, Volume VIII*, ed. John Morley (London: Chapman and Hall, 1870).

Hay, William Delisle, *Three Hundred Years Hence; Or, a Voice from Posterity* (London: Newman and Co, 1881).

Hazlitt, William and Philip Roche, *A Manual of the Law of Maritime Warfare* (London: V & R Stevens and G.S. Norton, 1854).

Helps, Sir Arthur, *Conversations on War and General Culture* (London: Smith, Elder, and Co, 1871).

Hoke, Jacob, *Historical Reminiscences of the War; or Incidents which Transpired in and About Chambersburg* (Chambersburg, PA: M.A. Foltz, 1884).

Hopkins, Gerard Manley, "Balloons and Aeronauts: In Two Parts," *Once a Week: An Illustrated Miscellany, Volume 13, June to December 1865* (London: Bradbury & Evans, 1865).

Hough, William, *The Practice of Courts-martial, and Other Military Courts* (London: Parbury, Allen & Co ..., 1834).

Howard, Sir William, *The British Expedition to Crimea* (London: George Routledge and Sons, 1877).

Hozier, Henry, *The Seven Weeks' War: Its Antecedents and Its Incidents* (London: Macmillan, 1872).

Hozier, Henry, *The Franco-Prussian War: Its Causes, Incidents, and Consequences, Volume 2* (London: William Mackenzie, 1872).

James, W.H., "Balloon Warfare," *Mechanics' Magazine, Volume 4, July to December 1860* (London: Robertson, Brooman, & Co., 1861).

Jay, William, *War and Peace: The Evils of the First, and a Plan for Preserving the Last* (London: Thomas Ward and Co., 1842).

Jay, William, *A Review of the Causes and Consequences of the Mexican War* (Boston: Benjamin B. Mussey & Co., 1849).

Jefferson, Thomas, *Memoir, Correspondence, and Miscellanies, From the Papers of Thomas Jefferson, Volume 4* (Boston: Gray and Bowen, 1830).

Johnson, Willis Fletcher, *Life of William Tecumseh Sherman* (Philadelphia: Edgewood Publishing, 1891).

Jones, Sir John T., "Bombardment," in *Aide-mémoire to the Military Sciences, Volume 1* (London: John Weale, 1846).

Kent's Commentaries on the International Law (Cambridge: Deighton, Bell, and Co., 1866).

Ladd, William, "Extracts for the Circular Letter of the American Peace Society," in *Thirteenth Annual Report of the Committee of the Society for the Promotion of Permanent and Universal Peace* (London: Richard Clay …, 1829).

Ladd, William, "Changes in Public Opinion," in *The Essays of Philanthropos on Peace and War* (Portland: Shirley & Edwards, 1825).

Ladd, William, *An Essay on a Congress of Nations* (Boston: Whipple and Damrell, 1840).

Lancaster, William Joseph [Harry Collingwood], *The Log of the* Flying Fish*: A Story of Aerial and Submarine Peril and Adventure* (London: Blackie & Sone …, 1887).

Lang, Hermann, *The Air Battle: A Vision of the Future* (London: William Penny, 1859).

Langley, S.P., "The Flying Machine," *McClure's Magazine* 9:2 (1897); 647–660.

Lawrence, Thomas, *The Principles of International Law* (Boston: D.C. Heath & Co., 1895).

Leeds, Josiah, *The Primitive Christians' Estimate of War and Self-Defense* (New Vienna, OH: Peace Association of Friends in America, 1876).

Leroy-Beaulieu, Paul, *Contemporary Wars: 1853–1866; Statistical Researches Respecting the Loss of Men and Money Involved in Them* (London: London Peace Society, 1869).

Le Queux, William, *The Great War in England in 1897* (London: Tower Publishing, 1894).

Lieber, Francis, *Manual of Political Ethics, Volume 2* (Boston: Charles C. Little and James Brown, 1839).

Lieber, Francis, *Instructions for the Government of Armies of the United States in the Field, Originally Issued as General Orders No. 100, Adjutant General's Office, 1863* (Washington DC: Government Printing Office 1898).

Lincoln, Abraham, *The Life and Public Services of Abraham Lincoln* ed. Henry Raymond (New York: Derby and Miller, 1865).

Livermore, Abiel Abbot, *The War with Mexico Reviewed* (Boston: The American Peace Society, 1850).

Lorimer, James, *The Institutes of the Law of Nations; a Treatise of the Jural Relations of Separate Political Communities* (Edinburgh: Blackwood and Sons, 1883).

Macdougall, Patrick, *The Theory of War: Illustrated by Numerous Examples from Military History, Volume 1* (London: Longman, Brown, Green Longmans, & Roberts, 1858).

Macqueen, John Frasier, *Chief Points in the Laws of War and Neutrality, Search and Blockade* (London: William and Robert Chambers, 1862).

Maine, Henry, *International Law* (London: John Murray, Albemarle Street, 1888).

Maine, Henry, "The Modern Laws of War," in *International Law: A Series of Lectures Delivered Before the University of Cambridge, 1887* (New York: Henry Hold and Company, 1888).

Mance, John, *The Soldier Spiritualized; or National, Compared with Spiritual Warfare* (London: Partridge and Co., 1859).

Manning, William Oke, *Commentaries on the Law of Nations* (London: Macmillan, 1875).

Manual of Military Law (London: H.M. Stationery Office …, 1899).

María de Pando, José, *Elementos del Derecho Internacional* (Madrid: Imprenta de Algeria y Charlain …, 1843).

Marmont, Marshal, *The Spirit of Military Institutions* trans. Frank Schaller (Columbia, SC: Evans and Cogswell, 1864).

Mercer, Cavalié, *Journal of the Waterloo Campaign: Kept Throughout the Campaign of 1815* (London: William Blackwood and Sons, 1870).

"Mexican Narrative of the Events at the Heroic City of Vera Cruz, While Besieged by the American Army," *Niles' National Register*. 8 May 1847.

Migne, Jacques Paul, *Theologiae cursus completes* (Paris: Apud Editores, in via Gallice Dicta, 1839).

Mill, James, *Essays* (London: J. Innes, Oxford Street, 1823).

Mill, James, "Law of Nations," *Supplement to the Encyclopedia Britannica* (London: J. Innes, 1825).

Mill, John Stuart, "A Few Words on Non-Intervention" [1859], *New England Review* 27:3 (2006); 252–264.

Mill, John Stuart, *The Contest in America* (Boston: Little, Brown and Company, 1862).

Moffat, W. Graham and John White, *What's the World Coming To?* (London: Elliot Stock, 1893).

Moltke, Helmuth von, *Moltke on the Art of War: Selected Writings* ed. Daniel Hughes (London: Greenhill Books, 1893).

Moody, Loring, *A History of the Mexican War* (Boston: Bela Marsh, 1848).

Moynier, Gustave, *Help for Sick and Wounded*. trans. John Furley (London: John Camden Hotten …, 1870).

Nys, Ernest, *Les Origines Du Droit International* (Harlem: Erven F. Bohn 1894).

Odell, Samuel, *The Last War, or the Triumph of the English Tongue* (Chicago: Charles H. Kerr & Co., 1898).

Ollier, Edmund, *Cassell's History of the United States, Volume 3* (New York: Cassell Petter & Galpin, 1874).

Orman, Charles, *The Art of War in the Middle Ages: A.D. 378–1515* (Oxford: B.H. Blackwell, 1885).

Paley, William, *The Principles of Moral and Political Philosophy* (Boston: West and Richardson, 1815).

Paltock, Robert, *The Life and Adventures of Peter Wilkins, among the Flying Nations in the South Seas* (London: J. Limbird …, 1833).

Pasley, Sir Charles, *Rules Chiefly Deduced from Experiment, for Conducting the Practical Operations of a Siege, Part II* (London: John Weale, 1842).

Phillimore, Sir Robert, *Commentaries upon International Law, Volume 3* (London: Butterworths, 1873).

Piercy, M.J., *Catechism of the Diocese of Paris* (London: Richardson and Son, 1850).

Plarr, Victor, "Henry Tracy Coxwell," in *Men and Women of the Time: A Dictionary of Contemporaries* (London: George Routledge and Sons, 1899).

Pollard, Edward A., *The Last Year of the War* (New York: Charles B. Richardson, 1866).

Pollock, Frederick, *An Introduction to the History of the Science of Politics* (London: Macmillan, 1890).

Polson, Archer, *Principles of the Laws of Nations* (Philadelphia: T. & J.W. Johnson & Co., 1860).

Pond, George E., *The Shenandoah Valley in 1864, Volume 11* (New York: Scribner, 1883).

Pope, John, "General Orders, No 5," in John Codman Ropes, *The Army Under Pope, Volume 4* (New York: Charles Scribner's Sons, 1881).

Proceedings of the International Conference on Aerial Navigation (New York: The American Engineer and Railroad Journal, 1894).

Project of an International Declaration concerning the Laws and Customs of War (Brussels. 27 August 1874). https://ihl-databases.icrc.org/ihl/INTRO/135.

Read, Hollis, *Foot-prints of Satan; Or, The Devil in History* (New York: E.B. Treat ..., 1872).

Reed, Samuel Rockwell, *The War of 1886, Between the United States and Great Britain* (Cincinnati: Robert Clarke & Co., 1882).

The Rebellion Record: A Diary of American Events with Documents, Volume 11 ed. Frank Moore (New York: D. Van Nostrand, 1868).

Reddie, James, *Inquires in International Law* (Edinburgh: William Blackwood & Sons, 1842).

Report of the Secretary of War, Volume 1 (Washington DC: Government Printing Office, 1892).

Rickaby, Joseph, *Moral Philosophy or Ethics and Natural Law* (London: Longman, Green & Co., 1888).

Ripley, George and Charles Dana (eds.), "Bombardment," in *The New American Cyclopædia, Volume III* (New York: Appleton and Company, 1873).

Risley, John S., *The Law of War* (London: A.D. Innes & Co., 1897).

Ropes, John Codman, *The Army under Pope, Volume 4* (New York: Charles Scribner's Sons, 1881).

Salverda, J.G.W. Fijnje van, *Aerial Navigation* trans. George Waring (New York: D. Appleton and Company, 1894).

"Science in War," *Gunton's Magazine, Volume 16, January–June 1898* (New York: The Gunton Company, 1898).

Scoffern, John, *Projectile Weapons of War and Explosive Compounds* (London: Cooke & Whitley ..., 1852).

Scott, Sir Walter, *Minstrelsy of the Scottish Border, Volume 1* (Edinburgh: James Ballantyne ..., 1803).

Scott, Sir Walter, *The Life of Napoleon Buonaparte, Emperor of the French, Volume 2* (Exeter: J. & B. Williams, 1832).

Senior, Nassau Williams, *Historical and Philosophical Essays* (London: Longman, Green, Longman, Roberts, & Green, 1865).

Serviss, Garett, *Edison's Conquest of Mars* [1898] (Auckland, New Zealand: The Floating Press, 2012).

Sheridan, Philip, *Personal Memoirs of P.H. Sheridan, General, United States Army, Volume 1* (London: Chatto & Windus, 1888).

Sidgwick, Henry, *The Elements of Politics* (London: Macmillan, 1897).

Sidgwick, Henry, *A Memoir* (London: Macmillan, 1906).

Sidgwick, Henry, "The Inductions of Ethics," in *Lectures on the Ethics of T.H. Green, Mr. Herbert Spencer, and J. Martineau* (New York: Cambridge University Press, 2012).

Snow, Freeman, *International Law: A Manual* (Washington, DC: Government Printing Office, 1898).

Southey, Robert, *Sir Thomas More; Or, Colloquies on Progress and Prospects of Society* (London: John Murray …, 1829).

Spencer, Herbert, "Letter V" in *The Proper Sphere of Government* (London: W. Brittain …, 1843).

Spencer, Herbert, *The Principles of Ethics, Volume 1* [1897] (Indianapolis: Liberty Fund, 1978).

Spencer, Herbert, *The Principles of Ethics, Volume 2* [1897] (Indianapolis: Liberty Fund, 1978).

Spinks, Thomas, *The Ecclesiastical and Admiralty Reports, Volume 1* (London: A. & G.A. Spottiswoode, 1855).

Stebbins, Rufus, "Address," *The Advocate of Peace, July and August, v. 15–18, 1854–1858* (Washington DC: American Peace Society, 1859).

Stockton, Frank, *The Great War Syndicate* (New York: Dodd, Mead, & Company, 1889).

Straith, Hector, *A Treatise on Fortification* (Croydon: William Annan, 1833).

Sumner, Charles, *The True Grandeur of Nations* (Boston: J.H. Eastburn, 1845).

Tennant, Charles, *Utilitarianism Explained and Exemplified in Moral and Political Government* (London: Longman, Green, Longman, Roberts, & Green, 1864).

Tennyson, Alfred Lord, *The Works of Alfred Tennyson* (Philadelphia: Gebbie & Company, 1881).

Thomas, John P., *A Treatise of Universal Jurisprudence* (London: Longman …, 1828).

Thring, Theodore, *A Treatise on the Criminal Law of the Navy* (London: V.& R. Stevens and Sons, 1861).

Tomes, Robert, *The War with the South: A History of The Late Rebellion, Volume 3* (London: Virtue & Yorston, 1867).

Tompson, Henry Steuart, *Catechism on the Army Discipline and Regulations Acts, 1879. Courts-martial* (London: W. Mitchell & Company, 1880).

Tone, William Theobald Wolfe, *Essay on the Necessity of Improving Our National Forces* (New York: Kirk and Mercein, 1819).

Twiss, Sir Travers, *The Laws of Nations Considered as Independent Political Communities* (Oxford: Clarendon, 1863).

Upham, Thomas, *The Manual of Peace* (New York: Leavitt, Lord, and Co., 1836).

Upton, Francis Henry, *The Law of Nations Affecting Commerce During War* (New York: John S. Voorhies, 1863).

Vallecillo, Don Antonio, *Ordenanzas de S.M. el Régimen, Disciplina, Subordinación y Servicio* (Madrid: Emprenta de los Señores Andrés y Deas, 1850).

Verne, Jules, *Five Weeks in a Balloon: A Voyage of Exploration and Discovery in Central Africa* (London: Sampson Low, Marston, Low & Searle, 1874).

Verne, Jules, *The Clipper of the Cloud* (London: Low, Marston, Searle & Rivington, 1887).

Victor, Orville, *The History, Civil, Political, and Military of the Southern Rebellion, Volume IV* (New York: James P. Torrey, 1864).

Voss, Julian von, *Ini: Ein Roman aus dem ein und zwanzigsten Jahrhundert* (Berlin: Karl Freidrich Amelang, 1810).

Walker, Thomas Henry, *The Science of International Law* (London: Cambridge University Press, 1893).

The War Correspondence of the Daily News (London: Macmillan, 1871).

The War of the Rebellion: A Compilation of the Official Records of the Union and Confederate Armies, Series 1, Volume 1 (Washington DC: Government Printing Office, 1880).

The War of the Rebellion: A Compilation of the Official Records of the Union and Confederate Armies, Series 1, Volume 3 (Washington DC: Government Printing Office, 1892).

The War of the Rebellion: A Compilation of the Official Records of the Union and Confederate Armies, Series 2, Volume 4 (Washington DC: Government Printing Office, 1899).

Ward, Captain B.R., *Manual of Military Ballooning* [1896] (East Sussex: Rediscovery Books, 2006).

Waterloo, Stanley, *Armageddon: A Tale of Love, War, and Invention* (Chicago: Rand McNally & Co. 1898).

Wellington, Duke of, *A Memoir of Field-Marshal the Duke of Wellington, Volume 1* (London: A. Fullarton and Co. 1814).

Wells, H.G., *The Time Machine* (London: J.M. Dent & Sons Publishers, 1895).

Wells, H.G., *The War of the Worlds* (London: William Heinemann, 1898).

Whelpley, Samuel, *Letters Addressed to Caleb Strong, Esq. Late Governor of Massachusetts* (Providence, RI: Miller and Hutchens, 1818).

"When War Shall Be No More," *The Argosy, Volume 18, April to September 1894* (New York: Frank A. Munsey & Company, 1894).

Whewell, William, *The Elements of Morality: Including Polity, Volume 2* (London: John W. Parker, 1845).

Wildman, Richard, *Institutes of International Law, Volume 1* (London: William Benning & Co., 1849).

Wilkinson, Henry, *Engines of War* (London: Longman, Orme, Brown, Green, and Longmans, 1841).

Wilmot, Colonel Eardley, "Military Law in Germany," *Journal of the Royal United Service Institution, Volume 35* (London: Harrison and Sons, 1891).

Winthrop, William, *Military Law, Volume 1* (Washington, DC: W.H. Morrison, 1886).

"The Wise and Unwise Proposals for Altering the Code of Maritime Warfare," *The Economist* 20:966 March 1, 1862.

Wolseley, Garnet, *The Soldier's Pocket-Book for Field Service* (London: Macmillan, 1871).

Woosley, Theodore D., *To the Study of International Law* (New York: Charles Scribner, 1871).

Wright, Henry C., *Defensive War Proved to be a Denial of Christianity and of the Government of God* (London: Charles Gilpin, 1846).

Young, Arthur, *A Digest of Maritime Law Cases, 1837–1860* (London: Horace Cox, 1865).

Young, Philip, *History of Mexico: Her Civil Wars, and Colonial and Revolutionary Annals* (Cincinnati: J.A. & U.P. James, 1847).

The Twentieth Century

Abbot, Willis John, *Aircraft and Submarines: The Story of the Invention, Development and Present-Day Uses of War's Newest Weapons* (New York: G.P. Putnam's Sons, 1918).

Ader, Clément, *Military Aviation* [1909] trans. Lee Kennett (Maxwell Air Force Base, AL: Air University Press, 2003).

"Aeroplane as Scouts; Not Fighters," *Popular Mechanics*, July 1901.

"Aerial Navigation" Part XI, Articles 313 to 320, in *Papers Relating to the Foreign Relations of the United States: The Paris Peace Conference, 1919, Volume 13* (Washington DC: Government Printing Office, 1947); 642–646.

"Aerial Torpedo Boat," *Popular Mechanics* (May 1907).

"Air Service Tactical Application of Military Aeronautics," 5 January 1919, Maxwell AFB, Ala., USAF Historical Research Agency, document 167.4–1, 3.

"Airship Safely Bombards City, Knabenshue Drops Missile on City Hall," *Los Angeles Herald* December 18, 1908.

Alexander, John, *The Conquest of the Air: The Romance of Aerial Navigation* (New York: A. Wessels Company, 1902).

Anderson, Clinton C., *The War Manual, Volumes 1–2* (London: T.F. Unwin, 1916).

Angell, Sir Norman, *The Menace to Our National Defense* (London: H. Hamilton, 1934).

"Annual Report of the Secretary of War," *War Department Annual Reports, 1919, Volume I* (Washington DC: Government Printing Office, 1920).

Anscombe, G.E.M., "The Justice of the Present War Examined," [1939] in *War in the Twentieth Century: Sources in Theological Ethics* ed. Richard Miller (Louisville, KY: Westminster, 1992).

Anscombe, G.E.M., *Mr. Truman's Degree* (Oxford: [n.p.], 1957).

Anscombe, G.E.M., "War and Murder," in *Nuclear Weapons: A Catholic Response* [1961] ed. Walter Stein (New York: Steed and Ward, 1962).

Arnold-Forster, W., "The Political Background of the Disarmament Conference," in *Disarmament in 1932: The Year-Book of the Disarmament Information Committee* (Geneva: Published at the Office of Disarmament, 1932).

Ashmore, Edward B., *Air Defence* (New York: Longmans, 1929).

Aston, Sir George, *Sea, Land, and Air Strategy* (London: John Murray, 1914).

Bacon, Gertrude, *Balloons Airships and Flying Machines* (New York: Dodd, Mead & Company, 1905).

Bacon, John Mackenzie, *The Dominion of the Air: The Story of Aerial Navigation* (London: Cassell and Co., 1902).

Baden-Powell, B., "How Airships are Likely to Affect War," *Royal United Services Institution Journal* 54:387 (1910); 555–581.

Baker, Joseph, *The Laws of Land Warfare* (Washington, DC: Government Printing Office, 1919).

Baker, Joseph and Louis McKernan, *Selected Topics Connected with the Laws of Warfare as of August 1, 1914* (Washington DC: Government Printing Office, 1919).

Baldwin, Stanley, "The Bomber will Always Get Through," *The Parliamentary Debates, House of Commons, Volume 270* (London: H.M. Stationery Office, 1932).

Barclay, Sir Thomas, *Law and Usage of War: A Practical Handbook of the Law and Usage of Land and Naval Warfare and Price* (Boston: Houghton Mifflin Company, 1919).

Baschab, Charles Reinhard, *A Manual of Neo-Scholastic Philosophy* (London: B. Herder Book Company, 1937).

Basic Field Manual, Rules of Land Warfare (Washington, DC: Government Printing Office, 1940).

Bassett, John Spencer, *The Lost Fruits of Waterloo: Views on a League of Nations* (New York: Macmillan, 1919).

Baty, Thomas and John Morgan, *War: Its Conduct and Legal Results* (London: John Murray, 1915).

Beamish, Richard and Francis March, *America's Part in the World War: A History of the Full Greatness of Our Country's Achievements* (Chicago: John C. Winston Company, 1919).

Bernhardi, General Friedrich von, *Germany and the Next War* [1911], trans. Allen Powles (New York: Longmans, Green, and Co., 1914).

Birkhimer, William, *Military Government and Martial Law* (London: Kegan Paul, Trench, Trubner & Co., 1914).

Blanchon, G., *The New Warfare* (London: George Harrap & Co, 1917).

Bloch, Ivan, *The Future of War* (Boston: The World Peace Foundation, 1914).

"'Blow-up' By Airship: Sky Pilot Knabenshue Gets Permission to Make an 'Attack' on the City Tonight," *Los Angeles Times* December 17, 1908.

Blyth, James, *The Aerial Burglars* (London: Ward, Lock & Co., 1906).

Bodart, Gaston and Verson Kellogg, *Losses of Life in Modern Wars* (Oxford: Clarendon, 1916).

Boeckel, Florence, *Between War and Peace: A Handbook for Peace Workers* (New York: Macmillan, 1928).

Bogard, Ernest, *Direct and Indirect Costs of the Great War: Preliminary Economic Studies of the War, of the Carnegie Endowment for International Peace* (New York: Oxford University Press, 1919).

"Bomb-Dropping for Aeroplanes," *Flight* December 18, 1914.

"Bomb-dropping from Airships," *Antiaircraft Journal, Volume 33* (Fort Monroe, VA: Coast Artillery School Press, 1910).

Bordwell, Percy, *The Law of Law Between Belligerents: A History and Commentary* (Chicago: Callaghan & Co., 1908).

Boureut-Aubertot, J., *Les Bombardements Aériens* (Paris: Les Presses Universitaires de France, 1923).

Brittain, Vera, *England's Hour* [1941] (New York: Continuum, 2005).

Brittain, Vera, *Seeds of Chaos: What Mass Bombing Really Means* (London: Bombing Restriction Committee, 1944).

Brittain, Vera, *Massacre by Bombing: The Facts Behind the British-American Attack on Germany* (New York: Fellowship of Reconciliation, 1944).

Brown, James, "Pan American Cooperation in Aeronautics," *Journal of Air Law and Commerce* 9:3 (1938); 468–502.

Bruce, Eric Stuart, "The Use of Balloons in War," *The Aeronautical Journal* 6:23 (1902); 43–44.

Bruce, Eric Stuart, *Aircraft in War* (London: Hodder and Staughton, 1914).

Butler, Geoffrey and Simon Maccoby, *The Development of International Law* (New York: Longmans Green & Co, 1928).

Campbell, M.V., *Practical Bombing as Applied by the Canadian and British Armies* (Detroit: Bartlett Publishing, 1917).

Capper, Colonel J.E., "The Military Aspect of Dirigible and Aeroplanes," *Antiaircraft Journal* 33 (1910); 147–168.

Carreras y Artau, Joaquín, *Doctrinas de Francisco Suárez Acerca del Derecho de Gentes y Sus Relaciones con el Derecho Natural* (Gerona: Carreras, 1921).

Choate, Joseph, *The Two Hague Conferences* (Princeton: Princeton University Press, 1913).

Churchill, Winston, *The World Crisis, 1915* (London: Thornton Butterworth Limited, 1927).

Claxton, William, *The Mastery of the Air* (Glasgow: Blackie and Son, Limited, 1915).

Cobbett, Pitt, *Leading Cases of International Law* (London: Sweet and Maxwell, 1924).

Colbey, Elbridge, "Aërial Law and War Targets," *The American Journal of International Law* 19:4 (1925); 702–715.

Coles, Robert, *The Struggle for Empire: A Story of the Year 2236* (London: Elliot Stock …, 1900).

Collins, Francis Arnold, *The Air Man: His Conquests in Peace and War* (New York: The Century Co., 1917).

"Commission on the Responsibility of the Authors of the War and on Enforcement of Penalties" *The American Journal of International Law* 14:1/2 (1920); 95–154.

Conference on the Limitations of Armament, Washington November 12, 1921–February 6, 1922 (Washington D.C. Government Printing Office, 1922).

"Convention Relating to the Regulation of Aerial Navigation, 1919," *Journal of Air Law and Commerce* 1:1 (1930).

Cooper, John Cobb, "The International Air Navigation Conference, Paris, 1910," *Journal of Air Law and Commerce* 19:2 (1952); 127–143.

David, Evan, *Aircraft: Its Development in War and Peace and Its Commercial Future* (New York: Charles Scribner's Sons, 1919).

Davis, George, *The Second Peace Conference, The Rules of War on Land, Working Memoranda* (Washington, DC: Government Printing Office, 1907).

Davis, Malcolm, "Public Opinion and Disarmament," in *Disarmament in 1932: The Year-Book of the Disarmament Information Committee* (Geneva: Office of Disarmament, 1932).

de Valles, Arnaldo, *Premier Congrès Juridique Internacional de La Réglementation de la Locomotion Aérienne* (Paris : L. Larose et L. Tenin, 1911).

Dickinson, Edwin, *The Equality of States in International Law* (Cambridge, MA: Harvard University Press, 1920).

Disarmament in 1932: The Year-Book of the Disarmament Information Committee (Geneva: Office of Disarmament, 1932).

Dodeman, Charles, *Bombe Silencieuse* (Tours: Maison Alfred Mame et Fils, 1916).

Douhet, Giulio, *The Command of the Air* trans. Dino Ferrari [1921] (Washington DC: Air Force History and Museums Program, 1998).

Doyle, Arthur Conan, *Great Britain and the Next War* (Boston: Small, Maynard & Co., 1914).

"Draft Disarmament Convention: Proposal Text Prepared in Light of the Modifications Adopted in the First Reading on June 8, 1933. Geneva, September 22, 1933," in *Disarmament and Security: A Collection of Documents, 1919–55* (Washington DC: Government Printing Office, 1956).

Draft Rules of Aerial Warfare, The Hague, February 1923. https://wwi.lib.byu.edu/index.php/The_Hague_Rules_of_Air_Warfare.

Dumas, Samuel and Knud Otto Vedel-Petersen, *Losses of Life Caused by War* (Oxford: Clarendon, 1923).

Dutton, Arthur, "War Too Humane," *The Pacific Monthly* 16 (1906).

Edmunds, Sterling, *The Lawless Law of Nations* (Washington, D.C.: John Byrne, 1925).

Edmunds, Sterling, "The Laws of War: Their Rise in the Nineteenth Century and Their Collapse in the Twentieth," *Virginia Law Review* 15:4 (1929); 321–349.

Ellinwood, Ralph, *Behind German Lines: A Narrative of the Everyday Life an American Prisoner of War* (New York: Knickerbocker Press, 1920).

Ellis, W.E., "Aerial-Land and Aerial-Maritime Warfare," *The American Journal of International Law* 8:2 (1914); 256–273.

Emery, Henry, *Some Economic Aspects of War* (Washington DC: Government Printing Office, 1914).

Ennis, William, *Flying Machines Today* (New York: D. Van Nostrand Company, 1911).

"Estimates Presented by Major Baird," HC Deb February 1918 vol 103 cc957–1068. https://api.parliament.uk/historic-hansard/commons/1918/feb/21/estimates-presented-by-major-baird#S5CV0103P0_19180221_HOC_386.

Fagg, John, "Mission Accomplished," in *The Army Air Forces in World War II: Europe, Argument to V-E Day, January 1944 to May 1945* eds. Wesley Craven and James Cate (Chicago: University of Chicago Press, 1951).

Ferris, Richard, *How to Fly; Or, The Conquest of the Air* (London: T. Nelson, 1910).

Finch, George, "James Brown Scott, 1866–1943," *The American Journal of International Law* 38:2 (1944); 183–217.

Foch, Marshal, *The Principles of War* [1903] trans. Hillarie Belloc (New York: Henry Holt and Company, 1920).

Foch, Ferdinand, *The Principles of War* (New York: The H.K. Fly Company, 1918).

Folks, Homer, *The Human Costs of the War* (New York: Harper & Brothers, 1920).

Ford, John C., "Morality of Obliteration Bombing," *Theological Studies* 5:3 (1944); 261–309.

Fradkin, Elvira, *The Air Menace and the Answer* (New York: Macmillan, 1934).

Frank, Waldo, *Virgin Spain: The Drama of a Great People* (New York: Boni & Liveright, 1926).

Frothingham, Arthur, *Handbook of War Facts and Peace Problems* (New York: National Security League, 1919).

Fuller, J.F.C., *The Reformulation of War* (London: Hutchison and Co., 1923).

Fuller, J.F.C., *Towards Armageddon: The Defence Problem and its Solution* (London: Lovat Dickson, 1937).

Garner, James, *The German War Code* (Urbana, IL: University of Illinois, 1918).

Garner, James, *International Law and the World War* (New York: Longmans, Green and Co, 1920).

Generals and Other Officers, *Franco-German War, 1870–71* trans. J.F. Maurice (London: Swan Sonnenschein and Co, 1900).

The German War Book trans. J.H. Morgan (London: John Murray, 1915).

German White Book Concerning the Responsibility of the Authors of the War trans. Carnegie Endowment for International Peace (New York: Oxford University Press, 1924).

"Germany's Air Madness," in *European War Pamphlets, Volume 55* (New York: [n.p.], 1921).

Gilchrist, Harry L., *A Comparative Study of World War Casualties from Gas and Other Weapons* (Washington DC: Government Printing Office, 1928).

Graham-White, Claude and Harry Harper, *The Aeroplane in War* (Philadelphia: Lippincott, 1912).

Grahame-White, Claude and Harry Harper, *Aircraft in the Great War: A Record and Study* (Chicago: A.C. McClurg, 1915).

Grahame-White, Claude and Harry Harper, *Air Power: Naval, Military, Commercial* (New York: Frederick A. Stokes Company, 1917).

Great Britain War Office, *Statistics of the Military Effort of the British Empire During the Great War, 1914–1920* (London: H.M. Stationery Office, 1922).

Grebler, Leo, *The Cost of the World War to Germany and Austria–Hungary* (New Haven CT: Yale University Press, 1940).

Greene, Francis, "Recent Changes in the Art of War," *Proceedings of the New York State Historical Association* 15 (1916); 156–175.

Greer, Rowen Allen, *International Aerial Regulations* (Washington DC: Government Printing Office, 1926).

Griffith, George, *The Lord of Labour* (London: F.V. White, 1911).

Groves, P.R.C., "The Influence of Aviation on International Relations," *Journal of the Royal Institute of International Affairs* 6:3 (1927); 133–152.

Guarini, Emile, "Free and Dirigible Balloons," *Scientific American* 90:23 (1904). 436.

Hall, Cyril, *Modern Weapons of War, By Land, Sea and Air* (London: Blackie and Son, 1915).

Hall, Norris, "Science in War," in *The Next War* (Cambridge, MA: Harvard University Press, 1925).

Harbord, Major General James, "Review: *The Great Delusion. A Study of Aircraft in Peace and War* by 'Neon,'" *Naval Institute Proceedings*, 53:293 (1927). 899.

Harmon, Benjamin F., "The Past and Future of Defense against Aircraft," *The Coastal Artillery Journal* 63:5 (1925). 555–556.

Harris, Arthur, *Bomber Offensive* [1947] (Barnsley: Pen & Sword Books, 2005).

Hart, Henry Liddell, *Paris, or The Future of War* [1925] (New York: Garland Publishing, 1972).

Hart, Henry Liddell, *The Remaking of Modern Armies* (Boston: Little, Brown, and Company, 1928).

Hartman, Sylvester J., *A Textbook of Logic: A Normative Analysis of Thought* (New York: American Book Company, 1936).

Hawthorne, Major H.L., "Balloons and Dirigibles in War," *Antiaircraft Journal* 32:2 (1909); 107–116.

Hazeltine, H.D., "Recent Growth on Aerial Law," *Flight International* 10:1 (1919), 295–375.

Hearne, R.P., *Aerial Warfare* (London: John Lane …, 1909).

Hearne, R.P., *Airships in Peace and War* (London: John Lane …, 1910).

Hershey, Amos, "History of International Law Since the Peace of Westphalia," *The American Journal of International Law* 6:1 (1912); 30–69.

Higgins, A. Pearce, *War and the Private Citizen: Studies in International Law* (London: P.S. King & Son, 1912).

Higgins, A. Pierce, *Non-combatants and the War* (London: Oxford University Press, 1914).

Hilderbrandt, Alfred, *Airships Past and Present* trans. W.H. Story (New York: D. Van Nostrand Company, 1908).

Hoernes, Hermann, *A Compendium of Aviation and Aereostation, Balloons, Dirigibles, and Flying Machines* (London: Charles Griffin & Company …, 1911).

Holland, Thomas, *Letters to* The Times *upon War and Neutrality* (New York: Longmans, Green, and Co., 1921).

Hornaday, William T., *A Searchlight on Germany: Germany's Blunders, Crimes and Punishment* (New York: American Defense Society, 1917).

Houard, Georges, "Long-Distance Bombardment from Air-Planes and American Aviation," *The Living Age* Eighth Series, Volume XII (1918).

Hull, William, *The Two Hague Conferences and Their Contributions to International Law* (Boston: Ginn & Company, 1908).

Hyde, Charles Cheney, *Land Warfare* (Washington, DC: Government Printing Office, 1918).

Hyde, Charles Cheney, *International Law: Chiefly as Interpreted and Applied By the United States, Vol. 2* (Boston: Little, Brown, and Co., 1922).

Imperial Conference, 1923: Summary of Proceedings in *Appendix to the Journals of the House of Representatives of New Zealand, 1923, Volume I* (Wellington NZ: [n.p.], 1924).

Irwin, Will, *The Next War* (New York: P. Dutton, 1921).

Jackman, W.J. and Thos Russell, *Flying Machines: Construction and Operation* (Chicago: Charles C. Thompson, 1912).

Jackson, Thomas, "The Lot of the Innocent Bystander in War," *The Public Health Journal* 7:5 (1916); 276–281.

James, William W. and Benjamin W. Wickling, *Injuries of the Jaws and Face: With Special Reference to War Casualties* (London: John Bale & Staples, 1940).

Johnson, Thomas, "Underground Fortresses Guard France from Invasion," *Popular Science* 129:4 (1936). 14–15, 118–119.

"The Kaiser at Prayer," *The Literary Digest*, 1 June 1918.

Keating, Joseph, "The Ethics of Bombing," *The Catholic Mind*, XXXVI (1938); 269–281.

Kuhn, Arthur, "International Aerial Navigation and the Peace Conference," *The American Journal of International Law* 14:3 (1920); 369–381.

Lancaster, F.W., *Aircraft in Warfare: The Dawn of the Fourth Arm* (London: Constable and Company, 1916).

Lascelles, Captain E., "The Airship and Flying Machines in War," *The Calvary Journal* 5 (1910); 208–212.

Laws of War: Bombardment by Naval Forces in Time of War (Hague IX); October 18, 1907. http://avalon.law.yale.edu/20th_century/hague09.asp.

Lea, General Homer, "The Aeroplane in War: Some Observations on a Military Delusion," *Harper's Weekly* 54 (1910); 8–9.

Lee, Blewett, "Sovereignty of the Air," *The American Journal of International Law* 7:3 (1913); 440–496.

Le Queux, William, *The Invasion of 1910: With a Full Account of the Siege of London* (Toronto: Macmillan, 1906).

Littlefield, Walter, "War Casualties of All the Nations," in *The European War:* The New York Times *Current History, Volume XVIII* (New York: New York Times Company, 1919).

Les Lois De La Guerre Continentale (Paris : Librairie Payot, 1916).

"Los Angeles Attacked from Airship," *Popular Mechanics* (February 1909).

Lougheed, Victor, *Vehicles of the Air: A Popular Exposition of Modern Aeronautics with Working Drawings* (Chicago, IL: Reilly and Britton, 1910).

Manisty, Herbert, "Aerial Warfare and the Laws of War," *Transactions of the Grotius Society* 7 (1921); 33–41.

Marshall, Logan, *Horrors and Atrocities of the Great War* (Philadelphia, PA: P.W. Ziegler, 1915).

Marsh, Col. W. Lockwood (ed.), *International Air Congress, London, 1923: Report* (London: Offices of the Royal Aeronautical Society, 1923).

Martin, Franklin Henry, *The World War* (Doubleday Doran, 1933).

Maude, F.N., "The Bombardment of Open Towns," *The Contemporary Review* 114:275 (1918); 271–78.

Maxim, Hudson, *Defenseless America* (New York: Hearst's International Library Co., 1915).

McDonald, Walter, *The Principles of Moral Science* (Dublin: Browne and Nolan, 1903).

McKenna, Charles, *Francis de Victoria, Founder of International Law* (Washington, DC: Catholic Association for International Peace, 1930).

McReavy, Lawrence L., "Reprisals: a Second Opinion," *Clergy Review*, 20 (1941). 131–138.

Mercier, Désieré, *A Manual of Modern Scholastic Philosophy, Volume 2* (London: K. Paul, Trench, Trubner & Company, 1917).

Miaja de la Muela, Adolfo, *Internacionalistas Españoles del siglo XVI, Francisco Suárez (1512–1569)* (Valladolid: Cuesta, 1932).

Miles, Major Sherman, "War in the Third Dimension," *The North American Review* 223:833 (1926–27); 594–605.

"Military Aeroplanes," *Antiaircraft Journal* 37: (1912); 101.

Military Airplane Data Prepared for the Select Committee of Inquiry into Operations of the United States Air Services, Volume 5 (Washington DC: Government Printing Office, 1925).

Miller, Henry, *The Paris Gun: The Bombardment of Paris by the German Long Range Guns and the Great German Offensives of 1918* (New York: Jonathan Cape & Harrison Smith, 1930).

Ministère de la Guerre: Données de statistiques relatives à la guerre 1914–1918 (Paris: Imprimerie Nationale, 1922).

"Minutes of a Meeting of the Supreme War Council Held in M. Pichon's Room at the Quai d'Orsay, Paris, March 12th, 1919 at 3 p.m." in *Papers Relating to the Foreign Relations of the United States: The Paris Peace Conference 1919, Volume 4* (Washington DC: Government Printing Office, 1943).

Mitchell, Charles, *Aerial Navigation in Warfare* (Toronto: [n.p.], 1911).

Mitchell, Thomas, *Casualties and Medical Statistics of the Great War* (London: H.M. Stationery Office, 1931).

Mitchell, William, *Our Air Force: The Keystone of National Defense* (New York: E.P. Dutton & Company, 1921).

Mitchell, William, *Winged Defense: The Development and Possibilities of Modern Air Power – Economic and Military* (New York: G.P. Putnam's Sons, 1925).

Moore, John Bassett, *A Digest of International Law, Volume VI* (Washington, DC: Government Printing Office, 1906).

Morgenthau, Hans, "Death in the Nuclear Age," *Commentary* Sept. 1, 1961. http://www.commentarymagazine.com/article/death-in-the-nuclear-age/.

Mowbray, Jay H., *Conquest of the Air by Airship and other Flying Machines* (Philadelphia: National Publishing Company, 1910).

Mowbray, Jay H., *Thrilling Achievement of 'Bird Men' with Flying Machines* (Philadelphia: National Publishing Company, 1911).

Muller, Hollis, *Manual of Military Aviation* (Menasha, WI: George Banta Publishing, 1918).

Müller-Meiningen, Ernst, *Who are the Huns? The Law of Nations and Its Breakers* trans. R.L. Orchelle (Berlin: Georg Reimer Publisher, 1915).

Nagle, William, *Morality and Modern Warfare: The State of the Question* (Baltimore, MD: Helicon Press, 1960).

Naiden, Captain Earl, *Air Service* (Leavenworth, KS: The General Service Schools Press, 1920).

"Naval and Military Aeronautics," HC Deb 02 August 1909 vol 8 cc1564–617. https://api.parliament.uk/historic-hansard/commons/1909/aug/02/naval-and-military-aeronautics.

Naval War College, *International Law Situations with Solutions and Notes* (Washington DC: 1912).

Neumann, Georg Paul, "The Possibility of Making Use of Balloons and Motor Airships in the Navy," *United Services Institution Journal* 52:369 (1908); 1502–1517.

Neumann, Georg Paul, *The Air Force I Knew, 1914–1918: Memoirs of the Imperial Air Force in the Great War* [1921] trans. J.E. Gurdon, ed. Bob Carruthers (Barnsley: Pen and Sword Aviation, 2014).

Nurick, L., "The Distinction between Combatant and Noncombatant in the Law of War," *The American Journal of International Law* 39:4 (1945); 680–697.

Office of the Adjutant General of the Army, *Summary of Casualties Among Members of the American Expeditionary Forces during the World War, 1919* (Washington DC: Government Printing Office, 1920).

Oppenheim, Lassa, *International Law, a Treatise* (London: Longman, Green, and Co., 1905).

Phillimore, G.G., "Bombardments," *Problems of the War* 15 (1915); 59–65.

Phillipson, Colman, *International Law and the Great War* (London: T. Fisher Unwin, 1915).

Picciotto, C.M., "Some Notes on Air-Warfare," *British Institute of International and Comparative Law* 15:2 (1915); 150–155.

Pierce, W.O'D., *Air War: Its Technical and Social Aspects* (London: Watts & Co., 1937).

Ponsonby, Arthur, *Falsehood in War-Time: Containing an Assortment of Lies Circulated throughout the Nations during the Great War* (London: E.P. Dutton & Company, 1928).

Porter, J.B., *International Law: Having Particular Reference to the Laws of War on Land* (Fort Leavenworth, KS: Press of the Army Service Schools, 1914).

"The Position of Irregular Troops under the Laws of War," *The Law Times Volume 138* (London: Field & Queen, 1915).

Practical Warfare: Chapters on Armies and Navies in Action (London: E. Nash, 1915).

Premier congrès du Comité juridique international de l'aviation (Paris: A. Pedone, 1912).

Protection of Civilian Populations Against Bombing from the Air in Case of War, Unanimous resolution of the League of Nations Assembly, September 30, 1938.

Quindry, Frank, "Aerial Bombardment of Civilian and Military Objectives," *Journal of Air Law and Commerce* 2:4 (1931); 474–509.

"Radio and Aerial: Commission of Jurists to Consider and Report upon the Revision of the Rules of Warfare – General Report," *International Law Documents* (Washington D.C.: Government Printing office, 1926).

Raleigh, Walter, *The War in the Air: Being the Story of the Part Played in the Great War by the Royal Air Force, Volume 1* (London: Clarendon, 1922).

Raven, Charles Earle, *War and the Christian* (New York: Universal Christian Council for Life and Work, 1938).

Report of the International Commission to Inquire into the Causes and Conduct of the Balkan Wars (Washington, DC: The Carnegie Endowment for International Peace, 1914).

A Report on the Characteristics, Scope of Utility, Etc. of Railway Artillery, Volume 1 (Washington DC: Government Printing Office, 1921).

Richards, Erle, *International Law: Some Problems of the War* (Oxford: Clarendon, 1915).

Ross, John Elliot, *Ethics from the Standpoint of Scholastic Philosophy* (New York: Devin-Adair Company 1938).

Rotch, A. Lawrence, *The Conquest of the Air; Or, The Advent of Aerial Navigation* (New York: Moffat Yard and Company, 1909).

Royse, M.W., *Aerial Bombardment and the International Regulation of Warfare* (New York: H. Vinal, 1928).

Ryan, John K., *Modern War and Basic Ethics* (Washington DC: Catholic University of America Press, 1933).

Ryan, John K., "Unlimited War and Traditional Morality," *The Modern Schoolman* 22:1 (1944); 24–32.

Salvaneschi, Nino, *Let Us Kill the War: Let Us Aim at the Heart of the Enemy!* (Milan: Art Edition of Bianco e Nero, 1917).

Sand, Peter, Jorge de Sousa Freitas, and Geoffrey Pratt, "An Historical Survey of International Air Law Before the Second World War," McGill *Law Journal* 7:1 (1944); 24–42.

Santos-Dumont, Alberto, *My Airships* (London: Grant Richards, 1904).

Schuman, Frederick, *Europe on the Eve: The Crisis of Diplomacy, 1933–1939* (New York: A.A. Knopf, 1939).

Scott, James Brown, *The Classics of International Law* (Washington DC: Carnegie Institution of Washington, 1907).

Scott, James Brown, "Bombardment by Naval Forces," *The American Journal of International Law* 2:2 (1908); 285–294.

Scott, James Brown, *The Hague Conventions and Declarations of 1899 and 1907* (New York: Oxford University Press, 1915).

Scott, James Brown, "Preface," in Francisci de Victoria, *De Indis et De Iure Belli Relectiones* (Washington DC: Carnegie Institution of Washington, 1917).

Scott, James Brown, "A Single Standard of Morality for the Individual and the State," *Proceedings of the American Society of International Law at its Annual Meeting* 26 (1932); 10–29.

Scott, James Brown, "The Seventh International Conference of American States," *The American Journal of International Law* 28:2 (1934); 219–230.

Scott, James Brown, *The Spanish Origin of International Law: Francisco de Vitoria and His Law of Nations* (Oxford: Clarendon, 1934).

Scott, James Brown, *The Catholic Conception of International Law* (Washington DC: Georgetown University Press, 1934).

Scott, James Brown, *Law, the State, and the International Community: A Commentary on the Development of Legal, Political, and International Ideals* (New York: Columbia University Press, 1939).

Scott, George and James Garner, *The German War Code* (Washington DC: US Committee on Public Information, 1918).

Scully, Joseph, "The Ethical Implications of Atomic Energy," in *The Implications of Atomic Energy* (Brooklyn: St. John's University, 1950).

"The Seeming Triumph of the Mono-plane," *Journal of United States Artillery*, Volume 34 (Fort Monroe, VA: Coast Artillery School Press, 1910).

Seversky, Alexander, *Victory Through Air Power* (New York: Simon & Schuster, 1942).

Sherman, William, *Air Warfare* [1926] (Maxwell Air Force Base, Alabama: Air Force Press, 2002).

Shiel, M.P., *The Purple Cloud* (London: Chatto and Windus, 1901).

Slater, Thomas, *Questions of Moral Theology* (New York: Benziger Brothers, 1915).

Slessor, Sir John Cotesworth, *Air Power and Armies* (London: Oxford University Press, 1936).

Spaight, J.M., *War Rights on Land* (London: Macmillan, 1911).

Spaight, J.M., "Reviewed Work: *Aerial Bombardment and the International Regulation of Warfare.* by M.W. Royse," *Journal of the Royal Institute of International Affairs* 7:6 (1928); 432–33.

Spaight, J.M., *Bombing Vindicated* (London: Geoffrey Bles, 1944).

Stockton, Charles, *Outlines of International Law* (New York: Charles Scribner & Sons, 1914).

Stone, Colonel. F.G., "The Limitations of Aerial Bombardment by International Law," *Flight* 11 (1909); 7–16.

Stone, Julius, *Legal Controls of International Conflict* (London: Stevens, 1954).

Stowell, Ellery, *Intervention in International Law* (Washington DC: John Byrne 1921).

Sykes, Sir Frederick Hugh, *Aviation in Peace and War* (London: Edward Arnold, 1922).

K.V.R.T. "Aerial Warfare and International Law," *Virginia Law Review* 28:4 (1942); 516–527.

Talbot, Frederick A., *Aeroplanes and Dirigibles of War* (Philadelphia: J.B. Lippincott, 1915).

Thomson, Christopher, *Air Facts and Air Problems* (London: John Murray, 1927).

Thone, Frank, "No Flying Devastation for Cities," *Science News Letter* 21:584 (1932); 386–387.

The Times History of the War, Volume 11 (London: The Times Printing House Square, 1917).

"To Make Night Trip in Airship, Knabenshue to Ascend at Midnight, Will Demonstrate How City Can Be Destroyed," *Los Angeles Herald* December 17, 1908.

"Torpedo Airship Controlled by Wireless is the Latest Invention," *New York Times* May 22, 1910.

"Torpedoes of the Air: Bomb-Droppers Directed by Wireless," *Illustrated London News*, September 6, 1913.

Townsend, Captain G.L., "The Use and Effect of Flying Machines on Military Operations," *Infantry Journal* 7 (1910); 244–258.

Training and Operations Group, Operations Division, Air Service, *Pursuit and Combat Training Regulations* (Washington DC: Government Printing Office, 1920).

Trelles, Camilo Barcia, *Francisco de Victoria, Fundador del Derecho Internacional Moderno* (Cuesta: Talleres Tipográficos, 1928).

Trenchard, Hugh, "Air Ministry," *Supplement to the London Gazette* 1 January 1919.

Tryon, James, "The Regulation of War," *The Yale Law Journal* 20:7 (1911); 535–547.

Tucker, Robert, *The Law of War and Neutrality at Sea* in *International Law Studies*, NavPers 15031, 1955, Volume 50 (Washington DC: Government Printing Office, 1957).

Tulloch, T.G., "The Aerial Peril," *The Nineteenth Century and After, Volume 65, January to June 1909* (New York: Leonard Scott Publication Co., 1909).

Turner, C.C., *The Struggle in the Air, 1914–1919* (London: Edward Arnold, 1919).

Union of the American Republics, *Bulletin of the Pan American Union* (Washington DC: US Government Printing Office, 1928).

United Nations War Crimes Commission, *History of the United Nations War Crimes Commission and the Development of the Laws of War* (London: H.M. Stationery Office, 1948).

"The U.S. Aerial Bombing Experiments on Warships," *Flight* September 15, 1921.

US Air Service Tactical School, Langley Field, VA, *Bombardment* (Washington DC: Government Printing Office, 1926).

United States, Department of the Army, *The Law of Land Warfare* (Washington DC: Department of Army, 1956).

Valentine, E. Seton and F.L. Tomlinson, *Travels in Space: A History of Aerial Navigation* (London: Hurst and Blackett, 1902).

van der Leeuw, H.J., *A Discourse on the Results of Aerial Navigation in the Year 1918*, trans. E. Halliday (Rotterdam: D.J.P. Storm Lotz, 1909).

Vanderpol, Alfred, *La Guerre devant le christianisme, suivi d'une traduction du « De Jure Belli » de François de Victoria* (Paris: A. Tralin, 1912).

Vanderpol, Alfred, *La doctrine scolastique du droit de guerre* (Paris: A. Pedone, 1919).

Verne, Jules, *Master of the World* in *The Works of Jules Verne* ed. Charles Horne (New York: Vincent Parke and Company, 1911).

Walker, Frederick, *Aerial Navigation* (London: Crosby, Lockwood and Son, 1902).

War Department: Office of the Chief of Staff, *Rules of Land Warfare* (Washington DC: Washington Government Printing Office, 1914).

War Office, *Manual of Military Law* (London: H.M. Stationery Office, 1914).

Wash, Major C.H., "Revue de L'Armee de L'air," in *Military Review* 15:57 (1935); 110–112.

Wells, H.G., *The War in the Air* (London: George Bell and Sons, 1908).

Wells, H.G., *The Way the World is Going* (London: Ernest Benn Limited, 1928).

Westermarck, Edward, *The Origin and Development of the Moral Ideas, Volume 2* (London: Macmillan, 1908).

Westlake, John, *International Law, Part I* (New York: Cambridge University Press, 1910).

Westlake, John, *International Law, Part II* (London: Cambridge University Press, 1913).

Wherry, William, *Aerial Warfare in International Law* (New York: New York University School of Law, 1939).

Wigmore, John H., *Select Cases on the Law of Torts, with Notes, and a Summary of Principle* (Boston: Little, Brown, & Co., 1912).

Williams, Paul, "Legitimate Targets in Aerial Bombardment," *The Journal of International Law*, 23:3 (July 1929). 570–581.

Williams, R.F., "Developments in Aerial Law," *University of Pennsylvania Law Review* 75 (1926); 139–153.

Willmore, John Selden, *The Great Crime and Its Moral* (London: Hodder and Stoughton, 1917).

Wilson, George, *Handbook on International Law* (St. Paul: West Publishing Company, 1910).

Wilson, H.W., *Convicted Out of Her Own Mouth: The Record of German Crimes* (New York: George H. Doran Company, 1917).

Winfield, Percy, "Aircraft Attacks," *The Law Magazine and Review* (London: Butterworths, 1915), 257–271.

"Wireless Control of Airships," *Flight* 6 February 1909.

Woodhouse, Henry, *Textbook of Military Aeronautics* (New York: The Century Co., 1918).

Woodhouse, Henry, *Textbook of Aerial Laws* (New York: Frederick A. Stokes Company 1920).

Wright, Herbert, *Francisco Suárez: Address in Commemoration of His Contribution to International Law and Politics* (Washington DC: Catholic University of America Press, 1933).

"'Zeppelin' Armed with Aerial Torpedoes," *Popular Mechanics* (October 1909).

Zimmerman, Alfred, "International Law and Social Consciousness," *Transactions of the Grotius Society* 20 (1934); 25–44.

Index

Printed in the United States
by Baker & Taylor Publisher Services